Fundamentals of Chemistry in the Laboratory

Fundamentals of Chemistry in the Laboratory

SECOND EDITION

Kurt Irgolic
Larry Peck
and
Rod O'Connor

with illustrations by

Paul Glenn

TEXAS A & M UNIVERSITY

Harper & Row, Publishers
New York, Hagerstown, San Francisco, London

To Bill Cook and Art Martell, with thanks.

Our special appreciation is extended to Mrs. Candy Gonzales for the manuscript preparation and to Steve Bales for the art photography.

Sponsoring Editor: Wayne E. Schotanus
Project Editor: Ralph Cato
Designer: Rita Naughton
Production Supervisor: Will C. Jomarrón
Compositor: Typothetae Book Composition
Printer and Binder: The Murray Printing Company
Art Studio: J & R Technical Services Inc.

Fundamentals of Chemistry in the Laboratory, Second Edition

ISBN 0–06–043219–5

Contents

Preface

The main goal of the first-year chemistry laboratory must be to help the student learn and understand some fundamental aspects of chemistry. To do this it is necessary to present some of the more common laboratory techniques and operations useful in the investigation of chemical systems. It is desirable, we believe, also to use in the laboratory some of the concepts, information, and problem-solving skills previously studied in the nonlaboratory part of a chemistry course.

If the laboratory experience stopped here, it would omit a very important aspect of the science of chemistry: the *useful applications* of chemical knowledge and skills.

It is our intention in this laboratory text to present some of the fundamentals of experimental chemistry within the framework of familiar and interesting chemical systems to illustrate these applications.

To gain the maximum benefit from the learning experiences of a chemical laboratory program, we suggest the importance of:

1. Understanding, *in advance of your laboratory period,* the concepts and procedures involved in the laboratory work to be performed.
2. Knowing and using *all* safety precautions necessary to work properly in the chemical laboratory.
3. Employing good laboratory techniques to perform work efficiently, while maintaining clean and neat working areas.
4. Maintaining an accurate laboratory notebook with complete records of experimental work, observations, data, and calculations.
5. Following up your laboratory work by a careful study of the results of your investigations.

Many of the things that chemists do require sophisticated (and expensive) equipment and the combination of many different skills and quite extensive knowledge. Rarely do practicing chemists, especially in a research field, have a "cookbook" to follow. Before they can design investigations and plan their experimental work, however, they must have learned the fundamental techniques and concepts required. Many of these are useful also in fields other than those of professional chemists. Agriculture, biology, environmental science, medical sciences, and oceanography (among others) are areas in which chemical laboratory skills are essential.

It is our hope that you will find this laboratory program a useful and interesting part of your study of the experimental sciences. Your suggestions for improvement of the program would be most welcome.

Kurt Irgolic Larry Peck Rod O'Connor Paul Glenn

Laboratory Safety

A chemical laboratory is not necessarily the safest of all places in which to work. Neither, however, is it something to be afraid of. With proper understanding of what you are doing, careful attention to safety precautions, and adequate supervision, you will find the chemical laboratory a safe learning situation.

Laboratory accidents can be classified in two general categories: those caused by your own negligence and those outside your control. Since both are possibilities, it is important that you learn both safety and emergency procedures. If you know and follow safe working practices, you will pose no hazard to yourself or others. If, in addition, you can use proper procedures for coping with emergencies, you may help save yourself or others from serious harm during an unexpected accident.

Every laboratory system will have some unique features for the prevention of accidents or for handling emergencies. It is important to you that you become thoroughly familiar with the special safety aspects of your own laboratory area. The following are some general precautions and procedures applicable to any chemical laboratory.

SAFETY

1. **WEAR APPROVED EYE PROTECTION AT ALL TIMES.**
   The authors have witnessed accidents, such as ammonia splattering and explosion of small glass containers, which have caused permanent eye damage that would have been prevented by the wearing of simple laboratory goggles.
2. **NEVER EAT, DRINK, OR SMOKE IN A CHEMICAL LABORATORY.**
   Tiny amounts of some chemicals may be toxic. Many solvents are easily ignited by a lighted cigarette.
3. **NEVER WORK IN A CHEMICAL LABORATORY WITHOUT PROPER SUPERVISION.**
   Your best protection against accidents is the presence of a trained, conscientious supervisor who is watching for potentially dangerous situations and who is capable of handling an emergency properly.
4. **NEVER PERFORM AN UNAUTHORIZED EXPERIMENT.**
   Many "simple" chemicals form explosive or toxic products when mixed.
5. **NEVER PIPET BY MOUTH AND NEVER INHALE GASES OR VAPORS.**
   Always use a mechanical suction device for filling pipets. Reagents may be more caustic or toxic than you expect. If you must sample the odor of a gas or vapor, use your hand to waft a small sample toward your nose.

6. **EXERCISE PROPER CARE IN HEATING OR MIXING CHEMICALS.**
   Be sure of the safety aspects of every situation in advance. For example, never heat a liquid in a test tube that is pointed toward you or another student. Never pour water into a concentrated acid. Proper dilution technique requires slowly pouring the concentrated reagent into water, while stirring to avoid localized overheating.
7. **BE CAREFUL WITH GLASS EQUIPMENT.**
   Cut, break, or fire polish glass only by approved procedures. When inserting a glass rod or tube through a rubber or cork stopper, lubricate the glass and the stopper, protect your hands with a cloth towel, and use a gentle twisting motion.

EMERGENCY PROCEDURES

1. **KNOW THE LOCATION AND USE OF EMERGENCY EQUIPMENT.**
   Find out where the safety showers, eyewash spray, and fire extinguishers are located. If you are not familiar with the function of any emergency equipment, ask your instructor for a lesson.
2. **DON'T UNDER-REACT.**
   Any chemical contact may be potentially hazardous, especially if near the eyes. In case of contact with a chemical reagent, wash the affected area *immediately* and *thoroughly* with water and notify your instructor. In case of a splatter of chemical over a large area of your body, don't hesitate to use the safety shower. Don't hesitate to call for help in an emergency.
3. **DON'T OVER-REACT.**
   In the event of a fire, don't panic. Small, contained fires are usually best smothered with an asbestos pad or damp towel. If you are involved in a fire or serious accident, don't run. Ask for help immediately and keep calm. Quick and thorough dousing under the safety shower can minimize damage. Be prepared to help someone else involved in an accident, calmly and efficiently, but don't get in the way of your instructor when he or she is answering an emergency call.

These precautions and procedures are not all you should know and practice in the area of laboratory safety. One of the most important safety rules is to *know what you're doing*. Read experimental procedures before coming to the laboratory, taking special note of potential hazards and indicated safety precautions.

Take the time to find out all the safety regulations for your particular course. Follow these meticulously. Remember that unsafe laboratory practices endanger you and your neighbors.

If you have any questions regarding safety or emergency procedures, discuss them with your instructor. Then sign and hand in the following safety agreement.

Safety Agreement

I have studied and agreed to follow the safety regulations as required for this course. I have located all emergency equipment and now know how to use it. I understand that I may be dismissed from the laboratory for failure to comply with stated safety regulations.

Signature ______________________________

Print name ______________________________

Date ______________________________

Course ______________________________

Section ______________________________

Instructor ______________________________

Safety Agreement

I have studied and agreed to follow the safety regulations as required for this course. I have located all emergency equipment and now know how to use it. I understand that I may be dismissed from the laboratory for failure to comply with stated safety regulations.

Signature ______________________________

Print name ______________________________

Date ______________________________

Course ______________________________

Section ______________________________

Instructor ______________________________

Math Skills Review

[BASED ON MATERIAL FROM *FUNDAMENTALS OF CHEMISTRY*, ROD O'CONNOR, HARPER & ROW (1974)]

Students rarely find chemical concepts difficult to master, but the mathematical applications of chemical principles are often troublesome. Sometimes the difficulty lies in the area of mathematical manipulations, which were once familiar but have grown "rusty" through disuse. More frequently it is the solving of "word problems" that is the major hindrance to learning chemistry.

The level of this text does not require a high degree of mathematical sophistication, but reasonable skill in problem solving is necessary. The following sections will illustrate the type of mathematics required and will enable you to identify those areas requiring some review or further study.

The general areas of competence may be roughly classified as follows:

A. Algebraic manipulations.
B. Dimensional conversions (using the "unity factor" method).
C. Significant figures.
D. Manipulation of exponents.
E. Scientific notation.
F. Logarithms.
G. Problem solving (converting verbal problems into mathematical formulas for solution).

In addition, some facility with using a slide rule is highly desirable.

Try the following Pre-Test, and check your answers at the end of this section. If you miss any questions, review the topics indicated, then try the Post-Test. If you are still having trouble, consult your instructor for additional help.

PRE-TEST

A. Algebraic Manipulations

1. $x + 6 = 10$
$x =$ ______

2. $7 - x = 24$
$x =$ ______

3. $2x + 8 = 4$
$x =$ ______

4. $\frac{x}{6} = \frac{1}{4}$
$x =$ ______

5. $\frac{3}{x} = \frac{2}{7}$
$x =$ ______

6. $\frac{5x}{12} = \frac{9}{16}$
$x =$ ______

B. Dimensional Conversions

1. 4 ft = ______ in.
2. 264 ft = ______ mile
3. 2 in. = ______ yard
4. 16 tons = ______ lb
5. 15 pints = ______ gal
6. 1 ml = ______ liter
7. 1 cm = ______ m
8. 1 dm = ______ m
9. 1 kg = ______ g
10. 1 megaton = ______ tons
11. 1 in. = ______ cm
12. 1.00 lb = ______ g
13. 1.00 qt = ______ liter
14. 1 cm = ______ Å
15. 212°F = ______ °C
16. 100°C = ______ °K
17. 2.0 ft = ______ dm
18. 5.0 km = ______ miles
19. 125 lb = ______ kg
20. 0.0020 in. = ______ Å
21. 37.0°C = ______ °F
22. −40°F = ______ °K

C. Significant Figures

Report each answer with the proper number of significant figures:

1. 0.20 lb × 125 ft = ______ ft lb
2. 0.1364 ft × 12 in./1 ft × 2.54 m/1 in. = ______ cm
3. 2 × 13.7 mg = ______ mg

D. Manipulation of Exponents

1. $10^2 \times 10^5 =$ ______

2. $\frac{10^5}{10^2} =$ ______

3. $\frac{10^4}{10^7} =$ ______

4. $10^{-2} \times 10^5 =$ ______

5. $\frac{10^4}{10^{-7}} =$ ______

6. $(10^4)^3 =$ ______

7. $\frac{(10^4)^2}{(10^4)^5} =$ ______

8. $\sqrt{10^{16}} =$ ______

E. Scientific Notation

Write each quantity in scientific notation:

1. 125 cm = ______ cm
2. 0.125 cm = ______ cm
3. 1000.0 cm = ______ cm
4. 0.0020 cm = ______ cm
5. 1,000,000 cm = ______ cm

F. Logarithms (Base 10)(See Log Table, Appendix B)

1. log 100 = ______
2. log(0.001) = ______
3. log 2.00 = ______
4. log 20.0 = ______
5. log 0.0200 = ______
6. $\log(2.0)^2 =$ ______
7. $\log(0.2)^2 =$ ______
8. $\log \sqrt{2.0} =$ ______
9. $\log \sqrt[3]{9.0} =$ ______
10. $\log\left(\frac{13 \times 0.47}{0.028}\right) =$ ______
11. log ______ = 6.000
12. log ______ = −4.000
13. log ______ = 0.6990
14. log ______ = 3.6990
15. log ______ = −0.6990
16. log ______ = −1.3980

G. Problem Solving

1. A gross of assorted nails weighs 1.1 kg. What is the average mass of a nail in this assortment in grams? ______
2. A manufacturing concern plans to distribute nut/bolt/washer sets. Each set will consist of one 18.5 g nut, one 31.0 g bolt, and two 5.50 g washers. The sets will be sold in 1-gross packages. What will be the net mass of each package? ______
3. What is the maximum mass of nut/bolt/washer sets, as described in Problem 2, that could be prepared from 8.00 kg of nuts, 18.2 kg of bolts, and 12.2 kg of washers? ______
4. The nut/bolt/washer set manufacturer described in Problem 2 finds that his assembly line is not entirely efficient. Some components are lost and some sets are improperly assembled and must be discarded at the inspection site. On the average, when the assembly process begins with 50.0 kg of nuts and the equivalent numbers of bolts and washers, only 138 kg of complete sets pass the final inspection. What percentage of the theoretical production actually passes inspection? ______

REVIEW

The following discussions represent simple pragmatic approaches to the types of mathematical operations most frequently encountered in an introductory chemistry course. They are not intended as mathematical theory, but rather as a review or introduction of the how-to-do-it of solving simple problems.

Section A: Algebraic Manipulations

The key to solving simple algebraic equations containing a single unknown (e.g., $x + 6 = 10$) is to realize that the equation represents a statement that two quantities are equal. Thus any mathematical operation can be applied to one side of the equation if it is also applied to the other. It is then necessary to determine what operations are required for the side containing the unknown (to "isolate" the unknown) and to apply these operations to both sides of the equation. This principle is illustrated by the following solutions to the problems in part A of the Pre-Test.

EXAMPLE 1

$$x + 6 = 10$$

SOLUTION

To isolate x, it is necessary to subtract 6:

$$(x + 6) - 6 = (10) - 6$$
$$x = 4$$

EXAMPLE 2

$$7 - x = 24$$

SOLUTION

To isolate x, it is necessary to subtract 7, then multiply by −1:

$$(7 - x) - 7 = (24) - 7$$
$$-x = 17$$
$$(-x)(-1) = (17)(-1)$$
$$x = -17$$

EXAMPLE 3

$$2x + 8 = 4$$

SOLUTION

To isolate x, it is necessary to subtract 8, then divide by 2:

$$(2x + 8) - 8 = (4) - 8$$
$$2x = -4$$
$$\frac{(2x)}{2} = \frac{(-4)}{2}$$
$$x = -2$$

EXAMPLE 4

$$\frac{x}{6} = \frac{1}{4}$$

SOLUTION

To isolate x, it is necessary to multiply by 6:

$$\left(\frac{x}{6}\right)(6) = \frac{1}{4}(6)$$
$$x = \frac{6}{4}$$
$$x = 1.5$$

EXAMPLE 5

$$\frac{3}{x} = \frac{2}{7}$$

SOLUTION

First, convert the unknown term from a denominator to a numerator position (by multiplying by x):

$$\left(\frac{3}{x}\right)x = \left(\frac{2}{7}\right)x$$
$$3 = \frac{2x}{7}$$

Then, to isolate x it is necessary to multiply by 7 and divide by 2; that is, multiply by $\frac{7}{2}$:

$$(3)\left(\frac{7}{2}\right) = \left(\frac{2x}{7}\right)\left(\frac{7}{2}\right)$$
$$\frac{21}{2} = x$$
$$x = 10.5$$

EXAMPLE 6

$$\frac{5x}{12} = \frac{9}{16}$$

SOLUTION

To isolate x, it is necessary to multiply by 12 and divide by 5 (i.e., multiply by $\frac{12}{5}$):

$$\left(\frac{5x}{12}\right)\left(\frac{12}{5}\right) = \left(\frac{9}{16}\right)\left(\frac{12}{5}\right)$$
$$x = \frac{108}{80}$$
$$x = 1.35$$

Section B: Dimensional Conversions

Most scientific measurements are expressed in metric system units.* Until these units are universally accepted, it is necessary to convert from one measuring system to another. A large number of conversion factors have been tabulated in such sources as the *Handbook of Chemistry and Physics* (Chemical Rubber Company), but most conversions required in an introductory study of chemistry can be made by using the simple factors given in Table 1 and the metric system prefixes in Table 2. Conversions among the Celsius (centigrade), Fahrenheit, and Kelvin ("absolute") temperature scales are given in Table 3.

Table 1 Common conversion factors

Approximate English-metric equivalents
1.00 pound (lb) = 454 grams (g)
1.00 quart (qt) = 0.946 liters (l)
Exactly defined equivalents
English
1 mile = 5280 feet (ft)
1 yard = 3 feet (ft)
1 foot (ft) = 12 inches (in.)
1 ton = 2000 pounds (lb)
1 pound (lb) = 16 ounces (oz)
1 gallon (gal) = 4 quarts (qt)
1 quart (qt) = 2 pints (pt)
1 pint (pt) = 16 ounces (fluid)(oz)
Metric
1 cubic centimeter (cm^3) = 1 milliliter (ml)
1 angstrom unit (Å) = 1×10^{-8} centimeter (cm)
1 micron (μ) = 1×10^{-6} meter (m)
English-metric equivalents
1 inch (in.) = 2.54 centimeters (cm) (exactly)

Table 2 Common metric system prefixes

mega = million (10^6)
kilo = thousand (10^3)
deci = tenth (10^{-1})
centi = hundredth (10^{-2})
milli = thousandth (10^{-3})

Table 3 Temperature scale conversions

$°F = (\frac{9}{5})(°C) + 32°$
$°C = (\frac{5}{9})(°F - 32°)$
$°K = °C + 273°$

Unity Factors

Any quantity can be multiplied by unity (one) without changing its value. A unity factor employs this to change the dimensions of a quantity without changing the value of the quantity, although the numbers used to express the value may be changed. If someone asked you the number of inches in 2 yards, you would reply, "72." The value of the length was not changed; that is, it is the same length whether you call it 2 yards or 72 inches (or 6 feet). You probably made this conversion almost intuitively. All the unity factor method does is to provide a systematic procedure for making such conversions in cases a bit too complicated for the "intuitive approach."

Since any fraction whose numerator and denominator are equivalent has the value of unity, you can make unity factors from any dimensional equivalents, such as those given in Table 1 or implied by the prefix meanings of Table 2, for example,

Equivalents	Corresponding unity factors
1 in. = 2.54 cm	$\frac{1\text{ in.}}{2.54\text{ cm}}$ or $\frac{2.54\text{ cm}}{1\text{ in.}}$
1 mile = 5280 ft	$\frac{1\text{ mile}}{5280\text{ ft}}$ or $\frac{5280\text{ ft}}{1\text{ mile}}$
1 kg = 1000 g (kilo = thousand)	$\frac{1\text{ kg}}{1000\text{ g}}$ or $\frac{1000\text{ g}}{1\text{ kg}}$

Unity factors are valuable in problems of dimensional conversion and, as we shall see later, in a variety of other types of problem situations. The method simply requires setting up the problem in fractional form, using dimensions for all quantities involved. The quantity to be converted is multiplied

*For information concerning the new "SI units," consult your instructor.

by appropriate unity factors until all dimensions can be canceled except for those desired in the answer. In multistep conversions it is well to have a systematic sequence; typically, numerator dimensions are converted first, then denominator units.

EXAMPLE 7

Convert 60 miles per hour (60 miles hr^{-1}) to feet per second. (Remember that "per" means "divide by.")

STEPWISE SOLUTION

$$\frac{60 \text{ miles}}{1 \text{ hr}} = \underline{\qquad} \text{ ft sec}^{-1}$$

First, convert the numerator unit (miles ⟶ feet), using a unity factor expressing the feet-miles equivalence, in a form such that the "miles" will cancel:

$$\frac{60 \not{\text{miles}}}{1 \text{ hr}} \times \frac{5280 \text{ ft}}{1 \not{\text{mile}}} = \frac{(60 \times 5280) \text{ ft}}{1 \text{ hr}}$$

Next, convert the denominator unit (hour ⟶ seconds), using a set of unity factors expressing hour ⟶ minutes ⟶ second equivalences, in proper form for dimensional cancellation:

$$\frac{(60 \times 5280) \text{ ft}}{1 \not{\text{hr}}} \times \frac{1 \not{\text{hr}}}{60 \not{\text{min}}} \times \frac{1 \not{\text{min}}}{60 \text{ sec}} = \frac{(60 \times 5280) \text{ ft}}{(60 \times 60) \text{ sec}}$$

The combined solution is

$$\frac{60 \not{\text{miles}}}{\not{\text{hr}}} \times \frac{5280 \text{ ft}}{\not{\text{mile}}} \times \frac{1 \not{\text{hr}}}{60 \not{\text{min}}} \times \frac{1 \not{\text{min}}}{60 \text{ sec}} = \frac{60 \times 5280 \text{ ft}}{60 \times 60 \text{ sec}} = 88 \text{ ft sec}^{-1}$$

Note that if any factors had been omitted or included "upside down," the necessary cancellation of units would not have worked.

Once the dimensions desired for the answer are obtained, the numbers are placed in scientific notation for calculation of the final answer. It is important that an "estimation" step be included as a check on the reasonableness of an answer. The estimation step involves rounding all numerical quantities to the nearest whole number and performing indicated operations to determine the order of magnitude expected for the final answer.

EXAMPLE 8

Convert 62.4 pounds per cubic foot (62.4 lb ft^{-3}) to units of grams per cubic centimeter.

SOLUTION

First, convert all dimensions:

$$\frac{62.4 \text{ lb}}{1 \text{ ft}^3} \times \frac{454 \text{ g}}{1.00 \text{ lb}} \times \frac{1 \text{ ft}^3}{(12 \text{ in.})^3} \times \frac{(1 \text{ in.})^3}{(2.54 \text{ cm})^3} = \frac{(62.4 \times 454) \text{ g}}{[(12)^3 \times (2.54^3)] \text{ cm}^3}$$

Second, convert to scientific notation (see Section E):

$$\frac{(6.24 \times 10^1 \times 4.54 \times 10^2) \text{ g}}{[(1.2 \times 10^1)^3 \times (2.54)^3] \text{ cm}^3} = \frac{(6.24 \times 4.54 \times 10^3) \text{ g}}{[(1.2)^3 \times (2.54)^3 \times 10^3] \text{ cm}^3}$$

Third, estimate the order of magnitude:

$6.25 \approx 6$
$4.54 \approx 5$
$1.2 \approx 1$
$2.54 \approx 3$

$$\frac{6 \times 5 \times 10^3}{(1)^3 \times (3)^3 \times 10^3} = \frac{30 \times 10^3}{27 \times 10^3} = \sim 1$$

Hence, the final answer should be closer to 1 than to, for example, 0.1 or 10.

Finally, perform the calculation (e.g., using a slide rule):

$$\frac{(6.24 \times 4.54 \times 10^3) \text{ g}}{[(1.2)^3 \times (2.54)^3 \times 10^3] \text{ cm}^3} = 1.00 \text{ g cm}^{-3} \qquad \text{(to three-place accuracy)}$$

Section C: Significant Figures (See also Appendix C.)

The numerical value of a measured quantity is limited by both the precision and the accuracy of the measurements involved. Precision is concerned with how close together the results of a series of measurements of the same quantity are, whereas accuracy is determined by how close a measured value is to the "true" value (usually compared to an exactly defined standard).

For example, suppose someone wishes to know the time at which an event took place. He might consult his wristwatch. However, if for some reason he wants to be more certain of the time, he might check his watch against his alarm clock or even use the telephone to dial "time and temperature." The chances are that he would arrive at three different values, and he might choose to average these values. However, if the three values were widely different (poor precision), he might wish to consult a number of other timepieces. If he can obtain several independent readings that differ only slightly, then he feels a certain confidence in accepting the average value. If the knowledge of the time were really important, then he might telephone the National Bureau of Standards and set ("calibrate") his watch against an accepted standard. This procedure would guarantee an improvement in accuracy, whereas any improvement in precision alone would simply increase the confidence in the measured value. Ten very close readings from ten different electric clocks during a power failure would give good precision, but poor accuracy.

Measurements in science should, whenever possible, use measuring devices that have been calibrated against accepted standards and repetitive determinations whose precision indicates minimized errors in the use of the devices. There is still another limitation on the accuracy of a reported value; this involves the calculations made with measured quantities. Although pure numbers (dimensionless) and exactly defined whole-number conversion factors (e.g., 1 ft = 12 in., or 1 in. = 2.54 cm) do not limit the number of meaningful figures in a computed quantity, other factors do. Common sense suggests that one cannot improve the accuracy of a measured value simply by performing mathematical operations. The accuracy of a computed value (number of significant figures reported) cannot exceed that of the least accurately known quantity used in the calculation. In most problem situations at the introductory level, it is routine to use "slide rule accuracy"; that is, answers are reported with the number of significant figures that can be read from a simple slide rule. One must, however, be aware of the limitations described above and apply them when necessary. (See also Sections D and E for exponential manipulations.)

EXAMPLE 9

The diameter of a cylinder is estimated, using a yard-stick, to be 1.1 in. The diameter of a second cylinder is measured carefully with a calibrated micrometer and found to be 0.8417 ± 0.0002 in. What value can meaningfully be reported as the sum of the diameters of the two cylinders?

SOLUTION

$$\begin{array}{r} 1.1 \\ \underline{0.8417} \\ 1.9417 \end{array}$$

But the less accurately known quantity is measured only to the nearest 0.1 in., so the sum can be reported only as approximately 1.9 in.

EXAMPLE 10

How many samples of 25.0 mg each could be prepared from approximately 2 lb of a crystalline salt?

SOLUTION

$$\frac{1\ \text{sample}}{25.0\ \text{mg}} \times \frac{10^3\ \text{mg}}{1\ \text{g}} \times \frac{454\ \text{g}}{1\ \text{lb}} \times \frac{2\ \text{lb}}{1} = \underline{\qquad\qquad}\ \text{samples}$$

Estimating

$$\frac{10^3 \times 5 \times 10^2 \times 2}{2.5 \times 10^1} = 4 \times 10^4$$

$$\frac{10^3 \times 4.54 \times 10^2 \times 2}{2.50 \times 10^1} = 3.63 \times 10^4$$

However, the number of significant figures is limited by the measured quantity "approximately 2 lb." (Note: not by the exact figures "1 sample," "10^3 mg," "1 g," or "1 lb.") Thus the answer can be reported only as "approximately 4×10^4 samples."

The proper use of scientific notation clearly indicates the number of significant figures.

Suggestion: To convince yourself that the above solution is correct, remember that "approximately 2 lb" would generally indicate a weight between 1.5 and 2.5 lb (i.e., values in this range would "round off" to 2 lb). Calculate the number of samples if 1.5 lb were available and if 2.5 lb were available. Then explain why rounding to one digit was reasonable for this example.

Section D: Manipulation of Exponents

Scientific calculations are frequently handled most conveniently by expressing quantities in scientific notation (see Section E). Such operations require simple manipulation of exponents, usually exponents of 10. When the same base (e.g., 10) is involved, the following rules apply:

1. When the operation involves multiplication, add the exponents algebraically; for example,

$$10^3 \times 10^2 = 10^{(3+2)} = 10^5$$
$$10^4 \times 10^8 \times 10^{-5} = 10^{[4+8+(-5)]} = 10^7$$

2. When the operation involves division, subtract the divisor exponent from the numerator exponent; for example,

$$\frac{10^5}{10^3} = 10^{(5-3)} = 10^2$$
$$\frac{10^7}{10^{11}} = 10^{(7-11)} = 10^{-4}$$
$$\frac{10^8}{10^{-2}} = 10^{[8-(-2)]} = 10^{10}$$
$$\frac{10^6 \times 10^4}{10^3 \times 10^9} = 10^{[6+4-(3+9)]} = 10^{-2}$$

3. When the operation involves roots or powers, divide the exponent by the root number or multiply the exponent by the power number, respectively; for example,

$$\sqrt{10^4} = (10^4)^{1/2} = 10^{4/2} = 10^2$$
$$\sqrt[5]{10^{20}} = (10^{20})^{1/5} = 10^{20/5} = 10^4$$
$$(10^2)^3 = 10^{2\times 3} = 10^6$$
$$(10^{-7})^4 = 10^{-7\times 4} = 10^{-28}$$

[Note: To take "roots" for cases in which the exponent is not a multiple of the "root number," use logarithms (see Section F).]

The operations described apply only to cases in which the same base number is involved. To illustrate this point, show that $2^3 \times 10^2$ is not equal to $(2 \times 10)^5$.

Section E: Scientific Notation

It is inconvenient to work with the standard form of very large or very small numbers. An exponential notation (see also Section D) is commonly employed to simplify calculations involving such numbers. The procedure for converting from standard form to scientific (exponential) notation* follows.

Step 1: Reset the decimal point so that there is a single digit (not zero) to its left, for example,

$$0.00044 \longrightarrow 0.0004_{\wedge}4 \longrightarrow 0.0004_{\odot}4$$

Step 2: Count the number of digits between the original and new decimal point positions to determine the magnitude of the exponent of 10, for example,

1 2 3 ④
$0.0004_{\odot}4$
10^4

*In scientific notation, the only zeros recorded are those that represent actual measured quantities. Zeros used in standard form merely to indicate numerical magnitude are omitted.

Step 3: Determine the sign for the exponent of 10 by the direction from the new to the original decimal point position (left is negative, right is positive), for example,

0.0004⊙4

$\longleftarrow$

10^{-4}

Thus,

$0.00044 = 4.4 \times 10^{-4}$

EXAMPLE 11

Assume that the approximate diameter of the solar system, to the nearest million miles, is 7,300,000,000. Write this in scientific notation.

SOLUTION

Step 1:

(understood location of decimal point)

7,,300,000,000⊙

Step 2:

1 2 3 4 5 6 7 8 ⑨

7⊙300,000,000.

10^{9}

Step 3:

7⊙300,000,000.

$\longrightarrow$|

(Direction indicates positive exponent.)

Since the measured quantity is known only to the nearest million, the last six zeros (indicating order of magnitude) are omitted. Thus 7,300,000,000 miles (to the nearest million) is written as 7.300×10^{9} miles.

EXAMPLE 12

Write 0.00020 in. in scientific notation.

SOLUTION

Step 1:

0.0002⊙0

Step 2:

1 2 3 ④

0.0002⊙0

10^{4}

Step 3:

0.0002⊙0

|$\longleftarrow$

(Direction indicates negative exponent.)

The four zeros preceding the 2 are not measured quantities but only represent order of magnitude. The final zero does indicate a measurement to the nearest 0.00001 in. Thus 0.00020 in. is written as 2.0×10^{-4} in.

Section F: Logarithms

Chemical calculations normally use base 10 logarithms. These logarithms are exponents of 10 and, as such, are manipulated as described in Section D. Logarithms consist of a whole number (characteristic) followed by a decimal number (mantissa) or zero. The mantissa is found from a logarithm table, such as the one in Appendix B.

1. To find the logarithm of a number:
   a. Write the number in scientific notation (Section E), for example,

   $$343 = 3.43 \times 10^2$$

   b. Look up the logarithm (mantissa) of the number preceding the power of 10 in a log table. (Note: All such mantissas are decimal numbers, although most log tables do not show the decimal point.) For example,

   $$\log 3.43 = 0.5353$$

   c. Add this to the exponent of 10. For example,

   $$\log 343 = 0.5353 + 2 = 2.5353$$

In some cases the log will be a negative number. This is perfectly legitimate, although other forms are more frequently used. Operational uses of logs in many chemical calculations are simpler using negative logs. For example,

$$\log 0.0200 = ?$$

Step a: $0.0200 = 2.00 \times 10^{-2}$
Step b: $\log 2.00 = 0.3010$
Step c: $\log 0.0200 = 0.3010 + (-2) = -1.6990$

(Entire logarithm is negative)

2. To find a number from its logarithm:
   a. Be sure the log is written in a form such that the decimal part (mantissa) is positive. (That is how the logs are conventionally tabulated.) For example, what number has the logarithm -4.3010? Add the log to the next largest whole number, then subtract that number. For example,

   $$(-4.3010 + 5) - 5 = 0.6990 - 5$$

   b. Look up the mantissa in the logarithm part of the table and find the number to which it most closely corresponds. For example, the number 5.00 corresponds to 0.6990 in the mantissa part of the table.
   c. Write this number as the first part of a scientific notation. Then use the whole number in the log as the exponent of 10. For example,

   from log table: $0.6990\ (-5) \rightarrow 5.00 \times 10^{-5}$

If the original log is positive, follow steps **b** and **c** above. For example, what number has the logarithm 11.0828?

from log table: $11.0828 \rightarrow 1.21 \times 10^{11}$

3. To multiply or divide, using logs:
   a. Write the operation in fractional form. (Use scientific notation.) For example,

   $$\frac{(7.1 \times 10^{-3}) \times (1.48 \times 10^{2}) \times 2.73}{8.1 \times 6.02 \times 10^{23}}$$

   b. Combine all powers of 10 into a single power of 10 to appear in the numerator (see Section D). For example,

   $$\frac{7.1 \times 1.48 \times 2.73}{8.1 \times 6.02} \times 10^{(-3+2-23)} = \frac{7.1 \times 1.48 \times 2.73}{8.1 \times 6.02} \times 10^{-24}$$

   c. Obtain the sum of the logs of all numbers in the numerator. Subtract from this the sum of the logs of all numbers in the denominator. For example,

Numerator	Denominator	
log 7.1 = 0.8513	log 8.1 = 0.9085	1.4578
log 1.48 = 0.1703	log 6.02 = 0.7796	−1.6881
log 2.73 = 0.4362	sum: 1.6881	−0.2303
sum: 1.4578		

d. Find the number that has this logarithm (Section F, 2). Combine it with the power of 10 found in step **b** above. For example,

$(-.2303 + 1) - 1 = .7697 - 1$

from log table: $.7697 - 1$ → 5.88×10^{-1}

Then

(step **b**)
$5.88 \times 10^{-1} \times 10^{-24} = 5.88 \times 10^{-25}$

Note that if any of the numbers in the calculation are measured quantities or other than exactly defined conversion units, round off the final answer to the proper number of significant figures (see Section C).

4. To find roots and powers, using logs: Since logs are exponents of 10, roots and powers are found in the same way as in any other exponential situation (Section D).

Roots

$\sqrt[3]{1728} = ?$

a. Find the log of the number.

$\log(1728) = \log(1.728 \times 10^3) = 3.2375$

b. Divide the log by the "root number."

$\log \sqrt[3]{1728} = \frac{1}{3}(3.2375) = 1.0792$

c. Find the antilog (the number that has this logarithm).

$\sqrt[3]{1728} = \text{antilog } 1.0792 = 1.2 \times 10^1$

Powers

$(25)^2 = ?$

a. Find the log of the number.

$\log(25) = \log(2.5 \times 10^1) = 1.3979$

b. Multiply the log by the power number.

$\log(25)^2 = 2 \times 1.3979 = 2.7958$

c. Find the antilog.

$(25)^2 = \text{antilog } 2.7958 = 6.25 \times 10^2$

Section G: Problem Solving

Some people seem to have an intuitive grasp of how to approach a problem to obtain a rapid and efficient solution. Although there is little doubt that some are more gifted than others in this respect, it is a talent that can be acquired. Proficiency in this skill, as in all skills, improves with practice.

When a mechanically inept customer takes his car to a mechanic with the complaint that it is getting poor gas mileage and makes a "funny rumbling sound," this poses a problem for the mechanic. Although the talented mechanic might not express it this way, his approach to this problem involves certain distinguishable phases:

1. Identification of what he wants to find out.
2. Recognition of what information is given that might be useful.
3. Identification of other information needed, but not given (i.e., information that must be obtained from memory and from other sources).
4. Selection of a method of attack on the problem.
5. Manipulation of information and tools to achieve a solution.
6. Checking that the solution is probably satisfactory.

For the mechanic, these steps might involve

1. Find out what is causing the poor gas mileage and unusual engine noise.
2. Given: Poor mileage + "funny rumbling sound."
3. Other information: Memory of previous experience with similar problems plus application of various suitable test instruments and other pertinent observations.
4. Selecting parts to be replaced, adjustments to be made, and tools to be used.
5. Fixing the engine.
6. Test driving.

This system is a general one for solving all kinds of problems. It is the basis for most scientific research, and it is the key to solving all sorts of mathematical problems. The more problems you solve, the easier it becomes, until you, like the competent mechanic, can solve problems without formally thinking out each step of the method. Until the stage of "intuitive" proficiency is reached, however, it is worthwhile to proceed slowly and systematically through the various phases of problem solving.

Let us illustrate the approach by a "case study" of the first three problems given in Section G of the Pretest.

PROBLEM 1

A gross of assorted nails weighs 1.1 kg. What is the average mass of a nail in this assortment, in grams?

SOLUTION

Step 1: Find mass per nail (in grams).
Step 1: Given 1.1 kg per gross (of nails).
Step 3: Needed, not given: How many items per gross? How many grams per kilogram?
Answers (from memory): "144 per gross," "1000 g per kilogram."
Step 4: Method of attack
Divide mass by number to find average mass of each. Use unity factor method, including kilogram to gram conversion.
Step 5: Manipulation

$$\frac{1.1\text{ kg}}{\text{gross}} \times \frac{1\text{ gross}}{144} \times \frac{1000\text{ g}}{1\text{ kg}} = \frac{1100\text{ g}}{144}$$

Estimate

$$\frac{1.1 \times 10^3}{1.44 \times 10^2} \approx \frac{1 \times 10^3}{1 \times 10^2} \approx 10$$

Solve

$$\frac{(1.1 \times 10^3)\text{g}}{1.44 \times 10^2} = 0.764 \times 10^1\text{g} \qquad (7.6\text{ g, to two-place accuracy})$$

Step 6: Check
a. Dimensions canceled properly.
b. Math check

$$7.6\text{ g} \times \frac{1\text{ kg}}{10^3\text{ g}} \times \frac{144}{1\text{ gross}} = 1.1\text{ kg/gross}$$

PROBLEM 2

A manufacturing concern plans to distribute nut/bolt/washer sets. Each set will consist of one 18.5-g nut, one 31.0-g bolt, and two 5.50-g washers. The sets will be sold in 1-gross packages. What will be the net mass of each package?

SOLUTION

Step 1: Find net mass of a 1-gross package of the sets.
Step 2: Given 1 nut/set at 18.5 g/nut; 1 bolt/set at 31.0 g/bolt; 2 washers/set at 5.50 g/washer; 1 gross per package.

Step 3: Needed, not given:

How many items per gross?

Answer (from memory): 144 per gross

What does "net mass" mean?

Answer (from memory): mass of package contents (i.e., 1 gross of the sets)

Step 4: Method of attack

Find the mass of each set and multiply by the number of sets per package. Use unity factor method where appropriate. (Note: Often there is no unique method of attack. Here, for example, we might have decided to find the masses of 1 gross of nuts, 1 gross of bolts, and 2 gross of washers, and then add these together.)

Step 5: Manipulation

$$\left(\frac{18.5\text{ g}}{1\text{ nut}} \times \frac{1\text{ nut}}{1\text{ set}}\right) + \left(\frac{31.0\text{ g}}{1\text{ bolt}} \times \frac{1\text{ bolt}}{1\text{ set}}\right) + \left(\frac{5.50\text{ g}}{1\text{ washer}} \times \frac{2\text{ washers}}{1\text{ set}}\right)$$

$$= \frac{[18.5 + 31.0 + (2)(5.50)]\text{ g}}{1\text{ set}} = \frac{60.5\text{ g}}{1\text{ set}}$$

$$\frac{60.5\text{ g}}{1\text{ set}} \times \frac{1\text{ gross set}}{1\text{ pkg}} \times \frac{144}{1\text{ gross}} = \frac{(60.5 \times 144)\text{g}}{1\text{ pkg}}$$

Estimate

$$6.05 \times 10^1 \times 1.44 \times 10^2 \approx 6 \times 1 \times 10^3 \approx 6000$$

Solve

$$\frac{(6.05 \times 1.44 \times 10^3)\text{ g}}{1\text{ pkg}} = 8.71 \times 10^3\text{ g/pkg} \qquad (8710\text{ g/pkg})$$

Step 6: Check

a. Dimensions canceled properly.

b. Arithmetic rechecked for accuracy.

PROBLEM 3

What is the maximum mass of nut/bolt/washer sets, as described in Problem 2, that could be prepared from 8.00 kg of nuts, 18.2 kg of bolts, and 12.2 kg of washers?

SOLUTION

Step 1: Find maximum mass of sets from given weights of components.

Step 2: Given 8.00 kg of nuts, 18.2 kg of bolts, 12.2 kg of washers.

Step 3: Needed, not given

Composition of a set

Answer: (Given in Problem 2)

1 nut + 1 bolt + 2 washers

Masses of components

Answer: (Given in Problem 2)

18.5 g/nut, 31.0 g/bolt, 5.50 g/washer

Mass of a set

Answer: (Found during solution to Problem 2)

60.5 g/set

Step 4: Method of attack

The word "maximum" in the problem suggests that there is some limitation other than just the sum of the masses of the components given. It must be recognized that we can no longer make sets when we run out of any one component. Thus the maximum mass of sets will correspond to the mass obtainable from the limiting component (i.e., the smallest mass of sets obtainable from the masses of each component available). Since we are dealing with masses (rather than "pure" numbers), some appropriate equivalents would be useful in constructing unity factors.

$$18.5\text{ g nuts} = 60.5\text{ g sets}$$
$$31.0\text{ g bolts} = 60.5\text{ g sets}$$
$$(2 \times 5.50)\text{ g washers} = 60.5\text{ g sets}$$

Parts-by-mass unity factors

$$\frac{60.5\text{ g sets}}{18.5\text{ g nuts}} = \frac{60.5\text{ sets}}{18.5\text{ nuts}} \qquad \text{(by mass)}$$

$$\frac{60.5\text{ g sets}}{31.0\text{ g bolts}} = \frac{60.5\text{ sets}}{31.0\text{ bolts}} \qquad \text{(by mass)}$$

$$\frac{60.5 \text{ g sets}}{11.0 \text{ g washers}} = \frac{60.5 \text{ sets}}{11.0 \text{ washers}} \quad \text{(by mass)}$$

Then we must calculate the mass of sets obtainable by using up all of each component available and select the *smallest* of these values as the limit on the mass of sets obtainable.

Step 5: Manipulation

From nuts

$$\frac{60.5 \text{ sets}}{18.5 \text{ nuts}} \times \frac{8.00 \text{ kg nuts}}{1} = 26.2 \text{ kg sets}$$

From bolts

$$\frac{60.5 \text{ sets}}{31.0 \text{ bolts}} \times \frac{18.2 \text{ kg bolts}}{1} = 35.5 \text{ kg sets}$$

From washers

$$\frac{60.5 \text{ sets}}{11.0 \text{ washers}} \times \frac{12.2 \text{ kg washers}}{1} = 67.1 \text{ kg sets}$$

Since our calculations show that we shall run out of nuts when 26.2 kg of sets have been constructed, the maximum mass of sets obtainable is 26.2 kg.

Step 6: Check

a. Dimensions cancel.

b. Arithmetic rechecked for accuracy.

c. Logic rechecked.

Each new problem poses its own unique challenges, and there is no "magic formula" for problem solving. The stepwise approach suggested should prove useful in establishing a systematic technique. Facility will improve with practice.

POST-TEST

A. Algebraic Manipulations

1. $x - 5 = 9$

$x =$ ________

2. $17 - x = 4$

$x =$ ________

3. $3x + 2 = -11$

$x =$ ________

4. $\frac{x}{5} = \frac{2}{9}$

$x =$ ________

5. $\frac{7}{x} = \frac{28}{9}$

$x =$ ________

6. $\frac{3x}{7} = \frac{9}{5}$

$x =$ ________

B. Dimensional Conversions

1. 9.00 in. = ________ cm

2. 0.122 qt = ________ ml

3. 196 kg = ________ ton

4. 1.00 g cm^{-3} = ________ lb ft^{-3}

5. $-40°F$ = ________ °C

6. 70°C = ________ °K

C. Significant Figures

Report each answer with the proper number of significant figures:

1. $\frac{12.72 \text{ g}}{3.0 \text{ cm}^3} =$ ________ g cm^{-3}

2. 0.0050 mile × 5280 ft $mile^{-1}$

= ________ ft

3. 5 × 90.800 g × 1.00 lb/454 g

= ________ lb

D. Manipulation of Exponents

1. $10^3 \times 10^5 \times 10^{-1} =$ ________

2. $\frac{10^3}{10^8} =$ ________

3. $\frac{10^{-3}}{10^{-7}} =$ ________

4. $(10^{-3})^5 =$ ________

5. $\sqrt[3]{10^6} =$ ________

6. $\frac{10^{-7} \times 10^4}{10^{-2} \times 10^5} =$ ________

E. Scientific Notation

Write each quantity in scientific notation:

1. 1063 cm = ________ cm

2. 0.017 cm = ________ cm

3. 195.0 cm = ________ cm

4. 0.020 cm = ________ cm

F. Logarithms (Base 10)

1. log 264 = ________

2. log 0.264 = ________ or ________

3. $\log(26.4)^3 =$ ________
4. $\log \sqrt[3]{0.0264} =$ ________ or ________
5. log ________ = 4.9030
6. log ________ = −5.6020

G. Problem Solving

1. If 3 dozen eggs cost $1.70, what is the cost per egg, in cents? ________
2. A particular solvent mixture requires three parts acetic acid, two parts water, and two parts *n*-butyl alcohol (by mass). What mass of acetic acid is required for preparation of 140 g of the mixture? ________
3. The speed of light is approximately 3.0×10^{10} cm sec^{-1} in vacuum. How long, in minutes, would it take for light traveling at this speed to go a distance of 193,000 miles? ________
4. Baking soda can be prepared from table salt, ammonia, and carbon dioxide. If experience shows that in a particular plant operation 1.00 g of table salt will produce 1.39 g of baking soda, using appropriate quantities of ammonia and carbon dioxide, what mass, in kilograms, of table salt would be needed to form 2.26 kg of baking soda? ________

ANSWERS TO PRE-TEST

A. 1. 4
2. −17
3. −2
4. 1.5
5. 10.5
6. 1.35

If you missed any, review Section A and then try part A on the Post-Test.

B. 1. 48
2. 0.0500
3. 0.056
4. 32,000
5. 1.875
6. 0.001
7. 0.01
8. 0.1
9. 1000
10. 1,000,000
11. 2.54
12. 454
13. 0.946
14. 10^8
15. 100
16. 373
17. 6.1
18. 3.1
19. 56.8
20. 5.1×10^5
21. 98.6
22. 233

If you missed any, review Section B and then try part B on the Post-Test.

C. 1. 25 (limited by two significant figures in 0.20 lb.)
2. 4.157 (limited by the measured value 0.1364 ft, not by the exact quantities 12 in./1 ft and 2.54 cm/1 in.)
3. 27.4 (not limited by the pure number 2)

If you missed any, review Section C and then try part C on the Post-Test.

D. 1. 10^7
2. 10^3
3. 10^{-3}
4. 10^3
5. 10^{11}
6. 10^{12}
7. 10^{-12}
8. 10^8

If you missed any, review Section D and then try part D on the Post-Test.

E. 1. 1.25×10^2
2. 1.25×10^{-1}
3. 1.0000×10^3
4. 2.0×10^{-3}
5. 10^6 or 1.000000×10^6 (number of significant figures is uncertain)

If you missed any, review Section E and then try part E on the Post-Test.

F. 1. 2
2. −3
3. 0.3010
4. 1.3010
5. −1.6990 (or $\bar{2}.3010$)
6. 0.6020
7. −1.3980 (or $\bar{2}.6020$)
8. 0.1505
9. 0.3181
10. [1.1139 + (−0.3279) − (−1.5528) = 2.3388]
11. 10^6
12. 10^{-4}
13. 5.00
14. 5.00×10^3 (5000)
15. 2.00×10^{-1} (0.200)
16. 4.00×10^{-2} (0.0400)

If you missed any, review Section F and then try part F on the Post-Test

G. 1. 7.6 g
2. 8710 g/pkg
3. 26.2 kg
4. 84.5%

If you missed any, review Section G and then try part G on the Post-Test.

ANSWERS TO POST-TEST

A. **1.** 14
2. 13
3. $-4\frac{1}{3}$
4. $1\frac{1}{9}$
5. $2\frac{1}{4}$
6. $4\frac{1}{5}$

B. **1.** 22.9
2. 115
3. 0.216
4. 62.4
5. -40
6. 343

C. **1.** 4.2
2. 26
3. 1.00

D. **1.** 10^7
2. 10^{-5}
3. 10^4
4. 10^{-15}
5. 10^2
6. 10^{-6}

E. **1.** 1.063×10^3
2. 1.7×10^{-2}
3. 1.950×10^2
4. 2.0×10^{-2}

F. **1.** 2.4216
2. -0.5784 or $\bar{1}.4216$
3. 4.2648
4. -0.5261 or $\bar{1}.4739$
5. 8.00×10^4
6. 2.50×10^{-6}

G. **1.** 4.72¢
2. 60 g
3. 0.017 min
4. 1.63 kg

If you missed any, except for a few accidental errors you can correct, consult your instructor for suggestions for additional study.

Investigation 1

DETECTING IRON, COPPER, AND LEAD IN ORES

INTRODUCTION: Iron, copper, and lead are among the most important metals of our modern technological society. Iron is the fourth most abundant element in the earth's crust (the top three, in order, are oxygen, silicon, and aluminum). Rarely found in nature as the free metal, iron is produced commercially mainly from its oxide ores, *hematite* (Fe_2O_3) and *magnetite* (Fe_3O_4). Iron *pyrite* (FeS_2) is an important industrial mineral, but its sulfur content makes it unsatisfactory for most types of smelting operations. More than 100 million tons of iron ores are processed annually in the United States, mostly for the production of steel.

Copper, of critical importance for the manufacture of electrical apparatus, wires, and corrosion-resistant pipes and equipment, is much less abundant than iron, constituting only about 10^{-4} percent, by weight, of the earth's crust. Although "native copper" (the free metal) is found in nature, the deposits are relatively rare and most copper is prepared from its sulfide ores, such as *chalcocite* (Cu_2S) and *chalcopyrite* ($CuFeS_2$). Other common ores include *cuprite* (Cu_2O), *azurite* [$Cu_3(CO_3)_2(OH)_2$], and *malachite* [$Cu_2CO_3(OH)_2$]. Malachite samples are often polished for use in jewelry. The depletion of high-grade ores in the United States, coupled with problems of obtaining guaranteed imports from such copper-rich countries as Chile, has necessitated the use of low-grade ores whose smelting is more difficult. As a result, copper prices have skyrocketed in the past few years and it is now economically competitive to recover copper from scrap metals. Nearly half of our current copper production is from salvaged metal waste.

Lead, even less abundant in nature than copper, occurs mainly as *galena* (PbS), from which nearly 2 million tons of lead are produced annually, about one-fourth in the United States. Some of the other common ores are *cerussite* ($PbCO_3$) and *wulfenite* ($PbMoO_4$). Lead is used mainly in the manufacture of "storage" batteries, cable coverings, pigments, and various alloys. Like copper, nearly half of the current lead production is from reclaimed metal scrap, since high-grade lead ores have also been seriously depleted.

During the late nineteenth and early twentieth centuries, mining interests were able to influence state and national legislation to develop extremely favorable "mineral rights" laws and legal shelters for massive and highly profitable exploitation of mineral deposits. Much of this legislation is still in existence and it has been only recently that public challenges to mining and smelter pollution have made some slight impact on the mineral industries.

It has not been too many years since the days when the prospector and the assayer, lured by rich finds of gold and silver, were common sights in the American West. Much of the history of the metal industries in this country can be traced to discoveries of profitable deposits of copper and other "baser" metals made during the search for silver and gold.

The mining industry has progressed a long way from the days of the old assay office to which mineral samples were brought for a laborious, and usually time-consuming, analysis. Today the modern industrial laboratory has a wide variety of analytical instruments for the rapid and accurate assay of ore samples. Simple wet-chemical tests have not, however, become obsolete. The field geologist may need to do a quick preliminary assay of promising mineral finds and the individual prospector or small mining company will usually employ the slower, but less expensive, wet chemical methods.

In this experiment you will investigate methods used for dissolving ore samples, preparing the solutions for analysis, and making simple qualitative tests for some of the more important metals.

Before starting this investigation, you should be familiar with the generally accepted distinctions between "physical" change and "chemical" change and with "working definitions" of *acids, bases,* and *salts.*

References

1. O'Connor, Rod, 1977. *Fundamentals of Chemistry* (2nd ed.). New York: Harper & Row (Unit 3).
2. Moeller, Therald and Rod O'Connor. 1972. *Ions in Aqueous Systems.* New York: McGraw-Hill.

SAFETY You will be working with some dangerous reagents. Wear safety goggles *at all times.* In case of accidental contact with chemicals, wash *immediately and thoroughly* with water and notify your instructor. Exercise special precautions when heating chemicals to avoid splattering.

OBJECTIVES

1. Be able to handle reagents properly, including all safety precautions for the use of nitric acid and other hazardous chemicals.
2. Be able to record and interpret observations on tests performed on solutions of known composition in a systematic way so as to design an analysis scheme for solutions of unknown composition.
3. Be able to use your analysis scheme to test unknown solutions, prepared from ore samples, for the presence of Fe^{3+}, Cu^{2+}, or Pb^{2+} ions (or for any combination of these).

PROCEDURES

Before you can analyze your ore sample, you must dissolve it in nitric acid. This process converts the metallic compounds in the ore into metal nitrates, which are readily soluble in water to give metal cations and nitrate anions. The clear solutions, which you will obtain as described in procedure C, contain Fe^{3+}, Pb^{2+}, or Cu^{2+} (or all of these) depending on the composition of your original ore sample. You must now add other chemical substances (reagents), which are supplied to you as aqueous solutions, to detect the presence of these ions.

You will find in your laboratory station a beaker, test tubes, and dropping bottles containing the reagents you need. There will also be available dropping bottles containing solutions of $FeCl_3$, iron(III) chloride; $Pb(NO_3)_2$, lead(II) nitrate; and $Cu(NO_3)_2$, copper(II) nitrate.

In order to analyze your solution, obtained by dissolving the ore, for Fe^{3+}, Pb^{2+}, and Cu^{2+}, you must first become familiar with the behavior of these ions. In part A you will test known solutions (i.e., solutions for which you know the ions present) with the following reagents:

1.	$NaOH$	sodium hydroxide, 6 *M* (strong base)
2.	NH_3	ammonia, 6 *M* (weak base)
3.	HCl	hydrochloric acid, 6 *M* (strong acid)
4.	H_2SO_4	sulfuric acid, 3 *M* (strong acid)
5.	$K_4Fe(CN)_6$	potassium hexacyanoferrate(II) (a salt)
6.	NH_4CNS	ammonium thiocyanate (a salt)
7.	K_2CrO_4	potassium chromate (a salt)

These reagents will be aqueous solutions. Carefully observe what happens when the reagents are added. Pay special attention to differences in the behavior of Fe^{3+}, Pb^{2+}, and Cu^{2+}. Write all your observations into your notebook, recording descriptions of precipitates or color changes (or both). Make an attempt to express the reactions by means of chemical equations. Consult your laboratory instructor if you have difficulties with this task.

When you have performed all the reactions, study your results and deduce from them a scheme that will allow you to prove the presence of Fe^{3+}, Cu^{2+}, and Pb^{2+} in a solution that might contain all three of these ions.

After you have developed such a scheme and had it approved by your instructor, you are ready to analyze unknown samples.

The following sections contain the detailed procedures that should help you to perform the analyses. Follow the sequence suggested.

A. Procedure for Testing the Known Solutions Containing Fe^{3+} or Cu^{2+} or Pb^{2+}

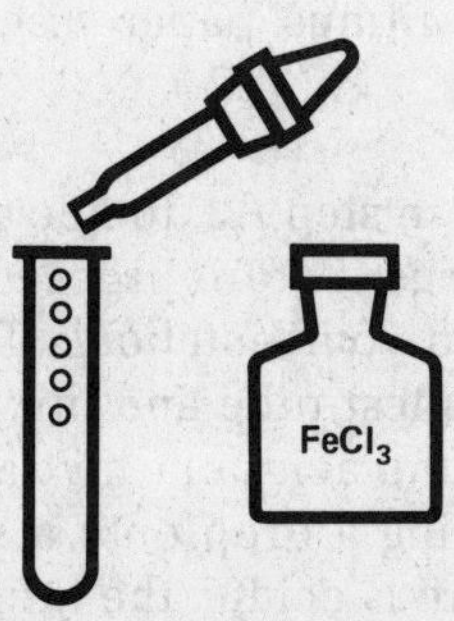

1. Place five drops of the solution containing Fe^{3+} into each of seven test tubes.

2. Add approximately $\frac{1}{2}$ ml (10 drops) of distilled water from the squeeze bottle to each test tube and agitate the mixture.

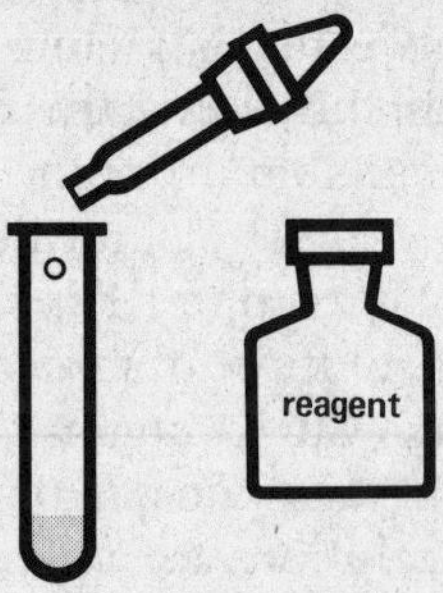

3. To each of the seven test tubes add, drop by drop, one of the seven reagents listed (i.e., a different reagent to each tube). Agitate after each addition. Keep adding reagent until on further addition of one drop no change is observed, or until you have added a maximum of six drops.

After the dropwise addition of the reagents, you should make one of the following observations:

No change in the appearance of the solution.
(If this is the case, make a note of this in your record book. No further tests are necessary for this solution.)
Formation of a precipitate.
(Continue as described in A4, A5, or A6, as appropriate.)
Formation of a colored solution.
(Continue as described in A4, A5, or A6, as appropriate.)

The reagents at your disposal are either acids, bases, or salts. If a precipitate or a colored solution was formed:

Continue with A4 if the reagent added was an acid.
Continue with A5 if the reagent added was a base.
Continue with A6 if the reagent added was a salt.

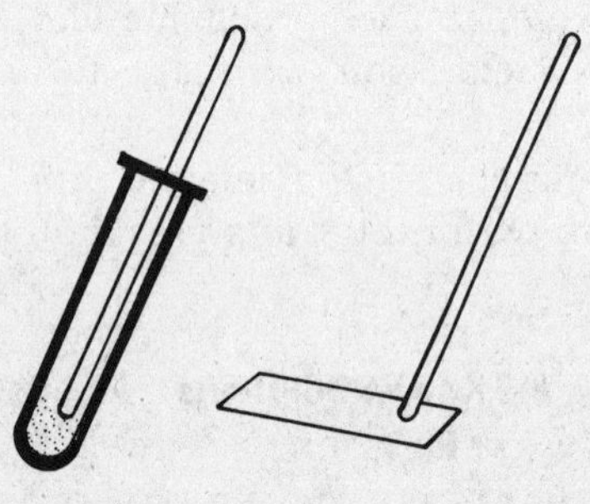

4. You have added an acid in step A3 to one of your solutions and observed a change. Before proceeding further, describe the observation in your notebook. Then add six drops of 6 *M* NaOH to this test tube and mix thoroughly. Record any observed changes. Add three more drops of the NaOH solution and mix again. Now check your mixture by dipping a clean stirring rod into the mixture and transferring a drop onto a strip of *red* litmus paper. If the mixture is basic, the litmus will turn blue. Should the solution not yet be basic, add three more drops of NaOH solution, stir, and repeat the litmus paper test. Record all observations.

5. You have added a base, in step A3, to one of your solutions and observed a change. Before proceeding further, describe the observation in your notebook. Then add six drops of 6 *M* HNO_3 to this test tube and mix thoroughly. Add three more drops of the acid and mix again. Check your mixture by transferring a drop onto a strip of *blue* litmus paper. If the mixture is acidic, the litmus will turn red. Should the solution not yet be acidic, add three more drops of HNO_3 solution, stir, and repeat the litmus paper test. Record all observations.

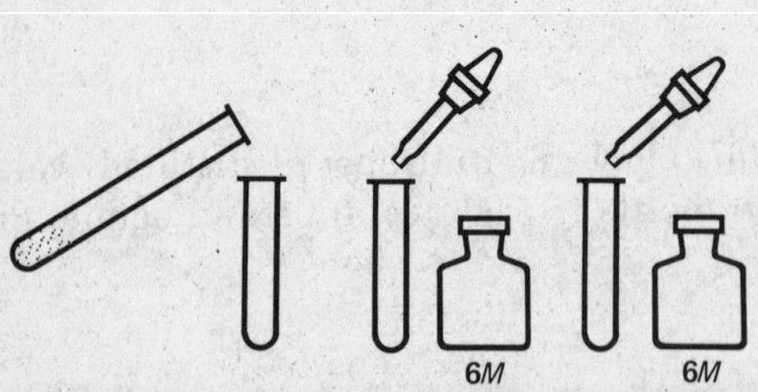

6. You have added a salt, in step A3, and observed a change. Mix the contents of the test tube thoroughly and record your observations in your notebook. Pour half of the mixture into a clean test tube. Add nine drops of 6 *M* NaOH to one of these test tubes and nine drops of 6 *M* HNO_3 to the other. Stir and test these mixtures for basicity (as in step A4) or acidity (as in step A5) using red or blue litmus, respectively. Record all observations.

Keep a detailed and accurate record of your observations. When you have finished the tests with Fe^{3+}, discuss your results with your instructor. He might suggest further experiments or ask you to work with the next ion. Wash the test tubes with tap water, using the test tube brush to loosen any material clinging to the walls. Rinse with distilled water from your squeeze bottle. Repeat procedures A1–A6 with the Cu^{2+} and Pb^{2+} solutions.

B. Reaction Scheme for the Identification of Fe^{3+}, Cu^{2+}, Pb^{2+}

Using the results obtained from procedure A, work out an analysis scheme that will allow the identification of each one of these three ions in a solution that might contain all three ions. Have your scheme checked by your instructor. If it is considered satisfactory, test your scheme on a known solution containing Cu^{2+}, Pb^{2+}, and Fe^{3+}.

Your scheme should be presented in the form of a flowchart, as illustrated by the following example for analysis of a mixture of silver ion (Ag^+), nickel ion (Ni^{2+}), and barium ion (Ba^{2+}). Chemical formulas may be omitted at the discretion of your instructor. Note that there are many different analysis schemes possible, but some are more efficient than others.

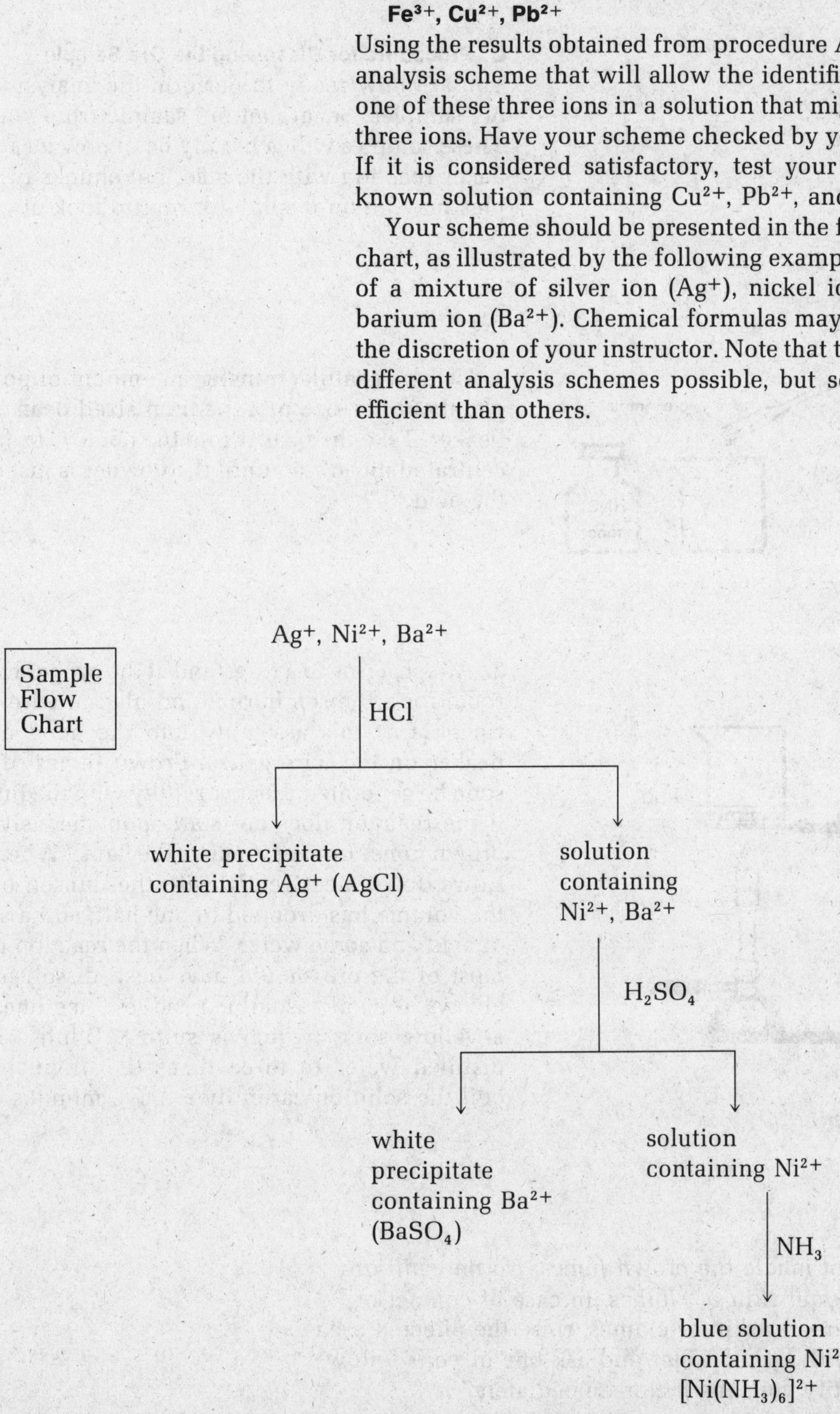

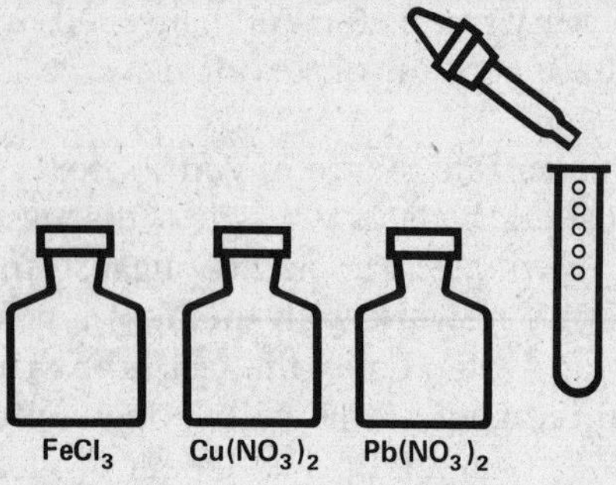

To prepare the known solution, add to a clean test tube five drops each of your Fe^{3+}, Cu^{2+}, and Pb^{2+} solutions. With this solution perform the tests outlined in your analysis scheme. When you have to separate a precipitate from a solution, use a centrifuge as described in Appendix A.

C. Procedure for Dissolving the Ore Sample

You are now ready to perform the analysis of unknown ore samples. Secure an ore sample from your instructor. These samples will generally be in powder form to insure faster reaction with the acid, but chunks of natural samples may be on display for you to look at.

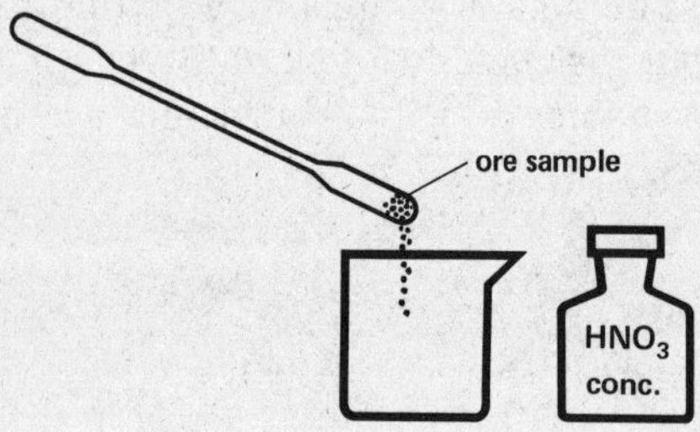

1. Using a spatula, transfer an amount of powdered sample about the size of a medium sized bean into a 100-ml beaker. Take the beaker into the hood. *Carefully* add concentrated nitric acid until the powder is just covered with the acid.

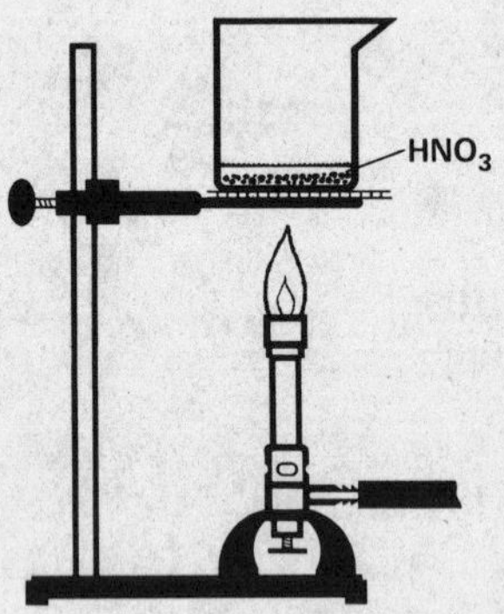

2. Attach a ring to a ringstand at the proper height to allow room for a Bunsen burner and place a wire gauze on the ring. Carry this assembly into the hood and put your beaker on the wire gauze. Brown fumes of NO_2 should soon be generated. Heat *carefully* with the Bunsen burner, if the reaction does not start spontaneously. *As soon as brown fumes appear,* remove the flame. When the reaction slows down, heat gently with the Bunsen burner. When the volume has dropped to one-half, add a small amount of acid and some water. When the reaction has subsided, most of the ore should have been dissolved. There will always be some insoluble residue, sometimes mixed with a yellow solid (which is sulfur). Dilute carefully with distilled water to three times the original volume and *boil* the solution carefully for five minutes.

Caution: *Do not inhale the brown fumes. Do not spill any nitric acid on your skin or clothes.* In case of contact of nitric acid with your skin or clothes, rinse the affected area immediately with *lots* of water and ask one of your fellow students to notify your instructor immediately.

3. Decant about 1 ml (20 drops) of the clear liquid from the solids into a test tube.

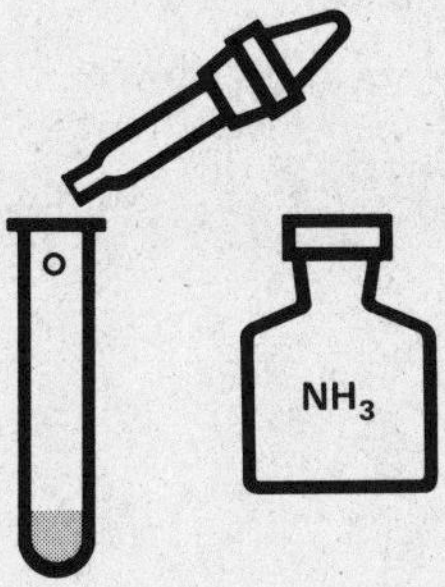

4. *Carefully* add ammonia, drop by drop, to the solution until a precipitate appears. Add 10 more drops of ammonia.

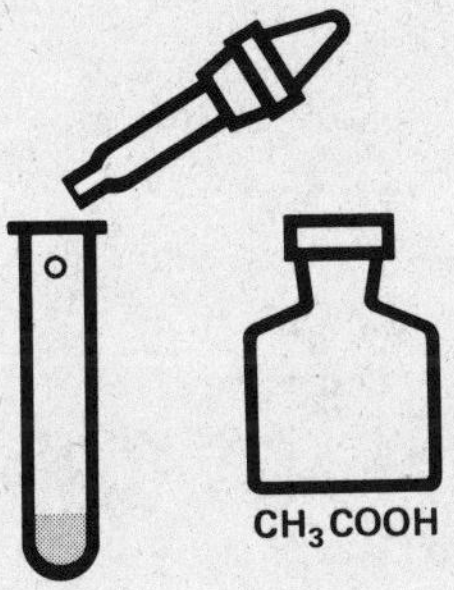

5. Drop into this mixture acetic acid, with careful swirling, until the solution becomes clear. It might take a minute or two for the solution to clear. Vigorous stirring will help. Should the solution still be cloudy add one drop of 6 *M* nitric acid. Note any observations. This clear solution will be analyzed for Cu^{2+}, Fe^{3+}, and Pb^{2+}.

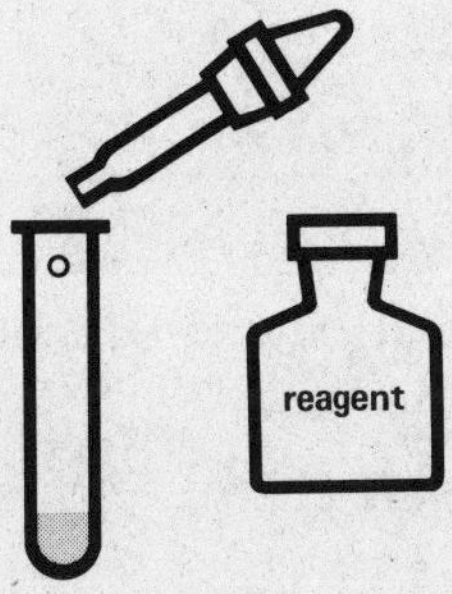

D. Testing for Fe^{3+}, Pb^{2+}, and Cu^{2+} in the Dissolved Ore Sample

Place five drops of the clear solution obtained in procedure C into a clean test tube and perform those reactions that will allow you to identify Fe^{3+}, Pb^{2+}, and Cu^{2+}. (See procedure B for reaction scheme.)

LABORATORY REPORT Using the observations made during this experiment, as recorded in your laboratory notebook, complete the Report Form and submit it to your instructor for grading. Do not use the Report Form as a laboratory notebook.

GRADE / INSTRUCTOR'S INITIALS

Report Form

FOR INVESTIGATION 1:

Detecting Iron, Copper, and Lead in Ores

NAME: ______________________ ID NO.: ______________________

LAB SECTION ______________________ INSTRUCTOR ______________________

Approved Analysis Scheme

Fe^{3+}, Cu^{2+}, Pb^{2+}

Analysis Results

Unknown identification code ______________________

Check ions found ☐Fe^{3+} ☐Cu^{2+} ☐Pb^{2+}

Describe how you applied the approved analysis scheme as shown above to identify the cations in your unknown.

______________________ ______________________
DATE SIGNATURE

Results of tests of procedure A for Fe^{3+} (Describe observations by brief notes or chemical equations, or both)

Reagent	Addition of reagent	Addition of 6 M HNO_3	Addition of 6 M NaOH
NaOH			X
NH_3			X
HCl		X	
H_2SO_4		X	
$K_4Fe(CN)_6$			
NH_4CNS			
K_2CrO_4			

Results of tests of procedure A for Cu^{2+} (Describe observations by brief notes or chemical equations, or both)

Reagent	Addition of reagent	Addition of 6 *M* HNO_3	Addition of 6 *M* NaOH
NaOH			X
NH_3			X
HCl		X	
H_2SO_4		X	
$K_4Fe(CN)_6$			
NH_4CNS			
K_2CrO_4			

Results of tests of procedure A for Pb^{2+} (Describe observations by brief notes or chemical equations, or both)

Reagent	Addition of reagent	Addition of 6 *M* HNO_3	Addition of 6 *M* NaOH
NaOH			X
NH_3			X
HCl		X	
H_2SO_4		X	
$K_4Fe(CN)_6$			
NH_4CNS			
K_2CrO_4			

Investigation 2

STOICHIOMETRY AND ECONOMICS

INTRODUCTION We live under the rule of certain natural laws. Among these are the laws of thermodynamics, which delineate for us some limitations of energy waste. Another law, "mass conservation," enables us to calculate the maximum amount of any chemical substance we can prepare from a fixed quantity of raw material, or the minimum amounts of chemicals required for preparation of a given quantity of a product. Stoichiometry deals with such calculations, as well as with the realities of chemical processes, which impose still further limitations.

The chemical industries are inherently linked to the stoichiometry of chemical systems and to the less precise, and sometimes unpredictable, "laws" of economics. If an industry is to produce some chemical substance—whether it be a cosmetic pigment, an insecticide, a detergent, or a birth control drug—the relationship between reactant requirements and product quantity can be calculated from appropriate stoichiometric considerations. However, if the industry is to survive in a free enterprise society (that is, if it is to operate profitably), then there are many variables that must be considered which are difficult to "calculate" with certainty. Unexpected changes in raw material or labor costs, for example, may force price increases to avoid financial loss; but if a competitive company has been able to avoid some of these increased costs, or to offset them by improved efficiency, the higher priced products may lose essential markets to the less expensive competition.

Large industries, through product diversification, can often absorb an economic "shock" in one area by shifting capital from more profitable ventures to provide a temporary cover for a period of economic readjustment. The small single-product industry may be unable to survive the same type of economic shock.

As consumers, we see primarily the retail price of a chemical product. Sometimes we wonder how a simple chemical, such as table salt, can be so expensive when "the ocean is full of it." Let us consider some of the factors that contribute to the eventual retail price of "table salt." There are, of course, various ways of obtaining sodium chloride. Some areas of the earth have enormous deposits of nearly pure salt, but these "salt mines" are not "typical" industries. Let us consider, instead, an alternative process. Evaporation of seawater leaves a residue of crude sodium chloride contaminated by small amounts of magnesium and calcium salts. Although these contaminants are not harmful to humans in the amounts found in "sea salts," they do absorb moisture and make the salt objectionable to the consumer who might wish his salt shaker to remain unclogged even on humid days. To obtain a purified salt, brine (seawater concentrated by evaporation) may be saturated with hydrogen chloride gas to reduce the solubility of sodium chloride to the extent that pure crystals precipitate. These are collected, washed lightly, dried, and ground to the desired particle size. Additives such as potassium iodide (for iodized salt) or moisture absorbents such as "sodium silico aluminate" (WHEN IT RAINS IT POURS®) may be blended in to improve marketability of the product.

Eventually, as the consumer, we pay 13 to 25 cents for a typical 26-oz (738-g) package of such salt. That is not very expensive compared to most prices today, but we still might wonder who gets the money, since the seawater is "free." Consider the following economic factors. Perhaps you can think of some others.

1. The chemical industry must pay for:
- **a.** The plant and equipment required to prepare salt from seawater.
- **b.** The chemical raw materials required (e.g., HCl, KI).
- **c.** The "nonchemical" materials needed (e.g., product packages, labels, shipping cartons).
- **d.** The shipping costs for delivery of the "free" seawater.
- **e.** The energy (perhaps as electric power) required to run the equipment and to light, heat, and air condition the plant. (Incidentally, salt plants often have an advantage over other chemical industries. They usually employ free sunlight for initial evaporation of the seawater to brine.)
- **f.** The people required to do the work (at various rates, depending on experience and knowledge).
- **g.** Warehousing costs to store the product until it is sold.
- **h.** Shipping costs to deliver the product to a "wholesale" distributor.
- **i.** "Miscellaneous costs," such as insurance, taxes, licensing fees, employee benefits, and equipment and plant maintenance and repair.
- **j.** Advertising to let people know the product is available, while suggesting reasons why they should buy it.
- **k.** The *money* needed to do all these things (as interest on loans, dividends on stock, etc.)

And, of course, the industry must make a fair profit to survive.

That still does not get the salt into your kitchen. There are other steps that add further costs.

2. The wholesale distributor must pay for:
- **a.** The plant and equipment needed.
- **b.** The chemical purchased from the basic industry.
- **c.** The energy required (electricity, truck fuel, etc.).
- **d.** The people needed to do the work.
- **e.** Shipping costs for delivery to the retail outlet.
- **f.** Miscellaneous costs (insurance, taxes, etc.).
- **g.** Advertising.
- **h.** Interest on loans, etc.

®Morton Salt Company, Chicago.

And, of course, the wholesale distributor must make a fair profit to survive.

There are, along the way, some other people who have to get paid. Their costs are generally "buried" in the total economy of the process, but we should not forget the trucker, the insurance man, the advertising executive, the academic consultant, the lobbyist, and the public official (whose salary depends on taxes collected). There are very many people whose income depends to some extent, directly or indirectly, on any given chemical industry. Now we come to the last stage before you get the salt.

3. The retailer must pay for:
- **a.** The store and equipment he needs.
- **b.** The chemicals purchased from the wholesale distributor.
- **c.** The energy required to light, heat, and air condition the store.
- **d.** The clerks and janitors and assistant managers and bookkeepers, etc.
- **e.** Miscellaneous costs (insurance, taxes, etc.).
- **f.** Advertising.
- **g.** Interest on loans, etc.

And, of course, the retailer must make a fair profit to survive.

Now, let us consider an all too common event: a sudden price increase at the level of the basic industry. This might involve a new labor contract at higher wages and improved worker benefits. It might involve an increased cost in raw materials or paper cartons, a new government requirement for added safety equipment or lower waste product emission, or (as we have seen quite recently) a sudden hike in electrical rates due to the "energy crisis." What can the industry do? Basically, there are three kinds of choices.

1. It can maintain its product's sales price by taking a smaller profit. (If it does, its stockholders will get smaller dividends, sell stock, etc.)
2. It can maintain its product's sales price by cutting some other cost factors. (Of course, if it "lays off" some workers, the productivity drops, sales decline, etc. If improved efficiency will help, that is fine, but that takes time and, usually, additional capital outlay.)
3. It can pass increased costs along to the wholesaler, who probably has to pass it on to the retailer, who probably has to pass it on to you.

Did you ever notice that?

Before starting this investigation, you should know how to calculate formula weights, how to make "weight-to-weight" stoichiometric calculations, how to calculate percentage yields, and how to relate "cost," "profit," and "selling price." In addition, you should be familiar with the use of a *triple beam balance* (Appendix A) or an alternate type of balance, as assigned by your instructor.

Save Our Environment
Warehouse
Buy X
Chemical Company
ZZ Pharmaceuticals
Buy ZZ Pharmaceutical
Shop at...
DRUGS

References

1. O'Connor, Rod. 1977. *Fundamentals of Chemistry* (2nd ed.). New York: Harper & Row (Unit 4).
2. Selections from *Education Publications Available from the Chemical Industry*. Issued by the Manufacturing Chemists Association, Inc., 1825 Connecticut Ave., N.W., Washington, D.C. 20009.

SAFETY Wear approved eye protection and be careful with hot equipment and chemicals. While heating your sample, beware of splattering.

OBJECTIVES

1. Be able to design and perform a simple experiment efficiently and safely, including selection of appropriate equipment.
2. Be able to make simple stoichiometric and economic calculations.

PROCEDURES

Prior to the laboratory period, you must go to a store and purchase a candle and a *small* bottle of Epsom salts ($MgSO_4 \cdot 7H_2O$). You need only 25 g of this salt. Check the appearance of the salt crystals. They must be transparent. If they appear opaque and white, they are not suitable for this experiment. Obtain a sales receipt showing only these two purchases. Bring the candle, Epsom salts and sales receipt with you to the laboratory. [It is permissible for a group of students to make a "joint venture" purchase and to share the Epsom salts and the use of the candle. However, each student must have at least 25 g of Epsom salts and a *copy* of the sales receipt. A Xerox or other copy is satisfactory if marked with "Joint venture purchase. For original sales receipt, see (student's name)".]

Now, here is the situation.

1. You have purchased your "raw material" (*Epsom salts*) and your "energy" (the *candle*).
2. You have made a capital investment in plant and equipment (when you paid your *laboratory fee*) which gives you access to the laboratory and a standard collection of equipment.
3. You are the "personnel" (chemist, engineer, shop foreman, worker, executive, marketing expert, salesman, etc.) You must keep a record of your time, by job, for the trip to the store and your laboratory working time, to the nearest one-quarter hour. Allow a labor cost of $2.00 per hour.
4. You are to design an experiment, select appropriate equipment, and carry out the conversion of Epsom salts to anhydrous magnesium sulfate ($MgSO_4$). The only limitations are these:
   a. The only energy (heat) source permitted is the candle.
   b. You start with approximately 25 g of $MgSO_4 \cdot 7H_2O$, weighed to the nearest 0.1 g.
   c. You must hand in to your instructor a satisfactory container ("decorated," if you like) of your final product, labeled with the *net* mass, to the nearest 0.1 g.
   d. You must "sell" your product to your instructor (for grade points, not money), using any "sales pitch" you

devise. He may bargain with you on such features as product yield and quality. You may wish to suggest that your product is a valuable laboratory drying agent.

e. Your instructor (acting as a "government safety inspector") must approve your setup and proposed procedure before you are allowed to begin. If he finds you without eye protection or following an unsafe laboratory practice, he will "close down the plant."

f. You must maintain "company records" (appropriate entries in your laboratory notebook).

5. We will tell you only that the conversion of $MgSO_4 \cdot 7H_2O$ to $MgSO_4 \cdot H_2O$ occurs around 150°C, while the remaining water is lost around 200°C. The rest of this experiment is left to your own ingenuity. Keep in mind that a clean plant (your laboratory work station) is a safe plant.

Note that your energy source will very likely generate by-products (soot and candle drippings) that will pollute your environment (the laboratory). In order for the next generation (the students using the laboratory after you) to enjoy an unpolluted environment, you have to leave the laboratory in its naturally clean condition: no candle drippings on the ringstand or the bench top, no soot on containers, and no soot on the ring and wire gauze.

In order to catch the candle drippings, place the candle on a paper towel spread over the base of the ringstand. Wipe sooty containers and rings clean with a dry paper towel after they are cool. If the soot does not come off this way, rub the sooty areas with moistened detergent powder and then rinse with water. Rub the wire gauze with moistened detergent on a paper towel. Rinse then with tapwater. Dry the wire gauze with the Bunsen flame.

Your instructor will represent the Environmental Protection Agency and levy penalties (in form of point deductions) against polluters.

CALCULATIONS AND REPORT

Using your sales receipt, time records, and notebook records of data and procedures, complete the entries and calculations specified on the Report Form. Maintain a copy of this information in your permanent notebook.

At the conclusion of your experimental work, your instructor might give you some typical stoichiometry problems to work. These will involve some more sophisticated processes, so be sure you have studied the subject of chemical stoichiometry prior to coming to the laboratory.

GRADE
INSTRUCTOR'S INITIALS

Report Form

FOR INVESTIGATION 2:
Stoichiometry and Economics

NAME: ____________________ ID NO.: ____________________

LAB SECTION ____________________ INSTRUCTOR ____________________

Costs

1. Raw material (attach purchase receipt)

Mass of $MgSO_4 \cdot 7H_2O$ actually used ____________ g

Cost of original package ____________ ¢

Net mass of original package ____________ g

Cost of mass actually used ____________ ¢

2. Energy (attach purchase receipt)

Candle (____________ ¢) × (fraction consumed) = ____________ ¢

(if shared: number sharing = ____________, cost per person = ____________ ¢)

3. Capital investment

(0.10) × (lab fee paid) = $____________

4. Personnel (round time to nearest $\frac{1}{4}$ hour: minimum $\frac{1}{4}$ hr per job)

Job	Hours	Cost @ $2/hr
trucker (trip to store)		
chemist/engineer (experiment design)		
worker, etc. (experiment operation)		
salesman (bargaining with instructor)		
other (describe)		

Total: $____________

Calculations

1. Mass of starting material ($MgSO_4 \cdot 7H_2O$) ____________ g

2. Theoretical yield of anhydrous $MgSO_4$ ____________ g

(OVER)

Show setup.

3. Net mass of packaged product ____________ g

4. Percent yield ____________ %

Show setup.

5. Purity (instructor's evaluation) ☐ Reagent grade ☐ Tech grade ☐ Unsatisfactory

6. Sales Price Collected $____________
(Your instructor will tell you the "monetary equivalent" of the grade points awarded for your product.)

7. Total cost (from reverse side) $____________

8. Profit (sales price − cost) $____________

9. Corporate taxes (28 percent of profit) $____________

10. Net profit after taxes $____________

11. Percent profit

$$\frac{\text{net profit}}{\text{sales price}} \times 100\%$$ ____________ %

12. Government Safety Inspection: ☐ OK ☐ penalty (No. points deducted) ______
Environmental Protection Agency Inspection: ☐ OK ☐ penalty (No. points deducted) ______

Discussion

(Attach separate sheet) Does the consumer get a "good deal" on the purchase of a box of salt?

Also attach pages showing your solutions to the stoichiometry problems assigned by your instructor.

______________ DATE ______________________ SIGNATURE

Investigation 3

ALUM FROM WASTE ALUMINUM CANS

INTRODUCTION The number of discarded (usually empty) beverage cans along the roadside attests to two popular American pastimes, thirst-quenching and littering. Aluminum is one of the most indestructable materials used in metal containers. The average "life" of an aluminum can is about one hundred years. The littering from nonreusable cans is itself a serious and costly problem. A more serious problem, the depletion of raw materials, will have to be solved in the near future. Although aluminum is the third most abundant element in the earth's crust, the supply of its *useful* ores is not inexhaustible and the processes for manufacturing aluminum from its ores require large amounts of electrical energy. Man must, therefore, find ways to "recycle" the materials he uses. A program for "recyling" aluminum cans is already in operation in several communities.

In this experiment you will explore a chemical process which could be used to transform scrap metal into a useful chemical compound. You will prepare an "alum," potassium aluminum sulfate dodecahydrate [$KAl(SO_4)_2 \cdot 12H_2O$], from waste aluminum. This alum is widely used in the dying of fabrics, in the manufacture of pickles, as a coagulant in water purification and waste-water treatment plants, and in the paper industry.

Although aluminum is a "reactive" metal, it reacts only slowly with dilute acids because its surface is normally "protected" by a very thin coating of aluminum oxide. Alkaline solutions (containing OH^-), on the other hand, dissolve the oxide layer and then attack the metal. Aluminum is converted by excess alkali to the tetrahydroxoaluminate(III) anion, which is stable only in basic solution. When acid is added slowly to a solution of this complex anion, initial reaction removes one hydroxide ion, causing the precipitation of the white, flocculent $Al(OH)_3$. If addition of acid is continued, the aluminum hydroxide is dissolved, eventually forming the hydrated aluminum cation $[Al(H_2O)_6]^{3+}$. Addition of alkali to the $Al(OH)_3$ precipitate will also bring about solution, by re-forming the complex $[Al(OH)_4]^-$. A hydroxide which can be dissolved by either acid or base is said to be *amphoteric.*

Alums are ionic compounds which crystallize from solutions containing sulfate anion, a trivalent cation (e.g., Al^{3+}, Cr^{3+}, Fe^{3+}) and a monovalent cation (e.g., K^+, Na^+, NH_4^+). Most alums crystallize readily as octahedra or cubes which, under the appropriate conditions, may grow to considerable size. The crystal lattice contains 12 water molecules, 6 of which are tightly bound to the trivalent cation, with the remaining 6 loosely bound to the sulfate anion and monovalent cation.

Before starting this investigation, you should know how to calculate formula weights, how to make "weight/weight" stoichiometric calculations, and how to calculate percentage yields. In addition, you should be thoroughly familiar with safety precautions and emergency procedures involving work with strong acids and bases.

References

1. O'Connor, Rod. 1977. *Fundamentals of Chemistry* (2nd. ed.). New York: Harper & Row (Units 3 and 4).
2. Fullman, R. L., *Scientific American*, offprint no. 260. San Francisco: W. H. Freeman.

SAFETY In this experiment you will be working with strong acids and bases. Wear *complete* eye protection at *all* times. Alkaline solutions are particularly hazardous to the eyes. In case of contact with any reagents, wash the affected area *immediately and copiously* with water and notify your instructor.

Exercise special care with flammable liquids and gases to avoid a fire.

OBJECTIVES

1. Be able to handle reagents properly, including all safety precautions for the use of acids and alkalis.
2. Be able to use a simple laboratory balance for weighing to ±0.05 g.
3. Be able to use properly the techniques of heating and transferring aqueous mixtures and of aspirator filtration.
4. Be able to make appropriate stoichiometric calculations, including percentage yield.

PROCEDURES

1. Bring to the laboratory an empty thin-walled aluminum beverage can. If you are unable to find one, ask your instructor.

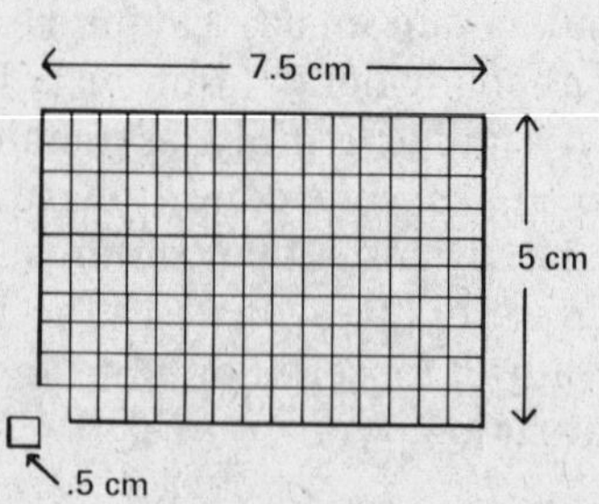

2. Using snips provided, cut from your can a piece of aluminum approximately 5 cm × 7.5 cm and scrape off any paint as completely as possible. Weigh the cleaned strip on a simple laboratory balance. You will need about one gram of aluminum. If your sample is too heavy, cut off any excess. If it is too light, cut and clean some more small pieces from a can. Weigh the final sample to the nearest 0.05 g and record this mass. Then cut your aluminum sample into small squares of about 0.5 cm edge length.

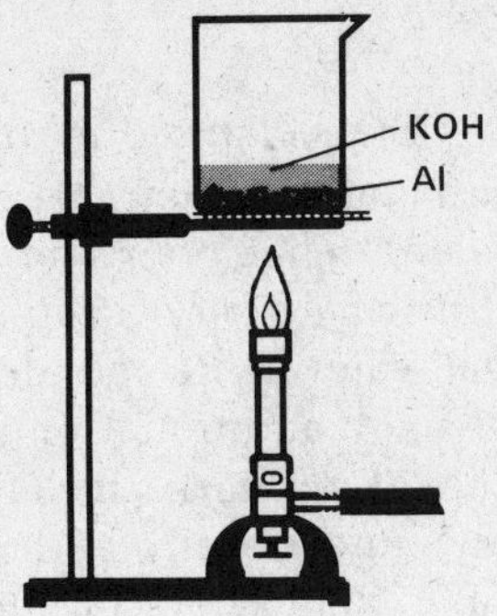

3. If possible, work at a fume hood. Place the aluminum clippings in a 250-ml beaker and add 50 ml of 1.4 *M* potassium hydroxide. Bubbles of hydrogen should soon form from the reaction between aluminum and aqueous hydroxide. Heat the beaker *gently* to speed up the reaction, preferably using a hot plate. (If a burner is used, be *very* careful. Hydrogen–air mixtures may ignite explosively.)

If the liquid level in the beaker drops to about half its original volume, add distilled water to reconstitute part of the volume. During the reaction the initially colorless mixture will turn dark gray. This is probably due to decomposition of residual paint or plastic lining. Note the periodic rise and fall of aluminum fragments during the reaction. (Suggest an explanation.) No more than 30 minutes should be required to complete the reaction. Final volume of the liquid should be about 60 percent of the original volume. The reaction is complete when hydrogen evolution ceases.

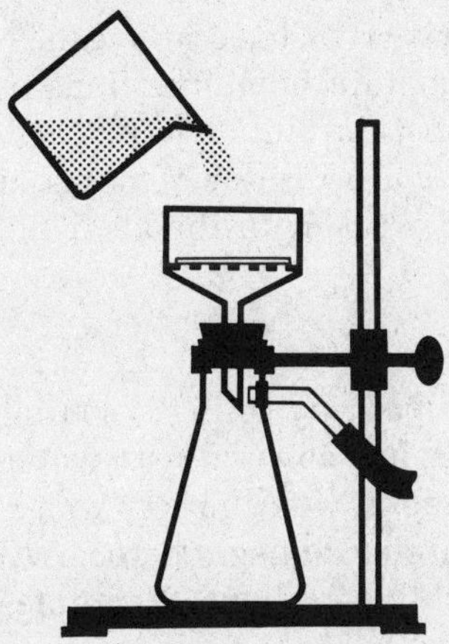

4a. The hot solution must be filtered, using an aspirator vacuum to remove any solid residue. Consult Appendix A if you are unfamiliar with the technique of vacuum filtration. Be sure that the filter flask is securely clamped.

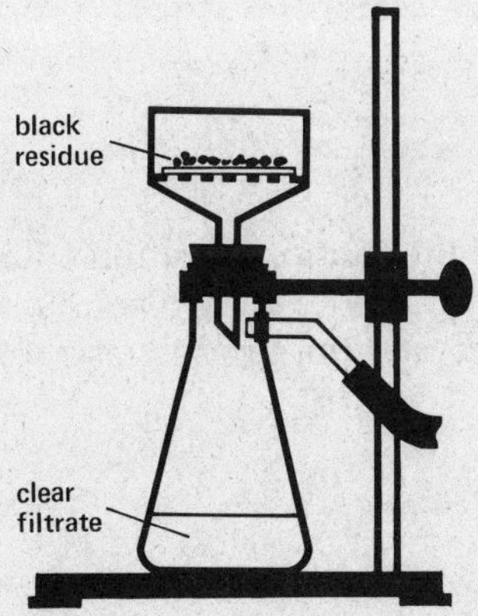

4b. The filtrate should be clear, with any dark residue left on the filter paper. Rinse the beaker twice with 5-ml portions of distilled water, pouring each rinse through the filter residue.

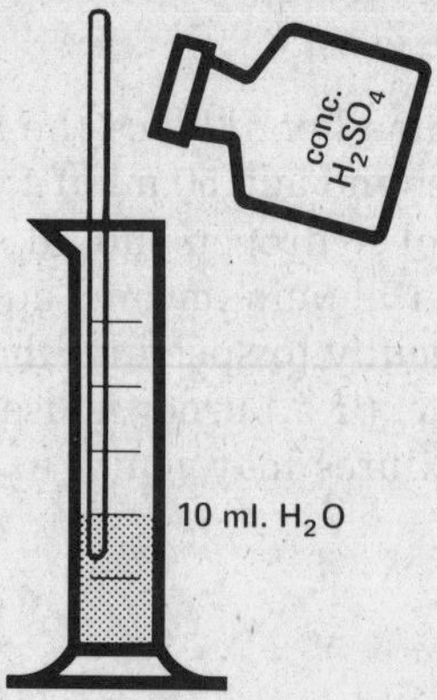

5a. Prepare 20 ml of 9 *M* H_2SO_4. Place 10 ml of distilled water in your 25 ml graduated cylinder. (Be sure your cylinder is *Pyrex* or equivalent.) Then pour *small* portions of concentrated (18 *M*) sulfuric acid *very carefully* from a small reagent bottle into the water in the cylinder, stirring with a glass rod after each small addition. Continue until the liquid level reaches the 20 ml mark on the cylinder. Stir the solution thoroughly, but carefully, and allow it to stand until cool. (The diluted acid will be quite hot!)

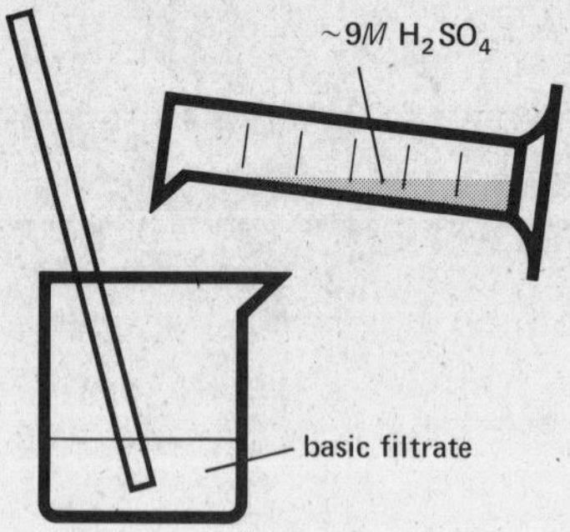

5b. Transfer the clear filtrate into a clean 250-ml beaker. Rinse the suction flask with 10 ml of distilled water and pour this also into the beaker. If the solution is not yet cool, place the beaker in a cooling bath. *Slowly and carefully, while stirring,* add 20 ml of 9 *M* H_2SO_4 to the cooled solution.

You will soon notice the appearance of a white precipitate of aluminum hydroxide. You will also notice heat generated by the neutralization reaction. Addition of more sulfuric acid will dissolve the $Al(OH)_3$. If necessary, warm the solution gently, while stirring, to complete the reaction. The final solution will contain potassium ions (from the KOH used), aluminum ions, and sulfate ions. If, after a few minutes of heating, any solid residue remains, filter the mixture as in step 4 and work with the clear filtrate.

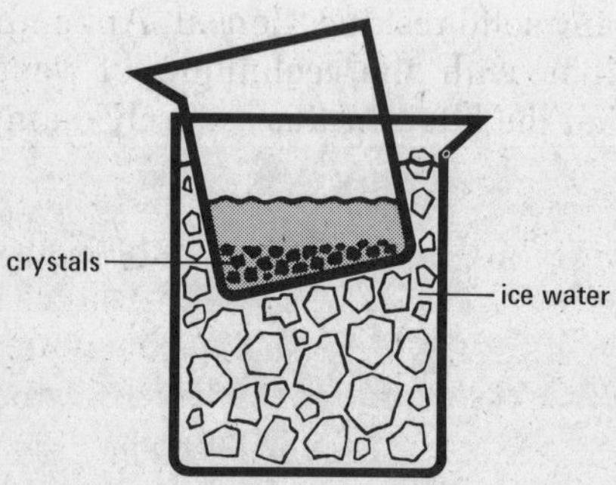

6. Prepare an ice–water bath by filling a 600 ml beaker half way with crushed ice, just covered with water. Set the reaction beaker (from step 5) into the ice–water bath to chill. Crystals of the alum should begin to form. Allow the mixture to chill thoroughly for about 15 minutes.

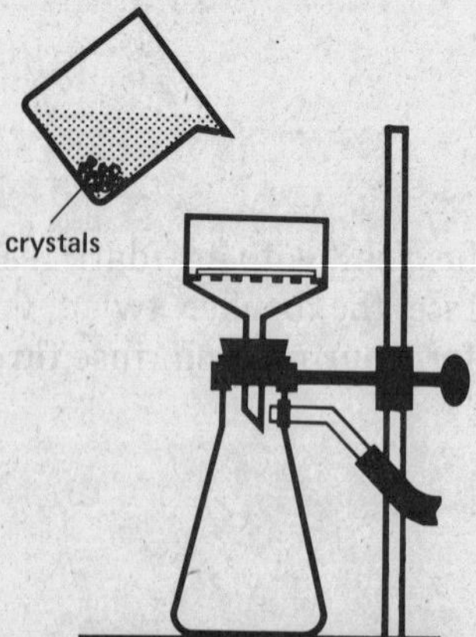

7. Clean and reassemble the suction filtration equipment. Filter your alum crystals from the chilled solution, transferring as much of the crystalline product as possible to the funnel.

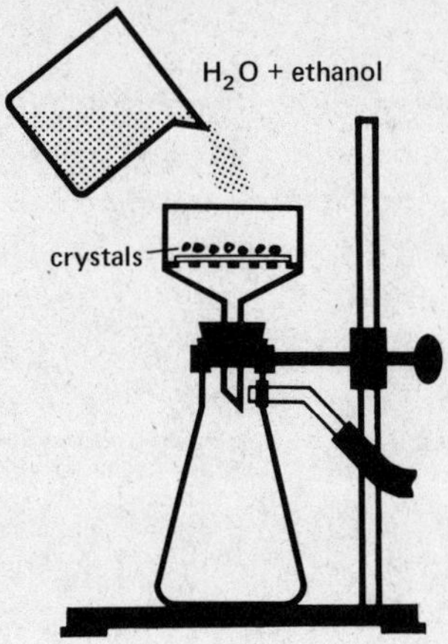

8. Mix 12 ml each of ethanol and water and chill the mixture for a few minutes in the ice–water bath. Use half this solution to rinse the remaining crystals from the beaker into the funnel. Rinse the beaker again with the second half of the solution. Use your spatula to distribute the crystals uniformly on the filter paper. Allow the aspirator to pull air through the crystals for about 10 minutes.

While the crystals are drying, weigh a large clean test tube and its stopper, to the nearest 0.05 g. Record this mass. Use your spatula to transfer all of the air-dried crystals from the filter paper into the test tube. Stopper the test tube and reweigh, recording the mass. *Calculate the mass of alum formed and the percentage yield, based on the weight of aluminum used. The complete sequence of reactions is summarized by the following equations:*

Reaction of aluminum with KOH:

$$2\ Al_{(s)} + 2\ KOH_{(aq)} + 6\ H_2O_{(\ell)} \longrightarrow 2\ K[Al(OH)_4]_{(aq)} + 3\ H_{2(g)}$$

Initial addition of H_2SO_4:

$$[Al(OH)_4]^-_{(aq)} + H^+_{(aq)} \longrightarrow Al(OH)_{3(s)} + H_2O_{(\ell)}$$

Reaction with more H_2SO_4:

$$Al(OH)_{3(s)} + 3\ H^+_{(aq)} \longrightarrow Al^{3+}_{(aq)} + 3\ H_2O_{(\ell)}$$

Precipitation of alum, on cooling:

$$K^+_{(aq)} + Al^{3+}_{(aq)} + 2\ SO_4{}^{2-}_{(aq)} + 12\ H_2O_{(\ell)} \longrightarrow KAl(SO_4)_2 \cdot 12\ H_2O_{(s)}$$

9. Show the test tube containing your alum to your instructor, reporting the mass of alum obtained and requesting a "product inspection." If your instructor considers your alum to be satisfactory, transfer the alum into the "Alum Storage Bottle" for later use. Otherwise, dispose of the alum as directed by your instructor.

REPORT Record all appropriate data, calculations, and observations in your permanent notebook. Use this record to prepare the brief Report Form.

GRADE	
	INSTRUCTOR'S INITIALS

Report Form

FOR INVESTIGATION 3:
Alum from Waste Aluminum Cans

NAME: ______________________ ID NO.: ______________________

LAB SECTION ______________________ INSTRUCTOR ______________________

Mass of aluminum used: ______________ g Technical grade ☐

Mass of alum obtained: ______________ g Reagent grade ☐

Approximate yield: ______________ % Instructor's initials ________

Present below the chemical equation or equations necessary for the yield calculation. Show the actual calculation by giving the numerical set up and the results.

Questions: (Answer on reverse side)

1. How many grams of alum could you have obtained from 20.0 g of aluminum on the assumption of the percentage yield you obtained in this investigation?
2. How many grams of hydrogen were released by the reaction of 1.00 g of aluminum with excess potassium hydroxide solution: assuming 100 percent efficiency?
3. How many moles of sulfuric acid would be required to make 4.74 g of $KAl(SO_4)_2 \cdot 12\ H_2O$, assuming 100 percent efficiency?

______________ ______________________

DATE SIGNATURE

Investigation 4

GROWING CRYSTALS

INTRODUCTION Sir W. H. Bragg, one of the first scientists to use x-rays for the study of crystal structures was fascinated by the intricate and orderly atomic patterns revealed by his investigations. He expressed this, in 1901, "The molecules are locked into crystalline structure when attachments are made at sufficient points, and the whole has the stability of an engineering structure."

Natural crystals have always held a fascination for man, whether they were precious stones, such as diamonds or rubies, or just a beautifully formed snowflake. Today we recognize, as did Bragg, that the regular geometric shapes of well-formed crystals reflect orderly three-dimensional networks of the atoms, ions, or molecules that constitute the crystal lattice.

Perfect crystals, in which the repetitive lattice pattern is completely uniform throughout the solid, are extremely rare. This is not surprising when you consider that a crystal is typically formed from a liquid (a solution or the molten solid) or a gas, in which the particles are quite randomly oriented. To initiate crystal growth, a tiny collection of particles must form in a highly ordered pattern. This process is called *nucleation* and it may occur spontaneously when particles of a pure substance collide with appropriate orientations and with low enough kinetic energies to permit the establishment of "permanent" interparticle attractions rather than "rebound collisions." More often, some foreign material provides a surface on which particles may be absorbed in orientations suitable for continued crystal growth. The nucleation of snowflakes, for example, often involves a "dust particle" as the center of adsorption for water molecules, and a string suspended in a supersaturated solution of a salt may provide nucleation sites as ions are adsorbed onto the fibers.

The tiny microcrystals initially formed may be perfectly ordered, but continued growth generally introduces imperfections. As the crystal grows, the surface areas for addition of more particles become continuously larger, thus requiring an ever-increasing availability of the proper particles, with appropriate orientations. When molecular crystals grow from the pure liquid (e.g., as pure water freezes) or gas (e.g., in the formation of iodine crystals from the vapor), there are no problems with imperfections from foreign particles, but it is most unlikely that uniform growth rates over all surface areas will continue. Nonuniform growth may result in crystal imperfections. Crystals grown from mixtures, such as salt solutions or impure liquids, may contain chemical imperfections from foreign particles trapped in crystal lattices in addition to physical imperfections resulting from irregular growth patterns.

The growing of nearly perfect crystals, or of crystals containing "planned imperfections," is an important aspect of modern *solid state technology*. Synthetic diamonds are grown for such uses as diamond-tipped drills; sodium chloride crystals are produced for use in the optics systems of infrared spectophotometers; and "doped" crystals of silicon or germanium (containing planned substitutional lattice defects) are prepared for solid state electronic equipment.

In this investigation you will have the opportunity of growing relatively pure alum crystals, using alum prepared from waste aluminum cans (Investigation 3), or of introducing planned "imperfections" by growing mixed crystals of alum and "chrome alum."

Before starting this investigation, you should be familiar with the characteristics of crystalline solids, the general types of crystal defects (imperfections) encountered during crystal growth, and the basic principles of growing crystals from supersaturated aqueous solution.

References

1. O'Connor, Rod, 1977. *Fundamentals of Chemistry* (2nd ed.). New York: Harper & Row (Unit 11 and Excursion 1).
2. Bunn, Charles, 1964. *Crystals: Their Role in Nature and in Science*. New York: Academic Press.
3. Holden, Alan and Phylis Singer. 1970. *Crystals and Crystal Growing*. New York: Doubleday.

OBJECTIVES

1. Be able to grow a crystal from a supersaturated aqueous solution.
2. Be able to make and record systematic observations over a long time period.
3. Be able to measure and describe geometric characteristics of a regular crystalline solid.

PROCEDURES

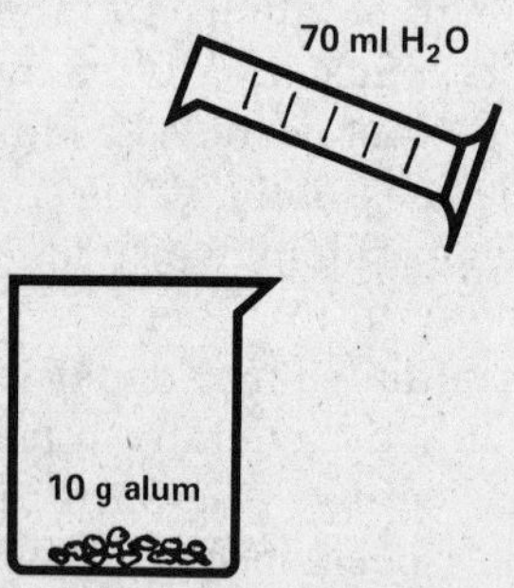

1. Weigh out at least 10 g of alum and place it in a clean 250-ml beaker. For each gram of alum used, add 7 ml of water. Heat the mixture to 50°–60°C and stir it until a clear solution is obtained. (If the mixture remains cloudy, allow the mixture to stand for a few minutes until the sediment has settled out, then carefully decant the clear solution into another clean 250-ml beaker.)

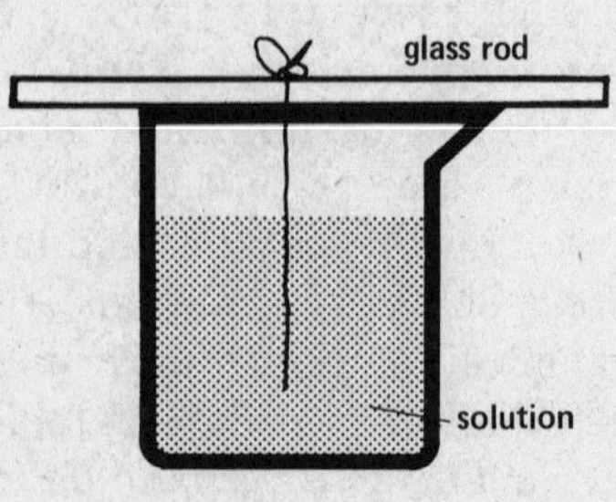

2. Tie a piece of thread to a glass rod so that when suspended in the solution, the thread will extend no more than $\frac{1}{2}$ in. below the surface of the solution. Smear stopcock grease on the part of the thread that will be *above* the solution to keep the solution from creeping up the thread. Put the rod in place on top of the container, cover with a paper towel held in place by a rubber band, use a pencil to punch a few holes (five or six) in the towel, and keep the beaker in a safe storage place. Crystals should have formed either on the string or at the bottom of the solution (or in both places) by the time you return for your next laboratory session.

3a. If crystals have formed on the string, crush off all but the best one. If not, decant the liquid from the crystal or crystals at the bottom of the container into another clean container. If more than one crystal grew, inspect them and pick one that shows good faces. Tie a fine thread to this "seed" crystal so that it may be suspended from the glass rod.

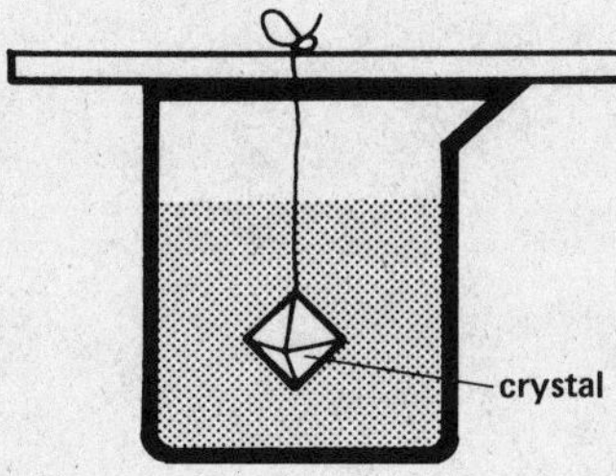

3b. Suspend the crystal in a *freshly* prepared solution of 10 g of alum in 70 ml of water as shown in the figure. This solution must be allowed to cool to room temperature before the crystal is introduced.

Note: It is important to cool your growing solution sufficiently before suspending your seed crystal in it, particularly if it was necessary to tie your seed crystal. If the solution has not been allowed to cool enough, it will not be saturated. It will start to dissolve the seed crystal and the seed crystal may slip out of the thread loop. You can get an indication if the solution is cool enough by carefully observing the portion of the solution near the suspended crystal. If the solution is less than saturated, the portion of the solution in contact with the crystal will dissolve some of the crystal. This will increase the density of that portion of the solution and it will flow to the bottom of the beaker. On careful observation this density current can be seen. If this happens, remove the seed crystal and allow the solution to cool further.

4. Repeat procedure 3b weekly until you have a nice crystal you want to keep. Weigh your crystal each week and describe its appearance (color, size, mass, shape). Your instructor will grade your efforts toward the end of the term. You may take your crystal with you after it has been graded. To preserve it and avoid its conversion to a white powder (by loss of water) cover it with a clear plastic spray available from your instructor.

If you would rather have a colored crystal, you may substitute a growing solution of "chrome alum," $KCr(SO_4)_2 \cdot 12H_2O$, for the *second* alum growing solution. Prepare your seed crystal as described above. Then prepare an alum solution as described above using 8 g of alum in 56 ml of distilled water. Prepare a second solution of *chrome* alum using 15 g per 25 ml of H_2O. The color of this solution will be a dark blue-green. Pour this solution slowly into the growing solution of ordinary alum, while holding the growing solution up to the light. Stop adding the chrome alum when you can still just see through the solution. Suspend your seed crystal in the solution as before. For subsequent growing stages, use ordinary alum solutions. This will give you a purple crystal inside a clear one.

REPORT Maintain regular notebook records of the appearance and mass of your crystal during successive growth stages. Record the first, second, and final weighings on your Report Form and answer the questions on the form. When you submit the final report, show your instructor your notebook records.

GRADE / INSTRUCTOR'S INITIALS

Report Form

FOR INVESTIGATION 4:

Growing Crystals

NAME: ______________________ ID NO.: ______________________

LAB SECTION ______________________ INSTRUCTOR ______________________

Final record due ______________
(Submit final crystal for instructor's inspection)

Record of Crystal Growth

	Date	Crystal mass (grams)
First measurement		
Second measurement		
Final measurement		

Describe the various stages of the crystal growth and sketch your final crystal on the back of this page. Does your crystal have any resemblance to one of the regular geometric solids? Which one? Measure the angles between the various faces. Are these angles characteristic of a regular geometric solid?

DATE ______________ SIGNATURE ______________________

Investigation 5

ISOLATION OF SUGAR FROM A NATURAL SOURCE

INTRODUCTION When most of us think of "sugar," we have in mind the common "table sugar" used in granular or cube form as a sweetener for our foods and beverages, or possibly the powdered "confectioner's sugar" often used in cooking. In scientific terminology, however, "sugar" is a *class* name applied to a large number of different chemical compounds, all of which have certain structural characteristics, such as a number of hydroxyl (OH) groups. Among the more common sugars are *glucose* (also known as *dextrose* or "grape sugar"), *fructose* (*levulose* or "fruit sugar"), *sucrose* ("cane sugar," "beet sugar," or "table sugar"), *maltose* ("malt sugar"), and *lactose* ("milk sugar"). The structural formulas of these sugars illustrate their chemical similarities and the molecular formulas, having the general relationship $C_x(H_2O)_y$, indicate why these sugars are *carbohydrates* ("hydrates of carbon"). It is obvious from the structural formulas, however, that such compounds are not really "hydrates" in the sense of simple hydrated salts such as $CuSO_4(H_2O)_5$. Although the ratio of hydrogen to oxygen in simple carbohydrates is the same as that in water, the carbohydrates (unlike hydrated salts) do not contain associated water molecules.

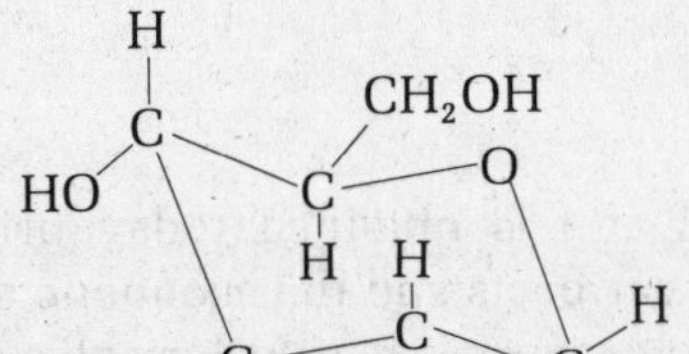

glucose ($C_6H_{12}O_6$)

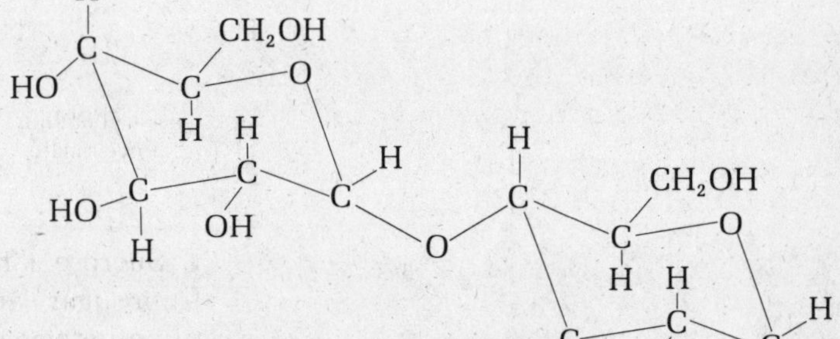

maltose ($C_{12}H_{22}O_{11}$)

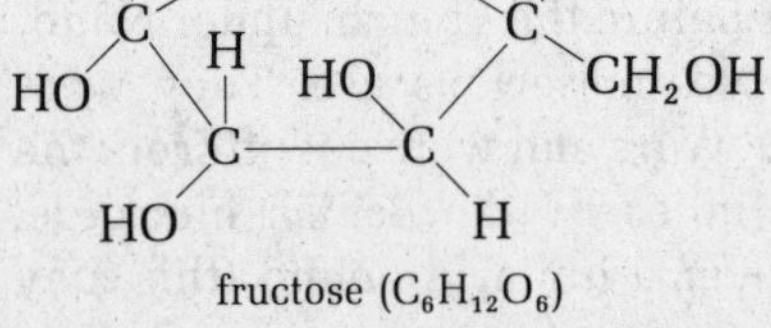

fructose ($C_6H_{12}O_6$)

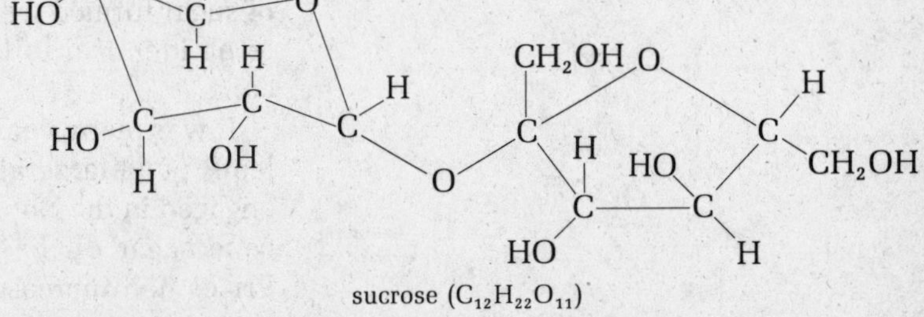

sucrose ($C_{12}H_{22}O_{11}$)

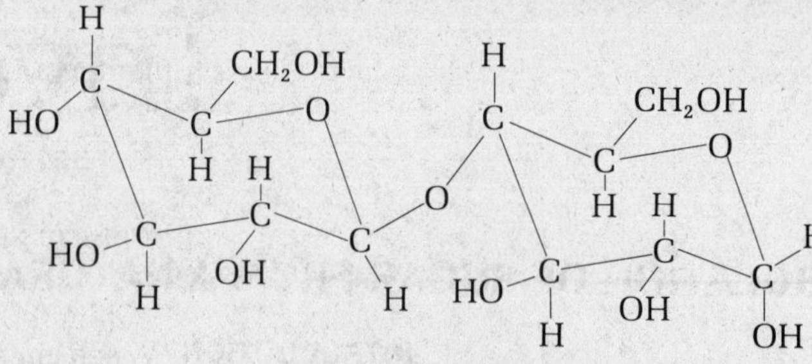

lactose ($C_{12}H_{22}O_{11}$)

The sweet taste commonly associated with sugars varies considerably among the simple carbohydrates. Although "sweetness" is a rather subjective property, it is generally considered that glucose (obtained mainly by hydrolysis of corn starch or potato starch and used generally as "powdered" or "confectioner's" sugar) is only about 75 percent as sweet as sucrose. Fructose (found in honey and many of the sweeter fruits, along with glucose and sucrose) is recognized as the sweetest of the common sugars, but two types of artificial sweeteners are even more effective. The cyclamates are ranked about 50 times sweeter than sucrose, and saccharin is some 9 times as sweet as the cyclamates. Unlike the carbohydrates, these sugar substitutes have no calories as foods.

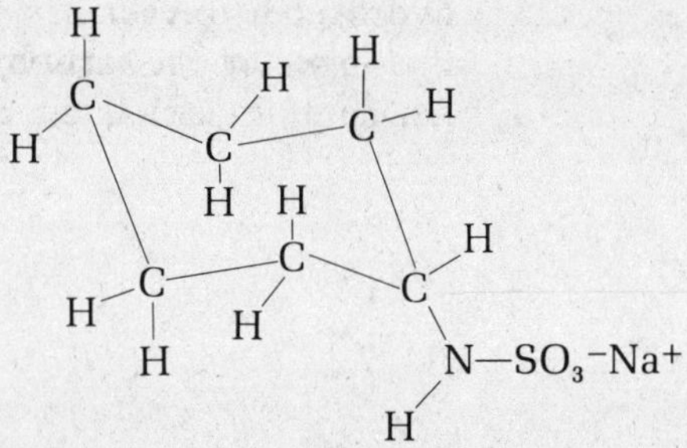

"sucaryl sodium"
a cyclamate

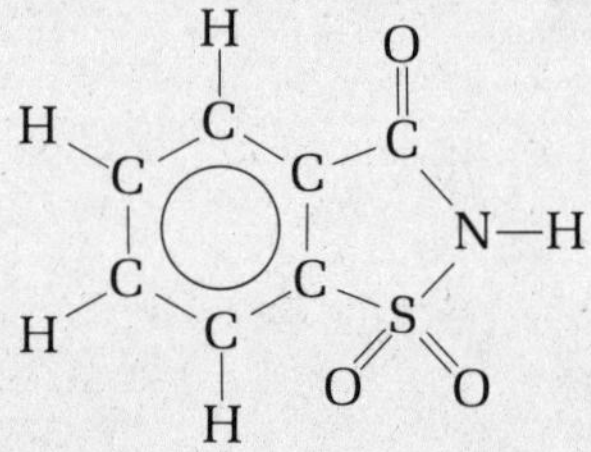

"saccharin"

Sucrose ("table sugar") is obtained today primarily from sugar cane or sugar beets and the economic competition between these two sources has, historically, been a major problem of the sugar industry. It was not until 1747 that a useful method of extracting sugar from beets was developed. The German chemist Andreas Marggraf sliced, dried, and pulverized the beet root. For each 8 oz of the powder, he added 6 oz of ethanol, and heated the mixture over a low flame. When the liquid began to boil, he quickly filtered the hot mixture. The filtrate was allowed to stand for several days in a corked flask, after which the crystals of sugar formed were collected. These crystals were found to be identical to the sugar crystals obtained from sugar cane.

It was many years later before the French, under Napoleon, put Marggraf's process to serious use. They were engaged in the Napoleonic Wars and were cut off from the cane sugar of the West Indies by the British blockade. Prices had increased to over a dollar a pound (this may seem familiar). Napoleon decreed that sugar beets be planted and built 334 beet sugar factories in 1812 and 1813. June 18, 1815, was Waterloo, not only for Napoleon, but also for the new sugar beet industry. With the end of the war and the lifting of the blockade, the West Indian cane producers poured their sugar into Europe. Within a year only one beet factory was still operating in France.

Generally, E. H. Dyer is credited with establishing the beet sugar industry in the United States on a successful basis in 1879 at Alvarado, California, where a plant still operates. (He may have been the second successful producer in California. A Spanish soldier who explored California between 1768 and 1772 wrote that the California Indians made molasses, candy, and sugar from a vegetable. A wild variety of sugar beet still grows in California).

The sugar beet manufactures sugar in its leaves by the process of photosynthesis. Carbon dioxide is absorbed from the air and water is taken in through the roots. Activated by the energy of the sun shining on the leaves, a sequence of complex chemical reactions converts the CO_2 and H_2O to sucrose ($C_{12}H_{22}O_{11}$), which is stored in the silvery white root.

Beets can withstand adverse weather better than most crops and are adaptable to widely varying climates, altitudes, and soils. Sugar beet production is an important part of the economy of 22 states, but the sugar content of any variety of beet will vary significantly depending on the growing conditions. It has been necessary to develop many different varieties of beets in order to have plants that would produce well in each growing area.

In this investigation you will extract sucrose from sugar beets. You may find it interesting to compare the procedures you use with those developed by Marggraf.

Before starting this investigation, you should be familiar with the nature and manifestations of intermolecular forces in pure substances and in mixtures. In addition you should be able to make stoichiometric calculations involving "percentage yield."

References

1. O'Connor, R. 1977. *Fundamentals of Chemistry* (2nd ed). New York: Harper & Row (Units 7, 9, and 11).
2. DeBey, H. J. 1969. *Introduction to the Chemistry of Life: Biochemistry*. Reading, Mass.: Addison-Wesley

SAFETY Do not taste any of the products prepared in this experiment. They are contaminated with methyl alcohol which can cause permanent damage to your central nervous system and your vision. Wear eye protection and use proper care in working with hot solutions.

OBJECTIVES
1. Be able to perform properly and safely the techniques of liquid extraction, solvent evaporation, and vacuum filtration.
2. Be able to prepare a crystalline compound.
3. Be able to make calculations of yield and percentage yield.

PROCEDURES

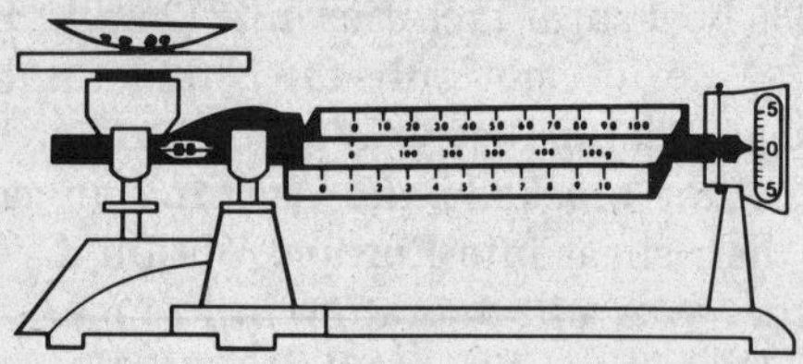

1. Obtain from your lab instructor a small piece of sugar beet. Weigh a clean dry 9-cm watch glass to the nearest 0.10 g. Add the beet sample to the watch glass and weigh again. You will need between 20 and 25 g of the sugar beet sample. If your sample is too heavy, cut off any excess. If it is too light, obtain additional sample from your instructor.

2. Transfer the beet sample to a clean mortar and crush the sample. You may find it easier if you first cut the sample into small pieces, using your spatula, before crushing it.

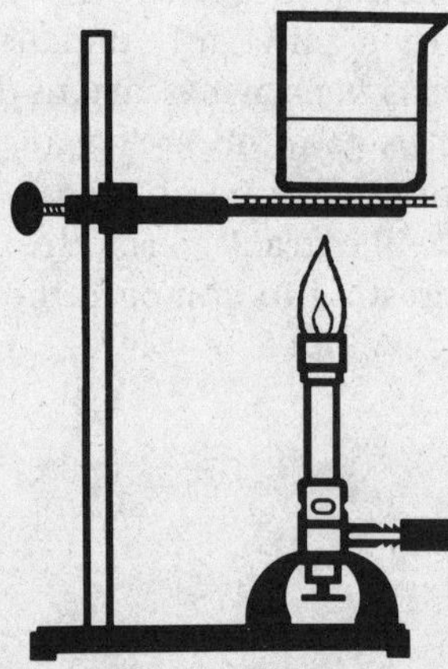

3. Transfer the crushed beet sample to a 250-ml beaker. Wash the watch glass, mortar, pestle, and spatula (and stirring rod if it was used to transfer the sample) with distilled water. Add the washes to the beaker containing the crushed beet sample. Add distilled water to the beaker until the total volume is approximately 100 ml. Place the beaker and its contents on a ring stand and gently boil the contents for 20 min.

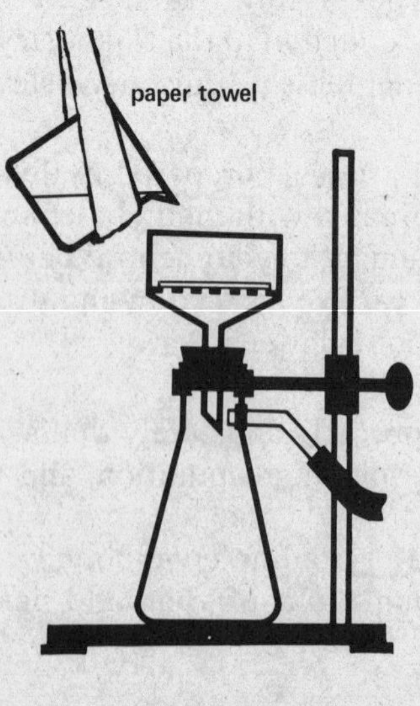

4. Using an aspirator, vacuum filter the hot solution to remove the fiber residue. Use a few milliliters of distilled water to rinse the beaker. Consult Appendix A if you are unfamiliar with the technique of vacuum filtration.

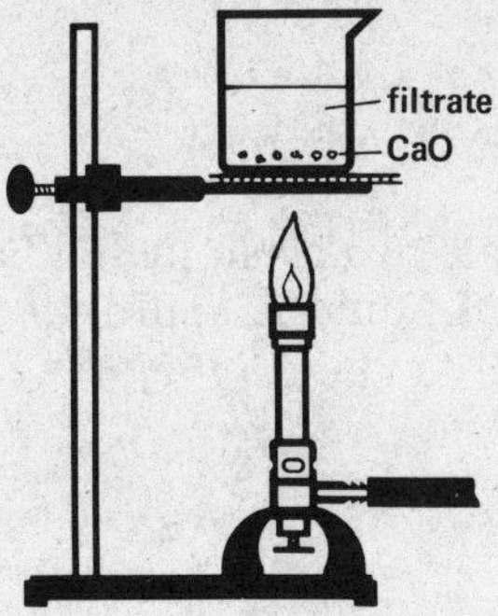

5. Transfer the clear filtrate to a clean 250-ml beaker. Rinse the filter flask with a small amount of distilled water and add this rinse to the beaker containing the filtrate. Add approximately 0.2 g (two spatula tips full) of calcium oxide to the filtrate and boil the mixture for five min.

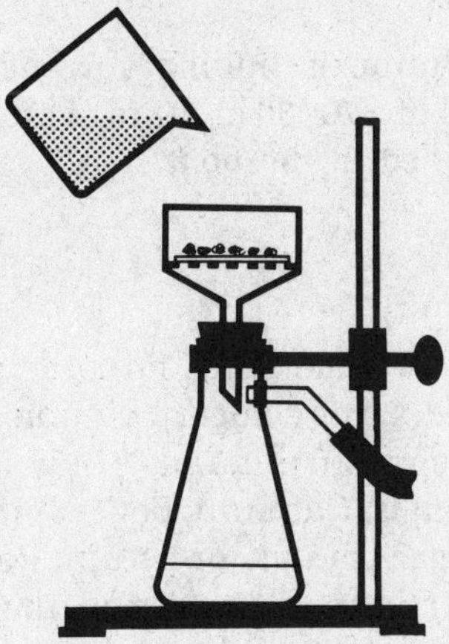

6. Filter the resulting mixture using the technique of vacuum filtration. Use a few milliliters of distilled water to rinse the contents of the beaker and to wash the residue retained on the filter paper.

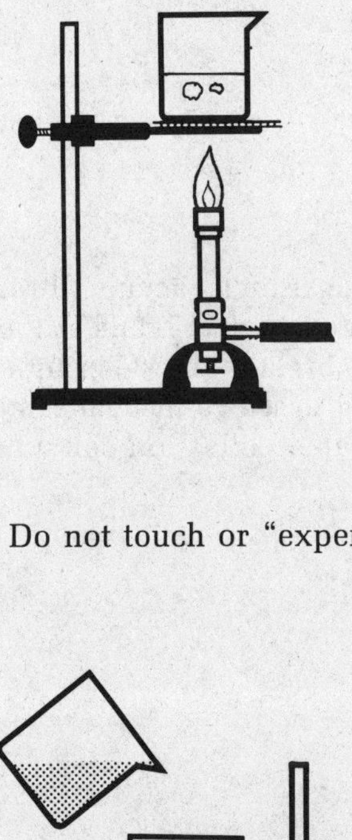

7. Transfer the clear filtrate to a clean 250-ml beaker. Add two small pieces of dry ice (each about the size of a small marble). Boil the solution for 10 min. At this point the volume of liquid must be less than 50 ml. If not, boil the solution until its volume is less than 50 ml.

Warning: Do not touch or "experiment" with "dry ice."

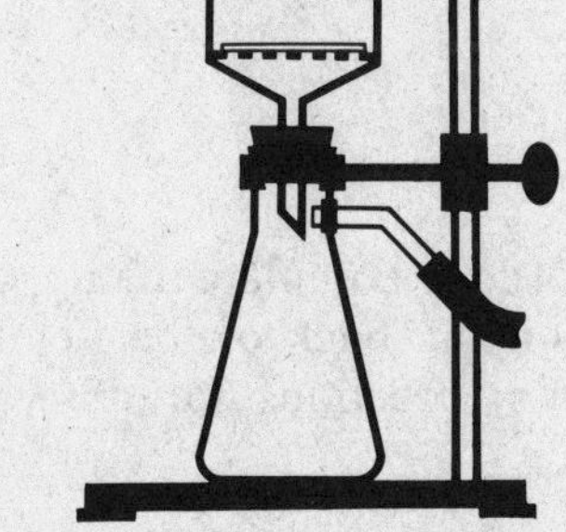

8. Filter the resulting mixture using the technique of vacuum filtration. Use a few milliliters of distilled water to rinse the contents of the beaker and to wash the residue retained on the filter paper.

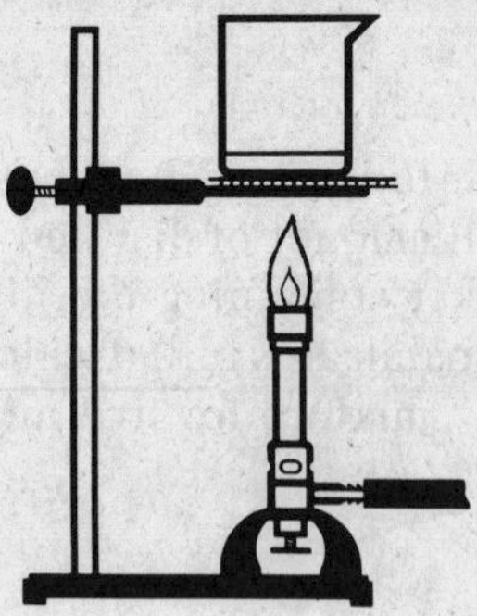

9. Transfer the clear filtrate to a clean 100-ml beaker. Boil gently until the volume of liquid is approximately 1 ml. (Do not allow the sample to char.)

While the above solution is boiling you may be requested to prepare 60 ml of a 70–30 mixture of ethanol and methanol, which is needed in the next step.

10. Add 10 ml of an ethanol–methanol mixture to the approximately 1 ml of the syrupy sugar solution. Stir the mixture thoroughly using a stirring rod. Allow the sugar to harden for a few minutes. *Carefully* pour off the liquid (do not pour off any sugar crystals or syrup). Repeat the addition of the alcohol mixture, the stirring, the settling, and the decanting until all of the hardened syrup is crystalline.

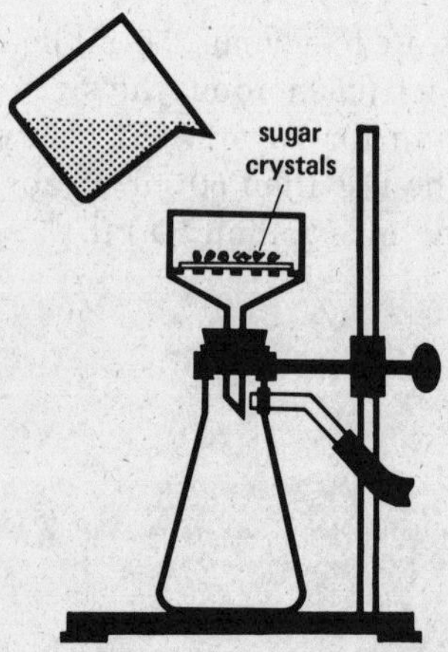

11. Filter the obtained sugar on a vacuum filtration apparatus and allow the crystals to dry. Transfer the obtained sugar to a weighed watch glass and determine its mass. Your instructor will want to see your sugar sample before you place it in a designated collection container.

Warning: *Do not try to taste your sugar.* It will be contaminated with methanol, which is very poisonous.

CALCULATIONS AND REPORT Use your notebook records to complete the Report Form. Show all calculations on the back of the Report Form. Follow your instructor's directions for answering the questions following.

QUESTIONS

1. When using the vacuum filtration apparatus, why must you disconnect the vacuum tubing before turning off the aspirator?
2. In this experiment, excess calcium ion was added to precipitate organic acids present. How was the excess calcium removed?
3. The formula of the sugar isolated (sucrose) is $C_{12}H_{22}O_{11}$. If you obtained 1.19 g of sugar from 22.38 g of beet, how large a beet sample would you need in order to isolate 1 mole of sugar? (Assume that you would obtain the same percentage yield in the larger run.)
4. Using the results of your experiment, how many tons of beets would be required for the preparation of 100 lb of sugar?
5. Assume that it only costs a penny an hour to supply fuel for the burner and that you boiled your sample for 1 hr and 15 min during this experiment. Using your results, what would the cost be for fuel alone for the production of 1 lb (454 g) of sugar?
6. (Optional) Look carefully at the structural formulas given in the Introduction for glucose, fructose, sucrose, maltose, and lactose. What similarities do you note? What differences?

GRADE
INSTRUCTOR'S INITIALS

Report Form

FOR INVESTIGATION 5:
Sugar from Sugar Beets

NAME: ______________________ ID NO.: ______________________

LAB SECTION ______________________ INSTRUCTOR ______________________

Mass of watch glass plus sample ______________ g

Mass of watch glass ______________ g

Mass of beet sample used ______________ g

Number of rinses of ethanol–methanol used ______________

Mass of sugar obtained ______________ g

Yield ______________ %

Description of Crystals Obtained

______________ DATE

______________ SIGNATURE

Investigation 6

THE SCIENCE OF SOAP

INTRODUCTION The "soap opera" is truly an American institution, having graduated from the early "Just Plain Bill" and "Helen Trent" of radio to the modern daytime television serials. Called "soap operas" because so many were sponsored by soap and detergent manufacturers, these dramas have provided many viewers with hours of escape from routine, while making their sponsors' slogans a part of our permanent culture.

It has not been so many years since the farm wife made her own soap in a big black kettle, using lard (rendered animal fats) and lye ($NaOH$). Now, with the ready availability and almost infinite variety of commercial products, soap making is almost a forgotten art.*

Natural animal fats and vegetable oils are rich in compounds called *glyceryl esters,* such as *glyceryl tristearate:*

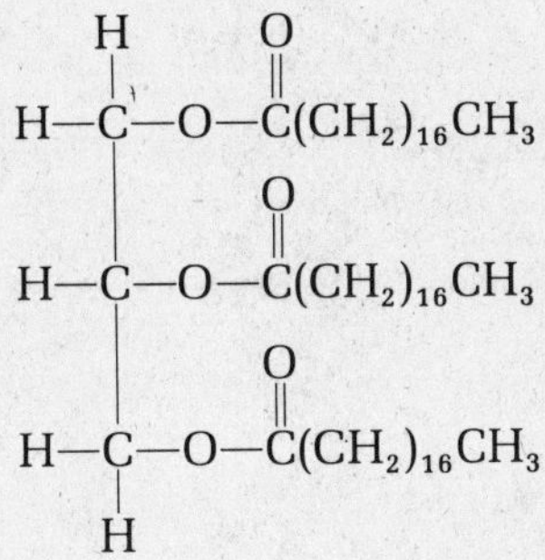

Esters may be thought of as having been formed by the combination of a *carboxylic acid* and an *alcohol,* with the elimination of water:

$$\underset{\text{ethyl alcohol}}{CH_3CH_2OH} + \underset{\text{stearic acid}}{HO\overset{O}{\overset{\|}{C}}(CH_2)_{16}CH_3} \rightleftharpoons \underset{\text{ethyl stearate (an ester)}}{CH_3CH_2O\overset{O}{\overset{\|}{C}}(CH_2)_{16}CH_3} + H_2O$$

In the glyceryl esters, the alcohol is the trihydroxy compound called *glycerol* (older name *glycerine*).

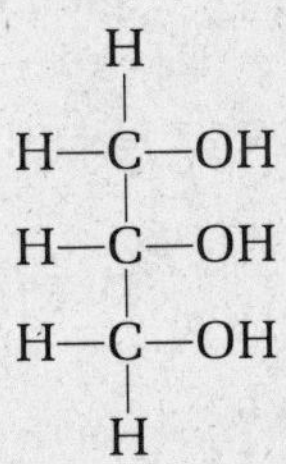

glycerol (1,2,3-trihydroxypropane)

*Some of the major restaurant chains still use waste cooking grease to make some of their own soap.

The carboxylic acid may be one of a number of long-chain acids, for example,

lauric acid	$CH_3(CH_2)_{10}CO_2H$
palmitic acid	$CH_3(CH_2)_{14}CO_2H$
stearic acid	$CH_3(CH_2)_{16}CO_2H$
oleic acid	$CH_3(CH_2)_7CH{=}CH(CH_2)_7CO_2H$
linoleic acid	$CH_3(CH_2)_4CH{=}CHCH_2CH{=}CH(CH_2)_7CO_2H$

Most glyceryl esters found in nature contain more than one kind of the various "fatty" acid moieties.

The formation of esters is a reversible reaction; that is, esters react with water to reform their "parent" alcohol and carboxylic acid(s). For example,

$$\underset{\text{ethyl stearate}}{CH_3CH_2O\overset{\overset{O}{\|}}{C}(CH_2)_{16}CH_3} + H_2O \rightleftharpoons \underset{\text{ethyl alcohol}}{CH_3CH_2OH} + \underset{\text{stearic acid}}{HO\overset{\overset{O}{\|}}{C}(CH_2)_{16}CH_3}$$

This "hydrolysis" reaction with water alone is typically fairly slow, but either strong acids (sources of H^+) or strong bases (e.g., OH^-) will act as *catalysts* to speed up the reaction. The base-catalyzed hydrolysis of esters is called *saponification*. This is the type of reaction we will be studying in this experiment.

Since one of the products of the hydrolysis of an ester is a carboxylic acid, the use of excess base in the hydrolysis mixture will result in the formation of a salt of the acid. For example,

$$\underset{\text{stearic acid}}{CH_3(CH_2)_{16}\overset{\overset{O}{\|}}{C}OH} + NaOH \longrightarrow \underset{\text{sodium stearate}}{[CH_3(CH_2)_{16}\overset{\overset{O}{\|}}{C}O^-]Na^+} + H_2O$$

The salts of long-chain carboxylic acids are called *soaps* and the water-soluble sodium or potassium salts are valuable as cleaning agents. Salts containing dipositive (e.g., Ca^{2+}, Mg^{2+}) or tripositive (e.g., Al^{3+}, Fe^{3+}) cations are insoluble in water. In "hard" water the presence of such cations results in the precipitation of insoluble soaps, familiar as "bathtub" ring.

The saponification of glyceryl esters, using aqueous sodium hydroxide, is the most common way of making a soluble soap. For example,

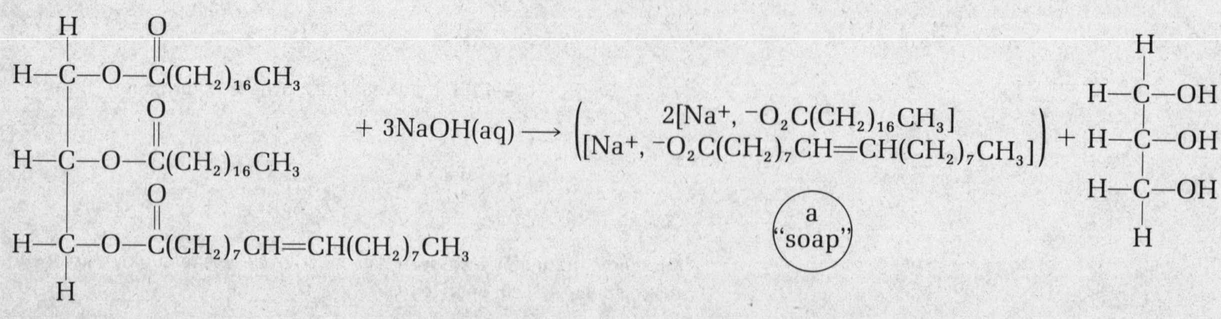

In "primitive" soap making, the glycerol was discarded with the waste residue solution. Modern industrial soap making recovers most of the glycerol, since it is itself a valuble commercial product. It is the principal ingredient of most skin lotions and, when esterified with nitric acid, it forms the powerful explosive "nitroglycerine" (glyceryl trinitrate).

In this experiment we shall perform a saponification reaction, and then study some of the characteristics of the soap prepared.

Before starting this investigation, you should be familiar with the properties of "hard water" and with some general characteristics of soaps and detergents. In addition, you should review the calculations of "weight-to-weight" stoichiometry and you must do the calculations required for part B.3 of this investigation.

References

1. O'Connor, Rod. 1977. *Fundamentals of Chemistry.* (2nd ed.) New York: Harper & Row (Units 4, 6, and 14).
2. Vold, M. J. and R. D. Vold. 1964. *Colloid Chemistry.* New York: Van Nostrand Reinhold.

SAFETY In this experiment you will be working with concentrated alkaline solutions. *Wear safety goggles at all times.* One drop of the hot alkali solution in the eye could cause blindness. In case of contact of any chemicals with the skin or clothing, wash immediately and copiously with water and notify your instructor. Be particularly careful to keep the alcohol used away from open flames.

OBJECTIVES

1. Be able to handle reagents properly, including all safety precautions.
2. Be able to measure, transfer, mix, and heat liquid reagents properly.
3. Be able to make reasonably accurate measurements of mass and volume.
4. Be able to make and record accurate observations and to interpret these in terms of ionic and molecular behavior.
5. Be able to calculate a percentage yield.

PROCEDURES

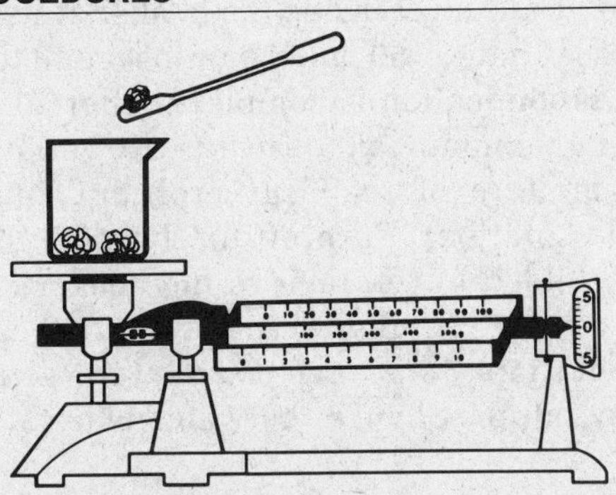

A. Preparation of a Soluble Soap

1. Weigh a clean, dry 400-ml beaker on the triple beam balance to the nearest 0.05 g. Advance the weights by 10 g and carefully transfer, with a spatula, enough fat to bring the balance beam to its equilibrium position. (You should have 9.5 to 10.5 g of fat in the beaker.) Record the actual mass transferred to the nearest 0.05 g.

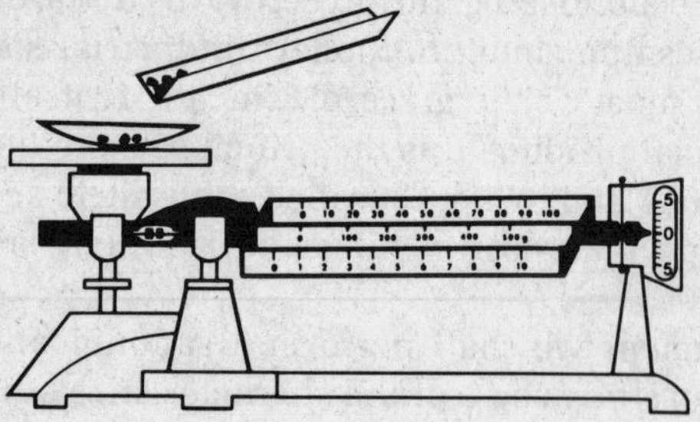

2. Using the triple beam balance, weigh out 2.5 g of sodium hydroxide pellets on a preweighed watch glass. (Caution: Avoid contact with these pellets.) Transfer all the pellets into a 125-ml Erlenmeyer flask containing 25 ml of distilled water. Swirl the flask until all the NaOH has dissolved. Then add 20 ml of 95-percent ethanol to this solution and swirl the mixture until it appears homogeneous. Pour this NaOH solution into the beaker containing the fat and mark the liquid level with a grease pencil on the outside of the beaker.

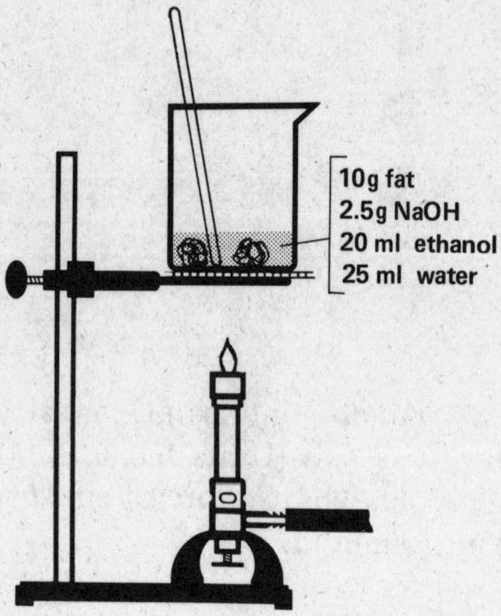

3. Before continuing work, place an asbestos square, large enough to cover your beaker, in a readily accessible location. Then place the beaker on a wire gauze supported on a ring clamp above a laboratory burner. Light and adjust the burner for a low cool flame and heat the mixture *gently*, while stirring it. (Caution: Ethanol is *very* flammable.) It is important to keep the mixture boiling very *gently*. Adjust the flame as necessary. If you observe foam rising in the beaker, you are heating too strongly. Remove the burner and reduce the flame size and the amount of air mixing with the gas. It is important to stir the mixture well during the heating process. Stirring will improve mixing for more rapid reaction and will prevent dangerous overheating and bumping. Do not permit the flame near the upper part of the beaker, where ethanol vapors might easily ignite.

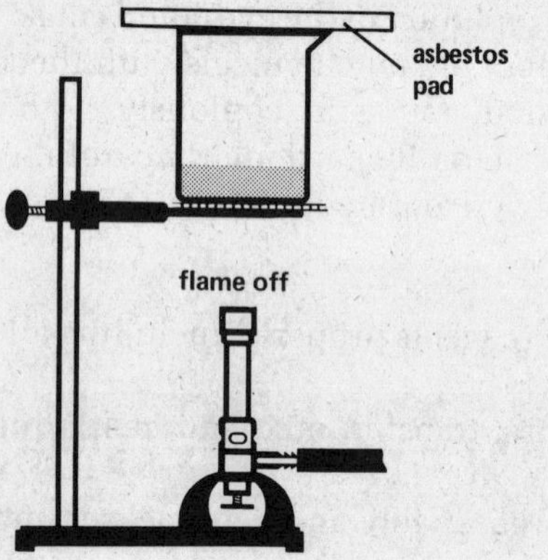

Warning: If your mixture should ignite, do not panic. *Turn off your burner* and cover the beaker with an asbestos pad. The beaker must be kept covered until all flames are gone. Notify the instructor.

4. Continue with gentle heating and stirring for 20 min. From time to time, squirt distilled water from your squeeze bottle around the inside of the beaker to replace liquid lost by evaporation and at the same time to rinse particles clinging on the inside of the beaker back into the reaction mixture. Keep the liquid level inside the beaker at its initial level of approximately 50 ml. Foaming commonly occurs when the saponfication is complete. After 20 min your beaker should contain an off-white to yellowish viscous mass (a mixture of soap, glycerol, and excess sodium hydroxide solution). Turn off the burner. Using beaker tongs or a folded towel, remove the beaker carefully from the ring support and set it into a 600-ml beaker half-filled with cold tap water. Let the mixture cool to below room temperature. Change the water when it becomes warm.

5. Weigh 16 g of sodium chloride into a 250-ml Erlenmeyer flask, using the triple beam balance. Add 50 ml of distilled water to the salt, stopper the flask, and shake the stoppered flask until most of the sodium chloride has dissolved. Should the solution be cloudy, filter it by vacuum filtration (see Appendix A).

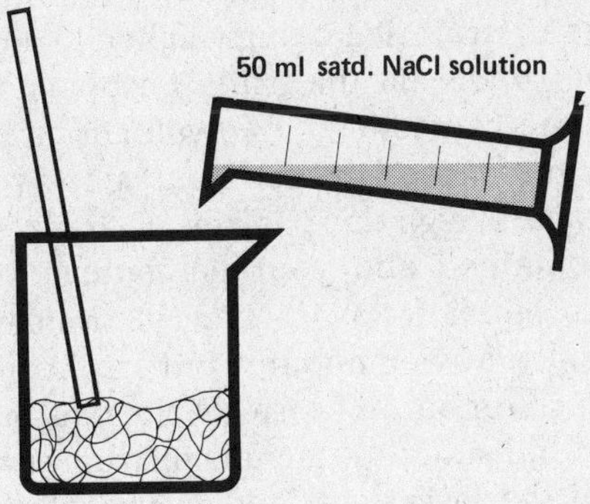

6. After the soap has cooled, pour into the beaker the aqueous sodium chloride solution just prepared. (This reduces the solubility of the soap in the aqueous phase. The process is called "salting out" of the soap.) Stir the mixture thoroughly with a spatula. The excess sodium hydroxide and the glycerol are dissolved, leaving a purified soap. It is extremely important that you mix the soap with the sodium chloride solution *very thoroughly*. Press the soap against the beaker wall, breaking up any chunks floating in the mixture. If this step is not properly executed, sodium hydroxide and glycerol will be trapped in the soap particles and will interfere with subsequent experiments to be performed with your soap.

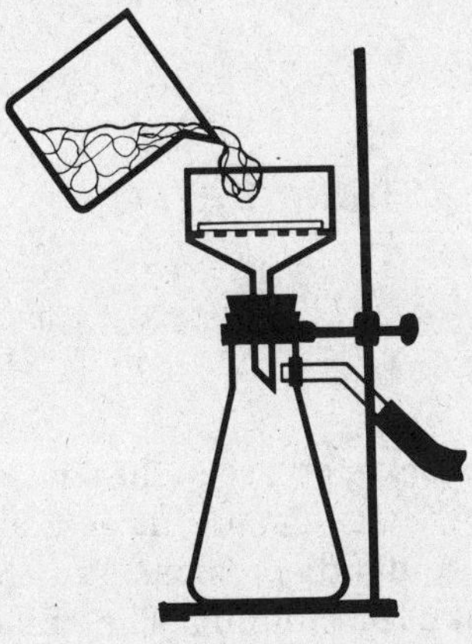

7. Set up the aspirator filtration equipment as instructed (Appendix A). Place a filter paper in the funnel and moisten it with distilled water. Turn on the water aspirator and *decant* the supernatant liquid through the funnel. Then transfer the soap into the funnel, proceeding as described in Appendix A for vacuum filtration.

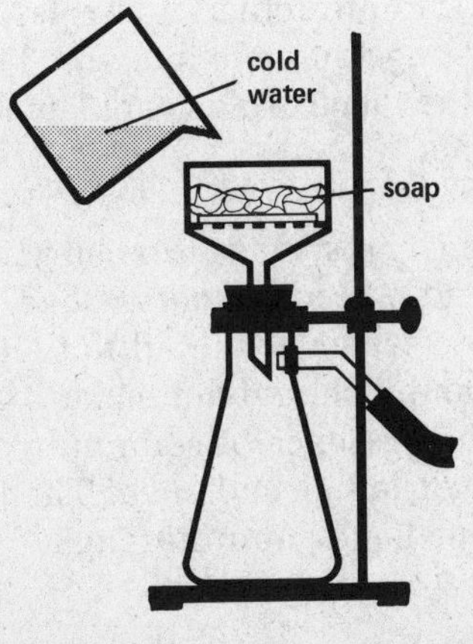

8. Add a few chunks of ice to 50 ml of distilled water in a 100-ml beaker and allow the water to chill. Use half of the cold water (no ice) to rinse any remaining soap from the reaction beaker into the funnel (while the aspirator is shut off), using a spatula to scrape soap from the interior of the beaker. Stir the soap—ice water mixture in the funnel into a slurry, using your spatula. (Be careful not to tear the filter paper.) Then turn on the aspirator to filter the mixture. Allow the aspirator to pull air through the soap for about 10 min to help dry it. When the soap is dry, remove a small sample for testing.

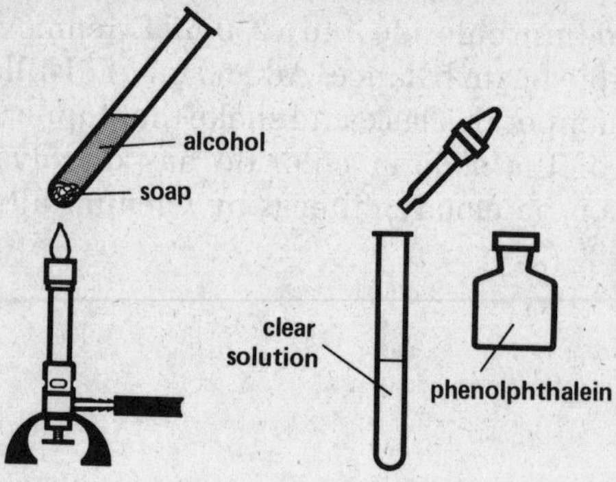

9. To check for residual sodium hydroxide in your soap, place half a spatula-tip full of the soap in a large, dry test tube and add 10 ml of 95-percent ethanol. Heat the mixture *carefully* over a low flame, with gentle swirling until the soap has dissolved. Then cool the test tube to room temperature and add three to five drops of phenolphthalein indicator. If the solution turns pink, you have too much NaOH left in the soap and it must be washed again with ice water until repeated tests show no pink color with phenolphthalein.

When your soap, shown to be free of significant contamination by NaOH, has air dried, it is to be weighed to ± 0.1 g. Weigh a clean paper towel on the triple beam balance. Place your soap in the center of this towel. Discard the filter paper. Then fold the towel around your soap sample to make a neat package. Weigh this package. Calculate the mass of the soap obtained and your percentage yield, assuming that the original fat was glyceryl tristearate. (The fat was not really this compound but its "formula weight" is close enough for our purpose.) Write your name on the soap package and submit it to your instructor after you have performed the tests described in procedure B.

B. Some Properties of Soaps

Soluble soaps owe their cleaning power to the affinity of the negatively charged carboxylate group for water and the affinity of the "hydrophobic" carbon chain for greases.

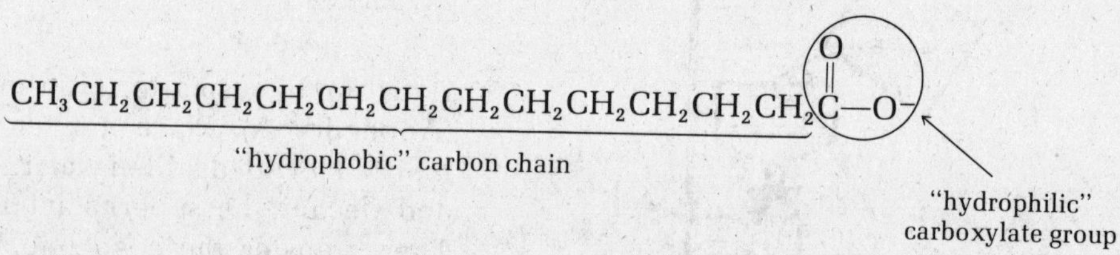

Soaps do not *dissolve* dirt or grease in the sense of forming a true solution, but they can cause these materials to form stable aqueous colloidal dispersions, thus removing such "dirt" from your skin or clothing. The greasy materials are broken up into submicroscopic globules containing soap ions. The hydrophobic region of the soap is dissolved in the grease droplet and the charged hydrophilic end sticks out into the water. Thus the globule is not only rendered more "attractive" to water, but the tendency for globules to agglomerate into an oil or grease film is diminished by the electrostatic repulsion of these "micelles" for each other.

1. Use some of the soap you just made to wash your hands. If the soap was prepared correctly and washed properly with salt water and cold water, it should not feel greasy (caused by excess fat) and it should not make your hands feel slick (caused by the presence of sodium hydroxide). Rinse your hands free of lather and soap. Then place a small dab of fat on your hands and thoroughly "grease" your hands. Try to wash off this fat with tap water alone. Then use your soap. Give an explanation for the observed results.

Soaps are not good cleaning agents for use in "hard" water. The dipositive (mainly Ca^{2+} and Mg^{2+}) and tripositive (mainly Al^{3+} and Fe^{3+}) cations responsible for water "hardness" form insoluble soaps. If, for example, the soap was sodium stearate and the "hardness" was due to Ca^{2+} and Fe^{3+}, then the insoluble soaps would have the formulas $Ca(C_{17}H_{35}CO_2)_2$ and $Fe(C_{17}H_{35}CO_2)_3$. Thus soap needed for cleaning is removed from solution. Synthetic detergents, mostly organic sulfonates, do not form insoluble salts with the ions in "hard" water.

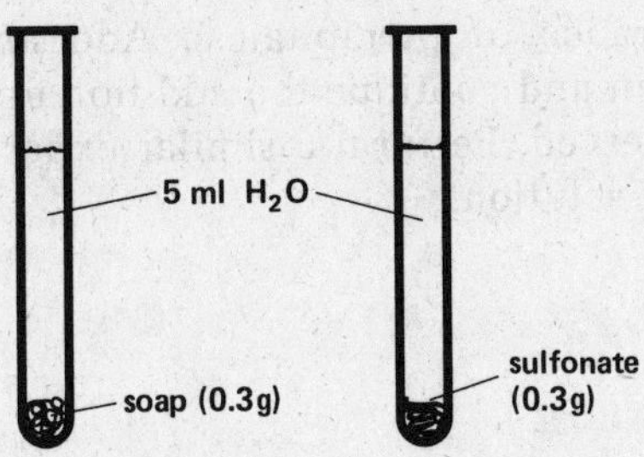

2. Prepare a clear dispersion of your soap by placing 0.3 g of the soap, weighed out on a triple beam balance, into a small test tube and adding 5 ml of distilled water. Warm and shake the mixture until you have an almost clear dispersion. Cool the dispersion by setting the test tube into a small beaker filled with cold tap water. Proceed similarly, using the same amounts of the appropriate materials, to prepare a dispersion of sodium dodecylbenzenesulfonate, which has the formula*

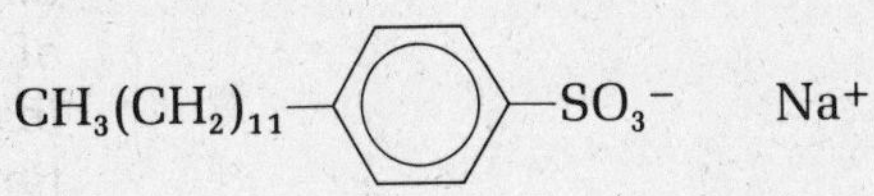

Calculations (To be completed prior to starting work)

The calcium chloride and magnesium chloride stock solutions to be used in the next procedure contain 10 mg ml^{-1} of $CaCl_2 \cdot 2H_2O$ or $MgCl_2 \cdot 6H_2O$, respectively. By definition, "hard" water contains at least 40 mg of Ca^{2+} liter^{-1}. Assuming that one drop is 0.05 ml, calculate (a) the volume of distilled water to be added to two drops of the stock $CaCl_2$ solution to obtain a solution containing 40 mg Ca^{2+} liter^{-1} and (b) the volume of distilled water to be added to three drops of the stock $MgCl_2$ solution to obtain a solution containing 60 mg Mg^{2+} liter^{-1}. Have your calculations checked by your instructor before proceeding further.

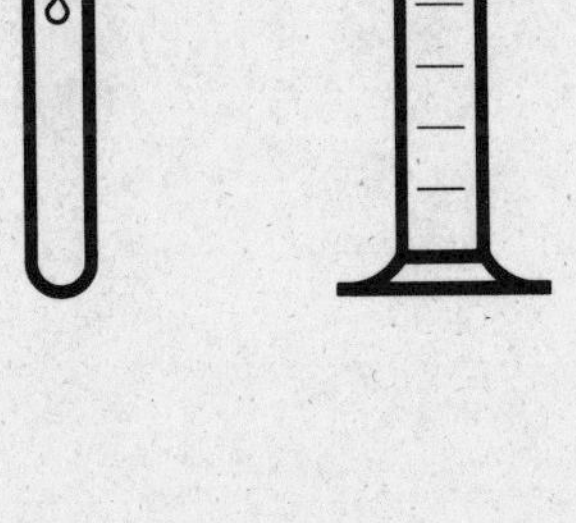

3. Prepare the solutions containing 40 mg liter^{-1} Ca^{2+} and 60 mg liter^{-1} Mg^{2+}, employing the stock dropping bottles, test tubes of appropriate size, and your graduated cylinder. (This procedure of making a dilute solution is not very precise. This is, however, acceptable for our purposes.)

* 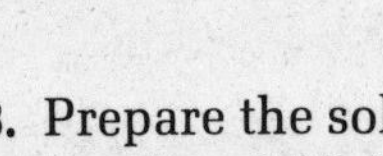

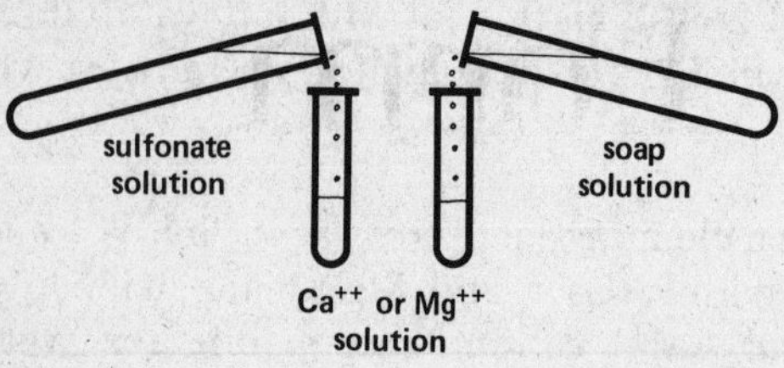

4. Divide the diluted Ca^{2+} solution equally between two small test tubes. Add to one a few drops of your soap solution and to the other a few drops of the sulfonate solution. Observe and record what happens. Repeat this procedure with the diluted Mg^{2+} solution.

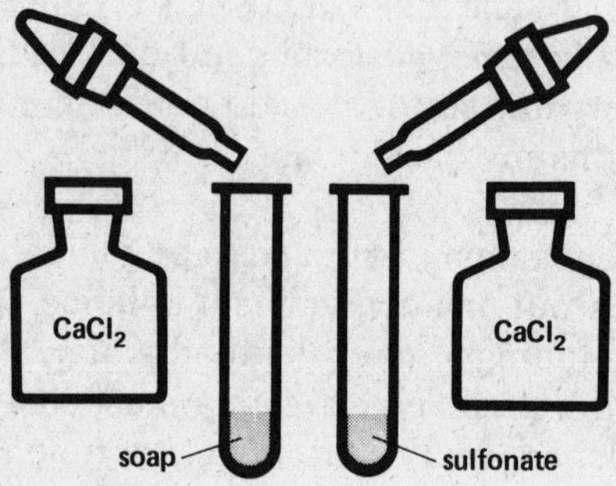

5. Place approximately 1 ml of your soap solution in a small test tube. Then add one drop of the stock Ca^{2+} solution (containing 10 mg ml^{-1} of $CaCl_2 \cdot 2H_2O$) to this test tube. Mix well and describe your observations (cloudiness, precipitates, appearance of precipitates). Add another drop of Ca^{2+} solution and continue the addition until no changes can be observed. Perform a similar experiment with your sulfonate solution.

REPORT Record in your permanent notebook all pertinent data, observations, and the calculation of your percentage yield of soap. Using these records, prepare and submit to your instructor the Report Form. Answer the following questions on a separate page and submit this page with your Report Form.

1. A dirty man takes a bath every day. He fills the bathtub with 100 liters of water, which contains 162 mg calcium hydrogen carbonate [$Ca(HCO_3)_2$] per liter. How many kilograms of soap will be used per month (30 days) to remove all the calcium ions as calcium stearate from the bath water? Assume that calcium stearate is completely insoluble in bath water.
2. What similarities exist between "soluble" soap and sodium dodecylbenzenesulfonate?
3. Why is dodecylbenzenesulfonate a better cleaning agent than soap in hard water? Will it also work in extremely hard water containing, for instance, 0.06 moles of Ca^{2+} $liter^{-1}$? Use your experimental results in answering this question.

GRADE

INSTRUCTOR'S INITIALS

Report Form

FOR INVESTIGATION 6:

The Science of Soap

NAME: ______________________ ID NO.: ______________________

LAB SECTION ______________________ INSTRUCTOR ______________________

Mass of fat used ______________ g

Mass of soap obtained ______________ g

Yield ______________ %

Show yield calculations below

Washing with Soap

Observations and explanations

Hard Water with Soap or Sodium Dodecylbenzenesulfonate

Volume of water to be added to stock $CaCl_2$ solution

______________ ml

Volume of water to be added to stock $MgCl_2$ solution

______________ ml

Show calculations on reverse side.

Observations: (Answer on reverse side)

Soap to hard water
Sulfonate to hard water

10 mg ml^{-1} $CaCl_2$ solution to soap

10 mg ml^{-1} $CaCl_2$ solution to sulfonate

______________ DATE ______________ SIGNATURE

Investigation 7

ANTIFREEZE MIXTURES

INTRODUCTION Most automobile engines employ a circulating water-cooling system to dissipate engine heat. Water, because of its relatively high heat capacity [1.0 cal g^{-1} $(°C)^{-1}$] and ready availability at low cost, makes a good coolant for the automobile radiator system, although some modern systems employ a water–ethylene glycol mixture for improved heat transfer.

The principal difficulty with water-coolant systems lies with the relatively high freezing point of water, 0°C (32°F). Thus automobiles would be limited to operation in moderately warm regions if water alone were to be used in their radiators. Fortunately, the freezing point of a solvent is lowered by a dissolved solute by an amount that can be calculated* from

$$\Delta t = \phi m k_{\text{f.p.}}$$

in which Δt is the change in freezing point of the solvent, ϕ is the number of dissolved *particles* per formula unit of solute, m is a concentration unit (*molality*) expressing the number of moles of solute per kilogram of solvent, and $k_{\text{f.p.}}$ is a molal freezing-point constant characteristic of the solvent. For water, $k_{\text{f.p.}}$ is -1.86°C kg (solvent) mole^{-1} (solvent particles).

It is important to note, for the purpose of the method used in this experiment, that the initial freezing point of an "ideal" solution is the same as the melting point of the frozen solution *if* the melting point is determined just when the final solid disappears. Since the molal freezing-point constant for water is not very large (compare, for example, -37.7 for camphor), a compound suitable for an antifreeze additive must be quite soluble in water to provide a solution whose concentration is sufficient to produce a significant lowering of the freezing point.

Calculate the initial† freezing point of the following solutions:

1. Saturated sodium chloride solution

37 g NaCl per 100 g H_2O, $\phi = 2$

2. Saturated "table sugar" solution

250 g $C_{12}H_{22}O_{11}$ per 100 g H_2O, $\phi = 1$

3. 50 percent ethylene glycol, by weight

100 g $C_2H_6O_2$ per 100 g H_2O, $\phi = 1$

*This equation represents an "ideal" calculation. Real systems are limited by such factors as the freezing point of the solute itself or by changes in solubility of the solute with temperature. The equation is still useful over a fairly broad range of concentrations.

†Since freezing normally involves the initial crystallization of the pure solvent, the remaining solution gradually increases in concentration.

Ideally, an antifreeze additive should be a liquid that is *miscible* (soluble in all proportions) with water. Otherwise evaporation of water from a loosely sealed radiator system could leave solute deposits that would clog the cooling system.

Obviously the antifreeze compound must be noncorrosive to metal or rubber to avoid damage to the system. Many years ago, alcohols (either methanol, CH_3OH, or ethanol, C_2H_5OH) were employed as antifreeze agents. The volatility of these compounds, however, was a problem in maintaining a radiator solution of proper solute concentration. Now we use a "permanent" antifreeze, usually ethylene glycol (1,2-dihydroxyethane), which will not be lost by evaporation.

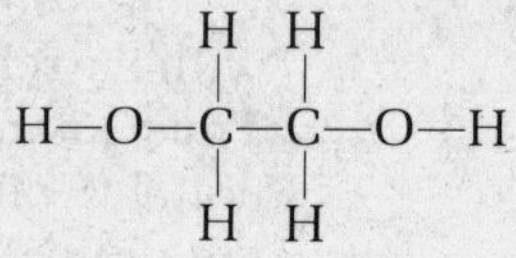

ethylene glycol

Ethylene glycol, the principal component of such well-known antifreeze preparations as Prestone and Zerex, has an ideal set of properties. It is noncorrosive to metal or rubber, perfectly miscible with water, high boiling (198°C), and liquid at room temperature for easy mixing (m.p., −13°C). In addition, water–ethylene glycol mixtures are quite effective as heat-transfer systems. Most commercial antifreeze preparations contain other additives, such as corrosion retardants.

In this experiment we will investigate some characteristics of water–ethylene glycol mixtures of potential value for use in automobile cooling systems.

Before starting this investigation, you should be familiar with the basic concepts of *colligative properties* of solutions and with the procedures described in the Calculations and Report section of this investigation. In addition, you must have completed the calculations assigned in the preceding Introduction before coming to the laboratory.

References

1. O'Connor, Rod. 1977. *Fundamentals of Chemistry* (2nd ed.). New York: Harper & Row (Units 9, 12, 13, and 15).
2. Eisenberg, D. and W. Kauzmann, 1969. *The Structure and Properties of Water.* New York: Oxford.

SAFETY Wear approved eye protection. Avoid contact with or inhalation of the vapors of the organic liquids used. Be particularly cautious with "dry ice": Contact with the skin for 2 sec or more can cause severe freeze burns. *Under no circumstances* should you "play" with "dry ice." Never seal "dry ice" in a *closed* system. Gas pressures may cause an explosion.

OBJECTIVES

1. Be able to use a laboratory balance and to measure liquid volumes accurately.
2. Be able to calculate liquid densities.
3. Be able to determine melting points accurately.
4. Be able to prepare and interpret data graphs.

PROCEDURES

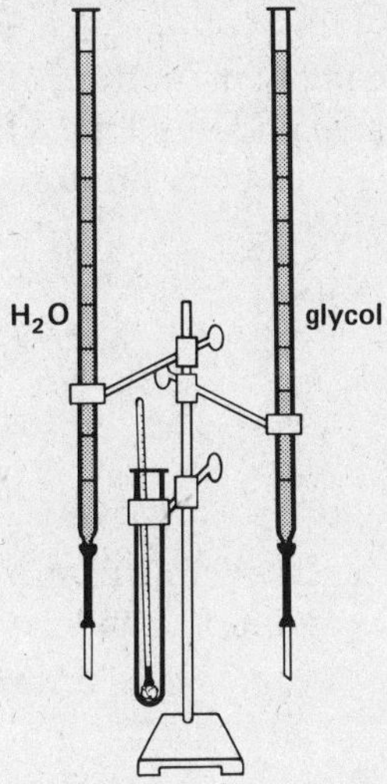

A. Density Determination

1. Obtain two burets and be sure that they are clean (Appendix A). Attach them with clamps to a ringstand. Rinse and fill one buret with distilled water. Rinse the other twice with 5-ml portions, then fill, with ethylene glycol. Label both burets. Then clamp a large test tube containing a wad of paper towel to the ringstand. Use this as a storage place for your thermometer when it is not in use. Be careful with your thermometer. It is both fragile and expensive.

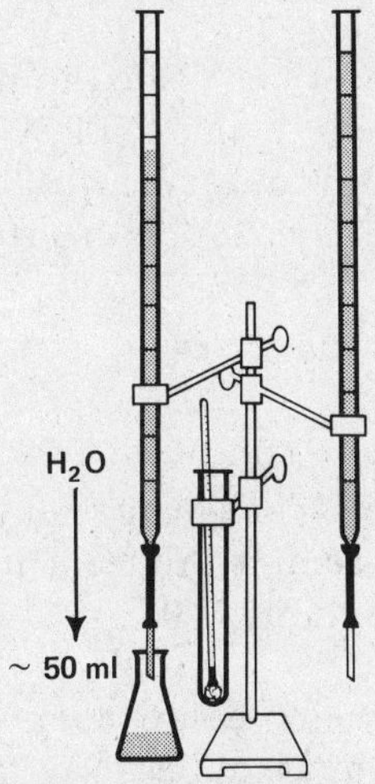

2. As a check on your technique, determine first the density of water. Weigh (Appendix A) a clean, dry 125-ml Erlenmeyer flask to the nearest 0.01 g. Drain just enough water from the buret into a waste flask to expel air from the buret tip and to drop the liquid level just below the zero mark. Read and record the volume to the nearest 0.1 ml. Then drain about 50 ml of water into the weighed flask. Record the buret reading and weigh the flask plus water to the nearest 0.02 g. Calculate the density of distilled water. (If your value differs from 0.99 g ml^{-1}, consult your instructor.) Discard the water and dry the flask, inside and out, with a clean towel.

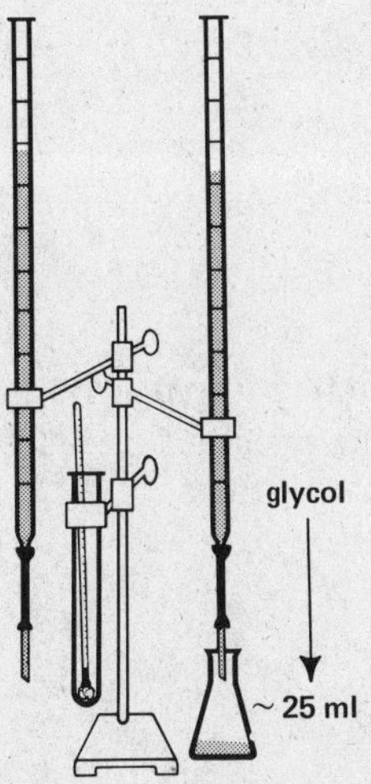

3. Reweigh the dried flask and measure into it from the buret about 25 ml of ethylene glycol, recording the volume to ±0.1 ml. Weigh the flask plus contents.

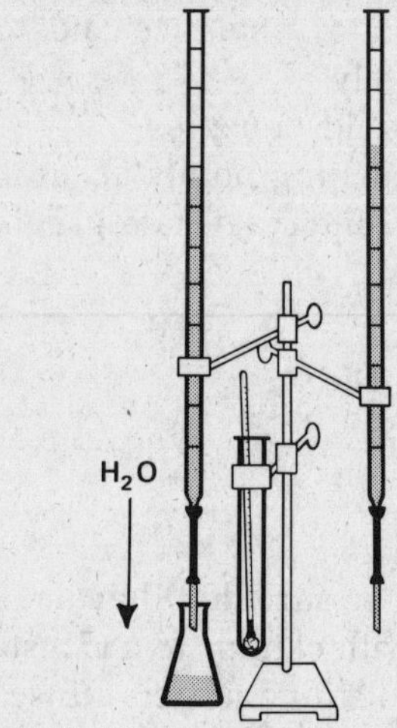

4. Now add to the glycol in the flask about 10 ml of the distilled water (recording the volume accurately) and reweigh. Continue the incremental addition of 10-ml portions of water, until you have added a total of ~50 ml of water. (Calculate the density of each mixture.)

Make a copy of the Density Data table from your Report Form into your permanent notebook and complete the entries in the table.

B. Calibration of a Thermometer

1. Secure a No. 6 rubber stopper drilled to hold a small test tube. Lubricate the stopper and the test tube with a few drops of glycol, protect your hands with a towel, and insert the test tube through the stopper nearly to the rim of the tube (Appendix A).

Form a team of four students. Each team member will prepare a different calibration sample. All team members will use four samples for procedures 3–9.

2. Fill the test tubes as directed below, then stopper and label the tubes. (Use only the tubes inserted into the stoppers.)

STUDENT 1 Crush some ice to a fine powder. Fill your test tube about halfway with the crushed ice. Then add about 1 ml of distilled water.

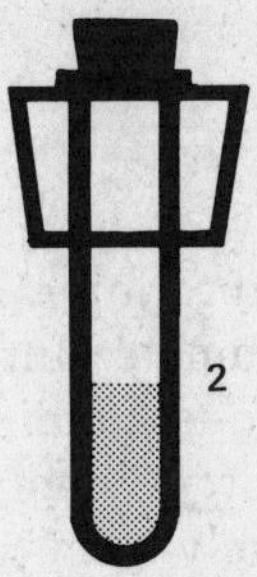

STUDENT 2 Pour 2 ml of pure glycol into your test tube.

STUDENT 3 Pour 2 ml of carbon tetrachloride into your test tube.

STUDENT 4 Pour 2 ml of chlorobenzene into your test tube.

3. Take your turn with the ice–water sample, inserting your thermometer *carefully* into the slush to a depth of about 3 cm. When the mercury column has stabilized, read and record the temperature. Remove the thermometer and wipe it dry. Store it in the test tube on the ringstand until you need it again.

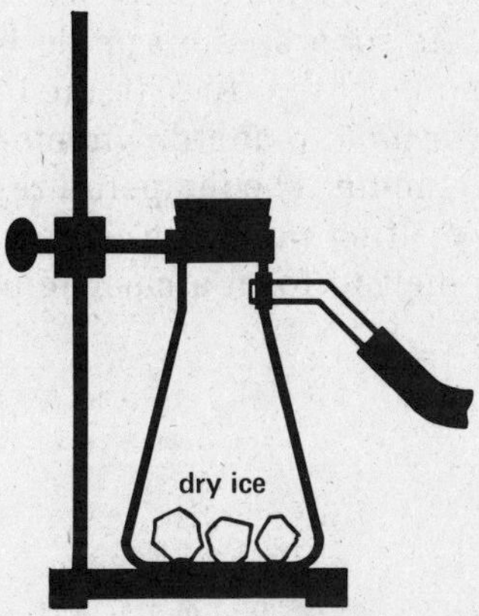

4. Clamp your suction flask to a ringstand. Then connect the flask to an aspirator. Place a few chunks of dry ice into the flask and stopper it.

Caution: Do not handle dry ice with your bare hands. Use a dry towel.

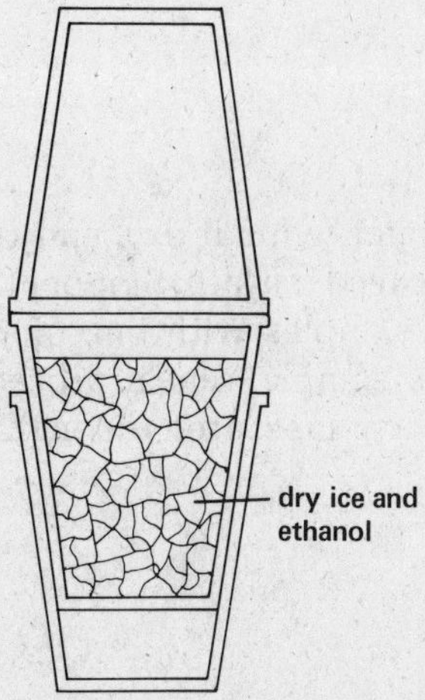

5. Obtain three styrofoam cups. Insert one cup into another and fill the upper cup nearly to the brim with powdered dry ice from the laboratory storage chest. Obtain about 60 ml of ethanol and pour it slowly onto the dry ice, stirring with a glass rod. When the very cold slurry evolves CO_2 only slowly, cover the mixture with the inverted third cup.

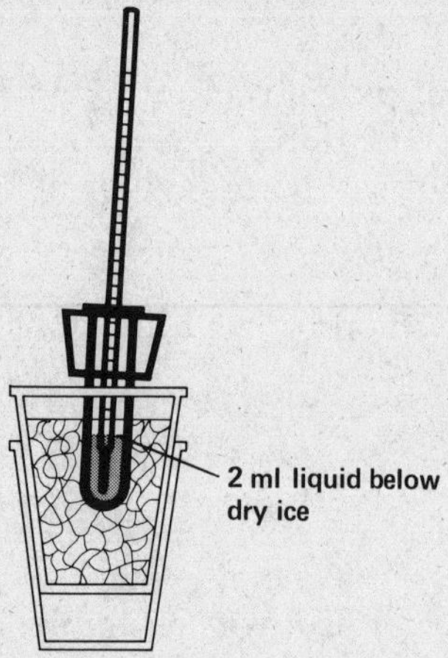

6. Obtain one of the liquid calibration samples and *carefully* slide a clean, dry thermometer into the test tube. Remove the cover of the dry ice–ethanol slurry and push the test tube down into the slurry, keeping it there until the sample has frozen completely. Then remove the test tube, cover the slurry, and. . . .

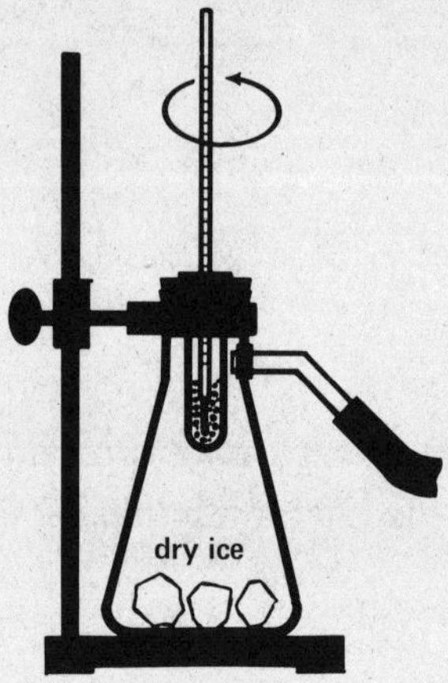

7. . . . place the test tube in the neck of the suction flask. Turn on the aspirator. As soon as the sample begins to thaw a little, stir the sample *carefully* with the thermometer, using a slow *circular* (*not* up-and-down) motion. Observe the thermometer column. The temperature will rise slowly. Record as the melting point the temperature at which the last crystals melt to form a clear liquid.

Relieve the vacuum by removing the hose carefully from its connection to the aspirator. *Then* turn off the aspirator.

8. Remove the thermometer, wipe it dry, and return it to its storage tube until you need it again. Stopper the sample test tube and exchange samples with one of your team members. Using the new sample, repeat procedures B6–B7. Continue until you have measured the melting points of all four samples.

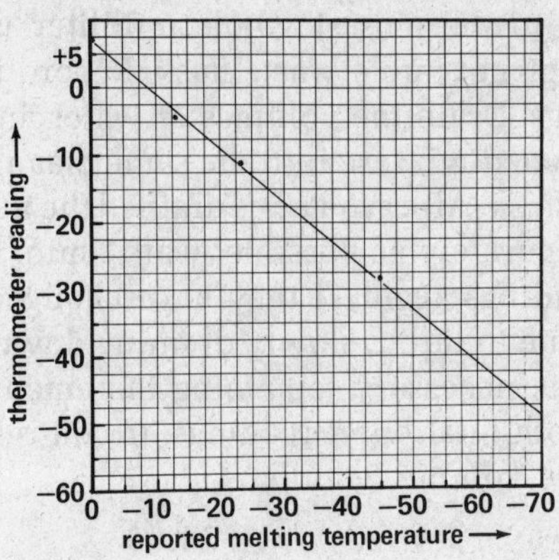

9. It is likely that your thermometer is not completely accurate. That is why you have just obtained some calibration data. In your notebook, plot a calibration graph using your experimentally determined melting points versus literature values (0°C for water, −13°C for ethylene glycol, −23°C for carbon tetrachloride, and −45°C for chlorobenzene). You will use this graph to correct temperature readings obtained in part C of this experiment. (Complete your calibration graph before starting Part C.)

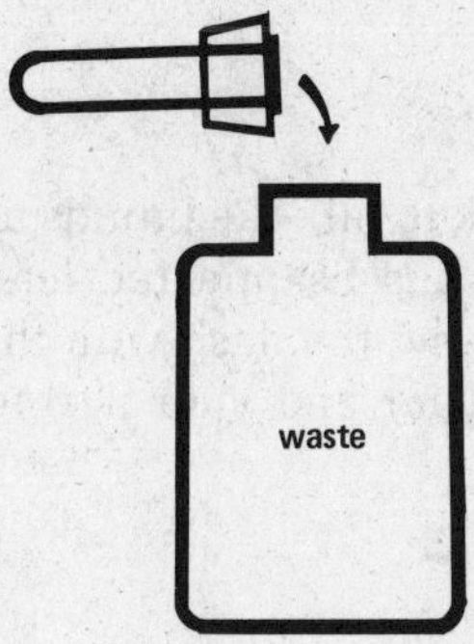

10. *After all team members have finished with all samples,* the ice–water mixture is to be poured down the drain. Each of the other samples is to be poured into *its own labeled* waste container. These liquids can be "recycled."

C. Melting Points of Glycol–Water Mixtures

You will work in pairs on procedures 1 and 2 of this part of the experiment. Your instructor will assign each pair one of the following glycol–water mixtures to prepare. Then each of you will determine the melting point of this mixture independently.

Table 7.1 Glycol–water mixtures

Mixture No.	Glycol (ml)	Water (ml)	Mixture No.	Glycol (ml)	Water (ml)	Mixture No.	Glycol (ml)	Water (ml)
1	18.2	80.0	5	56.0	38.4	9	76.7	15.6
2	36.4	60.0	6	63.6	30.0	10	79.1	13.0
3	45.5	50.0	7	68.7	24.4	11	83.6	8.0
4	52.1	42.7	8	72.7	20.0	12	87.2	4.0

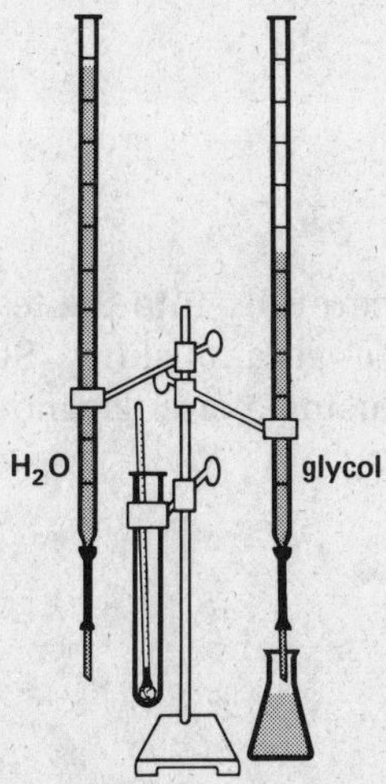

1. Fill the "distilled water" buret to above the top graduation, then drain water out carefully until the liquid level just reaches the zero mark. To conserve glycol, add just enough to the "glycol" buret to provide 10 ml more than the amount calculated for the glycol–water mixture assigned to you. (If more than 50 ml will be needed, you will have to fill the buret twice.)

Adjust the volume to the closest whole milliliter mark by draining a little glycol into a waste flask. Record both initial buret readings. Drain the volumes of water and of glycol for your assigned mixture into the same clean, dry 125-ml Erlenmeyer flask. After you have drained the necessary volumes, as accurately as possible, wait 1 min, then check the buret readings again. If they have changed by more than 0.1 ml (due to liquid slowly draining down the buret walls), add the necessary remaining increments to the flask. Record both final buret readings. If you refill a buret, keep accurate volume records.

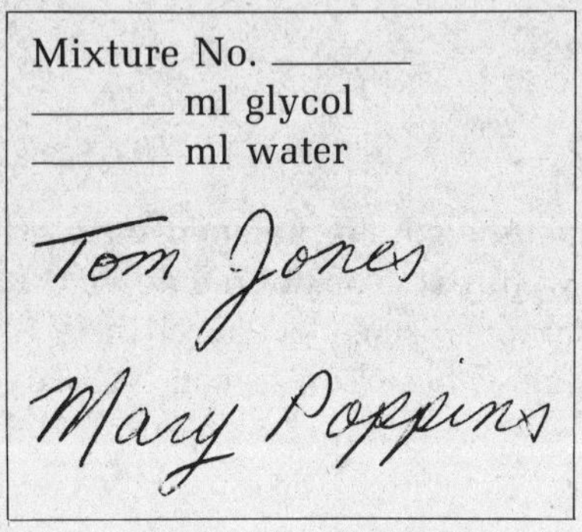

2. Carefully swirl the flask to mix the liquids until all evidence of density gradients has disappeared, leaving a clear homogeneous solution. Label the flask with the composition of the mixture and your and your partner's names. Stopper the flask.

The remaining steps are to be performed by each of you independently, but using portions of the same solution. You will use the equipment and general methods as in part B of this experiment.

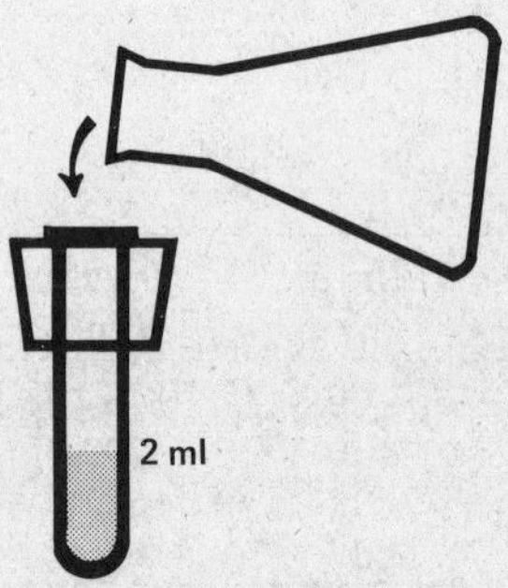

3. Check that all equipment is ready, that the test tube is clean and dry, and that you still have enough dry ice in the suction flask and styrofoam cups. Then transfer 2 ml of the glycol–water mixture into the test tube.

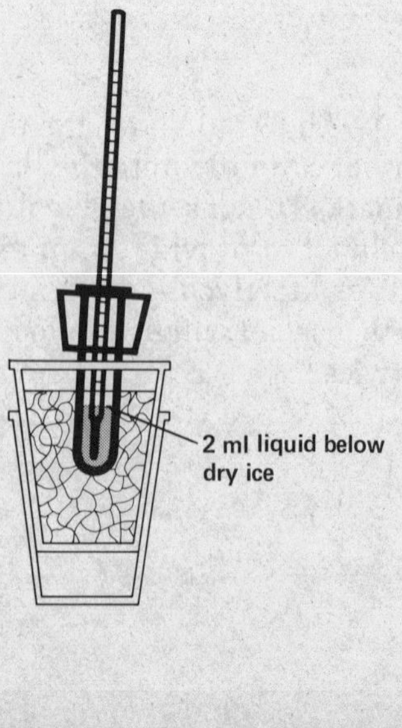

4. Slide the thermometer carefully into the test tube and place the tube in the dry ice–ethanol slurry. Stir the mixture slowly and *carefully*, using a circular motion. Avoid mixing in air.

5. Mixtures of high glycol content will become quite viscous. Crystallization is typically slow, usually beginning near the bottom of the test tube (where it is coldest). Check periodically for signs of crystal formation. Rotating the thermometer around its own axis in *gentle* contact with the bottom of the tube may facilitate crystallization if the mixture has supercooled.

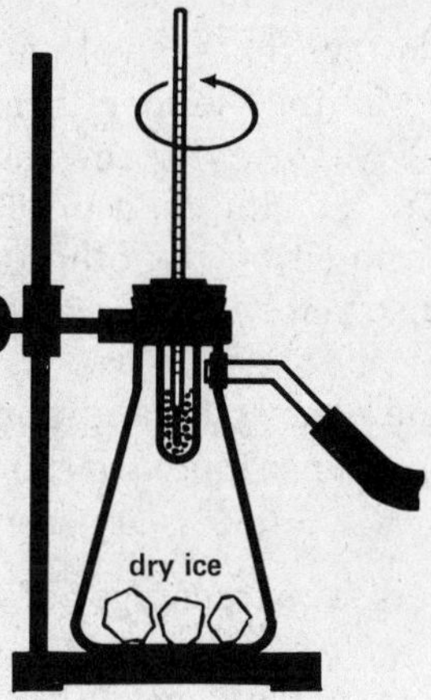

6. When the mixture has completely solidified, remove the test tube from the cooling bath, insert it in the top of the suction flask, and turn on the aspirator. When the solid begins to melt, stir slowly and carefully with a *circular* motion (*not* up-and-down), being careful to avoid stirring in air. (Air bubbles trapped in the viscous liquid cause a cloudiness that makes it difficult to determine the melting point accurately.) Record the temperature at which the last crystals disappear. Refreeze the mixture and repeat the melting point determination.

Use your thermometer calibration graph to correct the measured melting temperatures. Enter the temperatures in your notebook and enter your data in the *section* Report Form for inclusion in the total data collection. Secure a copy of the final records of this total collection from your instructor.

7. Pour *all* glycol–water mixtures into the labeled "waste glycol" container.

CALCULATIONS AND REPORT Note: Do not use the tables on the Report Form as your laboratory notebook. Enter all data into your notebook, perform the required calculations and *then* transfer the appropriate numbers to the Report Form.

Using your own experimental data, calculate the densities of the glycol–water mixtures to four significant figures (procedure A) and the percentage glycol by volume and mass. Then complete Table 7.2, Density Data, of your Report Form.

Definitions

$$\text{percent glycol by volume} = \frac{\text{volume of glycol}}{\text{volume of glycol} + \text{volume of water}} \times 100$$

$$\text{percent glycol by mass} = \frac{\text{mass of glycol}}{\text{mass of glycol} + \text{mass of water}} \times 100$$

Assume in these calculations that the volumes are additive. Complete Table 7.3, Melting Temperatures of Glycol–Water Mixtures. You have to enter the data of your own glycol–water mixture and the data of all the other mixtures that were determined by the other students in your section. Ask your instructor how these data will be made available.

Using Table 7.3 and the density data, prepare the following graphs on the graph paper that is part of the report section of this experiment.

1. Melting point (y) versus percentage of glycol by mass (x).
2. Melting point (y) versus percentage of glycol by volume (x).
3. Melting point (y) versus density of mixture (x).
4. Density (x) versus percentage of glycol by mass and by volume (y).

Note that Graphs 1 and 2 can be prepared using the data from Table 7.3, Melting Temperatures of Glycol–Water Mixtures. Graph 4 is constructed from data collected in Table 7.2, Density Data. Graph 3 requires that you read the densities of the mixtures, for which melting points are available, from graph 4. You may determine the densities using either the volume percentage or mass percentage composition of the mixtures.

Connect the points in all the graphs with a smooth line, employing French curves if necessary. Extrapolate the lines in graphs 1, 2, and 3 until they meet and determine the melting temperatures and composition (density, percentage by volume, percentage by mass) corresponding to this point of intersection. Enter these values into your Report Form.

On a separate page, to be attached to the Report Form, briefly defend or criticize the following statement made by a well-meaning salesman:

> Since a mixture containing about 60 percent glycol by mass will protect your radiator to around −50°C, I suggest that you fill your radiator with pure glycol to protect your cooling system to temperatures well below −50°C.

Also on a separate page, to be attached to the Report Form, explain how a measurement of the density of the liquid in your automobile radiator could be interpreted to provide an idea of how well your automobile's cooling system was protected against freezing; and work the following problem.

Problem

On arrival of the first cold front, an automobile owner weighed a 100-ml volumetric flask (110.21 g), and then filled it to the mark with a sample of the coolant (flask + coolant = 213.71 g). At what temperature will the coolant begin to freeze?

GRADE

INSTRUCTOR'S INITIALS

Report Form

FOR INVESTIGATION 7:

Characterization of Some Antifreeze Mixtures

NAME: ______________________ ID NO.: ______________________

LAB SECTION ______________________ INSTRUCTOR ______________________

Density Data

Mass of empty flask ______________ g

Table 7.2 Density data

Glycol (ml)		Water (ml)		Mass of		Total liquid volume*	Density ($g\ ml^{-1}$)	Percent glycol by	
approx.	actual	approx.	actual	flask + liquid	liquid			volume	mass
0	0	50							
25		0	0						
25		10							
25		20							
25		30							
25		40							
25		50							

*Assume that volumes are additive.

Sample Density Calculation

(Show setup for one of the glycol–water mixtures.)

Assigned glycol–water mixture (procedure C) ______________

Instructor's initials ______________

(OVER)

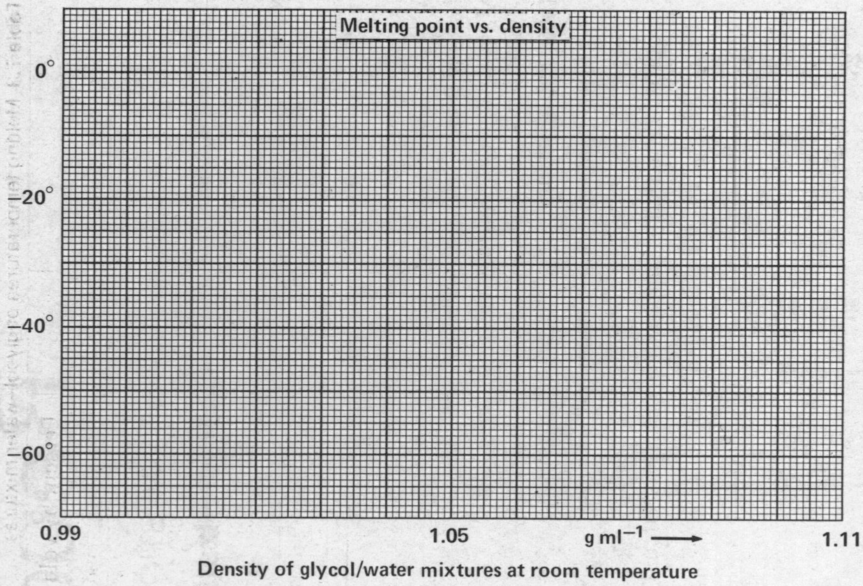
Melting point vs. density
0°
−20°
−40°
−60°
0.99
1.05
g ml^{-1} ⟶
1.11
Density of glycol/water mixtures at room temperature

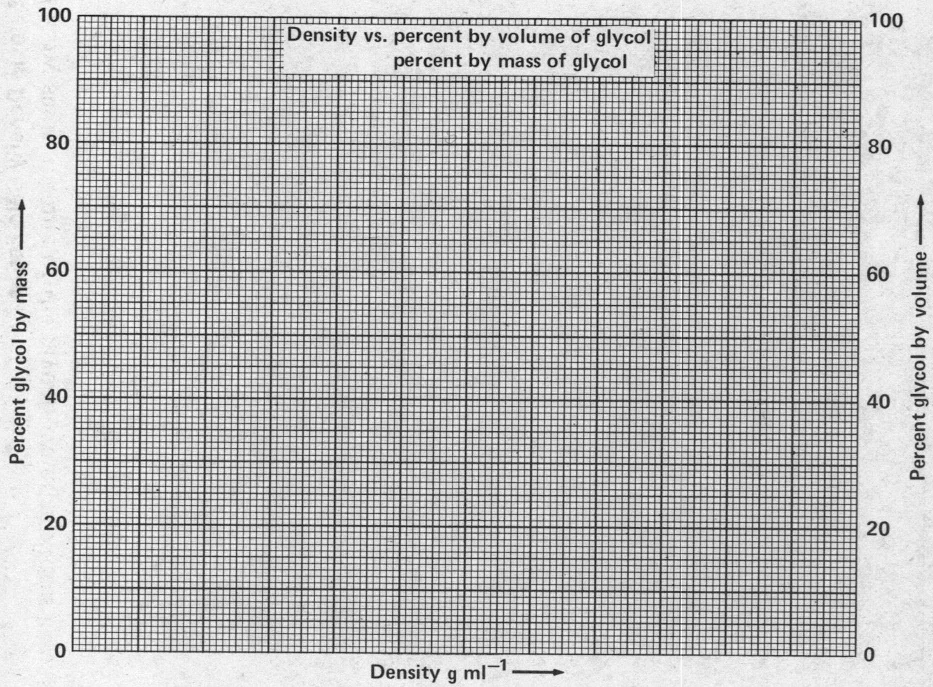
Density vs. percent by volume of glycol
percent by mass of glycol
100
80
60
40
20
0
Percent glycol by mass ⟶
Percent glycol by volume ⟶
Density g ml^{-1} ⟶

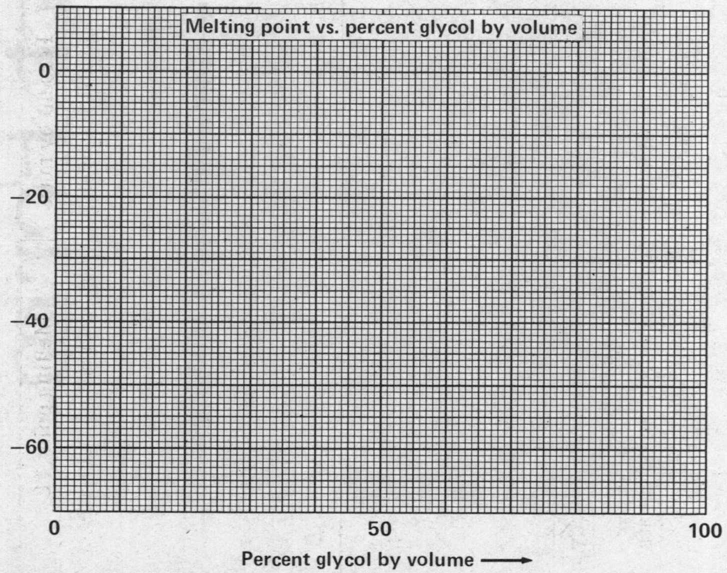
Melting point vs. percent glycol by volume
0
−20
−40
−60
0
50
100
Percent glycol by volume ⟶

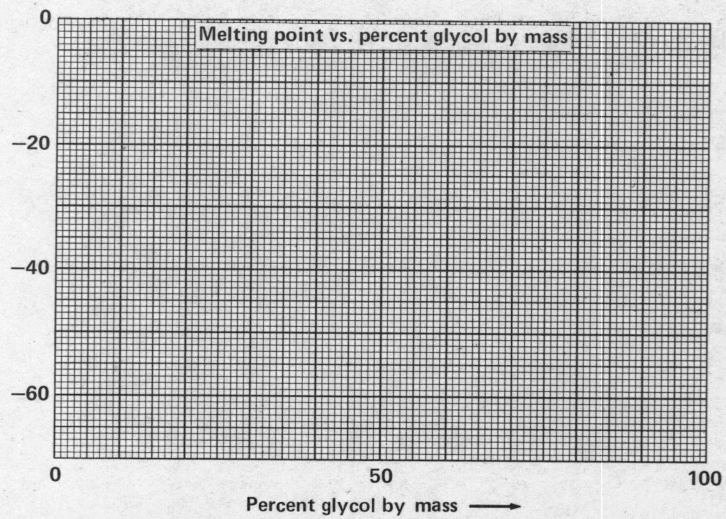
Melting point vs. percent glycol by mass
0
−20
−40
−60
0
50
100
Percent glycol by mass ⟶

Table 7.3 Melting temperatures of glycol–water mixtures

Mixture No.	Glycol (ml)	Water (ml)	Percent by volume		Percent by mass		Melting point, °C	
			glycol	water	glycol	water	experiment	corrected
1	18.2	80.0						
2	36.4	60.0						
3	45.5	50.0						
4	52.1	42.7						
5	56.0	38.4						
6	63.6	30.0						
7	68.7	24.4						
8	72.7	20.0						
9	76.7	15.6						
10	79.1	13.0						
11	83.6	8.0						
12	87.2	4.0						
13	2.0	0.0						
14	0.0	ice/water						

Sample Calculation

Use the mixture assigned to you to calculate the percent glycol by mass and volume.

Best mixture for automobile antifreeze ____________ % glycol by volume

____________ % glycol by mass

Density of this mixture ____________ g ml^{-1}

Initial freezing point of this mixture ____________ °C

____________ DATE ____________ SIGNATURE

Investigation 8

ALKA-SELTZER AND A LITTLE GAS

INTRODUCTION The "gastric juice" in the stomach contains hydrochloric acid, normally to the extent of 0.01–0.1 mole $liter^{-1}$. This acid is necessary to aid in the digestion of foods, and its secretion increases after a meal. An excess of hydrochloric acid (hyperacidity) may result from any of a number of causes, among them "overeating." The results are generally unpleasant and the remedy most commonly sought is some way of quickly neutralizing excess acid.

A popular home remedy used to be a tablespoon of baking soda ($NaHCO_3$), acting to consume excess acid:

$$HCO_3^-(aq) + H^+(aq) \longrightarrow H_2O(l) + CO_2(g)$$

Various commercial preparations are now more commonly employed, among them such alkaline compounds as magnesium oxide ("milk of magnesia"):

$$MgO(s) + 2H^+(aq) \longrightarrow Mg^{2+}(aq) + H_2O(l)$$

With any of these compounds, dosages must be carefully controlled to avoid reducing the stomach acidity too much, a condition producing undesirable physiological effects.

Alka-Seltzer uses a combination of sodium bicarbonate, citric acid, and calcium hydrogen phosphate to produce a solution capable of neutralizing excess stomach acid while avoiding the problem of excess alkalinity. One of the advertising slogans used with this product stressed this point—"Don't Over-Alkalize. Use Alka-Seltzer."

An Alka-Seltzer tablet "fizzes" when dropped into water because of the reaction between dissolved bicarbonate ion and citric acid to liberate carbon dioxide gas. There are several different "fizzing" preparations in addition to the "original" Alka-Seltzer. The components vary from one preparation to another. In some there is insufficient citric acid to consume all the bicarbonate, so that stomach acid may be neutralized by reaction with bicarbonate ion, hydrogen phosphate ion, or anions from the original citric acid. In other preparations an excess of the citric acid results in the formation of a buffered solution containing carbonic acid (mostly hydrated CO_2) and various anions of citric and phosphoric acids. The composition of the solution formed when Alka-Seltzer or similar multicomponent "fizzing" tablets dissolve in water is not simple, but the principle is one of providing a solution that will reduce excess acidity in the stomach to within a "comfortable" range, without risking excess alkalinity.

The history of Alka-Seltzer provides a classic study of the free enterprise system. The following summary is reprinted by permission*:

*Our appreciation is extended to Miles Laboratories, Inc. for their cooperation in supplying us with the compositions of some of their preparations and for their permission to quote from *Serving Needs in Health and Nutrition: The Story of Miles Laboratories, Inc.*

Great corporate enterprises rarely spring full-born from the brow of the founding father, and Miles Laboratories certainly is no exception. Actually, Miles was a small, rather static company for half a century until it began laying foundations in the 30's for worldwide leadership in several fields.

There were two key elements which enabled Miles to broaden its perspectives: innovation, supported by imaginative research, and secondly, the early establishment of orientation toward better health and quality of life as objective for that research.

The business, started by Dr. Franklin Miles at Elkhart, Indiana, in 1884, had humble origins. Dr. Miles, a physician with twelve years of medical training at Yale, Columbia, and the University of Michigan, began its affairs with products derived from his own medical practice and from his office dispensary. The products were so well received that a move to larger quarters was made three times during the first eleven years.

In the late 1920's the Company sought to modernize its formulations and focused its attention on effervescent technology, then still a somewhat novel form of pharmaceutical product. Strict humidity controls were required and Miles spent substantial sums on new equipment and securing experienced engineering staff to meet the wholly new requirements of manufacturing process and control. The wisdom of this research decision began to be realized later with the first process for ALKA-SELTZER, subsequently to become recognized as one of the most successful products ever introduced in any field.

The effervescent tablet that emerged as ALKA-SELTZER was based on a buffered antacid incorporating aspirin, bicarbonate of soda, and citric acid. It was found to be useful in several symptom areas. Indeed, its form was strange and at first the general public did not understand it.

"Hub" Beardsley and his brother Charles Beardsley, the new advertising executive, experimented with advertizing ALKA-SELTZER on radio, then a new medium. This was in 1932, the depths of the Depression, when most companies were investing little in new developments.

The Company achieved its first market penetration by sponsoring The Saturday Night Barn Dance on Station WLS in Chicago. The show was an immediate hit.

Many still recall with nostalgic affection the stars of the show—Uncle Ezra, Lulu Belle, Scotty, and the Hoosier Hot Shots. The Saturday Night Barn Dance ran for 14 years until the advent of television in 1947. Miles again was an early pacesetter on TV with The Quiz Kids, a highly rated information program, and with other shows.

THE STAFF IN FRONT OF DR. MILES MEDICAL COMPANY

The Company's reputation for new approaches to promotion was reinforced in post-war years by "Speedy," destined to become one of the best-known identifying symbols on television. Still later, TV audiences were to endorse another Miles innovation—the amusing soft-sell characterized by the "Stomach" commercial and catchy slogans such as "Try it, you'll like it," and "I can't believe I ate the whole thing."

Although ALKA-SELTZER was becoming established by 1936, management recognized the need for stronger research efforts that would increase the Company's scientific capability.

A major thrust of diversification has been in the area of chemicals. Miles bought the ingredients used in its products until it began the manufacture of aspirin in 1935. Today the Company is recognized as a major supplier of chemicals to the food and pharmaceutical industries, as well as a unique source of new research chemicals to investigators around the world.

A new process was perfected that has made Miles the world's second largest producer of citric acid, which is used extensively in many foods and beverages.

Another prime achievement in chemistry has been the development of biosynthesis technology for the production of complex substances of great importance in the food industry. One such product is fermentation production of the enzyme rennin used in place of calf rennet in cheese making. Other Miles' enzymes are being applied to the processing of baby foods, bread, tenderized beef, wine, and beer, to name just a few.

Early diagnosis of disease is essential in preventive medicine, and Miles is justifiably proud of its contributions in this field. Among other things, its research has developed new methods for measuring deviations in blood and urine biochemistry. These methods, although deceptively simple, are highly complex strip tests which have become standard in hospital laboratories and wards, as well as in doctors' offices throughout the world.

Perhaps the most exciting developments with global significance recently are beginning to appear in the field of nutrition. Perhaps a majority of the deaths in the world are related directly or indirectly to hunger because the malnourished are especially vulnerable to disease. There is an enormous potential market for new, low-cost food products based on vegetable protein. In 1970 Miles acquired Worthington Foods, a pioneer in textured vegetable protein and which currently is making excellent simulations of chicken, beef, ham, bacon, and sausages. Since 1974 Miles has marketed a second line of cholesterol-free food products under the MORNINGSTAR FARMS label. These are sold in supermarkets nationwide.

Miles' involvement in worldwide markets has been a major part of our diversification. The first foreign plant was started in Toronto in 1936, and overseas manufacturing operations were launched at Bridgend, Wales, in 1947.

Worldwide growth opportunities prompted the decision to make a public offering of the Company's stock in 1955 to help finance expansion of manufacturing facilities, mounting involvement in research, and to attract highly qualified professional managers.

Today, with 55 plants in 21 countries, and its products sold in more than 140 countries, Miles is a truly international concern with an innovative organizational structure. The U.S. based divisions have worldwide marketing responsibilities for their lines and provide professional inputs, through the local market managers, to fulfill these responsibilities. Overseas, each general manager has overall responsibility in his country or area for the optimal use of the local entity's resources and entrepreneurial direction. This structure with shared profit responsibility brings to bear the Company's full talents and resources in each market, combining the best of headquarters' expertise with skilled local management in each country.

Dr. Franklin Miles would be astonished, indeed, to learn that today products of the Company that bear his name are known in 140 countries around the globe.

In this experiment, we will utilize the reaction of Alka-Seltzer (or a similar preparation) with excess hydrochloric acid, so that all of the bicarbonate is converted to CO_2. Careful measurement of the CO_2 formed will be used to calculate the bicarbonate content of the original tablet.

Before starting this investigation, you should be thoroughly familiar with all types of "ideal gas law" calculations, including *Dalton's law of partial pressures*. In addition, you should know how to calculate the percentage composition of a mixture (by mass).

References

1. O'Connor, Rod. 1977. *Fundamentals of Chemistry* (2nd ed.) New York: Harper & Row (Units 4, 10, 19 and 20).
2. Compton, Walter Ames, 1973. *Serving Needs in Health and Nutrition*. New York: The Newcomen Society in North America.
3. Parsonage, N. G. 1966. *The Gaseous State*. Elmsford, N.Y.: Pergamon.

SAFETY You will be working with hydrochloric acid. Wear safety glasses at all times. Under *no circumstances* should the reaction of Alka-Seltzer (or a similar preparation) with acid or water be performed in a closed container. An explosion could result.

OBJECTIVES

1. Be able to transfer and mix hydrochloric acid solutions safely.
2. Be able to assemble and use apparatus for the collection and accurate measurement of gas generated by a chemical reaction.
3. Be able to make appropriate "gas law" calculations, including corrections for water vapor and gas solubility.
4. Be able to make appropriate calculations on the stoichiometry of the HCO_3^-/CO_2 system.
5. Be able to quickly weigh a solid on the analytical balance to ± 0.1 mg.

PROCEDURES

The contents of a two-tablet package are to be divided among four students. Students should work in pairs, as assigned by the instructor.

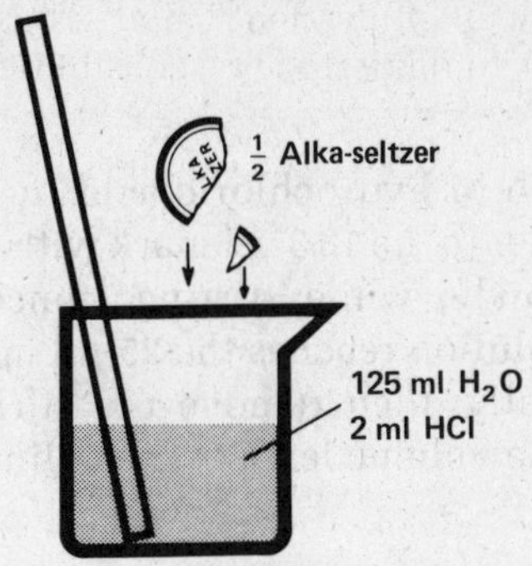

1. Each student pair takes half of their Alka-Seltzer tablet (or a similar preparation supplied by your instructor) and breaks off from this half some small fragments to determine the size of a tablet fragment that would weigh 0.2–0.3 g. Wrap the remainder of the tablet tightly in foil or plastic to prevent contact with atmospheric moisture. Then pour 125 ml of distilled water into a 250-ml beaker and add *carefully,* with stirring, 2.0 ml of concentrated (12 *M*) hydrochloric acid. Drop the half-tablet (including all fragments) into the solution and continue stirring until bubbling has ceased.

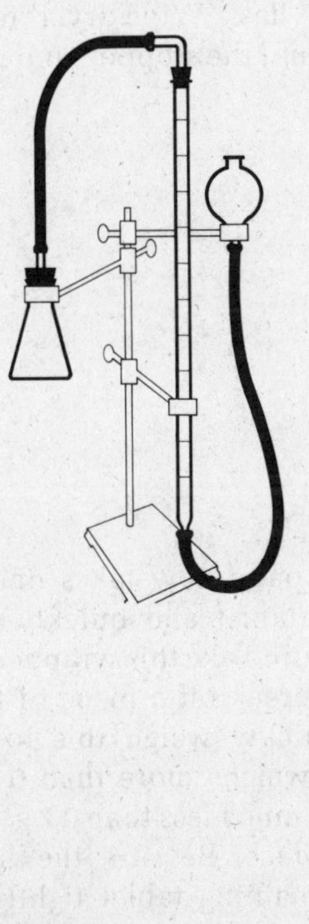

2. Attach a buret clamp parallel to the direction of the base of a ringstand and about 30 cm above the base. Attach two other clamps near the top of the ringstand at angles of about 45 deg to the direction of the first clamp. Connect one end of a 60 cm length of rubber tubing to the lower end of a 50-ml buret and fasten the buret securely in the center clamp so that the lower end of the buret is about 15 cm above the ringstand base. If the two required glass tube–rubber stopper units are not supplied already assembled, lubricate a size No. 00 one-hole rubber stopper and a previously prepared 90-deg bent glass tube and, protecting your hands with a towel, insert the glass tube *carefully* into the stopper until the end of the tube just protrudes below the stopper. Repeat this procedure with a straight piece of glass tube and a one-hole stopper to fit the 125-ml Erlenmeyer flask. [If you are not thoroughly familiar with the proper technique for inserting glass tubing through a stopper (Appendix A), consult your instructor *before* attempting it.] Now fasten the 125-ml flask securely in the upper left clamp, connect the two stoppers by another 60-cm length of rubber tubing, and insert the stoppers into the top of the buret and the flask, respectively. Finally, connect the tubing from the bottom of the buret to a leveling bulb and fasten the bulb securely in the right-hand clamp, being sure that the rubber tubing is not constricted by the clamp.

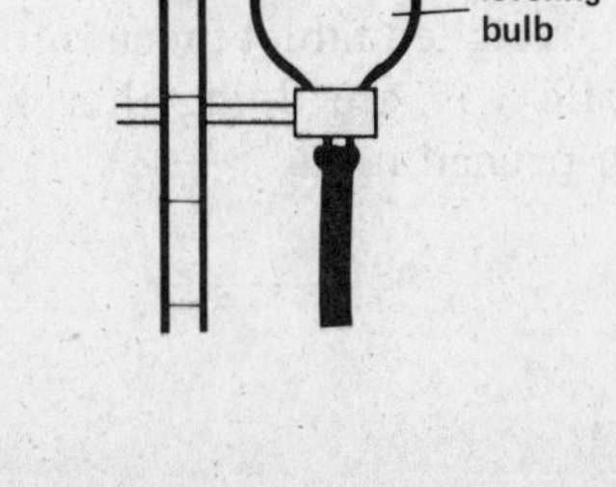

3. Remove the stopper from the flask. With the leveling bulb clamped securely, pour into it the solution from the 250-ml beaker (procedure 1). Loosen the clamp at the ringstand and alternately raise and lower the leveling bulb until all the air has been expelled from the lower tubing and no further bubbles rise in the buret. Clamp the bulb in its original position.

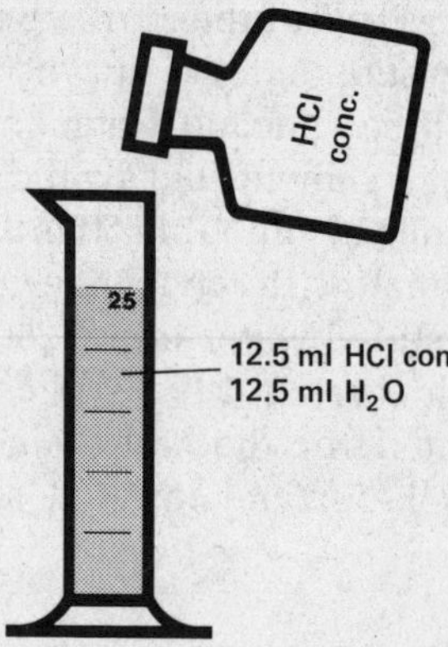

4. Prepare 25 ml of ~6 M hydrochloric acid by filling a 25-ml graduated cylinder to the 12.5-ml mark with distilled water and adding slowly, while stirring, concentrated (12 M) HCl until the solution reaches the 25-ml mark. Stir thoroughly, but carefully, then remove the stirring rod and read and record the volume level in the cylinder.

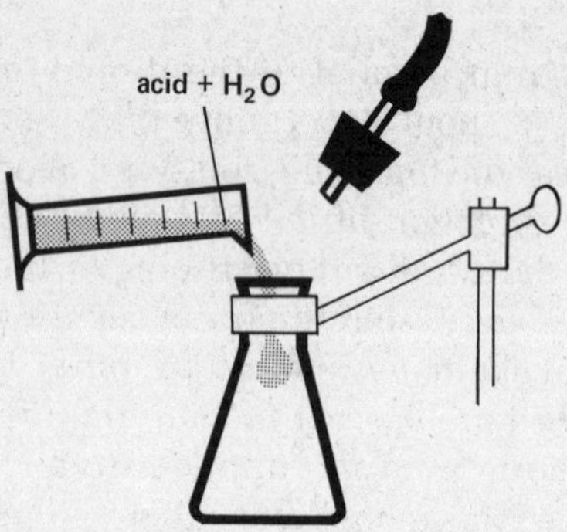

5. Carefully pour about 5 ml of the acid into the Erlenmeyer flask. Read and record the level of the acid remaining in the cylinder. From the difference in the two readings, calculate and record the volume of acid (to the nearest 0.5 ml) transferred to the flask. Wipe dry, if necessary, the inside neck of the flask and the stopper, and restopper the flask.

Caution: Plan the next steps carefully and perform them *quickly*. Alka-Seltzer deteriorates rapidly on exposure to moist air. Be sure your hands are dry.

6. Each of the student pairs now takes one-half of the remaining Alka-Seltzer tablet and quickly seals it in a wrapping of aluminum foil. Take this wrapped piece to the analytical balance and break off a piece of the size estimated at 0.2–0.3 g. Quickly weigh this to the nearest 0.0001 g. If the sample weighs more than 0.3 g, chip off small pieces. If it weighs much less than 0.2 g, start over by breaking off a larger piece. Be sure the final mass is recorded. Wrap the remaining tablet tightly. It will be used to break off three more pieces each weighing ~0.3 g.

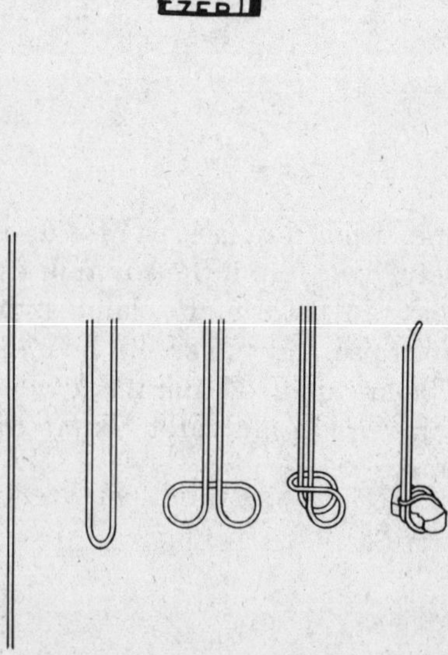

7. Using a thin thread about 20 cm long, quickly form a loop as shown, slip the weighed tablet piece into the loop, and pull it snug. While you are doing this, have your partner continue with procedure 8.

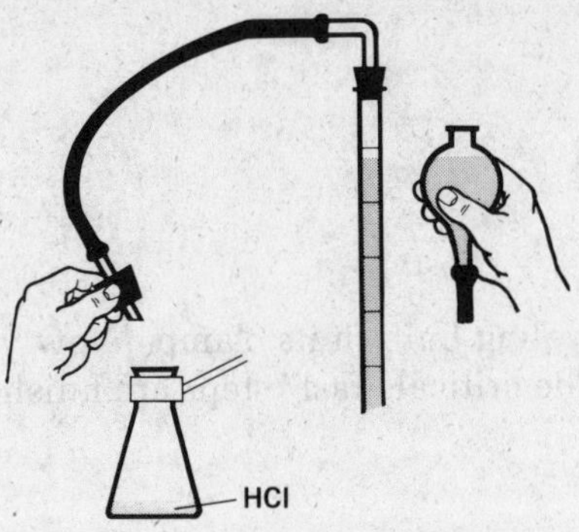

8. Your partner should open the Erlenmeyer flask and raise the leveling bulb until the liquid level in the buret is between the 0-ml and 2-ml marks.

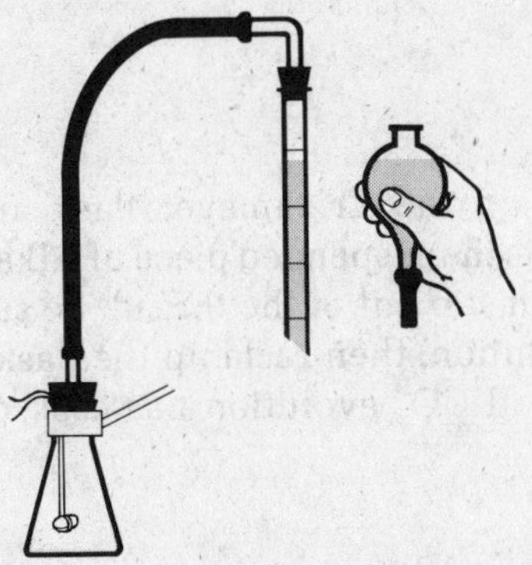

9. Then carefully suspend the tablet sample about 2 cm *above* the liquid surface, holding the thread in place, while slowly pushing the stopper snugly into the neck of the flask. Be sure the stopper fits tightly.

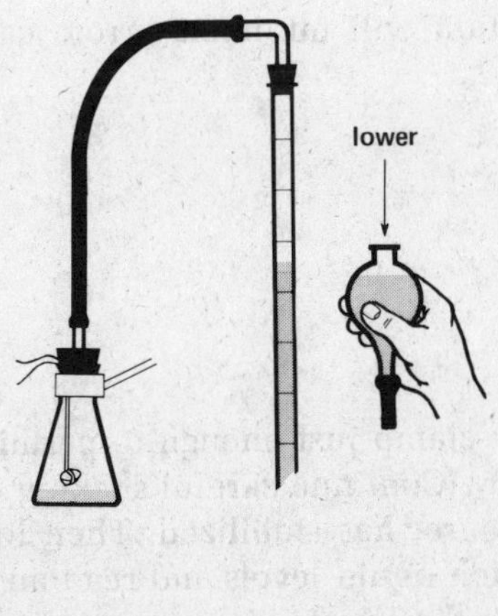

10. Now lower the leveling bulb to about 10 cm below its original position. If your system is gas tight, the liquid level in the buret should drop to somewhere *above* the level in the bulb. If this does not occur, check the system for leaks. (Check both stoppers carefully to be sure they fit snugly.) If you continue to have a gas leak, consult your instructor.

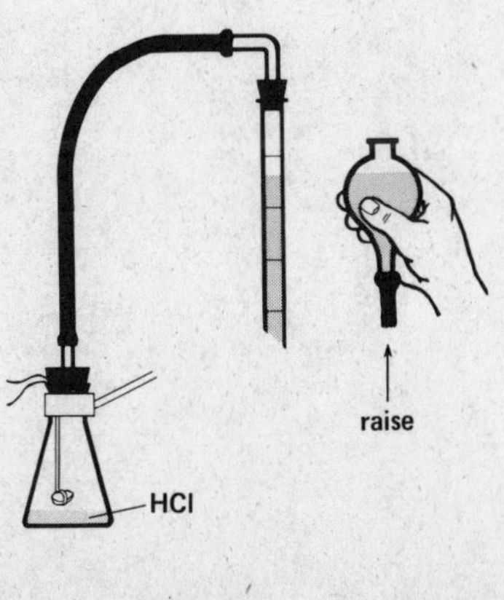

11. Now raise the bulb *until the two liquid levels match.* The buret reading should be between 1 ml and 5 ml. (If the level is below the 5-ml mark, your system was not gas tight and you will have to unstopper the buret and repeat procedures 5–11.) If the buret reading is in the correct range, record it to the nearest 0.1 ml.

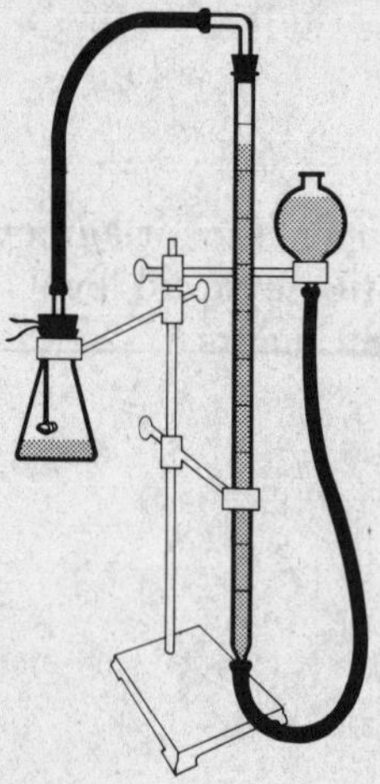

12. Replace the leveling bulb in its clamp. (Now you can slow down a bit. The critical "fast" steps are finished.)

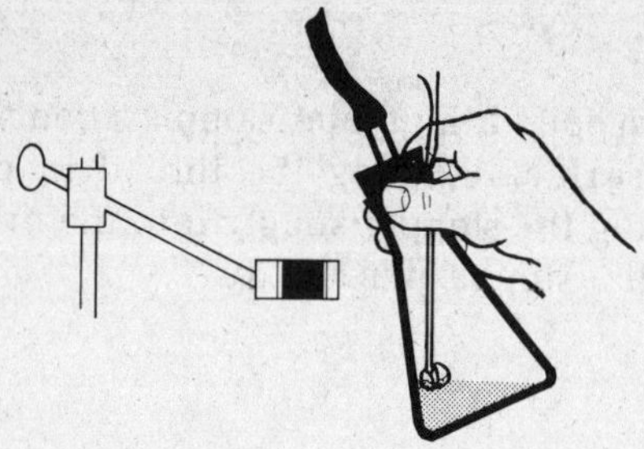

13. Carefully unclamp the Erlenmeyer flask and tilt it until the acid contacts the suspended piece of Alka-Seltzer. The sample will soon slip out of the thread. Be sure it has dropped into the solution, then reclamp the flask in position and wait until all CO_2 evolution has ceased.

It is advisable to keep the liquid levels in the buret and the bulb at the same height during gas evolution by lowering the bulb at the same rate at which the level in the buret drops. This precaution will minimize errors caused by leaks.

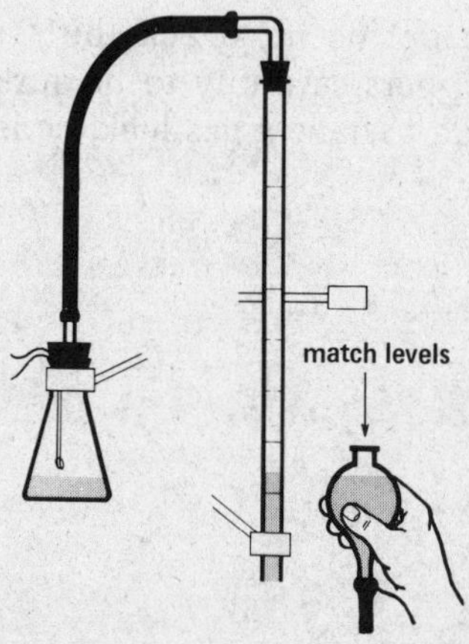

14. Loosen the flask clamp just enough to permit you to shake the flask gently. Continue careful shaking until the liquid level in the buret has stabilized. Then lower the leveling bulb to match liquid levels and read and record the liquid volume in the buret.

Caution: To check that there was no gas leakage during the reaction, raise the leveling bulb to its clamped position in Procedure 12. The liquid in the buret should not rise continuously. If it does, the results of this experiment are not valid and you will have to repeat your work, checking for better gas-tight seals.

Each pair should repeat this experiment until at least four good determinations of CO_2 volume are obtained. To do this, remove the stopper from the Erlenmeyer flask, discard the thread and solution, rinse the flask and stopper with distilled water, and wipe dry the stopper and flask neck. Then begin at procedure 5 (noting HCl solution volume before pouring any into the flask.) When you have completed a satisfactory set of experiments, consult your instructor for instructions for what to do with the apparatus.

ADDITIONAL DATA You will need to find the barometric pressure and the temperature of the laboratory. From the temperature you can determine the vapor pressure of water, using Table 8.1. You will also need to know the approximate amount of CO_2 dissolved in the aqueous solutions used during the experiment. Since we saturated the water in the buret–leveling bulb system with CO_2 (procedure 1), we may safely assume a negligible loss of CO_2 derived from the weighed Alka-Seltzer through absorption in the buret solution. We cannot, however, neglect CO_2 absorption by the acid solution in the flask. It has been determined, by experiments beyond the scope of our laboratory capabilities, that 0.80 ml of CO_2 gas (at 25°C and 760 torr) dissolves in each ml of 6.0 *M* HCl.

Table 8.1 Vapor pressure of water

C°	torr	C°	torr
15	12.8	26	25.2
16	13.6	27	26.7
17	14.5	28	28.4
18	15.5	29	30.0
19	16.5	30	31.8
20	17.5	31	33.7
21	18.6	32	35.7
22	19.8	33	37.7
23	21.1	34	39.9
24	22.4	35	42.2
25	23.8	36	44.6
		37	47.1

CALCULATIONS AND REPORT You should have four sets of data for mass of tablet piece and volume of CO_2 collected. Calculate the partial pressure of CO_2 in the buret from the barometric pressure and the vapor pressure of water at the laboratory temperature. Then correct your carbon dioxide volumes by adding a volume that corresponds to the amount of carbon dioxide dissolved in your hydrochloric acid. Keep in mind that the volumes to be added must correspond to the same pressure and temperature. From the corrected volumes of CO_2, the temperature, and the corrected pressure calculate the number of moles of CO_2 evolved from each of your Alka-Seltzer samples using the "ideal gas law." From this result, calculate the mass of HCO_3^- in each of the samples. Then calculate the percentage by mass of HCO_3^- in the original Alka-Seltzer tablet. Average the results (percentage by mass of HCO_3^-) and calculate the average deviation and range (see Appendix C).

Use your notebook records to complete the Report Form.

GRADE
INSTRUCTOR'S INITIALS

Report Form

FOR INVESTIGATION 8:

Alka-Seltzer and a Little Gas

NAME: ______________________ ID NO.: ______________________

LAB SECTION ______________________ INSTRUCTOR ______________________

Data

Barometric pressure __________ torr

Laboratory temperature __________ °C

Vapor pressure of water at the temperature of the lab __________ torr

Partial pressure of CO_2 in the buret __________ torr

	Experiment			
	No. 1	No. 2	No. 3	No. 4
Mass of tablet fragment, W				
Final buret reading, ml				
Initial buret reading, ml				
Volume of CO_2 collected, ml (V)				
Volume of 6 M HCl, ml				
CO_2 dissolved in 6 M HCl, ml (V')*				
$V + V'$				
Moles of CO_2				
Mass of HCO_3^-				
% HCO_3^- by mass				

Average percent HCO_3^- by mass __________

Average deviation __________ Range __________

*V' is the volume that the CO_2 dissolved in the hydrochloric acid would occupy at the laboratory temperature and the partial pressure of CO_2 in the buret.

Sample calculation for Experiment No. __________:

Partial Pressure of CO_2

Volume of CO_2 (expressed at the laboratory temperature and the partial pressure of CO_2) dissolved in 1 ml of 6 M HCl

(OVER)

Moles of CO_2

Mass of HCO_3^-

Percent HCO_3^- by mass

Average percent HCO_3^- by mass, average deviation

DATE ____________ SIGNATURE ____________________

Investigation 9

CRYSTAL MODELS, METALS, AND MINERALS

INTRODUCTION We have all seen many different crystalline materials, such as diamond or quartz, and we probably categorized them as crystals because of their regular geometric shapes and flat faces. That is a fairly superficial classification. A microscope slide has flat faces and a regular geometric shape, but it is not crystalline. There are some additional characteristics that distinguish between crystalline and noncrystalline (*amorphous*) solids. We know, for example, that a rough diamond can be "cut" to form a beautiful many-faceted gem by hitting the stone properly with appropriate tools. Of course, you have to know *where* to hit or the gem will be ruined. Expert diamond cutters are paid quite well for their work, not because they can "hit harder," but because they can judge the right places to cleave the rough stone to obtain a valuable gem. But not even the most expert diamond cutter can find a way of hitting a lump of glass to produce regular fracture patterns. The property that permits a pure solid to be fractured at appropriate locations to produce planar (flat) cleavage surfaces is a characteristic of most crystals. Amorphous solids do not possess this characteristic. There is still another familiar difference between crystalline and amorphous materials. Have you ever clutched a piece of ice tightly in your hand? As it warms up, the ice melts, but the lump does not get gradually softer. Contrast this with clutching a chunk of wax tightly. If you knead it in your hand, it gradually softens from the warmth of your hand. The wax is an amorphous solid and, as such, it softens and melts slowly over a broad temperature range. Ice, on the other hand, is crystalline. Pure crystalline substances melt, with little softening, over a small temperature range.

Many solids are not easily classified as either crystalline or amorphous on the basis of simple observations. In some cases the solids are mixtures of several different materials, some of which are crystalline. In other cases, such as the fibers of hair, neither classification is really appropriate.

Much of our modern technology finds important uses for crystals (such as those used in transistors) or amorphous materials (such as the special ceramic coatings for parts of space vehicles).

One of the things that chemistry tries to do is to explain properties of matter on the basis of chemical composition and atomic arrangement. It is suggested, for example, that the atoms of pure crystals are arranged in highly ordered patterns. Such uniform atomic arrays would account for the regular geometric shapes and planar cleavage patterns of crystals and, since such arrays would imply uniformity of interparticle forces, for the sharp melting points of pure crystals. An irregular atomic arrangement, on the other hand, would be consistent with the melting and fracture behavior of amorphous solids.

We may think of the regular three-dimensional pattern of the fundamental particles in a crystalline substance in terms of the orderly repetition of the *simplest* three-dimensional array characteristic of the overall lattice pattern. Such an arrangement of particles is called the *unit cell* and the geometric features of this *unit cell* are characteristic of the complete crystal lattice.

In this experiment we will work with physical models of some of the simpler crystals and relate these models to some real crystalline systems.

Before starting this investigation, you should be familiar with the general characteristics of crystalline substances, with the fundamental types of crystal lattice patterns, and with the concepts of close packing of spheres, lattice-point sharing, and interstitial sites. In addition, you should be able to calculate the number of particles assigned per unit cell on the basis of crystal lattice descriptions.

References

1. O'Connor, Rod. 1977. *Fundamentals of Chemistry* (2nd cd.) New York: Harper & Row (Unit 11 and Excursion 1).
2. Readings from *Scientific American*. 1973. *General Chemistry*. San Francisco: Freeman (Section IV).

SAFETY This experiment is not hazardous. However, if it is performed in a laboratory where reagents are set out, approved eye protection is required. Consult your instructor for the specific policy.

OBJECTIVES

1. Be able to construct models of hexagonal and cubic closest-packed lattices.
2. Be able to identify particles shared among unit cells in a crystal lattice and to determine the number of particles assigned to a particular unit cell.
3. Be able to identify tetrahedral and octahedral interstitial sites in a cubic closest-packed lattice and to relate these to the patterns of ionic arrangement in some simple crystalline salts.
4. Be able to draw three-dimensional representations for simple lattice patterns.

PROCEDURES

Work with a group of other students, as specified by your instructor, on procedures A-C1. Each group will be issued model-building sets.

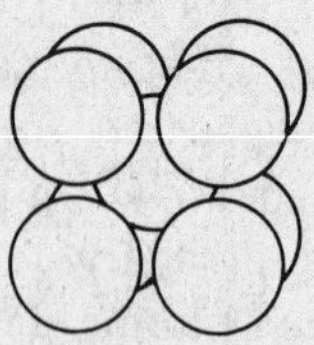

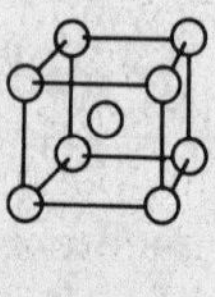

A. Close-Packing Patterns

The simplest types of unit cells (fundamental three-dimensional arrays) in crystals are those involving only monatomic particles. Thus crystal lattices of some of the common metals exhibit fairly simple packing patterns of the spherical metal atoms.

All the alkali metals and barium crystallize in the *body-centered cubic* pattern.

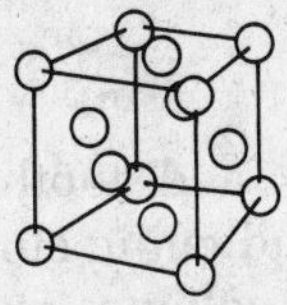

Calcium, strontium, nickel, silver, platinum, and gold are some of the metals crystallizing in the cubic closest-packed (face-centered cubic) pattern.

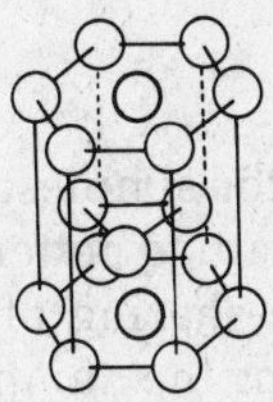

Magnesium, zinc, and cadmium form hexagonal closest-packed crystal lattices.

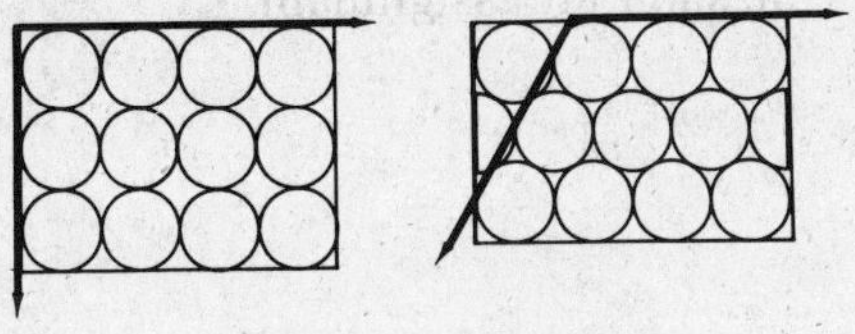

1. To see how these arrays are formed and to compare their "packing efficiencies," start first with a comparison of two-dimensional packing patterns. With pennies or similar sized disks build the two-dimensional arrays shown in the drawings. Use at least 12 disks for each array; however, a more extended array of 24 or more disks will make it easier to recognize the repeating pattern.

Sketch these patterns into your notebook. Then outline in your sketch the smallest unit containing whole disks or parts of disks (or both), which on movement along the x and y directions will yield repetitive areas that completely cover the two-dimensional space. Show your "two-dimensional unit cells" (the term applied to these smallest units) to your instructor before you proceed.

Then calculate the packing efficiencies of the two patterns in terms of the area required to accommodate one disk. To accomplish this you will have to calculate the area of your "unit cell" and find the number of disks in the "unit cell." Call the radius of the disks r. It is not necessary to introduce a numerical value for r to compare the packing efficiencies.

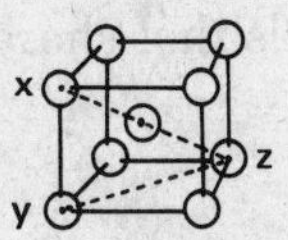

2. Now extend your study to three dimensions. Using Styrofoam spheres and toothpicks, construct a model of a body-centered cubic unit cell. Use this model to calculate the volume required to accommodate one sphere. The model and the drawings shown here should help you to visualize the geometric relations that are necessary for this calculation. Note that the distance separating the centers of the sphere x and the sphere z is equal to $4r$, with r designating the radius of a sphere. The distance between the centers of sphere x and sphere y is, however, not equal to 2r. Assume that the radius r of the spheres is expressed in centimeters.

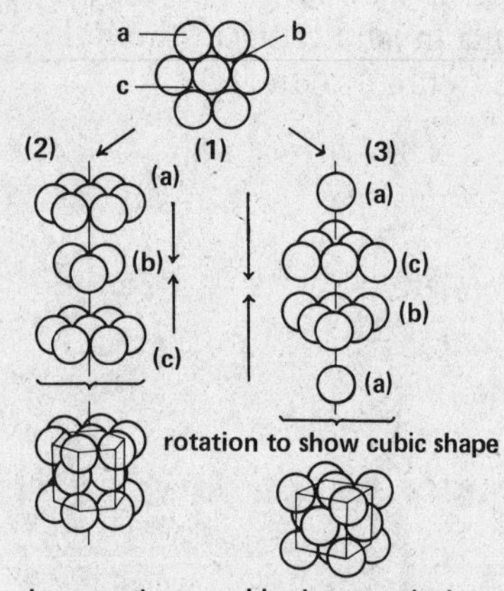

hexagonal closest packed cubic closest packed (face centered)

3. Using Styrofoam spheres and toothpicks, construct models of both hexagonal and cubic closest-packed unit cells. As above, calculate the volumes required to accommodate one sphere in each of the arrays. Use your models and the drawings to visualize the geometrical relations necessary for the calculations.

B. Lattice-Point Sharing

Remember that a unit cell is the simplest repetitive unit of a perfect three-dimensional lattice pattern. The repetition of the unit cell requires the sharing of particles among adjacent unit cells. In assigning the number of atoms "contained" by a single unit cell, this sharing must be considered. Only the part of an atom *not* common to other unit cells is "counted" for such an assignment.

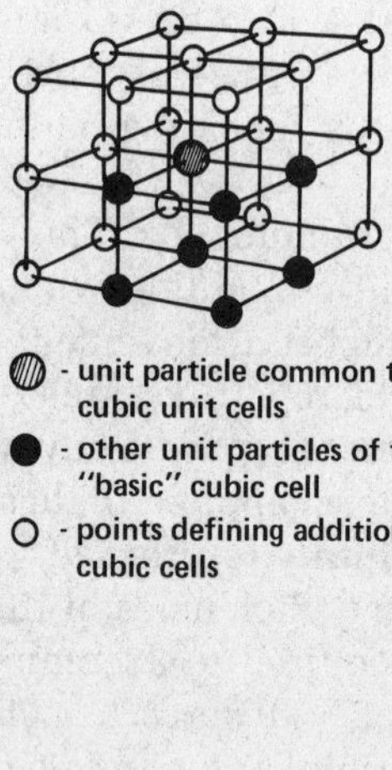

- unit particle common to eight cubic unit cells
- other unit particles of the "basic" cubic cell
- points defining additional cubic cells

1. Using toothpicks and Styrofoam spheres, construct a model of part of a lattice made up of repetitive *simple cubic* unit cells. Use the model to determine the maximum number of adjacent unit cells that would share a common corner. Draw a representation of this model in your notebook.

Calculate the fraction of a corner particle that should be assigned to a particular unit cell in the cubic system.

2. Take a model of a cubic closest-packed unit cell and use toothpicks and Styrofoam spheres to extend the unit cell into a larger repetitive lattice. Examine this lattice model to confirm that "corner sharing" is the same as in the case of a simple cubic system and to determine the number of adjacent unit cells that share a common face-centered particle.

Calculate the fraction of a face-centered particle that should be assigned to a particular unit cell in the cubic system.

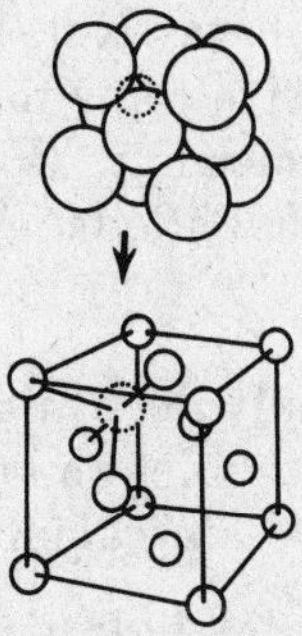

⁛ tetrahedral site (one under each corner)

C. Interstitial Sites

In both the cubic and hexagonal closest-packed lattices there are different types of "holes," called *interstitial sites.* You may recall that the tetrahedral pattern is characteristic of certain simple molecules, such as methane, in which four atoms surround a fifth. In a crystal lattice, the tetrahedral site is the hole symmetrically surrounded by four lattice particles. One such site occurs just behind each corner particle in a cubic closest-packed lattice. Since the tetrahedral sites are contained inside the imaginary cube defining the unit cell, eight such sites (under the eight corners) are counted for a face-centered cubic unit cell.

The octrahedral pattern consists of six particles symmetrically surrounding a seventh. Thus one octahedral site is located at the center of a face-centered cubic unit cell. The remaining octahedral positions are located at the centers of each edge of the unit cell.

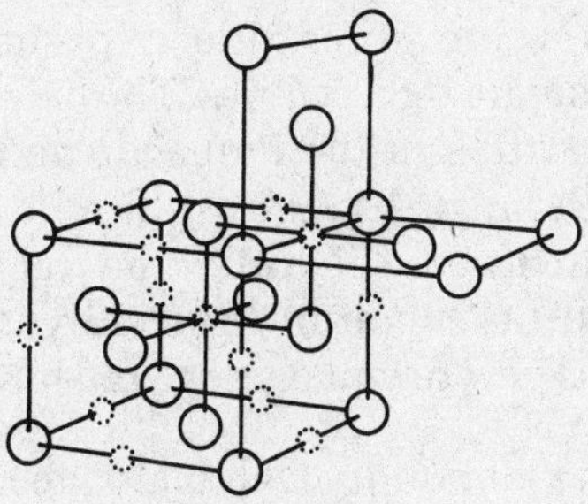

⁛ octahedral sites

In cubic closest-packed lattices of large ions (e.g., monatomic anions), there are two tetrahedral sites and one octahedral site per ion, which could be occupied by an ion of opposite charge if it were small enough to fit. Closest-packed arrays of anions are approximated in salts of the type M_2X (e.g., Li_2O, with Li^+ ions in all tetrahedral sites). If only half the available tetrahedral sites are occupied (e.g., ZnS, AgCl) or if cations occupy octahedral sites (e.g., LiCl, AgF), then salts of the type MX are formed.

1. Using a model of a cubic closest-packed unit cell, locate both *tetrahedral* and *octahedral* sites. Identify these by placing tiny Styrofoam spheres or small wads of aluminum foil in all such sites. Use drawings in your notebook to show why these locations are called "tetrahedral" and "octahedral," respectively and to show "sharing" of edge sites among adjacent unit cells.

Calculate the fraction of an edge-centered particle that should be assigned to a particular unit cell in the cubic system.

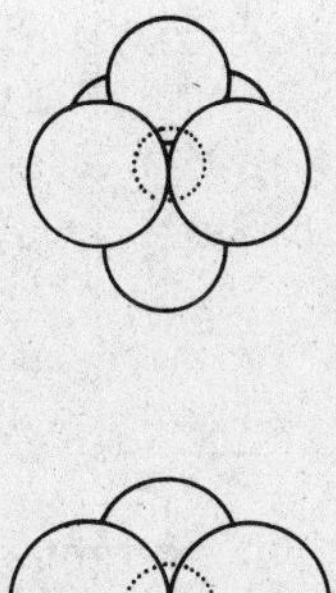

2. Partly assembled models of an octahedron and a tetrahedron constructed from 1.5 in. (3.8 cm) plastic spheres and an assortment of much smaller spheres are available in the laboratory. Complete the octahedron and the tetrahedron employing the available loose spheres. Note the hole present in the center of each of these geometric figures. How many spheres are necessary to create an octahedral or a tetrahedral hole? Which hole is larger?

3. Measure the diameter of the octahedral and the tetrahedral holes by placing the smaller spheres into the hole sites. When one of the small spheres fits snugly into one of the holes, check whether or not the loose large sphere required to complete the octahedron or tetrahedron still touches all four or three other spheres. Note the number of large spheres surrounding the small one in each of your arrangements. Measure, as accurately as possible, the diameters of the small spheres that filled the octahedral or tetrahedral holes.

4. Employing your experimentally determined diameters of the octahedral and tetrahedral holes and the size of the large spheres surrounding the holes, calculate:

  a. The ratio "radius of octahedral hole to radius of large sphere."
  b. The ratio "radius of tetrahedral hole to radius of large sphere."
  c. The ratio "octahedral hole radius to tetrahedral hole radius."

5. Instead of determining the radius ratios experimentally, you might want to calculate these values. The calculations involve only the application of the Pythagorean theorem on the appropriate right-angle triangles. It is a little tricky to find these triangles in the case of the tetrahedral arrangements of spheres. Your instructor might want you to perform these calculations before you come to the laboratory.

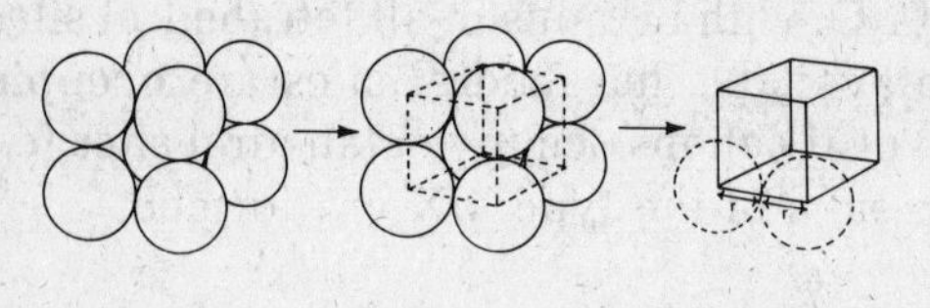

6. Before coming to the laboratory, calculate the radius of the hole found in the center of a cube when eight spheres of the same diameter are arranged at the corners of this cube. The spheres located next to each other touch each other. Calculate the ratio "radius of the eight-coordinated hole" to the radius of the large sphere.

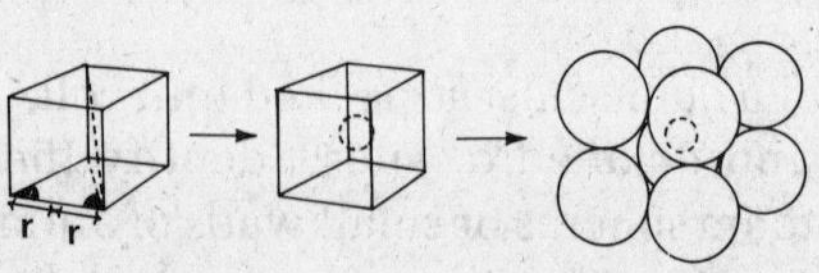

7. You will be given the formula of a compound that crystallizes in the cubic closest-packed arrangement. With the help of the ionic radii provided in Table 9.1 and the ratios calculated above decide whether the smaller ion will reside in the octahedral or tetrahedral holes.

Table 9.1 Some ionic radii (in Å, 1Å = 10^{-8} cm)

Li^+ (0.60)	Be^{2+} (0.31)	Al^{3+} (0.50)	Co^{2+} (0.72)	Zn^{2+} (0.74)	F^- (1.36)
Na^+ (0.95)	Mg^{2+} (0.65)	Sc^{3+} (0.81)	Ni^{2+} (0.69)	Cd^{2+} (0.97)	Cl^- (1.81)
K^+ (1.33)	Ca^{2+} (0.99)	Cr^{3+} (0.64)	Cu^+ (0.96)	Hg^{2+} (1.10)	Br^- (1.95)
Rb^+ (1.48)	Sr^{2+} (1.13)	Mn^{2+} (0.80)	Cu^{2+} (0.72)	Ag^+ (1.26)	I^- (2.16)
Cs^+ (1.69)	Ba^{2+} (1.35)	Fe^{2+} (0.75)	Fe^{3+} (0.64)	O^{2-} (1.40)	S^{2-} (1.84)

In order to solve this problem, you must be aware that the ions forming, for instance, the cubic closest-packed arrangement have the same electric charge and therefore repel each other. Generally, these ions can approach each other only until the distance between their nuclei equals two times the ionic radius. One then says that the ions "touch" each other. Under these conditions the "touching" ions generate a hole, the radius of which depends on the number of spheres surrounding this hole and on the radius of the spheres. You have experimentally determined (procedure C3) or calculated (procedures C5 and C6) the radii of tetrahedral, octahedral and eight-coordinated holes and deduced the appropriate radius ratios. A small ion of opposite charge can now reside in such a hole and hold together the otherwise highly volatile collection of ions of the same charge. Which type of hole the small ion will occupy is dependent on the radius of the small ion and the radius ratio "small ion to large ion." Experience has shown that generally the most stable crystalline structure, which is preferentially formed, is one in which the small ion fits snugly into the hole accessible to it or is somewhat larger. Arrangements in which the small ions "rattle" in their holes (i.e., the radius of the hole is larger than the radius of the small ion) are not very common. When the small ion becomes too large relative to a certain hole radius, it will expand its coordination number, that is, surround itself by a larger number of oppositely charged ions, and then generate for itself a hole of larger radius.

The following rules summarize this discussion:

a. A small ion will reside in a tetrahedral hole when the radius ratio of "small ion to large ion" is equal to the ratio of "radius of tetrahedral hole to radius of large ion" but smaller than the ratio "radius of octahedral hole to radius of large ion."

b. A small ion will reside in an octahedral hole when the radius ratio of "small ion to large ion" is equal to the ratio of "radius of octahedral hole to radius of large ion" but smaller than the ratio "radius of eight-coordinated hole to radius of large ion."

There are quite a few exceptions to these "rules."

8. You will be assigned a time to work alone with the model kit and asked to construct, within a specified time interval, a model of the unit cell of a common mineral of given unit cell description. Your instructor will inspect your model and ask you some questions about it. Some examples of possible minerals and of possible questions are given below. Record in your notebook a rough sketch of your model and the questions you were asked about it.

Some possible minerals	Some possible questions
Rock salt	"How many unit cells of this type in a crystal of the mineral would have a common edge?"
(Chloride ions in a face-centered cubic lattice with sodium ions in all the octahedral sites)	
Fluorite	
(Calcium ions in a face-centered cubic lattice with fluoride ions in all the tetrahedral sites)	"What is the empirical formula of this mineral and how was this determined from the unit cell description?"
Zinc blende	
(Sulfide ions in a face-centered cubic lattice with zinc ions in half the tetrahedral sites)	

REPORT Use your notebook records to complete the Report Form.

GRADE	
	INSTRUCTOR'S INITIALS

Report Form

FOR INVESTIGATION 9:

Crystal Models, Metals, and Minerals

NAME: ______________________ ID NO.: ______________________

LAB SECTION ______________________ INSTRUCTOR ______________________

A1. Two-Dimensional Packing Patterns

Sketch the two unit cells.

Area (cm^2) required for one disk in Pattern A:

______________ cm^2 $disk^{-1}$

Area (cm^2) required for one disk in Pattern B:

______________ cm^2 $disk^{-1}$

Which pattern can accommodate more disks per unit area? ______________

Show setups for calculations below.

(OVER)

A2. Three-Dimensional Packing Patterns
Body-Centered Cubic Unit

Volume (cm^3) required for one sphere: ____________________ cm^3 $sphere^{-1}$

Show calculations and required sketches below.

Cubic Closest-Packed Structure (Face-Centered Cubic Unit)

Volume (cm^3) required for one sphere: ____________________ cm^3 $sphere^{-1}$

Show calculations below.

Hexagonal Closest-Packed Structure

Volume (cm^3) required for one sphere: ____________________ cm^3 $sphere^{-1}$

Show calculations below.

Rank the body-centered cubic, the cubic closest-packed, and the hexagonal closest-packed structures in decreasing order of packing efficiency.

B and C1. Lattice-Point Sharing (Cubic Systems)

Particle location	Number of unit cells sharing particle	Fraction of particle counted per unit cell
corner		
edge-center (octahedral site)		
face-center		
body-center (octahedral site)		
tetrahedral site		

C4. Octahedral, Tetrahedral, and Eight-Coordinated Holes

Ratio "radius of octahedral hole to radius of large sphere"

Ratio "radius of tetrahedral hole to radius of large sphere"

Ratio "octahedral hole radius to tetrahedral hole radius"

C6. Ratio "Eight-Coordinated Hole Radius to Radius of Large Sphere"

C7.

Compound ____________________ has cubic closest packed structure.

Anion radius ____________________ Å

Cation radius ____________________ Å

Sites occupied by anion

Sites occupied by cation

C8. The mineral you were asked to construct a model of was ____________________

DATE ____________________ SIGNATURE ____________________

Investigation 10

ISOLATION OF A FATTY ACID

INTRODUCTION Historically, the term "fatty acid" was applied to any of a number of long-chain carboxylic acids obtained from the hydrolysis of animal fats. The salts of such acids are soaps, and the chemistry of soap preparation by the alkaline hydrolysis of the glyceryl esters from fats and oils has been discussed in Investigation 6.

Fatty foods are rich sources of energy and it is the fatty acids that provide most of this energy. The key to fatty acid metabolism was proposed in the early twentieth century by the German biochemist Franz Knoop, who showed that the breakdown of fatty acids in the body appeared to proceed by a series of reactions, each of which removed two carbon atoms from the carbon chain. Subsequent research has established the details of this complex stepwise process. It is the carboxyl group and the adjacent carbon atom that are removed first, with formation of a new fatty acid two carbons shorter than the original. Each successive step produces essentially the same result. The energy released by the overall process, and available for use by cellular processes, is on the order of 1000 kilocalories (kcal) per mole for a typical fatty acid.

The long-chain acids obtained from fats and oils are valuable commercial sources of many products in addition to soaps. The free acids may be obtained by acid hydrolysis of the glyceryl esters or by reaction of soaps with strong acids such as HCl. For example,

$$\underset{\text{sodium stearate (a simple soap)}}{[CH_3(CH_2)_{16}CO_2^-Na^+](aq)} + H^+(aq) \longrightarrow CH_3(CH_2)_{16}CO_2H(s) + Na^+(aq)$$

The long-chain acids, unlike their "cousin" acetic acid, are only slightly soluble in water. Since most soaps are mixtures of salts of different acids, the precipitated acids are also mixtures. Within recent years, techniques have been developed for separating these mixtures on a large scale by fractional distillation. One of the more important uses of these acids is in the production of long-chain alcohols which find a number of industrial applications. In the laboratory the conversion of acid to alcohol may be accomplished by lithium aluminum hydride reduction. For example,

$$4CH_3(CH_2)_{16}CO_2H + 3LiAlH_4 \longrightarrow 4H_2 + 2LiAlO_2 + [CH_3(CH_2)_{16}CH_2O]_4AlLi$$

$$\xrightarrow{H_2O,\ H^+} \underset{\text{1-octadecanol (an alcohol)}}{4CH_3(CH_2)_{16}CH_2OH} + Li^+ + Al^{3+}$$

The lithium aluminum hydride is too expensive to use on a commercial scale, but alternative methods are available and alcohols containing 12 to 18 carbons are produced in enormous quantities for use in preparing synthetic detergents. For example,

$$CH_3(CH_2)_{16}CH_2OH + H_2SO_4 \longrightarrow CH_3(CH_2)_{16}CH_2O{-}\overset{\displaystyle O}{\overset{\|}{\underset{\underset{\displaystyle O}{\|}}{S}}}{-}OH$$

octadecyl hydrogen sulfate

$$\Big\downarrow \; NaOH$$

$$CH_3(CH_2)_{16}CH_2{-}O{-}\overset{\displaystyle O}{\overset{\|}{\underset{\underset{\displaystyle O}{\|}}{S}}}{-}O^- \quad Na^+$$

sodium octadecyl sulfate
(a synthetic detergent)

In this investigation you will be preparing a fatty acid from a soap. Since the particular soap used is mostly the sodium salt of a single acid, you should be able to obtain a relatively pure product.

Before starting this investigation, you should be familiar with the general characteristics of acids and bases and with the relationships between molecule structure and water solubility. In addition, you must complete the calculations indicated in the procedures section of this investigation before coming to the laboratory.

References

1. O'Connor, Rod. 1977. *Fundamentals of Chemistry* (2nd ed.) New York: Harper & Row (Units 6, 7, 9, 11, 13, and 14).
2. Goldwhite, Harold. 1972. *An Introduction to Organic Chemistry*. Columbus, Ohio: Bobbs-Merrill.

SAFETY Wear approved eye protection at all times. Use caution in working with acids and be particularly careful in heating ethanol. *Ethanol is highly flammable.*

OBJECTIVES

1. Be able to perform safely and properly the techniques of dilution of concentrated acid, precipitation, filtration, and melting point determination.
2. Be able to isolate a fatty acid of good quality and in good yield.
3. Be able to make calculations involving solution stoichiometry, dilution, and percentage yield.

PROCEDURES

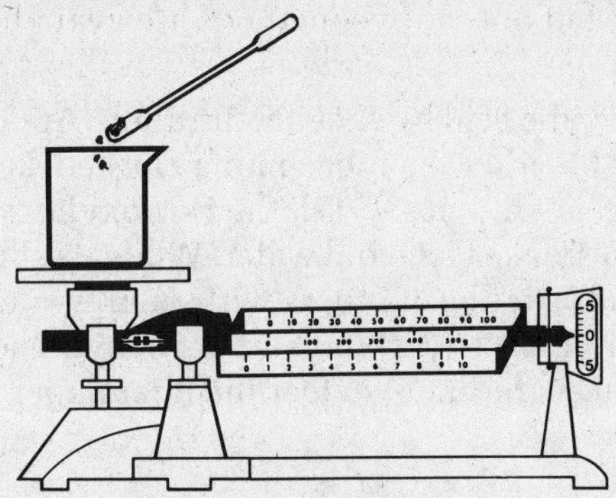

1. Weigh a clean 400-ml beaker and transfer into it approximately 10 g of the soap assigned by your instructor. Weigh the sample to the nearest 0.1 g and record the mass in your notebook.

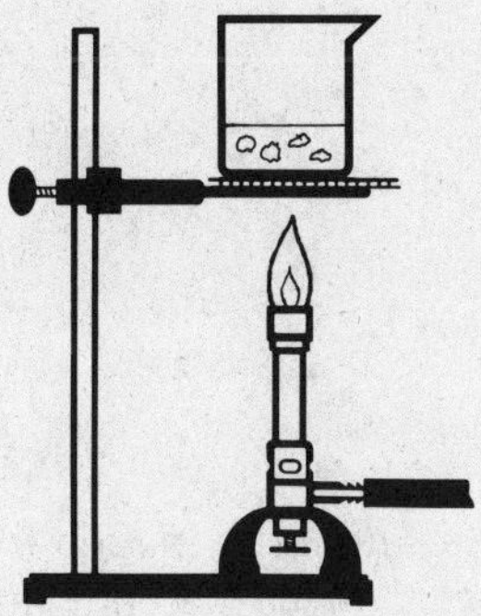

2. Add 100 ml of distilled water to the beaker containing your soap and heat the mixture until the soap is completely dispersed. If foaming becomes excessive, temporarily remove the burner and stir the mixture until foaming ceases.

Calculate the volume of 3 *M* HCl required to react completely with 10 g of soap, assuming the reaction to be represented by

$$CH_3(CH_2)_{16}CO_2Na + HCl \longrightarrow CH_3(CH_2)_{16}CO_2H + NaCl$$

Then calculate the volume of 12 *M* HCl required for preparation of twice the volume of 3 *M* HCl calculated for the reaction. Assuming volumes are additive, calculate the volume of water needed for this dilution. Have your instructor check your calculations before you proceed further.

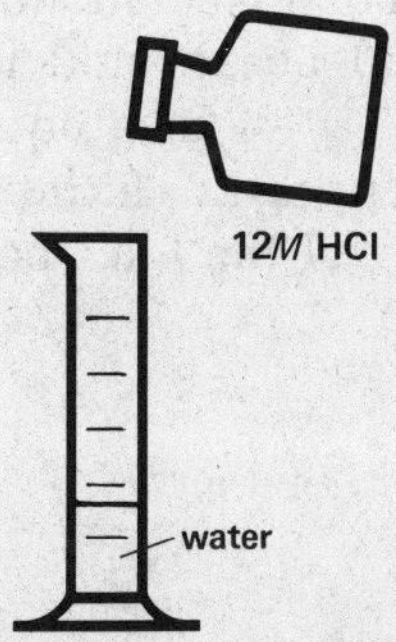

3. Prepare the volume of 3 *M* HCl calculated (twice the stoichiometric amount) by placing the calculated volume of water in your graduated cylinder and then carefully adding 12 *M* HCl, with stirring, until the total solution volume is obtained.

4. While stirring the hot soap dispersion, add the 3 *M* HCl. After all the hydrochloric acid has been added, add ice to the mixture and stir. Continue to add ice until the mixture has cooled to near 0°C (the last pieces of ice will melt very slowly).

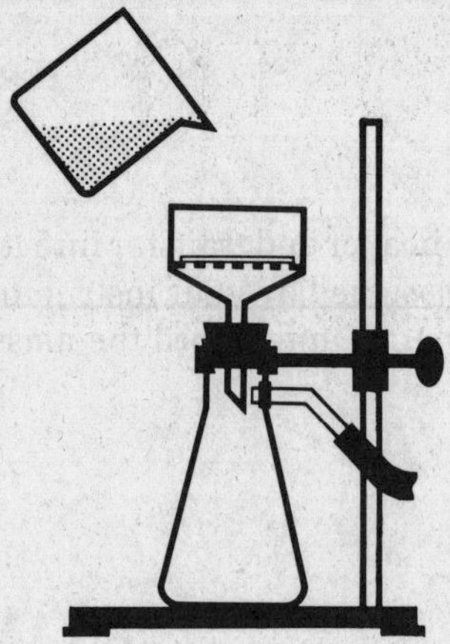

5. Isolate the carboxylic acid by vacuum filtration (Appendix A). To prevent the filter from becoming clogged, keep liquid in the funnel at all times. Wash the carboxylic acid with three 50-ml portions of ice-cold water. While washing the carboxylic acid, stir the mixture with your spatula. With your spatula crush any clumps of carboxylic acid that may form, being careful not to tear the filter paper.

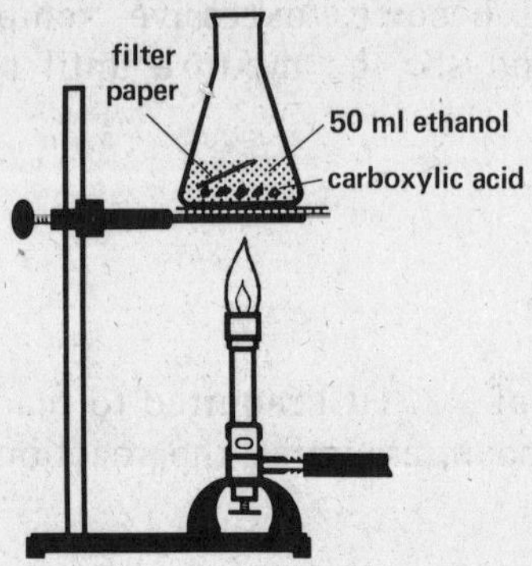

6. Transfer the precipitate and filter paper to a 250-ml Erlenmeyer flask. Add 50 ml of 95 percent ethanol and heat *very carefully* (*fire hazard*) until most of the acid has dissolved.

7. Fold a piece of filter paper of the appropriate diameter into quarters. Open one section of the folded filter paper and fit it into a powder funnel. Place the powder funnel with filter paper in a 250-ml beaker and filter the hot ethanol solution through it. Try to keep liquid in the filtering apparatus at all times during this filtration step. This will help to prevent clogging of the filter paper.

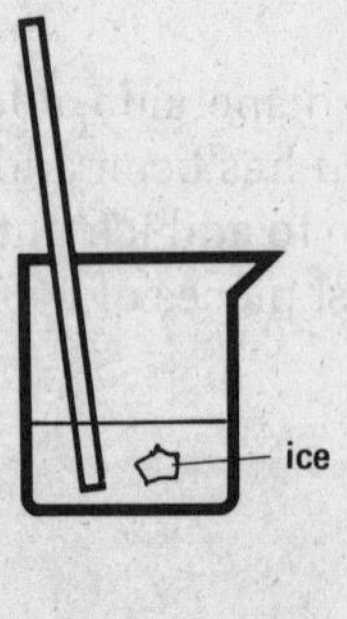

8. Add a piece of ice, about twice the size of a marble, to the clear alcoholic solution. Observe carefully. Record your observations in your notebook, along with a suggested explanation of these observations.

9. Place the beaker back on the ringstand and heat it *very carefully* with a *low* flame (*fire hazard*). Use your squeeze bottle to add distilled water to the warm alcoholic solution until the mixture becomes cloudy and remains cloudy on continued warming. Then add ethanol dropwise just until a clear solution is obtained. At that point, turn off the burner.

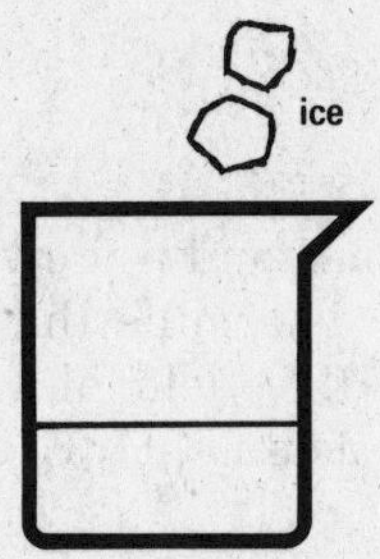

10. Remove the beaker from the ringstand and allow the solution to cool. Observe the solution as it cools and record your observations. After approximately 5 min of cooling, add, with stirring, small pieces of ice until the solution is approximately at 0°C.

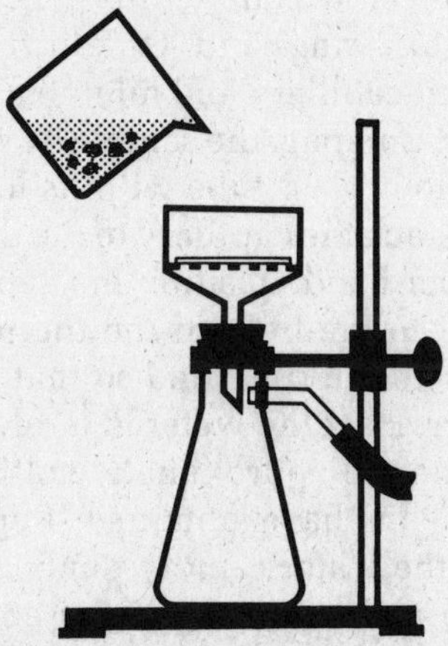

11. Separate the precipitated fatty acid by vacuum filtration. Again, keep the level of the liquid near the top of the funnel to prevent clogging.

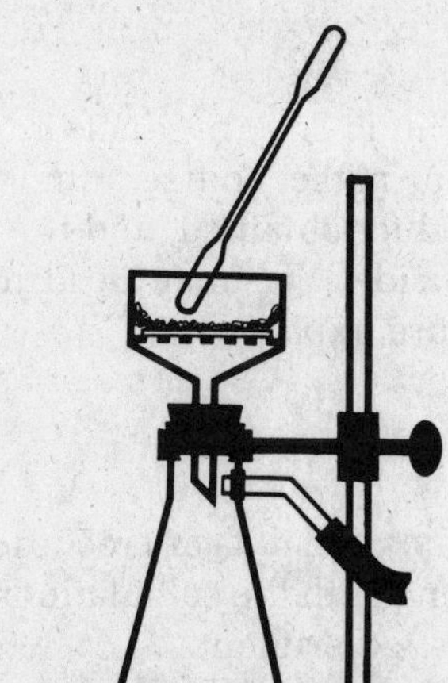

12. After all the liquid has been removed by filtration, press the collected carboxylic acid with your spatula. Continue to draw air through the funnel and to press the product for about 5 min.

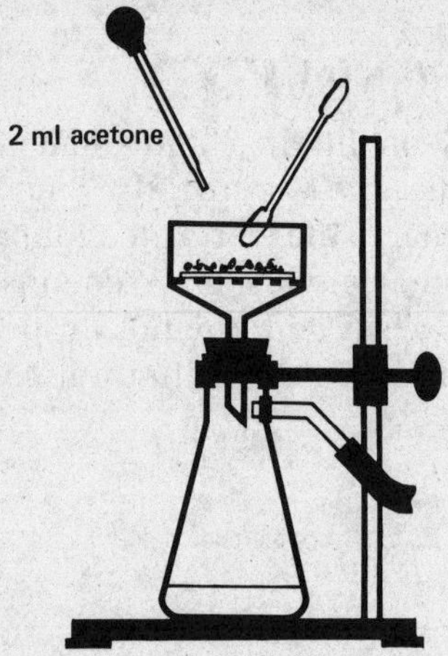

13. After 5 min of air drying disconnect the aspirator. Add 2 ml of acetone (not more) to the solid collected in the filter. With your spatula, mix the acetone and collected carboxylic acid until a uniform paste is obtained. Spread the paste evenly over the entire surface of the filter paper and turn on the aspirator. Draw air through your product until a white powder is obtained and you can no longer detect the odor of acetone. You will need to use your spatula to break up the product as it becomes dry.

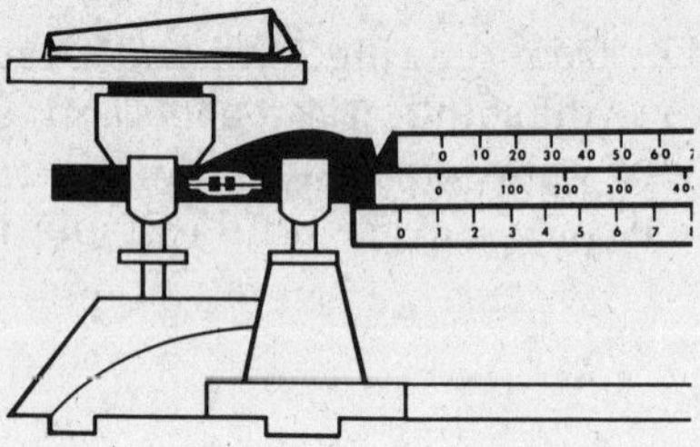

14. Use the triple beam balance and a preweighed piece of paper (half-page size) to determine the mass of the obtained product. Calculate the yield of your reaction assuming that the carboxylic acid is $CH_3(CH_2)_{16}CO_2H$.

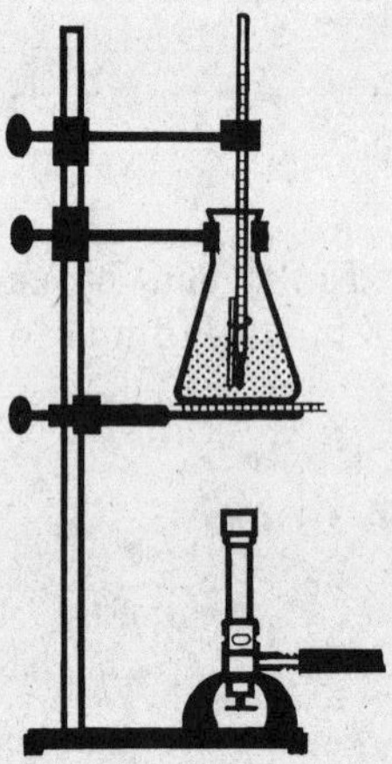

15. Fill a 125-ml Erlenmeyer flask three-quarters full with water. Clamp the flask to a ringstand. Obtain from your instructor a melting-point capillary and fill it to a depth of 2 mm with your product. Compact the sample by dropping the capillary through a long glass tube with its lower end against the bench top. Attach the capillary to the thermometer by means of a rubber band. Position the capillary so that the product is next to the bulb of the thermometer. Clamp the thermometer to the ringstand so that the bulb is immersed in the water. Heat the water at a rate of 5–10 deg min^{-1} and determine the approximate melting point of the carboxylic acid you have obtained. Repeat this determination, heating the water rapidly to about 10 deg below the approximate melting point of the acid. Then reduce the heating rate to 2 deg per minute and determine the melting point.

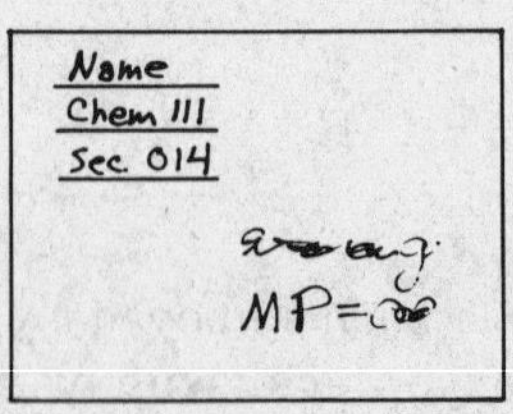

16. Fold the paper containing your product into an envelope shape and write your name, course number, section number, the mass of product obtained, and the product's melting point on the outside. You may need to use this carboxylic acid in a future experiment.

CALCULATIONS AND REPORT Be sure that all observations and data are recorded in your permanent notebook, along with the calculations required for various stages of this experiment.

Use this record to complete the Report Form for submission to your instructor.

GRADE / INSTRUCTOR'S INITIALS

Report Form

FOR INVESTIGATION 10:
Isolation of a Fatty Acid

NAME: ______________________ ID NO.: ______________________

LAB SECTION ______________________ INSTRUCTOR ______________________

Experimental values (Brand of soap used: ______________________)

Mass of soap used

Mass of soap + beaker ____________ g

Mass of beaker ____________ g

Mass of sample ____________ g

Volume of 3 *M* HCl used ____________ ml
Show all calculations on the back of this page.

Volume of 12 *M* HCl used ____________ ml
Show all calculations on the back of this page.

Mass of paper + acid ____________ g

Mass of paper ____________ g

Mass of acid ____________ g

Melting point of the acid obtained ____________ °C

Calculated Values

Theoretical yield ____________ g
Show all calculations on back of this page.

Percentage yield ____________ %
Show all calculations on back of this page.

Description of the product

DATE ____________ SIGNATURE ____________

Investigation 11

HEAT STORAGE FOR SOLAR HEATING

INTRODUCTION The "energy crisis" made its first real impact on the U.S. consumer in 1973, but a significant energy shortage had been clearly predicted years earlier. It has been suggested* that the announced goal of American independence from foreign energy sources would require very stringent energy conservation measures, rapidly accelerated consumption of U.S. fuel reserves, and virtual abandonment of environmental quality standards. This assessment may seem a bit overdramatic, but the facts on which it is based can hardly be disputed. We *do* have a very serious problem.

There are a number of alternative energy sources available. Fossil fuels (coal, oil, and natural gas) are the ones most commonly employed today for home and industrial use. Although the United States may soon begin to exploit the estimated 10 billion barrels of oil available from the North Slope of Alaska and the 55 billion barrels of oil estimated as easily recovered from the oil shale in Colorado, Utah, and Wyoming, it is still believed that the United States will have to import around 16-*million barrels of oil per day by* 1990. When you consider that conservative estimates suggest a U.S. demand for more than 150-billion barrels of crude oil over the next 15 years, it is easy to see why an "oil independence" will be difficult to achieve. Coal, on the other hand, is no problem. The United States has enormous deposits of coal. Although direct burning of coal poses significant environmental problems, advances in coal technology offer considerable potential for this important energy source. Nuclear energy, particularly with the development of breeder reactors, will undoubtedly ease the energy crisis, although there are a number of safety and environmental problems yet to be solved. Hydroelectric plants and geothermal power will make significant contributions to the nation's energy requirements, but these can contribute only a small percentage of the total need.

Among the more novel suggestions for developing new power resources are various ways of utilizing solar energy. Most of the heat and other radiation the earth receives from the sun is now "wasted." The problem is not one of insufficient availability, but rather one of finding efficient and economical ways of collecting and utilizing solar energy. Indirectly, we can use some of the heat from the sun to drive windmills (since it is primarily this heat that causes wind currents). More directly, solar radiation can be converted to electricity by appropriate cells, or the heat from the sun can be used to generate steam for turbines to drive electric generators. Until quite recently,

*Energy Policy in the U.S., David J. Rose, *Scientific American, 230,* 20–29 (1974).

research has been slow in the area of solar energy technology. With the present "crisis" atmosphere, the picture is changing somewhat. No reasonable suggestion for an energy resource can be ignored. In 1974, some 12-million dollars were earmarked for study of solar heat utilization in such areas as home space and water heating. It is this particular area that we will explore a bit in this experiment.

If heat from the sun is to be used over a prolonged period, including "sunless days" and nighttime, then we must have some way of storing that heat for release as needed. A number of alternatives have been suggested, among them the chemical system *sodium thiosulfate pentahydrate,* $Na_2S_2O_3 \cdot 5H_2O$. This compound has several advantages: As granular crystals or powder, it is easily transported and distributed into heat storage containers; it has a low enough melting point (48°C) to permit its liquefaction at temperatures easily achieved by solar heat "capture" equipment; and its heat of fusion is high enough to provide a good heat release per pound of chemical required. In this experiment you will determine this heat of crystallization for $Na_2S_2O_3 \cdot 5H_2O$.

The heat of crystallization of a liquid is numerically equal to the heat of fusion of the solid, but of opposite algebraic sign. The heat of crystallization refers to the heat *evolved* when a specified amount (typically 1 g) of a liquid crystallizes at its freezing point, whereas the heat of fusion measures the heat *absorbed* when a specified amount of the solid is liquified at its melting point.

Before starting this investigation, you should be familiar with the general principles of calorimetry and with enthalpy calculations involving temperature change and change of state. In addition, before coming to the laboratory you should *carefully* study the discussion at the conclusion of this investigation concerning how the experimental data should be handled and how graphs are to be prepared. Be certain that you understand this discussion or, if not, that you have prepared questions to ask of your instructor.

References

1. O'Connor, Rod. 1977. *Fundamentals of Chemistry* (2nd ed). New York: Harper & Row (Units 5 and 12).
2. Readings from *Scientific American.* 1973. *Chemistry in the Environment.* San Francisco: Freeman (Sections III and IV).

SAFETY Wear eye protection and exercise proper care in heating and in transferring hot equipment and reagents.

OBJECTIVES

1. Be able to measure temperature and time intervals accurately.
2. Be able to use a simple calorimeter properly.
3. Be able to make thermochemical calculations involving specific heat and heat of crystallization (or fusion).

PROCEDURES (Work in pairs on this experiment.)

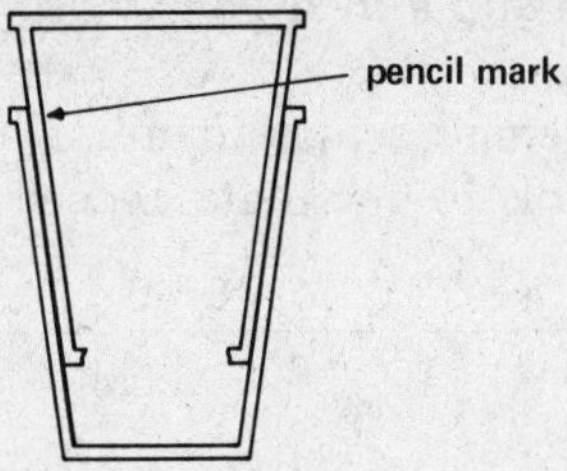

1. Carefully cut out the bottom from a Styrofoam cup. Push this cup firmly into a second (intact) cup and make a pencil mark 2 cm below the top rim of the inner cup. Weigh the cup assembly on a laboratory balance to the nearest 0.01 g. Place 200 ml of distilled water into a 250-ml beaker. Set the beaker aside for later use.

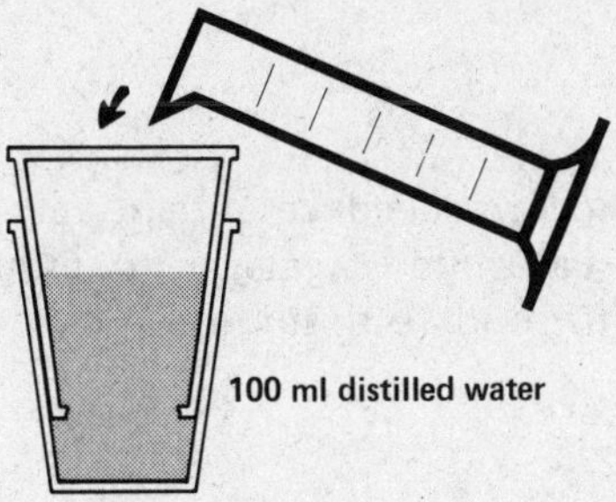

2. Measure 100 ml of distilled water into the cups. Weigh the cups containing the distilled water to the nearest 0.01 g.

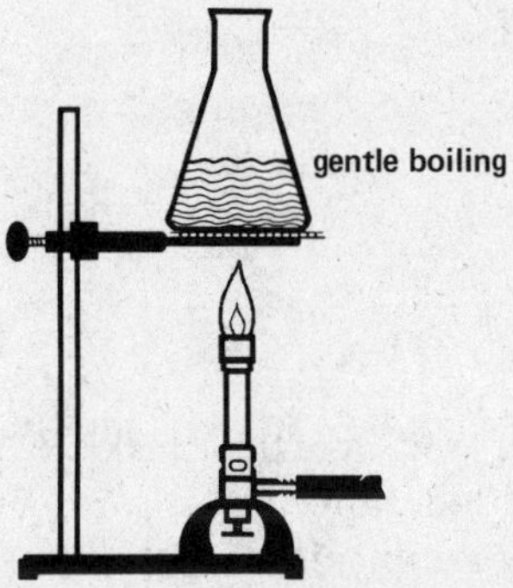

3. Pour about 80 ml of distilled water into a 125-ml Erlenmeyer flask and heat this water to boiling. Adjust the heat source so that the water is maintained at a *gentle* boil. Determine* the temperature of the boiling water from the graph of boiling point versus barometric pressure given at the conclusion of this investigation. Your instructor will announce the barometric pressure.

4. Be sure you can read a thermometer accurately. Use your thermometer (preferably one with 0.2 deg graduations) to stir the water in the Styrofoam cups, using a circular motion around the inside of the cups at a rate of about one circle per second. Practice reading and recording the temperature every minute for 4 min.

*Since the type of thermometer usually employed for this study does not read to 100°C, you must find the boiling point of water from a graph, rather than by direct measurement.

(Note that timing is critical in the next steps. Be sure that you and your partner have planned your work carefully for efficient cooperation.)

Before proceeding, prepare a table (similar to the following) in your notebook to facilitate record keeping.

Time (min.)	1	2	3	4	5	6	7	8	9	10	11	12
Temp (°C)												

5. Stirring as described in procedure 4, read and record the temperature of the calorimeter water every minute for 5 min. *Between the fifth and sixth minute. . . .*

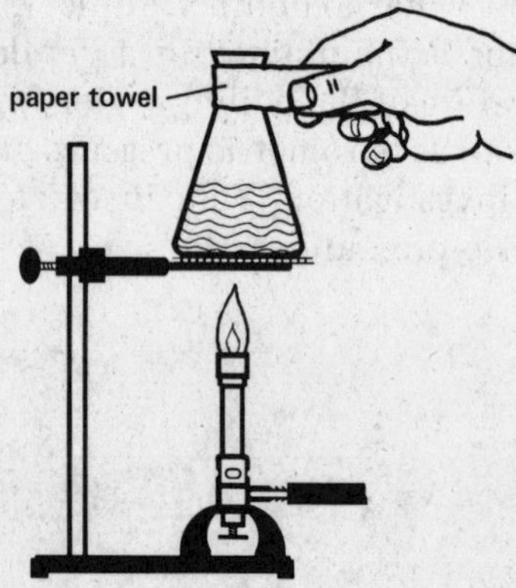

6. . . . have your partner remove the flask of boiling water, using a properly folded paper towel for insulation. *Exactly* at the start of the sixth minute. . . .

7. . . . boiling water is poured from the flask into the calorimeter until the liquid level just reaches the pencil mark. Continue timing and stirring, recording the temperature every minute. The temperature will rise quickly to a maximum and then begin to drop slowly. Keep stirring and recording temperatures until the *rate* of temperature change is constant (that is, until temperature *change* per minute is the same for 5 min). Weigh the cups containing the water to the nearest 0.01 g.

Now remove the thermometer, letting all water from the stem and bulb drain back into the cups. Dry the thermometer with a paper towel. Determine and record the mass of the calorimeter and water to the nearest 0.01 g.

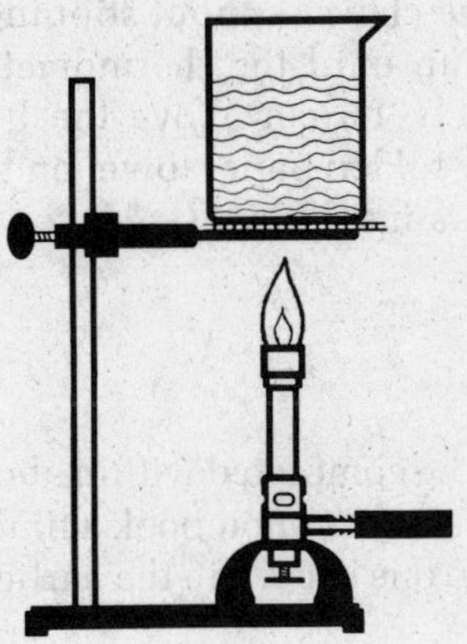

8. Pour 450 ml of water into a 600-ml beaker and heat the water to boiling. Adjust the heat to maintain a *gentle* boil.

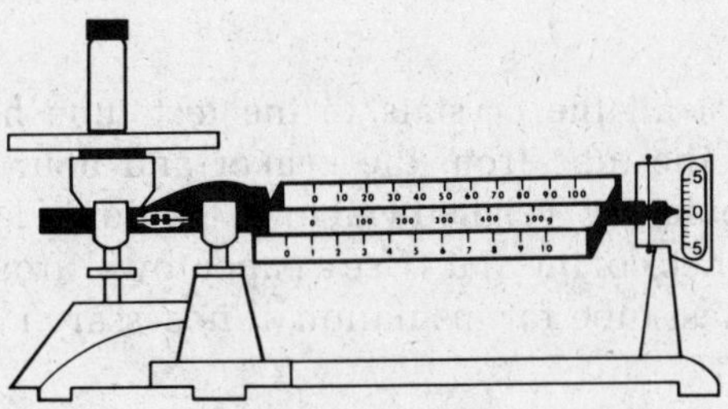

9. Weigh a clean, dry screw-cap vial (O.D. 2.5 cm, 7 cm height with cap) to the nearest 0.05 g. Record this mass.

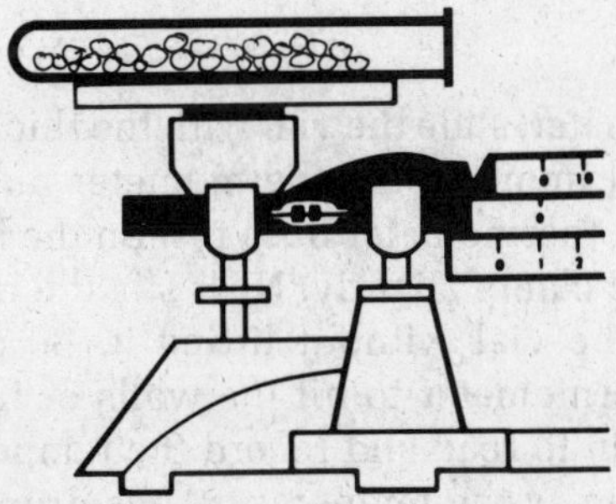

10. Weigh a 20-cm (length) test tube on the triple beam balance. Place about 41 g of $Na_2S_2O_3 \cdot 5H_2O$ in the test tube. Weigh the test tube containing the salt and record the data.

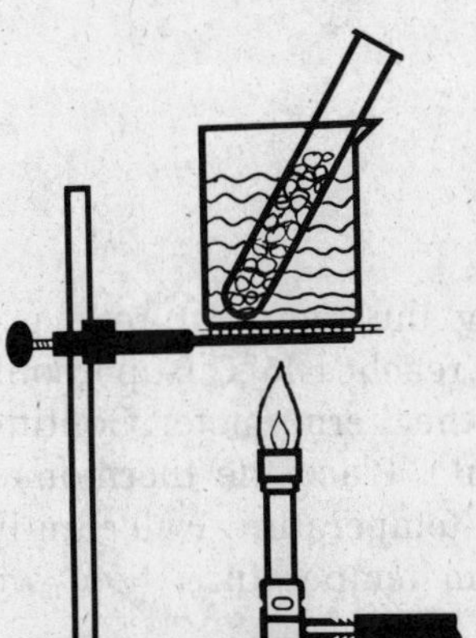

11. Set the test tube containing the sodium thiosulfate into the hot water in the 600-ml beaker. The crystals will soon begin to melt.

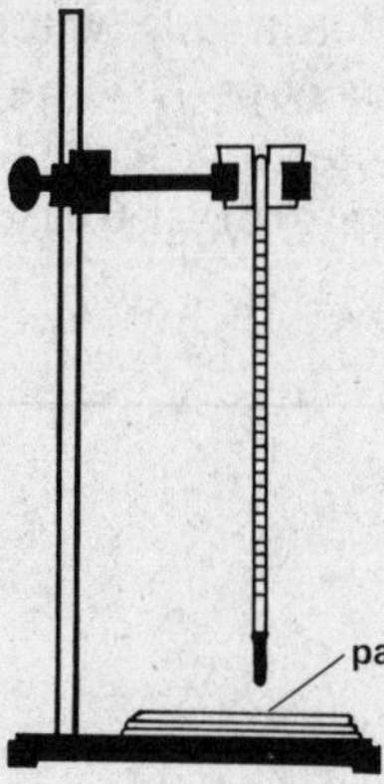

12. Carefully push the upper end of the thermometer (0.2 deg divisions) into a one-hole No. 6 rubber stopper until the thermometer end has almost reached the other side of the hole. Make sure that the thermometer is held tightly in the stopper and cannot slip out. Attach a clamp to a ringstand and open the claws of the clamp sufficiently to allow the stopper to rest in the claws without slipping through. Raise or lower the clamp until the thermometer bulb is suspended approximately 1.5 cm above the base of the ringstand. Place a twice-folded paper towel on the base of the ringstand to serve as insulation.

(Note: Steps 15–17 *must* be completed within about 10 min. Plan ahead now. Prepare in your notebook a time/temperature table similar to the one shown in the earlier note but extended to 40 min.)

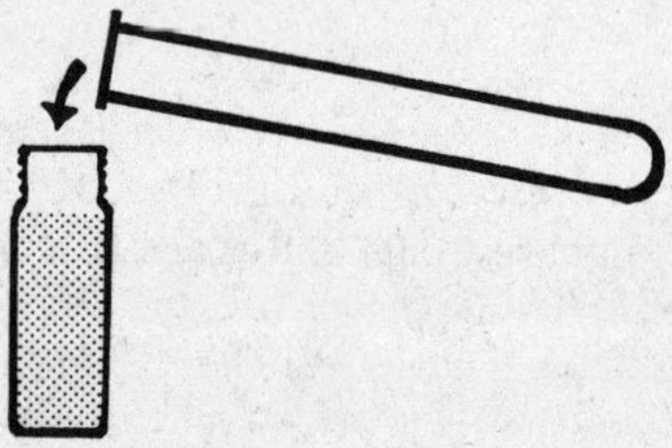

13. When almost all the crystals in the test tube have melted, remove the tube from the beaker and pour the melt into the previously weighed vial until the liquid level is just below the neck of the vial. (Use a paper towel around the neck of the test tube for insulation, if necessary.)

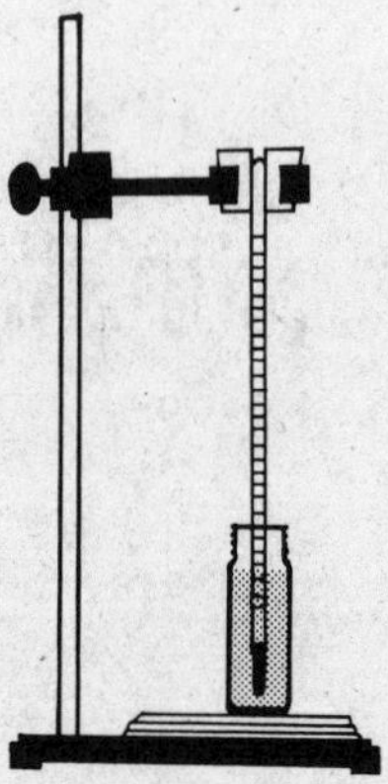

14. Lift the thermometer, slide the vial with the thiosulfate melt into place, and immerse the thermometer bulb into the melt. Should the thermometer bulb rest on the bottom of the vial, raise the clamp slightly. Now stir the melt by circularly moving the vial without lifting it. Be careful not to allow the thermometer to hit the walls or the bottom of the vial. Begin to read and record the temperature every minute as soon as the temperature has dropped to 48°C. This temperature is defined as that for "time zero."

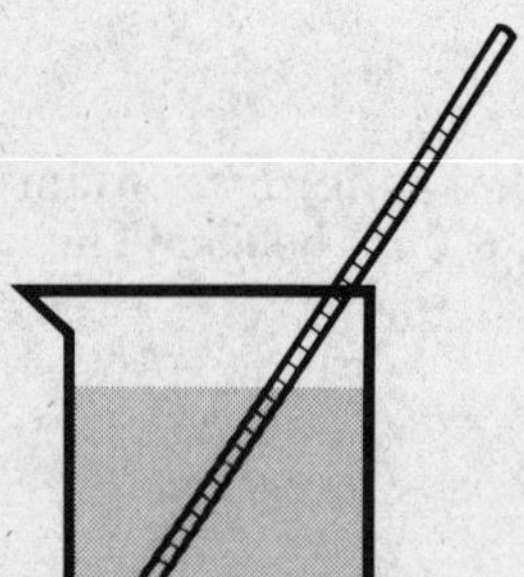

15. Continue recording the temperature every minute. When the temperature reaches 45°C (supercooled condition), carefully remove the thermometer. Continue recording each elapsed minute. Place the thermometer in the 250-ml beaker of room-temperature water until the thermometer registers room temperature, then wipe it dry and. . . .

16. . . . begin to stir the calorimeter water with the thermometer, reading and recording the temperature every minute for an additional 5 min. Between the additional fifth and sixth minutes . . .

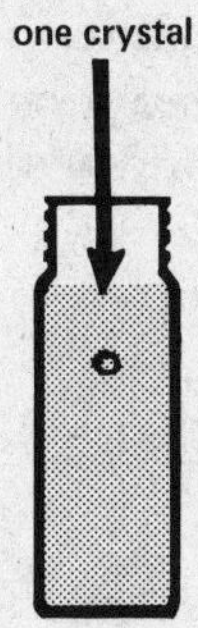

17. . . . have your partner drop one tiny crystal of thiosulfate into the vial and close the vial tightly. At the start of the sixth minute . . .

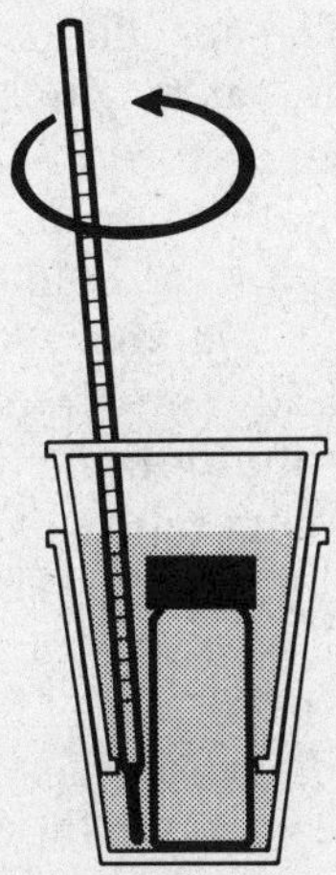

18. . . . slide the sealed vial carefully into the calorimeter. (The thiosulfate must still be liquid, except for the added "seed" crystal.) Stir the calorimeter water around the vial as described earlier, reading and recording the temperature every minute (estimating to 0.1°C). The temperature will slowly rise to a maximum and will then begin to drop. Continue stirring and recording the temperature until seven consecutive readings show a constant *rate* of cooling. (Be sure you do not unscrew the vial cap during the stirring.)

Note that if the salt began to crystallize before the vial was completely immersed in the calorimeter water, you will have to remove it (letting all water drain back into the calorimeter), tie a string around the vial's neck, and reheat the vial by submerging it in the hot water contained in the 600-ml beaker), and begin again with procedure 15.

19. When the cooling rate has remained constant for 7 min, remove the thermometer and vial and wipe the vial dry. Carefully unscrew the cap and wipe off any water adhering to the neck of the vial. Reseal the vial and weigh it to the nearest 0.01 g. Record this mass.

When you have finished, tie a piece of string around the neck of the vial and heat the vial in boiling water until the crystals melt. Pour the molten salt into the specified waste container, then wash and dry the vial.

CALCULATIONS AND REPORT Although you worked with a partner on this experiment, each one of you must carry out all the calculations required to obtain the heat of crystallization of $Na_2S_2O_3 \cdot 5H_2O$ from your raw data. Each one of you must record all the collected data in your own notebook. Use your notebook records to complete the Report Form for submission to your instructor. The following discussion should help you to understand the calculations and perform them more efficiently.

When hot and cold water are poured together, everyday experience tells you that the temperature of the mixture will be above the temperature of the cold water but below the temperature of the hot water. Such a temperature change implies that the hot water lost a certain amount of heat and the cold water gained heat. In the idealized case, when no heat is lost to the surroundings, the heat gained by the cold water is exactly equal to the heat lost by the hot water. This heat gain or loss can be easily calculated provided the following data are available:

The mass of the cold water.
The mass of the boiling water.
The temperature of the hot water.
The temperature of the cold water.
The specific heat of water.

The specific heat (also called the "heat capacity per gram, at constant pressure," and symbolized by c_p) is the amount of heat necessary to change the temperature of 1 g of a substance by 1°C. The specific heat for water is 1 cal g^{-1} deg^{-1}. For an experiment in which no heat is lost to the surroundings, the temperature of the water after mixing can be calculated from the data listed above. Consider, for instance, that a sample of cold water of mass M_{CW} and a temperature T_{CW} is mixed with a sample of boiling water of mass M_{BW} and temperature T_{BW}. Let us assume that the temperature after mixing is T_{MW}. Since it takes 1.0 calorie to raise the temperature of 1 g of water by 1°C, the number of calories necessary to raise the temperature of M_{CW} g of water from T_{CW} to T_{MW} can now be calculated. Similarly, the number of calories lost by M_{BW} g of hot water by having been cooled from T_{BW} to T_{MW} can be obtained. Equating the expression for the heat gained to the expression for the heat lost will produce an equation with the unknown T_{MW}.

Why is it then necessary to perform procedures 2–7 and experimentally determine T_{MW}? The unavoidable heat loss to the surroundings, which include the Styrofoam cups, the thermometer, and the air above the water and around the cup, forces you to do this additional experimental work. Since some of the heat lost by the boiling water (or other hot object) is used to warm the surroundings instead of the cold water, the temperature of the water after thermal equilibration will be lower than in the case of the idealized situation where no heat would be lost to the surroundings. If, for instance, the heat liberated by the crystallization of $Na_2S_2O_3 \cdot 5H_2O$ has to be calculated from the temperature increase of the calorimeter water, the results will be low, because the absorption of heat by the surroundings will keep the water temperature from rising to the theoretical value. In order to account for this heat loss, a correction factor called the calorimeter constant K is used in the calculations. The calorimeter constant is defined as the amount of heat taken up by the *surroundings* for each 1°C rise of the calorimeter water temperature. This constant is determined experimentally as described in procedures 2–7. In order to calculate the calorimeter constant, one again equates the heat gained with the heat lost. This time, however, the expression for the heat gained consists of two terms: the heat gained by the cold water *and* the heat gained by the surroundings. Since in this calculation the temperature of the water at thermal equilibrium is known from the experiment, the final equation contains the calorimeter constant K as the only unknown.

There is one more complication that must be considered. You will have noted in taking temperature readings of the calorimeter water, that the temperature of the water on addition of a hot object rises rather quickly, reaches a maximum, and then slowly decreases. Is the maximum temperature you read the temperature you should use in your calculations? The answer is no. The reason for this negative answer and the procedure to be used for determining the correct temperature follows.

Imagine cold water in the calorimeter. An object of higher temperature (water or a vial containing crystallizing $Na_2S_2O_3 \cdot 5H_2O$) is placed into the cold water. At the instant of addition the water in the calorimeter will be thermally inhomogeneous. The water closest to the hot object will have approximately the same temperature as the object, whereas the water close to the calorimeter walls will be cold. Therefore the water must be mixed by stirring. The transfer of heat from the inside of the vial to the water requires time. During this time the calorimeter water, which is warmer than room temperature, will lose heat. The opposing processes of heat transfer into the water and the heat loss from the water to the surroundings together with the time required for reaching thermal equilibrium make it impossible to measure directly the calorimeter water temperature expected for instantaneous heat transfer and mixing. This temperature can, however, be obtained by extrapolation.

Let us assume such instantaneous heat transfer and mixing took place on insertion of a hot object into the calorimeter water. At this moment the maximum possible water temperature would be measured. With passing time, heat would be lost to the surroundings and temperature readings at later times would give lower values. When temperature is plotted versus elapsed time, a straight line is obtained provided the temperature interval covered is small (curve *A*). In reality, for reasons discussed earlier, curve *B* is obtained. As illustrated in the figure, curve *B* joins curve *A*, and extrapolation of the linear portion of curve *B* to time zero, the time of mixing, will give the temperature that would be measured on *instantaneous* heat transfer and mixing.

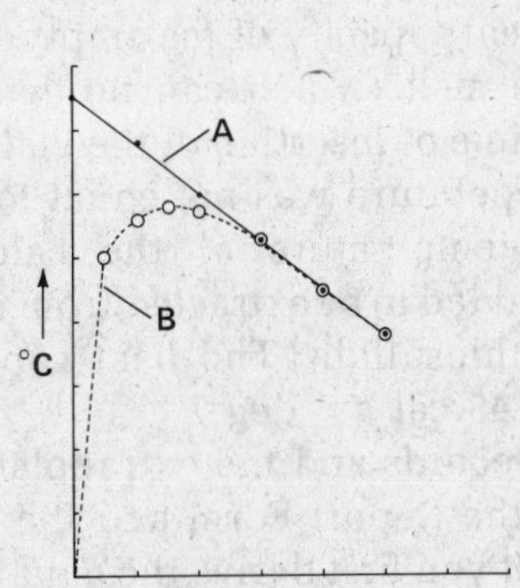

For the calculations ultimately leading to the heat of crystallization of $Na_2S_2O_3 \cdot 5H_2O$, the following four temperatures must be obtained:

1. The temperature of the calorimeter water after instantaneous heat transfer from the crystallizing melt.
2. The temperature of the calorimeter water after instantaneous mixing with the boiling water.
3. The temperature of the calorimeter water at the time the vial containing the melt is submerged.
4. The temperature of the calorimeter water at the time the boiling water is added.

The first two temperatures must be determined by extrapolation. Plot the measured temperature values on the temperature-time graph paper sheet. Parallel to the temperature axis draw an auxiliary axis intersecting the time axis at the point representing the time of mixing of the water samples or submersion of the vial. Connect those temperature points, representative of a constant rate of cooling, by a straight line and extend this line until it intersects the auxiliary temperature axis you have drawn into the graph. Read the temperature corresponding to the intersection point on the main temperature axis.

You could, of course, have read the other two temperatures (temperatures of the calorimeter water just before you added the boiling water or submerged the vial) directly from the thermometer. However, you were probably too busy preparing for the addition of the boiling water or the vial, checking time, or stirring the water. For this reason these two temperatures are also determined by extrapolation as described above.

The principles discussed earlier, which were used for the determination of the calorimeter constant, apply also to the calculation of the heat of crystallization of $Na_2S_2O_3 \cdot 5H_2O$. The heat of crystallization is defined here as the amount of heat liberated when 1 g of $Na_2S_2O_3 \cdot 5H_2O$ changes from the liquid to the solid state at the freezing point (melting point). The total amount of heat generated by the crystallizing thiosulfate, which is obtained as the product of the specific heat of crystallization and the mass of thiosulfate in the vial must be equal to the heat gained by the calorimeter water and the surroundings. The resulting equation with the specific heat of crystallization as the only unknown can easily be solved.

Our experiment has been accurate enough to permit you to determine the *approximate* crystallization heat of $Na_2S_2O_3 \cdot 5H_2O$. We will ignore a few of the "finer points" that would have to be considered in a more accurate determination, such as the heat capacity of the empty vial, and the temperature of the melt (which can be determined from your data) at the time of insertion of the vial into the calorimeter. If this temperature was not equal to the extrapolated maximum temperature of the calorimeter water, heat was either added to or extracted from the calorimeter water by the thiosulfate. The heat capacity of sodium thiosulfate is 0.45 cal g^{-1} deg^{-1}.

Use your notebook records and the extrapolated temperatures to complete the Report Form and the calculations. It is suggested that you first derive the equations for the calorimeter constant K and the specific heat of crystallization of $Na_2S_2O_3 \cdot 5H_2O$ (ΔH_c) using only those symbols defined in the Report Form. (If you use your own symbols,

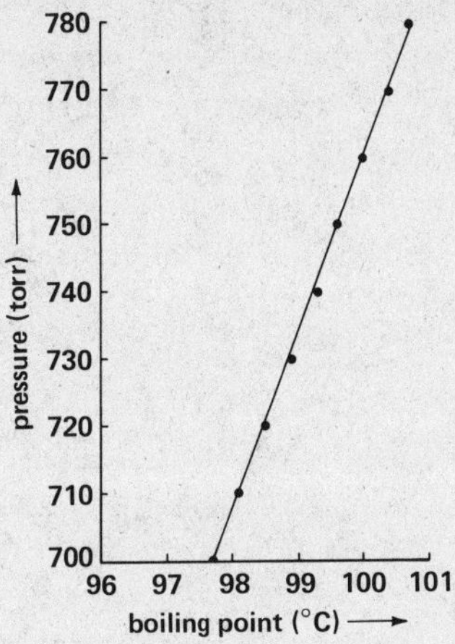

different from the suggested ones, your instructor will find it time consuming to check your calculations.) Insert numbers only into the finished equations.

Since the boiling point of water was not measured directly in this experiment, you should determine it from the graph shown, on the basis of the barometric pressure announced by your instructor.

GRADE / INSTRUCTOR'S INITIALS

Report Form

FOR INVESTIGATION 11:

Heat Storage for Solar Heating

NAME: ______________________ ID NO.: ______________________

LAB SECTION ______________________ INSTRUCTOR ______________________

Barometric pressure ______________ torr

Boiling point of water at this pressure ______________ °C

Directly Determined Data

Mass of empty calorimeter (M_C) (procedure 1) ______________ g

Mass of calorimeter (M_C) + (~100 ml) cold water (M_{CW}) (procedure 2)

______________ g

Mass of calorimeter (M_C) + cold water (M_{CW}) + boiling water (M_{BW}) (procedure 7)

______________ g

Mass of empty vial and cap (M_V) (procedure 9) ______________ g

Mass of empty vial and cap (M_V) and thiosulfate (M_{TS}) (procedure 19)

______________ g

Extrapolated Data

For the boiling water–cold water experiment

Temperature of cold water (extrapolated from data of procedure 5)

T_{CW} ______________ °C

Temperature of mixture (extrapolated from data of procedure 7)

T_{MW} ______________ °C

For the thiosulfate experiment

Temperature of cold water (extrapolated from data of procedure 16, graph)

$T_{CW}{}^{T}$ ______________ °C

Temperature of water after crystallization of thiosulfate (extrapolated from data of procedure 18, graph) $T_{MW}{}^{T}$ ______________ °C

Optional

Temperature of melt at time of insertion (extrapolated from data of procedure 14/15, $T_M{}^T$ ______________ °C)

Given

The heat capacity of water (C_P) is 1.0 cal g^{-1} deg^{-1}.

(OVER)

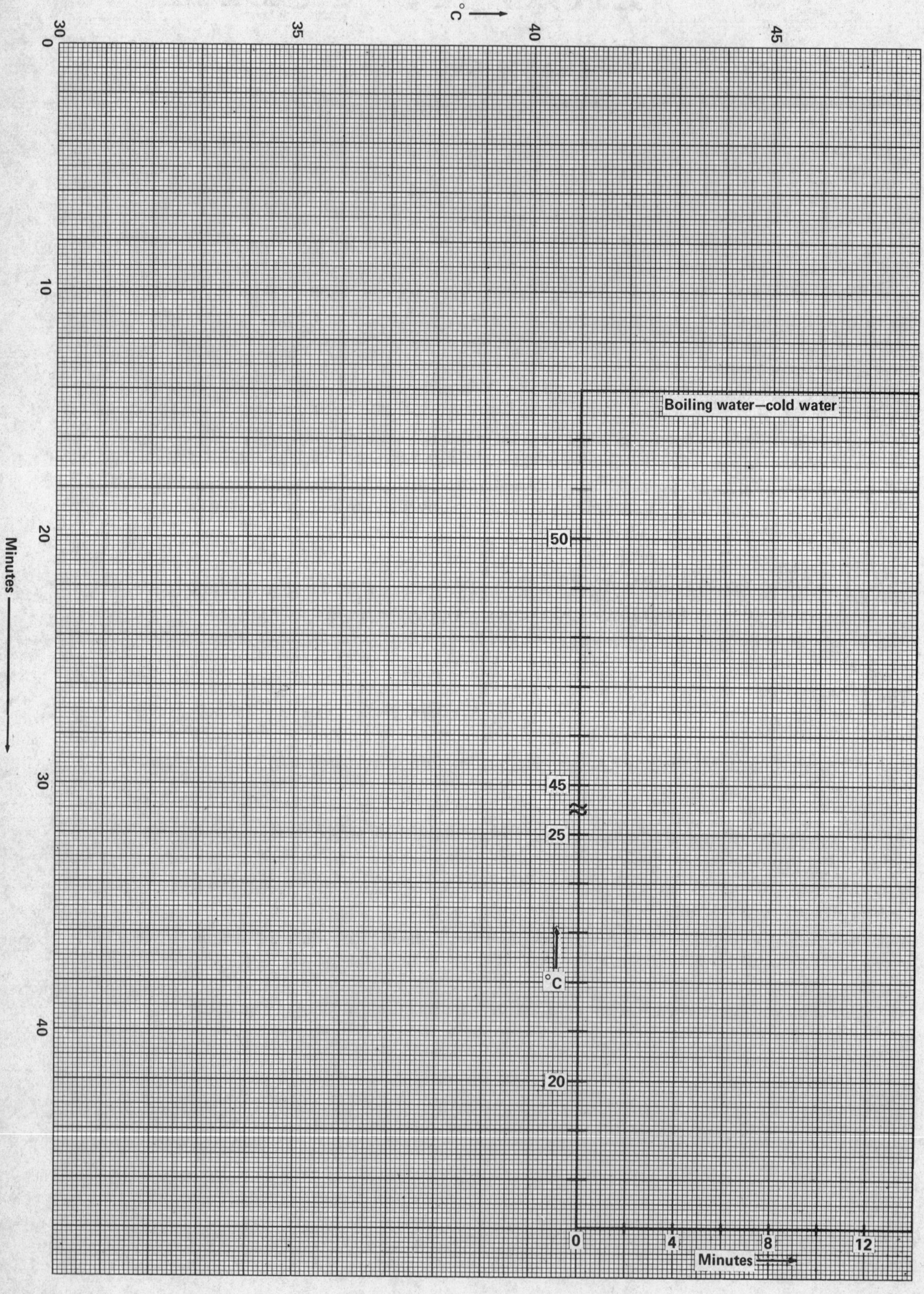
°C
30
35
40
45
Minutes
0
10
20
30
40
Boiling water–cold water
50
45
25
°C
20
0
4
8
12
Minutes

Calculations

For the boiling water–cold water experiment (give formulas not numerical values)

Heat lost by the boiling water ____________________

Heat gained by the cold water ____________________

Heat gained by the surroundings ____________________

Equation for calorimeter constant K =

Value for K ____________

Units for K ____________

For the thiosulfate experiment (give formulas not numerical values)

Heat lost by the crystallizing thiosulfate ____________________

Heat gained by the calorimeter water ____________________

Heat gained by the surroundings ____________________

Equation for specific heat of crystallization of thiosulfate (ΔH_C)

Value for ΔH_C ____________________

Units for ΔH_C ____________________

____________ ____________________

DATE SIGNATURE

Calculations

[illegible] water [illegible] ([illegible] numerical values)

[illegible]

Heat [illegible] by the [illegible]

Heat [illegible] by [illegible]

[illegible] the calorimeter [illegible]

[illegible]

[illegible] heat [illegible]

Heat lost [illegible]

Heat gained [illegible] the calorimeter [illegible]

[illegible] by [illegible]

[illegible]

Investigation 12

COPPER(II) AND ALGAE

INTRODUCTION All living systems require certain metal ions, such as Na^+, K^+, Mg^{2+}, Ca^{2+}, Cr^{3+}, Mn^{2+}, Fe^{3+}, Co^{2+}, and Cu^{2+}. These ions, and a few others, play critical roles in a number of essential biochemical processes. Some of these are shown in Table 12.1.

Table 12.1

Cation	Typical function
Na^+	principal cation in biological fluids *outside* cells
K^+	principal cation in biological fluids *within* cells
Mg^{2+}	component of some vital enzyme systems and of the chlorophyll systems of plants
Ca^{2+}	principal cation of teeth and bones
Cr^{3+}	involved in the function of insulin
Mn^{2+}	component of some important enzyme systems, among them the urea formation system in mammals
Fe^{2+} Fe^{3+}	involved in many vital enzyme systems, in oxygen transport and storage (hemoglobin and myoglobin), and in the ferredoxin system in photosynthesis
Co^{2+}	component of enzyme systems essential to amino acid metabolism and function of the genetic material (nucleic acids)
Cu^{2+}	in several enzymes and in oxygen transport system of invertebrates

Some of the other metal ions are quite toxic to most living systems. Both mercury poisoning (involving Hg^{2+}) and lead poisoning (by Pb^{2+}) are familiar examples.

There is still a great deal that we do not know about the behavior of metal ions in biological systems, although many current research programs are contributing significant new information. For example, it has recently been shown* that ceruloplasmin, a copper-containing protein in the blood, is essential in promoting the release of iron from the liver for incorporation into the hemoglobin of the red blood cells. This finding has cleared up a 50-year-old question as to why animals on a copper-deficient, but high-iron, diet develop anemia.

There is evidence that suggests a competition among metal ions for various "binding sites" in the complex molecules of living systems. In general, these metal ion–molecule coordination compounds will perform their allotted roles only if the proper metal is present. Metalloenzyme systems, for example, usually have very specific metal ion requirements. Many cases of metal ion poisoning

*Frieden, Earl. The Chemical Elements of Life. *Scientific American July, 1972.*

can be directly related to the substitution of a "foreign" metal ion for the one normally required by a particular biochemical system. The competitive nature of these processes suggests an interesting possibility: Too much of one of the essential metals might result in its substitution for a different essential metal in a critical metal-biochemical system. This is, indeed, the case.

As a result, it is necessary that the "diet" of any living organism contain not only the proper metals, but also the proper *range of concentrations* of these metals. Too little will result in a metal deficiency and too much may cause metal poisoning, sometimes with fatal results. We take advantage of this concept in a number of familiar systems.

For example, zinc(II) ion is essential to enzyme systems that help regulate the acidity of biological fluids and to other enzymes participating in the digestion of protein foods. Too much Zn^{2+}, however, inhibits the biological function of Cu^{2+}, apparently by competing for Cu(II) binding sites in certain proteins. Zinc oxide ointments are employed as topical agents for the treatment of certain fungal infections. Zinc oxide is not very soluble in water, so the small concentration of free Zn^{2+} absorbed through the skin is no problem to the person using the ointment. This concentration is, however, high enough to be toxic to the particular fungus. (Should the zinc oxide be *swallowed,* of course, by a human, its dissolution by the acid in the stomach could easily produce a concentration of Zn^{2+} toxic to the human.)

Another familiar system involves the use of copper(II) salts, usually as $CuSO_4$, to retard or prevent algae growth in swimming pools and decorative ponds. Algae require a certain concentration of Cu^{2+} to sustain their life processes, but a somewhat higher concentration is toxic, possibly because of the competitive replacement by Cu^{2+} of other essential metal ions. The concentration of Cu^{2+} toxic to most algae is well below the toxic limit for humans.

The concentration of very dilute solutions is often expressed employing "ppm" (parts per million) or "ppb" (parts per billion) rather than molarities. These concentration units are widely used in environmental sciences, which deal with very small quantities of hazardous materials in the air, in water, and in the soil. The units ppm (or ppb) when applied to solutions give the grams of *solute* per 1 million g (or 1 billion g respectively) of *solution*. Thus a solution containing 3 ppm $CuSO_4 \cdot 5H_2O$ has 3 g of this salt dissolved in 1×10^6 g of solution or 0.003 g in 1000 g of solution. The concentration unit "molarity" is defined as the number of moles of solute per 1 liter of solution. Since the ppm unit refers to a mass of solution and the molarity unit to a volume of solution, the conversion of one unit to the other involves the density of the solution. If the density is known, the mass can be converted to the volume using the formula

$$\text{mass} = \text{volume} \times \text{density}$$

When the density is exactly equal to one, the volume is numerically equal to the mass and molarity: ppm or ppm: molarity conversions are easily accomplished. Such conversions involve changes between *grams* of solute and *moles* of solute, as required.

In this experiment we will use the preparation of a 1-ppm $CuSO_4 \cdot 5H_2O$ algaecide solution as an exercise in getting acquainted with some of the skills and concepts involved in dealing with aqueous solutions.

Before starting this investigation, you should be thoroughly familiar with calculations involving ppm and molar concentration units, with mass-volume-density relationships, and with the techniques and mathematics of solution preparation and dilution. In addition, before coming to the laboratory, you must set up in your notebook the formulas for all calculations required in the Procedures section of this investigation. The actual numbers to be used in most of these calculations will be determined during your laboratory work.

References

1. O'Connor, Rod. 1977. *Fundamentals of Chemistry* (2nd ed). New York: Harper & Row (Unit 13).
2. Readings from *Scientific American*. 1973. *The Chemical Elements of Life*. San Francisco: Freeman (pp. 33–40).
3. Hale, Frank E. 1939. *The Use of Copper Sulfate in Control of Microscopic Organisms*. New York: Phelps-Dodge Refining Corp. (The "classic in the field," probably available now only through interlibrary loan.)

SAFETY Wear approved eye protection and exercise standard precautions in handling reagents.

OBJECTIVES

1. Be able to make accurate calculations of solute mass for a solution of specified concentration and volume and of volume quantities for specified dilutions.
2. Be able to weigh a solid chemical to the nearest 0.1 mg.
3. Be able to prepare and dilute solutions accurately.

PROCEDURES

You are to prepare a solution whose concentration of Cu^{2+} is sufficient to prevent the growth of the more common algae. In general, sufficient copper(II) sulfate pentahydrate is added to a water system to produce a concentration of about 1 ppm (1 part solute per million parts solution, by weight). You will prepare 100 ml of such a solution. This would require about 0.1 mg of $CuSO_4 \cdot 5H_2O$ and, although a typical analytical balance can weigh to 0.1 mg, the error in determination of such a small mass on an ordinary analytical balance might be as much as 100 percent. We will reduce the error by using a larger sample mass to form a more concentrated solution, then diluting this to the concentration desired. Since it is not necessary to have a concentration of *exactly* 1 ppm, we will not have to go through the rather tedious procedures of weighing the precise amounts of solute and solvent. Rather, we will prepare a solution based on solute mass and total solution volume and then calculate the final ppm concentration, which will be within the appropriate range for an effective algaecide.

1. You will need a clean 100-ml volumetric flask with a stopper. *Make sure the stopper fits tightly.* Place some tap water in a flask, stopper and invert it, securing the stopper with your fingers to prevent it from falling out. No water should leak out of the flask. If you detect any leakage, consult your instructor. Make sure that no foreign material is in the flask. Rinse it thoroughly with tap water and then with distilled water. Let all the water drain from the flask. The flask does not have to be dry inside.

Calculate the mass of $CuSO_4 \cdot 5H_2O$ required for the preparation of 100 ml of a 4.0×10^{-3} *M* solution of this salt. Have your instructor check your results before you proceed any further.

2. Weigh a clean, creased piece of *smooth* paper ($\sim5 \times \sim8$ cm) or aluminum foil on the analytical balance to the nearest 0.1 mg. Consult Appendix A if you are not familiar with an analytical balance. *Without spilling any of the blue crystals,* carefully add with your scoopula the amount of $CuSO_4 \cdot 5H_2O$ calculated above. Reweigh the paper and the $CuSO_4 \cdot 5H_2O$ to the nearest 0.1 mg. If this weighing shows that you added too much of the compound, remove a few crystals and place them into a beaker kept for this purpose next to the balance. If the mass is too light, add a few crystals more until the required mass is obtained. A final mass of $CuSO_4 \cdot 5H_2O$ up to 10 mg heavier than the calculated amount is acceptable.

3. Quantitatively transfer the $CuSO_4 \cdot 5H_2O$ into the volumetric flask. Rinse the funnel with distilled water, fill the flask half full with distilled water, and after removing the funnel, swirl until all crystals have dissolved. Should your solution remain cloudy upon shaking, add *one* drop of hydrochloric acid (6 *M*).

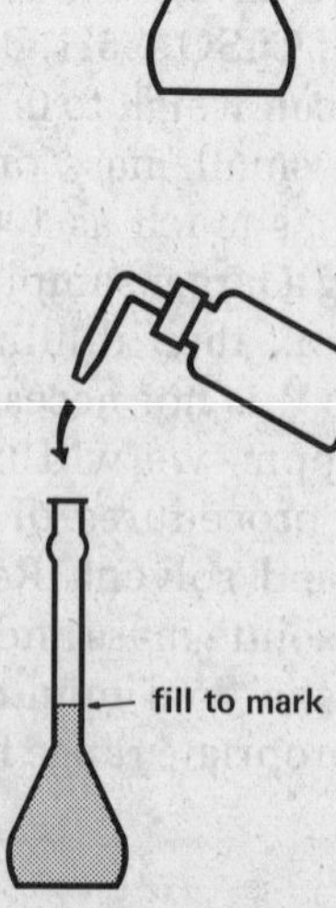

4. Then fill the flask to the mark, adding the last volume increment dropwise, stopper the flask, and mix the solution thoroughly by inverting and swirling the flask several times.

Calculate the molarity of the solution you have just prepared to three significant figures. Then calculate the volume of this solution required for the preparation of 100 ml of a 1.3×10^{-4} M solution. Have your instructor check your calculations before proceeding further.

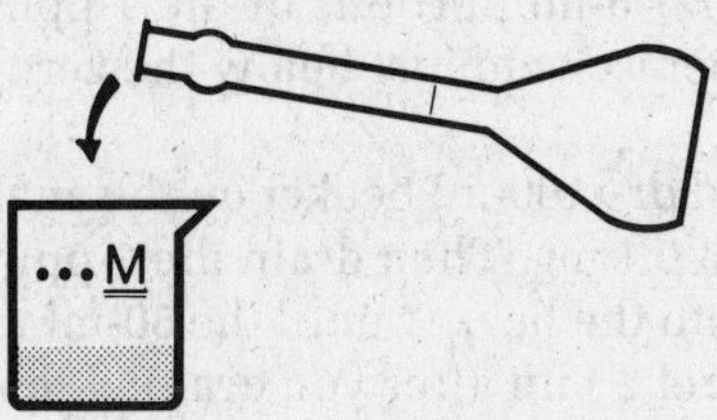

5. Pour the solution from the volumetric flask into a clean, dry 400-ml beaker labeled with the molarity calculated for the solution you have just prepared. Rinse a clean 50-ml buret twice with 10-ml portions of this solution, then fill it with the volume calculated for the preparation of 100 ml of a 1.3×10^{-4} M solution, plus an additional 5 ml. Record the liquid volume in the buret. Save the solution remaining in the beaker.

6. Rinse your volumetric flask thoroughly with distilled water. Drain the volume of solution calculated to prepare 100 ml of the 1.3×10^{-4} M solution as accurately as possible from the buret into the volumetric flask. Record the final buret reading. Fill the flask with distilled water as before, stopper and mix thoroughly by inverting and swirling the solution repeatedly.

Calculate the molarity of the solution you have prepared in procedure 6 to three significant figures. Since you do not know the density of a 1-ppm $CuSO_4 \cdot 5H_2O$ solution that you should prepare, proceed under the assumption that a solution containing 1.00 mg of $CuSO_4 \cdot 5H_2O$ per 1 liter of solution is equal in solute content to a 1-ppm solution containing 1.00 mg per 1000 g of solution. Calculate the molarity of a solution containing 1.00 mg of $CuSO_4 \cdot 5H_2O$ per 1 liter of solution. Finally, calculate the volume of the solution prepared in procedure 6 which is required to make 100 ml of a solution containing 1.00 mg of $CuSO_4 \cdot 5H_2O$ per 1 liter of solution. Check with your instructor before you proceed.

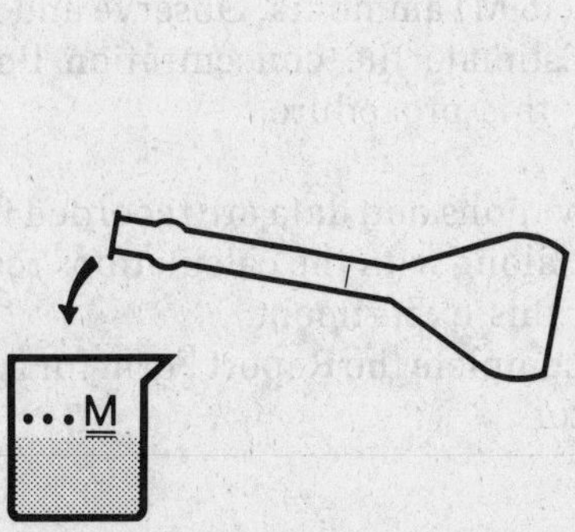

7. Empty the buret and rinse it three times with ~10-ml portions of the solution from the flask. Fill the buret to around the 40-ml mark with this solution, drain about 5 ml into a "waste solution" beaker, and pour the remaining solution from the flask into a clean, dry 250-ml beaker labeled with the solution molarity. Save this beaker containing the solution.

8. Rinse the volumetric flask thoroughly with distilled water. Record the buret reading, then drain the volume of solution calculated for preparation of the 1-ppm solution (from the buret into the flask). Record the final buret reading. Fill the flask to the mark with distilled water, stopper, and mix as before.

9. Rinse a buret thoroughly first with distilled water and then three times with ~5-ml portions of the 1-ppm solution and adjust the meniscus to just below the zero mark.

10. Weigh a *clean* and *dry* 100-ml beaker on the analytical balance to the nearest 0.1 mg. Then drain the 1-ppm solution from the buret into the beaker until the 50-ml mark is reached. Read the buret 2 min after you drained the buret to the 50-ml mark. Reweigh the beaker containing the solution to the nearest 0.1 mg. Make sure that no extraneous materials (water or dirt from the lab bench) adhere to the beaker during these operations.

Calculate the density of your final solution and its concentration in ppm. Obtain from a handbook or, if so instructed, determine experimentally the density of distilled water at the temperature of the laboratory. Compare the density of distilled water with the density of the solution.

Check the validity of the assumption made earlier that a solution containing 1.00 mg $CuSO_4 \cdot 5H_2O$ per 1 liter of solution is equal in solute content to a 1.00-ppm solution containing 1.00 mg of $CuSO_4 \cdot 5H_2O$ per 1000 g of solution. Proceed in the following manner: Since the density of a solution does not change significantly with small changes in concentration, use the density of your most dilute solution, which you have determined experimentally, as the density of a "real" 1.00-ppm solution. Calculate the molarity of a "real" 1.00-ppm solution and its $CuSO_4 \cdot 5H_2O$ content per 1000 g of solution. Compare these values with the molarity of the solution containing 1.00 mg $CuSO_4 \cdot 5H_2O$ in 1 liter of solution. Comment on the validity of the assumption in the Report Form under the heading "Molarity-ppm assumption."

11. Copper(II) ions form an intense blue colored tetraamine complex, $[Cu(NH_3)_4]^{2+}$, with ammonia, the formation of which can be used to detect Cu^{2+} in fairly dilute solutions. The aqueous (hydrated) Cu^{2+} ion has a characteristic light blue color. Do you note this in the three solutions you have prepared? Pour 1 ml of each of your three solutions into separate small test tubes. To each, add 1 ml of concentrated (15 *M*) ammonia. Observe and record your observations. Estimate the concentration limit for detection of Cu^{2+} by this procedure.

CALCULATIONS AND REPORT

Be sure that all observations and data are recorded in your permanent notebook, along with the calculations required for various stages of this experiment.

Use this record to complete the Report Form for submission to your instructor.

GRADE

INSTRUCTOR'S INITIALS

Report Form

FOR INVESTIGATION 12:

Copper(II) and Algae

NAME: ______________________ ID NO.: ______________________

LAB SECTION ______________________ INSTRUCTOR ______________________

Show setups on attached pages.

Calculated Values

Mass of $CuSO_4 \cdot 5H_2O$ needed for 100 ml of 4.0×10^{-3} *M* solution ______ g

Formula weight of $CuSO_4 \cdot 5H_2O$ ______ g mol^{-1}

Concentration of first solution ______ *M*

Volume of first solution needed to make 100 ml of 1.3×10^{-4} *M* solution ______ ml

Concentration of second solution ______ *M*

Volume of second solution needed to make 100 ml of 1.00 ppm solution ______ ml

Molarity of a solution containing 1.00 mg $CuSO_4 \cdot 5H_2O$ per liter ______ *M*

Concentration of your algaecide solution ______ *M*; ______ ppm

Laboratory temperature ______ °C

Density of water at this temperature ______ g ml^{-1}

Experimental Values

Paper + crystals ______ g

Paper ______ g

Crystals ______ g

First solution

Final buret reading ______ ml

Initial buret reading ______ ml

Volume added ______ ml

Second Solution:

Final buret reading ______ ml

Initial buret reading ______ ml

Volume added ______ ml

Density of your algaecide solution

Final buret reading ______ ml

Initial buret reading ______ ml

Volume added ______ ml

Beaker + solution ______ g

Beaker ______ g

Solution ______ g

Density of solution ______ g ml^{-1}

(OVER)

Molarity-ppm Assumption

"real" 1-ppm solution ____________ M ____________ mg $CuSO_4 \cdot 5H_2O$ per 1000 ml

1 mg per 1000-ml solution ____________ M ____________ mg $CuSO_4 \cdot 5H_2O$ per 1000 ml

Observations and Conclusions on Detection of Cu^{2+} (Procedure 11)

Estimate the concentration of Cu^{2+} that can just be detected with NH_3: ____________ M.

Remember that Cu^{2+} is only toxic to algae above a certain concentration range. Within another concentration range, Cu^{2+} is essential to algae growth. For certain algae, the optimum "growth concentration" is 2.0×10^{-8} M. Calculate for a 5000-gal swimming pool (18,920 liters):

a. The number of kilograms of $CuSO_4 \cdot 5H_2O$ to produce the optimum concentration for algae "bloom."

b. The number of kilograms of $CuSO_4 \cdot 5H_2O$ for a "lethal concentration" of 1.0 ppm.

Show setups on attached page.

Answers **a.** ____________ kg **b.** ________ kg

DATE ____________ SIGNATURE ____________

Investigation 13

WATER HARDNESS

INTRODUCTION It is estimated that water—as vapor in the atmosphere, liquid in underground deposits, and liquid or solid (ice and snow) on the surface of the earth—constitutes about 5 percent of the total mass of our planet. Although more than 80 percent of the earth's surface is covered by water, *less than* 3 percent of this is suitable for human consumption or agricultural use, and this percentage decreases slightly every day. Most of the surface waters of the earth (the oceans and seas) contain dissolved salts that make them unfit for drinking, washing, or irrigation. In addition, many of our lakes, rivers, and streams have become so polluted by human and industrial wastes that these critical sources of "fresh" water are fast becoming inadequate for human needs. Underground water is estimated to be more than 40 times that in lakes and streams, but there are significant problems in "tapping" it for wide-scale use. The largest source (more than 80 percent) of fresh water is, unfortunately, also difficult to obtain conveniently—the ice and snow of the polar regions, northern glaciers, and high mountain snowpacks.

The problems of obtaining water pure enough for human use, and in sufficient quantity to meet increasing demands, are enormous. We have not yet even halted the pollution of natural waters, except in isolated cases, so the situation gets worse every year.

When compared to the other difficulties with our water supplies, the problem of water hardness* (due to the presence of ions such as Mg^{2+}, Ca^{2+}, Al^{3+}, and Fe^{3+}) seems a minor one. It is not, however, insignificant. Although moderately hard water is generally suitable for drinking or agricultural use, it does represent some difficulties. The cations responsible for water hardness, leached from minerals in the earth or added in inorganic chemical wastes, form insoluble salts with ordinary soaps. As a result, effective cleaning with hard water requires one of four alternatives:

a. Use of *excess* soap so that "hard-water cations" are removed by precipitation of their insoluble soaps.

b. Removal of "hard-water cations" prior to use of the water for cleaning, by use of an ion-exchange "water softener" or a precipitating agent such as "washing soda" (Na_2CO_3).

c. Use of a synthetic detergent, not precipitated by "hard-water cations," instead of a soap.

d. Use of some chemical additive such as the "phosphates" or "silicates," to complex with "hard-water cations."

*Generally considered as equivalent to a Ca^{2+} concentration exceeding 40 ppm.

All of these alternatives present some economic or ecological problems. Until 1965, for example, the accumulation of nondegradable synthetic detergents in natural waters threatened to destroy aquatic ecosystems. Since that year, federal regulations require that all detergents must be of the biodegradable type. Even with that safeguard, there are many areas in which detergents are dumped into lakes and streams at rates greater than those of their removal by the natural degradation systems. "Phosphates" and other additives have been linked to increasing rates of algae growth. Water "softening" by ion exchange or addition of precipitating agents is generally "safe" from an ecological standpoint, but these methods are not inexpensive. When you consider that in the United States alone, nearly 100-*billion gallons* of water *per day* are used for some type of washing, you can see why the problems of hard water are not insignificant.

To determine the best way of handling a hard-water problem in a specific water supply, in terms of both economic and ecological criteria, it is necessary to know the "degree of hardness" of the water; that is, we must have a way of determining the concentration of dipositive and tripositive ions in the water. This experiment is concerned with a simple method for measuring "water hardness."

Before starting this investigation, you should be familiar with the characteristics of "hard" and "soft" water, with the general nature of soaps and synthetic detergents, with the principles of titration, and with the use of "ppm" as a concentration unit.

References

1. O'Connor, Rod. 1977. *Fundamentals of Chemistry* (2nd ed.) New York: Harper & Row (Units 6, 13, and 14).
2. Dreisbach, Dale, 1966. *Liquids and Solutions*. Boston: Houghton Mifflin.

SAFETY It is necessary in this experiment to heat a fairly large volume of ethanol. *This liquid is highly flammable.* Be extremely careful if a burner is used. (A steam bath is best. A hotplate, set on *low*, is satisfactory. An open flame is the last resort.) Wear eye protection at all times and be sure you are prepared to smother a fire if the ethanol should ignite.

OBJECTIVES

1. Be able to weigh and transfer a solid reagent accurately.
2. Be able to heat a flammable liquid safely.
3. Be able to read a buret and perform a titration accurately, using an indicator to detect completion of a reaction.

PROCEDURES

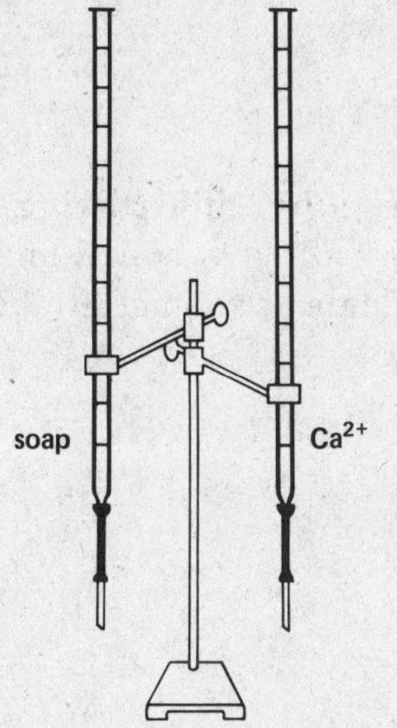

Obtain two 50-ml burets and check that they will drain without leaving water droplets adhering to the inside surface. If necessary, clean the burets as described in Appendix A. Attach the burets to a ringstand by buret clamps mounted at angles of about 45 deg to the direction of the ringstand base.

A. Preparation of Soap (Potassium Palmitate) Dispersion

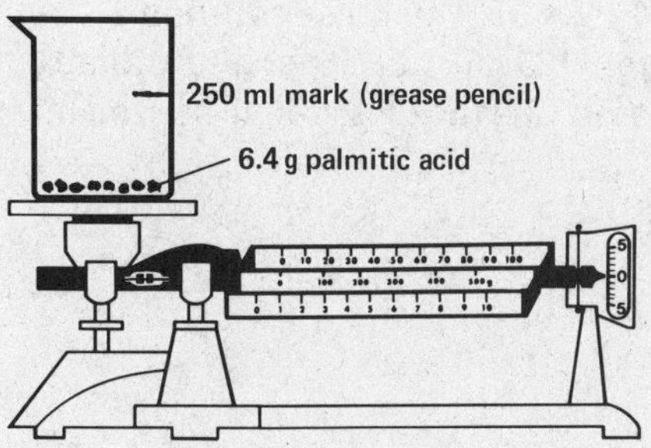

1. Measure 250 ml of distilled water into a 400-ml beaker and mark the resulting liquid level with a grease pencil on the outside of the beaker. Then pour the water into a clean container for later use and dry the inside of the beaker with a paper towel. Weigh this clean, dry 400-ml beaker on a laboratory balance to the nearest 0.1 g. Using a *clean* spatula or scoopula, carefully transfer 6.4 g of palmitic acid into the beaker. Record pertinent masses in your notebook. Remove the beaker from the balance.

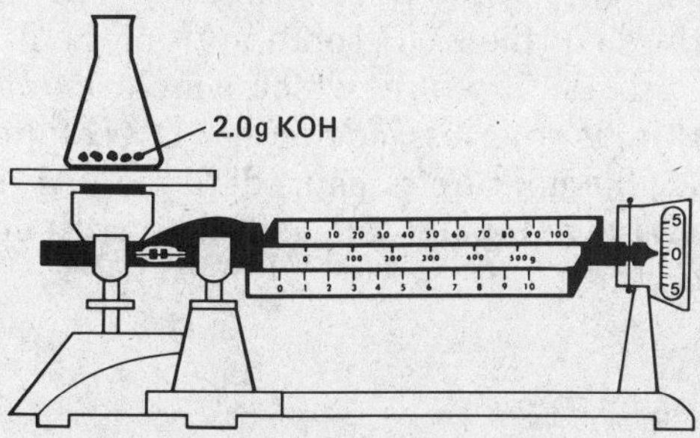

2. Weigh a clean, dry 125-ml Erlenmeyer flask to the nearest 0.1 g. Transfer into it about 2.0 g of potassium hydroxide pellets and reweigh to ± 0.1 g.

Caution: Avoid contact with these pellets. If you spill any, wipe them up immediately with a damp paper towel and dispose of them as directed by your instructor. In case of contact, wash immediately and thoroughly with water.

Remove the flask from the balance and add 25 ml of 95 percent ethanol.

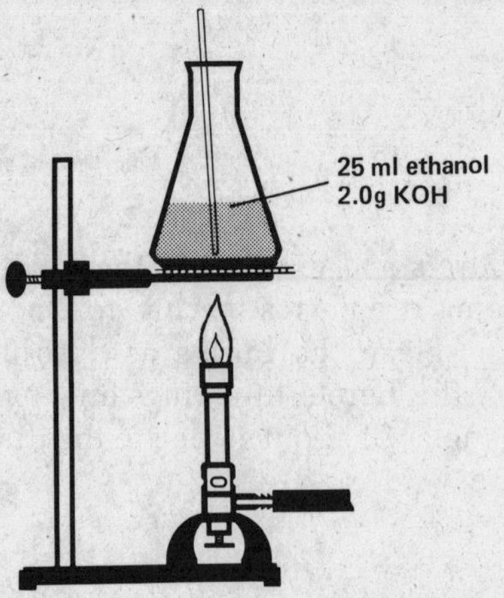

3. Warm the mixture *carefully,* while stirring, until all the KOH has dissolved. Do not allow to boil. Remove the flask from the vicinity of the flame as soon as all the potassium hydroxide has dissolved.

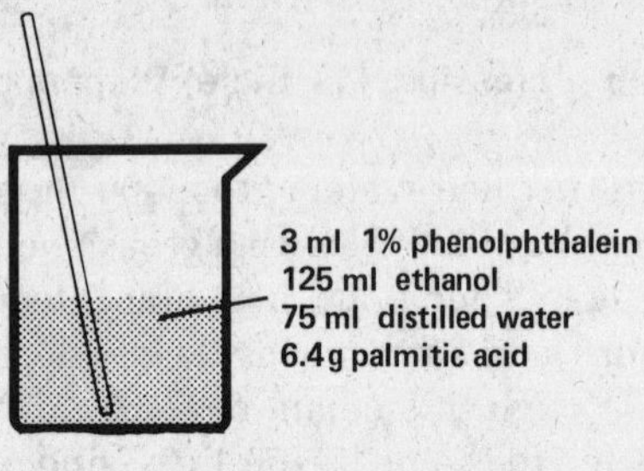

4. Into the 400-ml beaker containing the palmitic acid pour sequentially 3 ml of a 1-percent solution of phenolphthalein in ethanol, 125 ml of 95 percent ethanol, and 75 ml of distilled water. Use the glass rod to stir the mixture thoroughly.

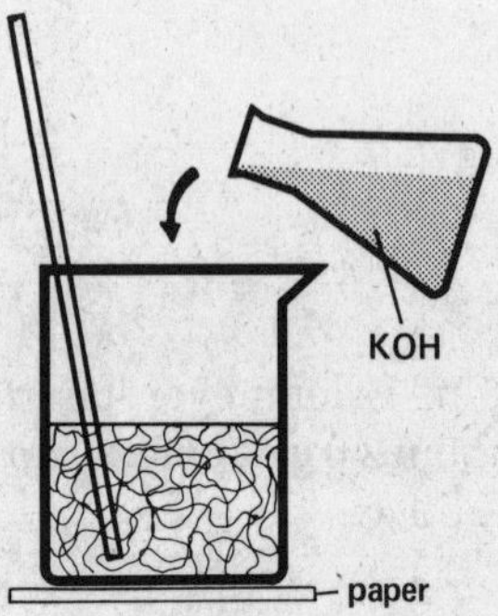

5. Place the 400-ml beaker containing the palmitic acid suspension on a sheet of white paper. Carefully pour not more than two-thirds of the KOH solution from the flask into the beaker. Stir the mixture, which should be dark red. (If your mixture is colorless, add more KOH solution, with stirring, until the mixture remains dark red.) At this point, some undissolved palmitic acid will still float around in the mixture.

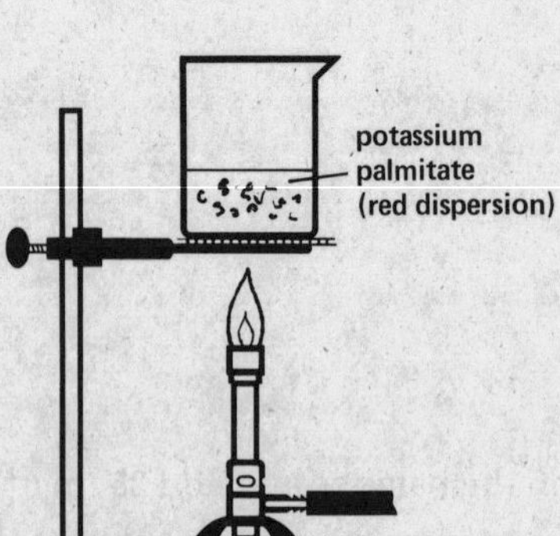

6. Warm the beaker containing the red dispersion *carefully,* while stirring, until the red color completely disappears. Then remove the beaker from the vicinity of the flame.

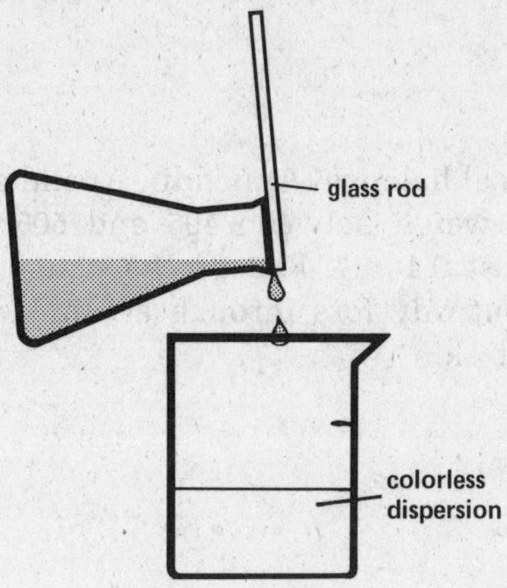

7. Then place a glass rod across the mouth of the flask containing the KOH solution, tilt the flask until liquid contacts the rod, and use this to add the KOH solution dropwise to the beaker, with stirring, *just* until the suspension in the beaker attains a permanent pink color that remains on continued stirring. At this point, if undissolved palmitic acid is still observed, repeat steps 6 and 7 until you obtain a clear potassium palmitate dispersion that is distinctly pink.

Should your dispersion become clear (no palmitic acid particles visible anywhere) but remain *dark red* in procedure 6 (caused by adding too much KOH solution in step 5), add one drop of 6 *M* HCl to the disperson. The addition of acid will cause some palmitic acid to precipitate. Heat the mixture until this precipitate dissolves. Should the dispersion still be red, add another drop of the HCl. Repeat this procedure until the solution becomes colorless. Then carefully repeat step 7.

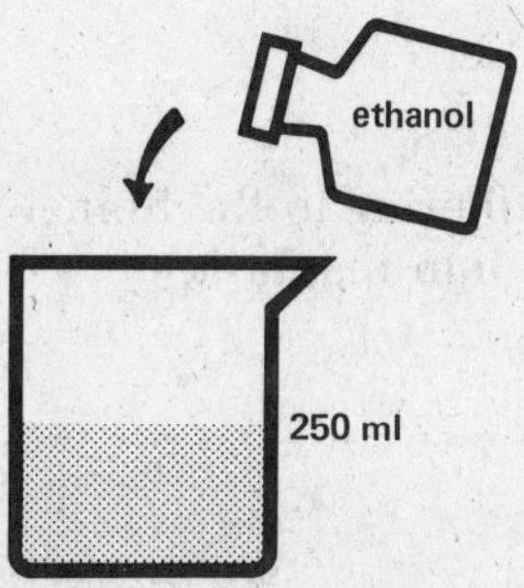

8. Set the beaker on an insulating pad, add 95 percent ethanol to the 250-ml mark on the beaker, and stir thoroughly. Pour the soap dispersion into a clean, dry 500-ml Erlenmeyer flask and stopper the flask *loosely*.

B. Standardization of the Potassium Palmitate Dispersion

The potassium palmitate is to be standardized against a solution containing a known concentration of Ca^{2+} ion. Your instructor will tell you whether the standard Ca^{2+} solution is available or whether you will have to prepare it. If it is available, clean and rinse (with distilled water) a 250-ml beaker and wipe it completely dry inside and out with a clean towel. Obtain in this beaker about 150 ml of the standard Ca^{2+} solution. If it is necessary for you to *prepare* the standard Ca^{2+} solution, follow carefully procedures B1–B5. Otherwise start at B6. First be sure that you are completely familiar with the use of a precision analytical balance, Appendix A.

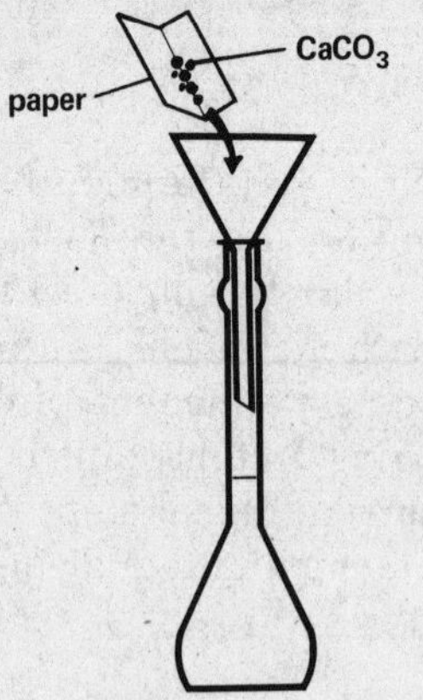

1. Using the analytical balance (Appendix A) and a piece of weighing paper, weigh between 495 and 505 mg of $CaCO_3$ to the nearest 0.1 mg. Record the actual mass. Transfer this *without any loss* through a funnel into a 100-ml volumetric flask.

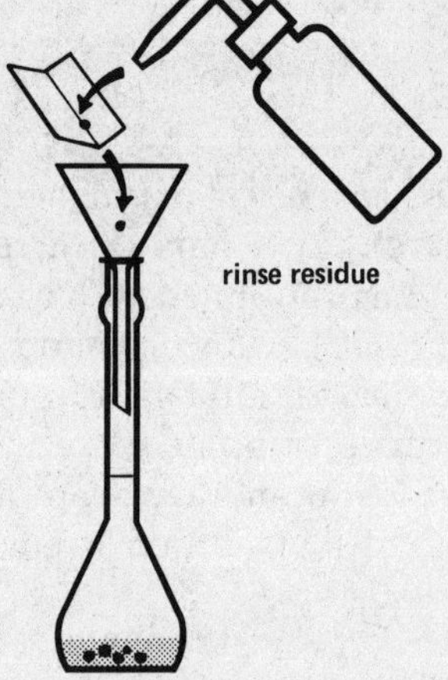

2. Rinse any residual particles from the weighing paper into the flask with a stream of distilled water from the squeeze bottle.

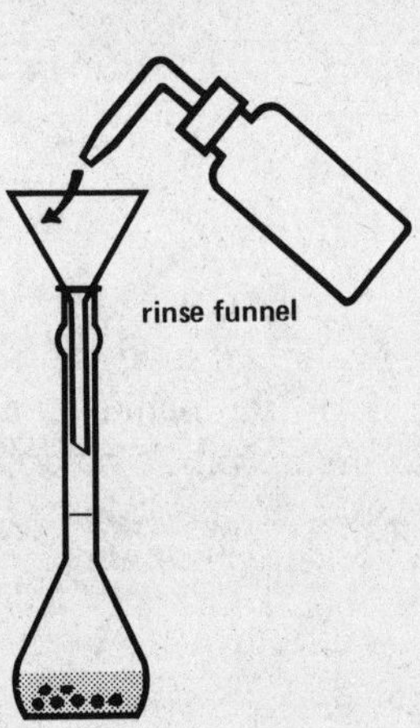

3. Rinse any material adhering to the funnel with about 10 ml of distilled water into the flask.

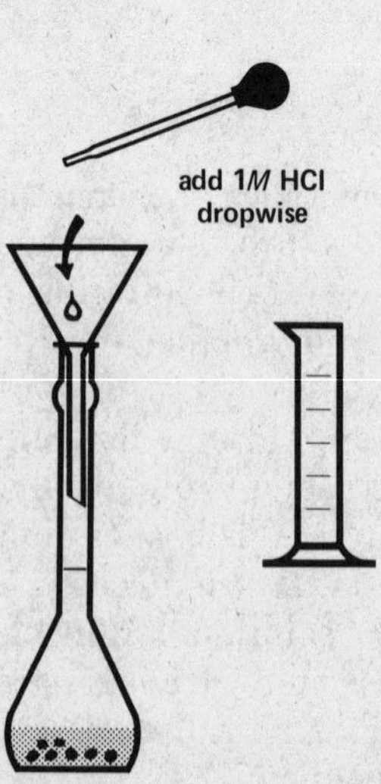

4. Prepare 16 ml of 1 *M* HCl by carefully adding, with stirring, 1.5 ml of 12 *M* HCl to 15 ml of distilled water contained in your graduated cylinder. Add 1 *M* hydrochloric acid to the flask, drop by drop, *slowly,* with swirling of the flask contents after each addition, until CO_2 evolution ceases. Continue this process just until *all* $CaCO_3$ has dissolved. This is critical.

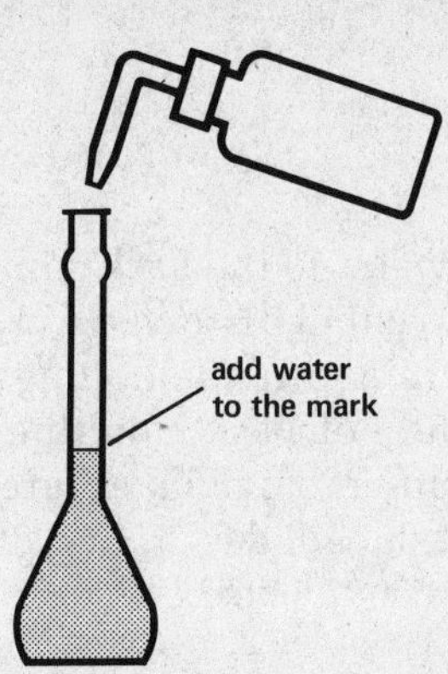

5. Now fill the volumetric flask *to the mark*, with distilled water, using a dropwise addition for the final increments. Stopper the flask tightly and shake thoroughly to mix the contents completely.

Now you are ready to standardize the potassium palmitate dispersion using a standard Ca^{2+} solution.

6. Prepare 6 ml of 1 *M* NaOH solution by adding 1.0 ml of your 6 *M* NaOH solution from your dropping bottle to 5 ml of distilled water contained in your graduated cylinder. Mix well and transfer the solution to a large test tube. Label the test tube "1 *M* NaOH." Clean the graduated cylinder with tap water and then rinse it with distilled water.

To prepare a 0.1 *M* HCl solution, place 11 ml of distilled water into your graduated cylinder. Add four drops (each drop has a volume of approximately 0.05 ml) of 6 *M* HCl from your dropping bottle and mix well. Store this solution in a large test tube labeled "0.1 *M* HCl."

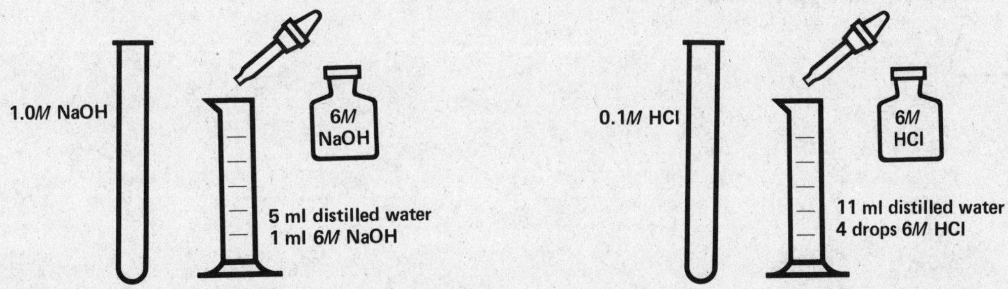

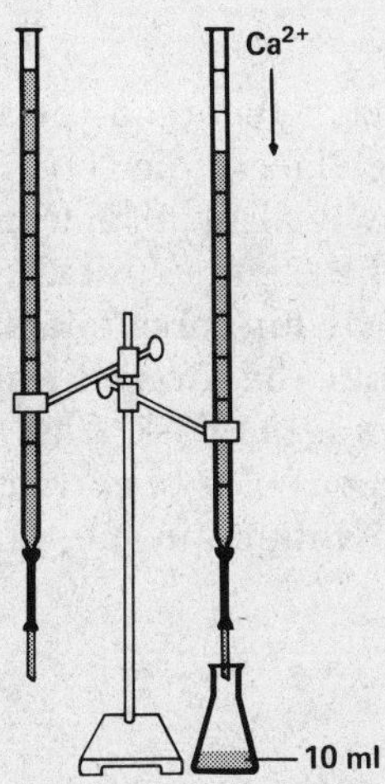

7. Rinse one buret twice with 5-ml portions, then fill to above the zero mark with the potassium palmitate dispersion. Rinse the second buret twice with 5-ml portions, then fill to above the zero mark, with standard Ca^{2+} solution. Allow just enough liquid to drain into a waste flask from each buret to expel any air from the buret tips and drop the liquid levels to slightly below the zero marks. Record the readings of both burets, then drain about 10 ml of the standard Ca^{2+} solution into a clean 125-ml Erlenmeyer flask. Record the buret reading after this addition.

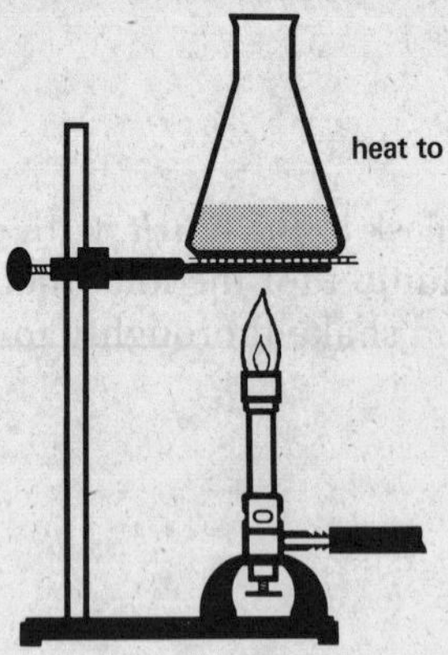

8. Add 50 ml of distilled water to the flask containing the standard Ca^{2+} solution. To expel dissolved CO_2, heat this solution to boiling and continue gentle boiling for 30 sec. Then carefully remove the hot flask from the ringstand and allow it to cool to around room temperature. To speed up the cooling process, set the hot flask into 200 ml of cold tap water contained in a 600-ml beaker.

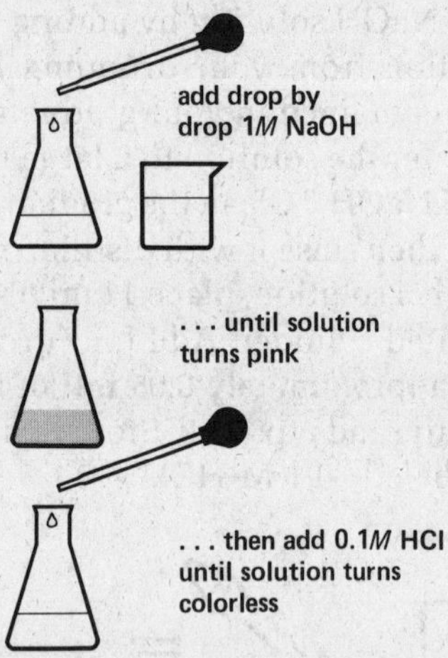

9. Add to the Ca^{2+} solution 10 drops of 1 percent phenolphthalein indicator. If the solution turns pink, add 0.1 *M* HCl, drop by drop with swirling, just until the color disappears. If the solution did not turn pink on addition of the indicator, add 1 *M* NaOH solution dropwise just until a permanent pink color forms. Then add 0.1 *M* HCl dropwise, with swirling, until the pink color just disappears.

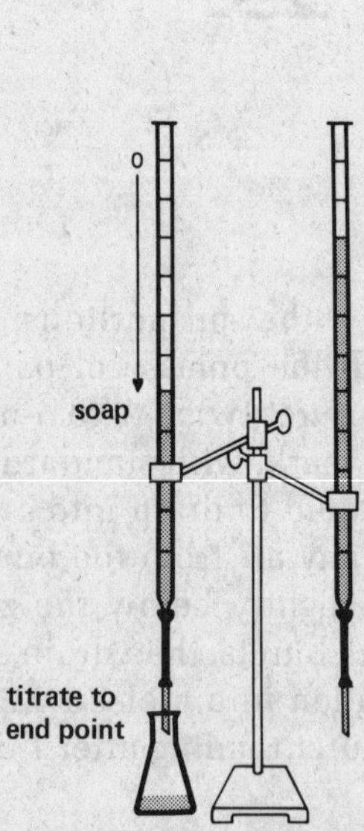

10. Now check that you have recorded the level in the potassium palmitate buret. Place a piece of white paper under the flask. Titrate slowly, dropwise, from this buret into the flask containing the Ca^{2+} solution, swirling the flask during the titration. A white precipitate of calcium palmitate will form. Continue the titration until a permanent pink color is observed in the flask. Check with your instructor that your endpoint color is correct, then read and record the final liquid volume in the buret.

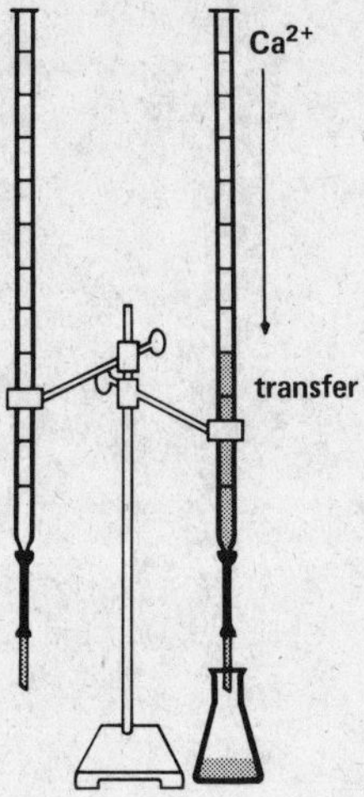

11. Repeat the titration twice, but using 35 ml of the standard Ca^{2+} solution with 25 ml of distilled water each time. Refill the burets before each titration. Use the information from your initial titration to estimate the volumes of potassium palmitate dispersion needed for these two titrations. Initial addition can be fairly rapid, but use a slow dropwise addition as you near the estimated titrant volume. Use the endpoint color from the first (10 ml Ca^{2+}) titration for comparison.

From the titrations in procedure 11 and the known composition of the standard Ca^{2+} solution, calculate the milligrams (Ca^{2+}) corresponding to 1.00 ml of your potassium palmitate dispersion. The formula for calcium palmitate is $Ca(C_{15}H_{31}CO_2)_2$.

C. Determination of Water Hardness

(Follow the appropriately modified procedures B8-B10.)

1. *Distilled water.* With your 25-ml graduated cylinder, measure 60 ml of distilled water into a clean, dry 125-ml Erlenmeyer flask and titrate this with your standarized potassium palmitate solution.
2. *Tap water.* Measure 60 ml of tap water into a clean, dry 125-ml Erlenmeyer flask and titrate as before.
3. *Unknown solution.* Obtain from your instructor a precisely measured volume (which will not be told to you) of an unknown solution in a clean 125-ml Erlenmeyer flask. Measure 50 ml of distilled water into the Erlenmeyer flask. Titrate this solution as before. Repeat with a second sample of the same unknown.

When you have finished your titrations, pour all solutions containing any ethanol into the specified "waste ethanol" container. Most of this can be "recycled" for use in other experiments.

CALCULATIONS AND REPORT

Calculate the "hardness" of the distilled water and tap water in ppm Ca^{2+}, and that of the unknown solution in terms of milligrams Ca^{2+}. (Parts per million Ca^{2+} represents as a good approximation milligrams Ca^{2+} per liter of the solution tested.) To do this, you need to use the equivalency (milligram Ca^{2+} corresponding to 1.00 ml of the dispersion) determined in your standardization of the potassium palmitate dispersion. (Note that we report hardness in terms of Ca^{2+} content even though other cations may be involved.)

Include all data and calculations in your permanent notebook and record those items requested also on the Report Form.

GRADE / INSTRUCTOR'S INITIALS

Report Form

FOR INVESTIGATION 13:

Water Hardness

NAME: ______________________ ID NO.: ______________________

LAB SECTION ______________________ INSTRUCTOR ______________________

Concentration of standard Ca^{2+} solution ______________ mg (Ca^{2+}) liter^{-1}

Titration Data **Unknown Number**__________

	Potassium palmitate dispersion titrant					
	Standard Ca^{2+} (trial 1)	Standard Ca^{2+} (trial 2)	Distilled water	Tap water	Unknown (trial 1)	Unknown (trial 2)
Initial volume						
Final volume						
Volume added						

Calculations

Show setups on reverse side.

1. Standardization of potassium palmitate (average of two trials)

______________ mg Ca^{2+} = 1.00 ml potassium palmitate dispersion

2. Water hardness determinations

a. Distilled water ______________ ppm Ca^{2+}

b. Tap water ______________ ppm Ca^{2+}

c. Unknown (average of two trials) ______________ mg Ca^{2+}

______________ ______________

DATE SIGNATURE

Investigation 14

MOLECULAR WEIGHTS OF SOME PHARMACEUTICAL CHEMICALS

INTRODUCTION A drug is defined in pharmacology as a chemical substance of known structure that has a therapeutic use. An enormous variety of synthetic drugs are now produced by the pharmaceutical industries. Some familiar examples follow.

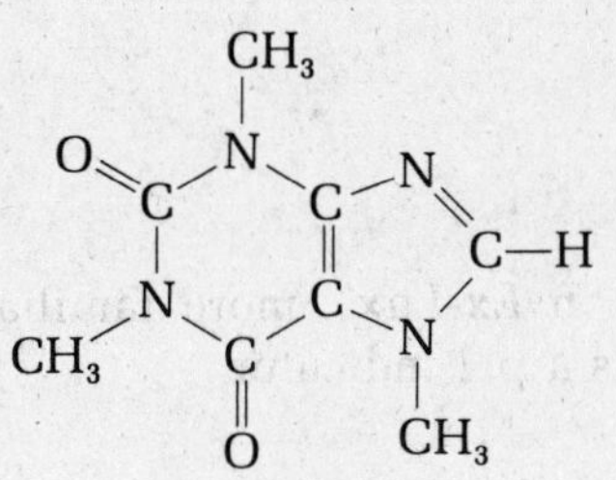
caffeine

An ingredient of coffee, tea, and various "cola" beverages; used as a stimulant drug in a number of nonprescription medications, such as *No-Doz* and *Anacin*.

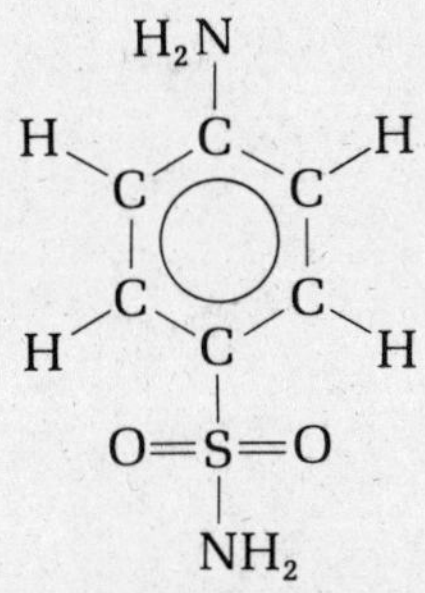
sulfanilamide

An antibiotic, one of the simpler *sulfa drugs*.

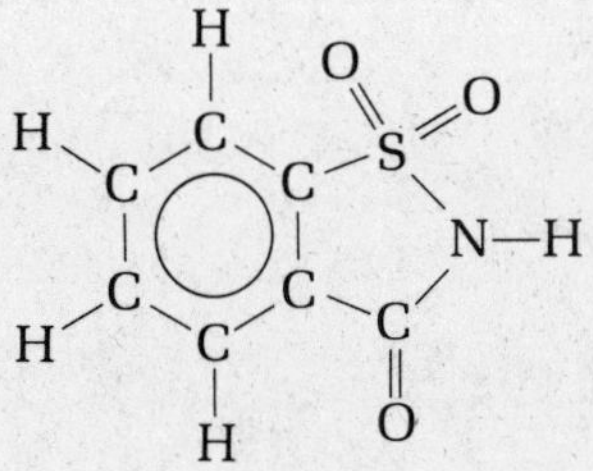
saccharin

A nonnutritive sweetener, used in many dietary foods and beverages.

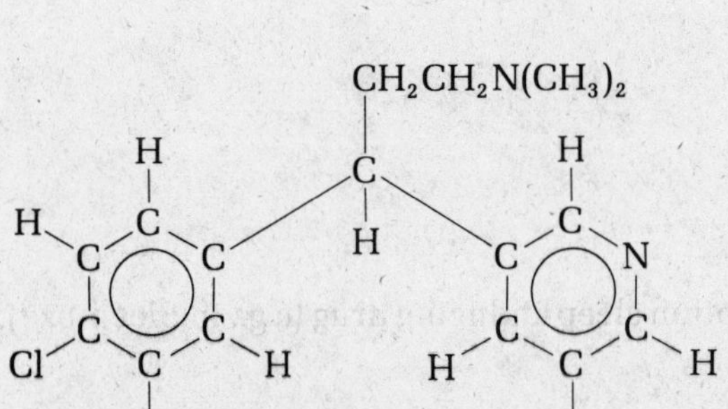
chlorpheniramine

A nonprescription antihistamine (e.g., an ingredient of *Coricidin*).

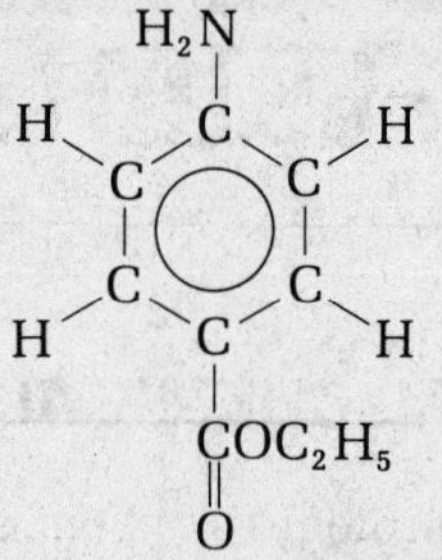

benzocaine

A local anesthetic, often used in topical ointments and solutions to reduce localized pain or irritation (e.g., in *Chiggerex* ointment).

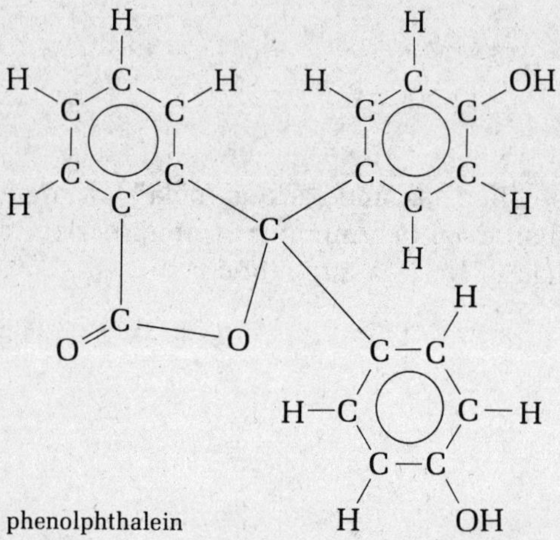

phenolphthalein

A laxative drug (e.g., in *Ex-Lax*), more familiar in the chemical laboratory as a pH indicator.

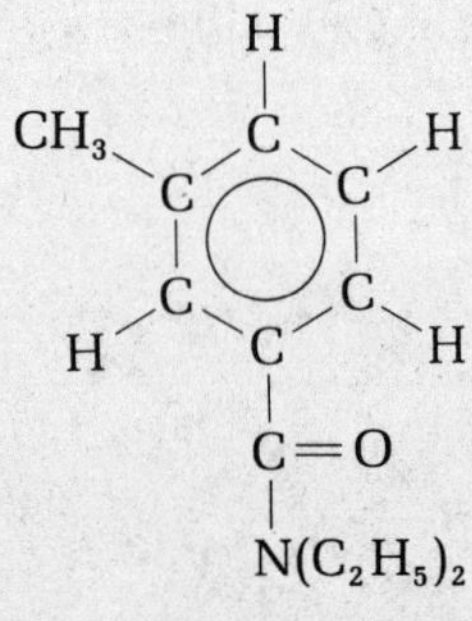

N, N-diethyl-*meta*-toluamide

An insect repellant (e.g., in *Off!*).

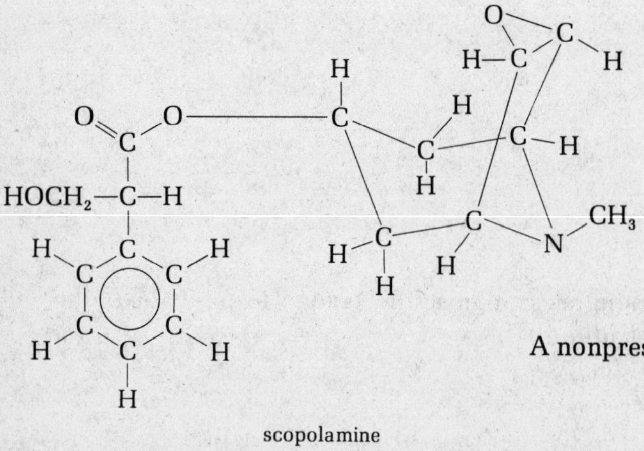

scopolamine

A nonprescription sleep-inducing drug (e.g., in *Sleep-eze*).

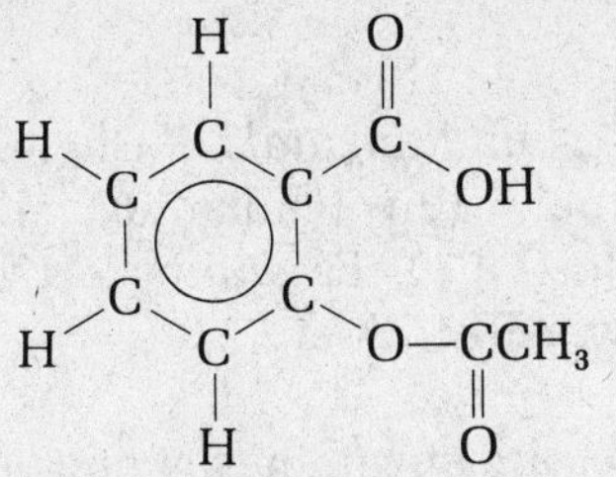

acetylsalicylic acid

Analgestic and antipyretic drug (*aspirin*), most commonly used as the sodium or calcium salt.

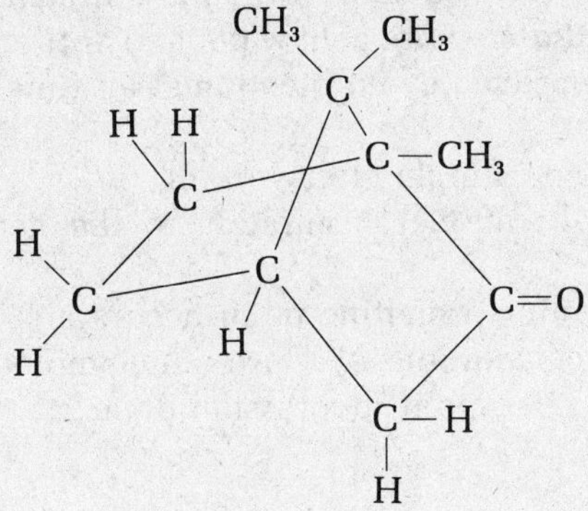

camphor

A stimulant and irritant drug of diverse applications ranging from mothproofing agents for fabrics to nasal decongestant preparations.

In spite of extensive research, we still know very little about the mechanisms of the actions of various drugs in the body. There are a number of theories, but none that have proved entirely satisfactory. This means that we cannot manufacture a drug for a specific purpose simply by selecting the structural features of a molecule needed for a particular biochemical function. Although we do know a few things about structure-function relationships, and more is being learned every day, this knowledge is not yet extensive enough to permit reliable drug design. Exhaustive testing of every new pharmaceutical chemical is necessary to determine its value as a drug, as well as possible adverse side effects that must be avoided or minimized.

Although some drugs, such as insulin, consist of very large molecules and others, such as cyclopropane (a general anesthetic), are quite small, most of the common drugs have molecular weights in the 100–400 range. In this experiment we will determine the approximate molecular weight of a familiar pharmaceutical chemical, using the melting-point depression of a camphor sample in which the "unknown" chemical has been dissolved.

Before starting this investigation, you should be familiar with the calculation of a molecular weight on the basis of the structural formula and with the methods used to determine approximate molecular weights from melting-point depression data. In addition, *before coming to the laboratory,* you must calculate the molecular weights for the "sample pharmaceutical chemicals" given in the Introduction to this investigation and for any compounds added to this list by your instructor.

References

1. O'Connor, Rod. 1977. *Fundamentals of Chemistry* (2nd ed.). New York: Harper & Row (Units 7, 8, and 15.).
2. Korolkovas, Andrejus. 1970. *Essentials of Molecular Pharmacology.* New York: Wiley.

SAFETY *Under no circumstances* are any of the chemicals used for this experiment to be *tasted, swallowed,* or *taken from the laboratory.* Many of them are toxic and all are hazardous for unsupervised use. Be careful also in working with the hot-oil bath. In case of contact with hot oil, soak the affected area *immediately* with cold water and notify your instructor. Wear approved eye protection at all times.

OBJECTIVES

1. Be able to do some simple glassworking.
2. Be able to weigh samples accurately to the nearest 0.1 mg.
3. Be able to determine a melting point accurately.
4. Be able to make calculations of approximate molecular weights from melting-point depression data.

PROCEDURES

1. Use a hot burner flame to seal one end of a 30-cm length of 4 mm (ID) soft glass tubing, rotating the tube slowly while heating.

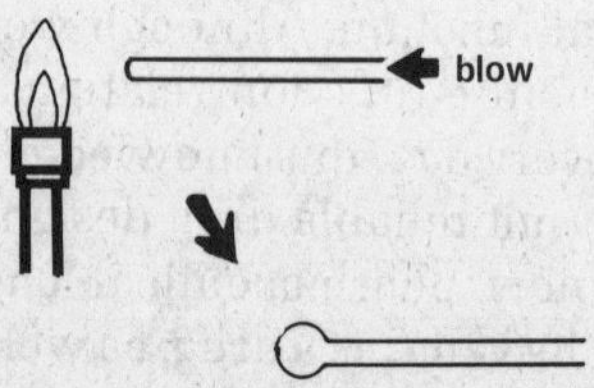

2. When the end of the tube has sealed, and while the glass is still soft, remove the tube from the flame and blow into the cool open end to form a bubble at the sealed end, about 1.5 cm in diameter.

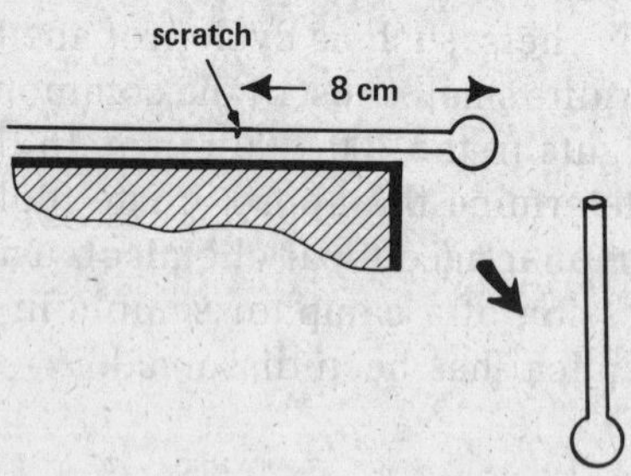

3. Place the hot tube on an asbestos pad to cool, or lean the tube against a ringstand base so that the hot end is above the base. (Do not place hot glass on the bench top.) When the tube has cooled, place it flat on the bench top with the bulb extending over the edge and scratch the tube with a file 8 cm from the sealed end. Carefully snap the tube at the scratch mark. Fire polish the end of the sealed tube.

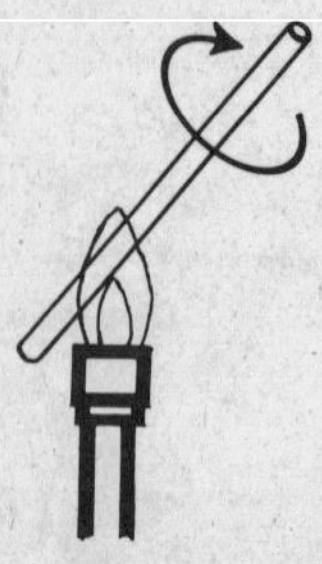

4. Hold the remaining unsealed tube at an angle along the edge of the hot inner cone of the burner flame, rotating the tube and moving it back and forth, to heat a 4-cm length near the middle of the tube until it is soft enough to bend of its own weight.

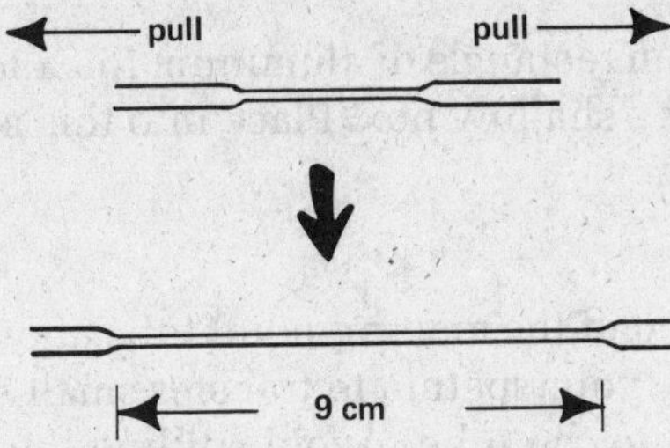

5. Then remove the tube from the flame and pull at both ends (not *too* strongly) to produce a capillary of 2 to 3 mm diameter at least 9 cm long.

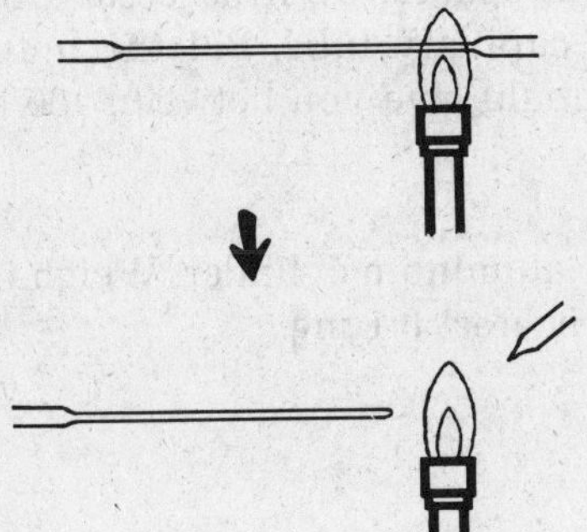

6. Heat one end of the capillary length and pull away the larger tubing. Then round off the end of the capillary neatly in the flame.

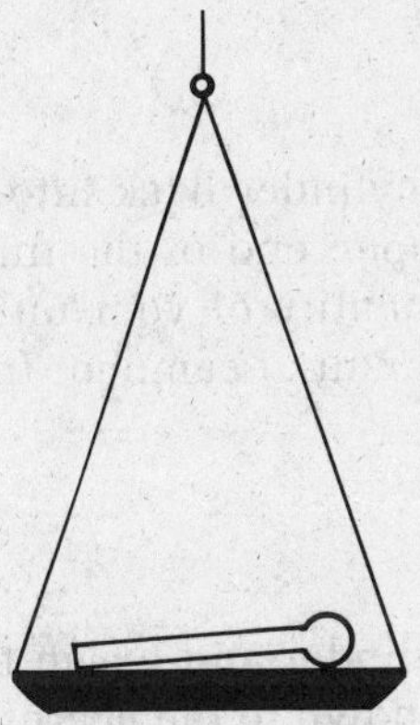

7. Be sure your melting-point tube (with the bulb) is cool. Then weigh it to the nearest 0.1 mg.

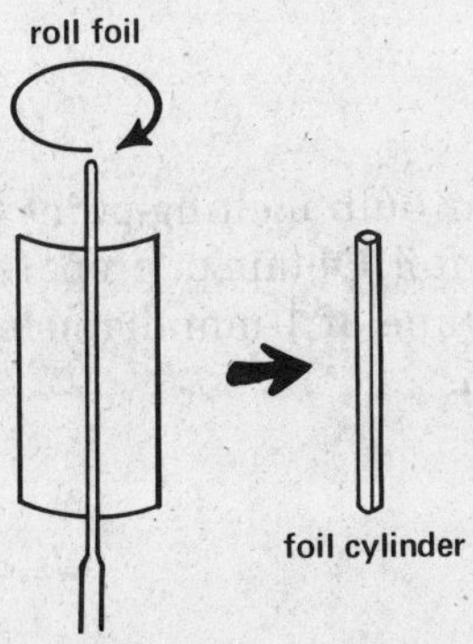

8. Cut a 9-cm × 2.5-cm rectangle of aluminum foil and fold it into a roll around the capillary tube.

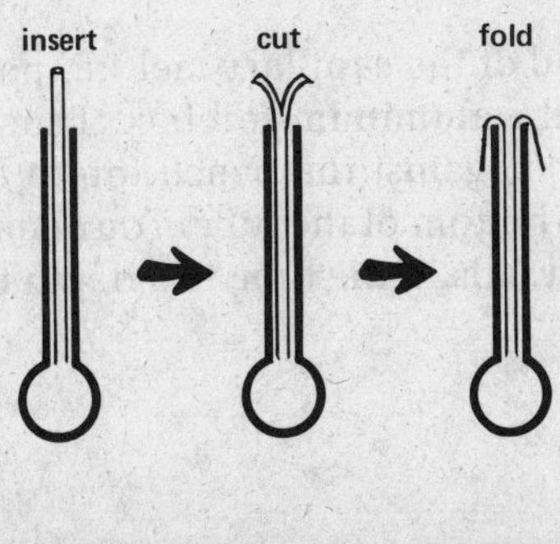

9. Insert the aluminum-foil cylinder into the melting-point tube until the end just extends into the upper part of the bulb. Cut the part extending from the tube in half and fold it over the edge of the tube.

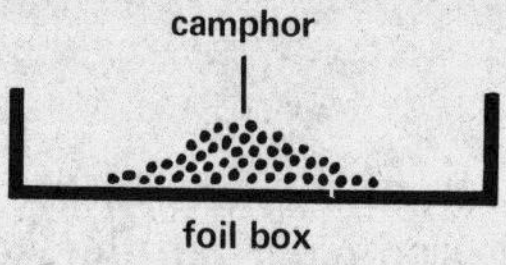

10. Cut a 9-cm × 7-cm rectangle of aluminum foil and fold up the edges to form a shallow box. Place into this a spatula load of camphor.

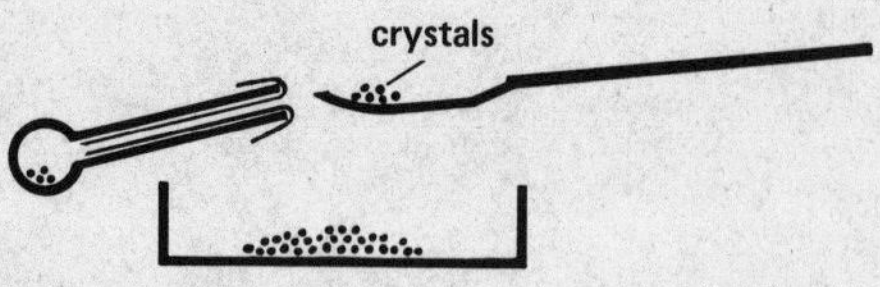

11. With the open end of the melting-point tube above the camphor sample, use your spatula to transfer small increments of camphor into the tube. Avoid spilling the camphor. If crystals adhere to the tube, tap the tube (*not* the bulb) gently with your spatula or, if necessary, push the crystals in with the capillary tube. Fill the bulb three-quarters full. This should give you between 100 and 300 mg of camphor.

Remove and save the aluminum cylinder. Weigh the tube and camphor to the nearest 0.1 mg.

12. Take a clean, dry watch glass to your instructor to obtain a sample of your crystalline unknown.

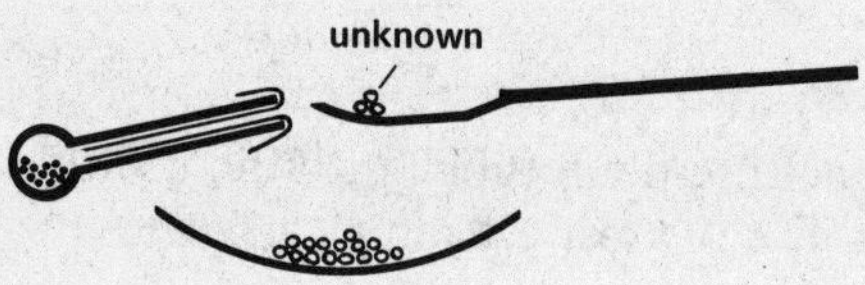

13. Insert the aluminum cylinder back into the melting-point tube. Holding the open end of the tube above the watch glass, transfer a quantity of your unknown equal to about one-tenth the amount of camphor (no more) into the tube.

Remove the aluminum cylinder and weigh the tube plus the camphor plus the unknown to the nearest 0.1 mg.

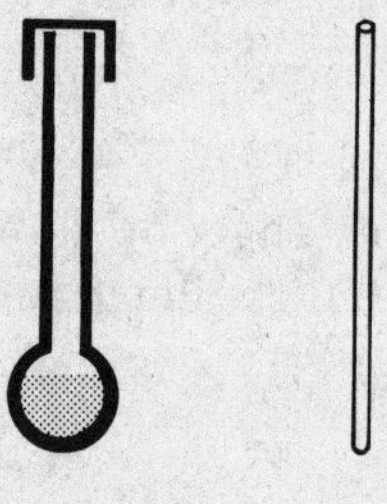

14. Cover the top of your bulb melting-point tube with a small piece of aluminum foil. Obtain or prepare a standard *capillary* melting-point tube of 1-mm diameter and 3-cm length sealed at one end.

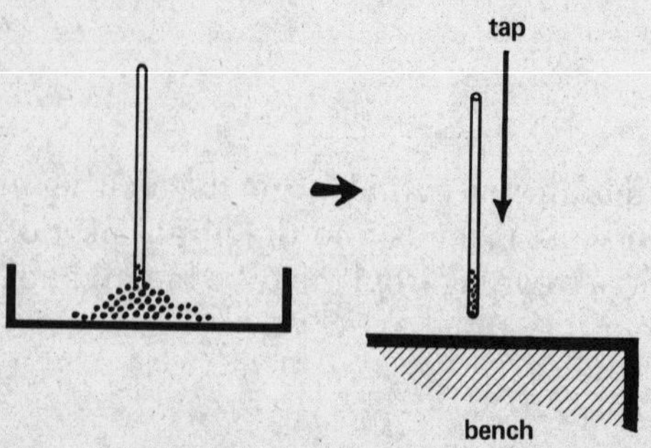

15. Press the open end of the capillary melting-point tube into the camphor in the aluminum-foil box. Then tap the closed end of the tube against the bench top to help the crystals slide into the bottom of the tube. You should have a compact mass of crystals in the tube to a depth of about 0.5 cm.

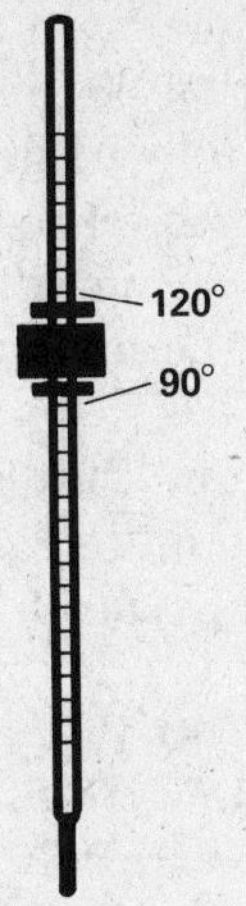

16. Slide onto a 260-deg thermometer, in sequence, a small piece of soft rubber tubing, a thick piece of vacuum hose, and another piece of soft rubber tubing. Position these pieces to cover the 90–120 deg region of the thermometer.

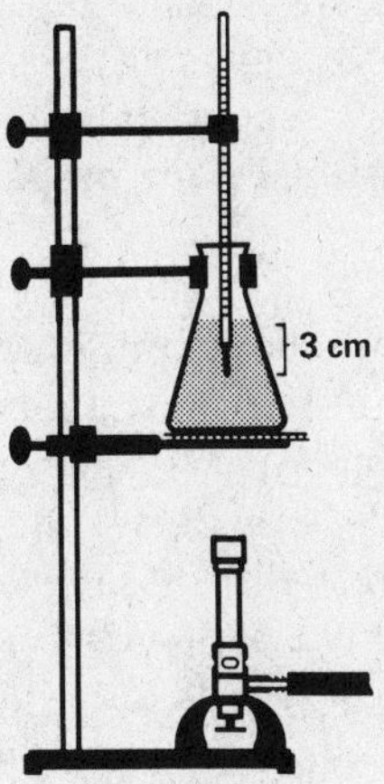

17. Clamp a 125-ml Erlenmeyer flask containing about 75 ml of heating oil so that the flask is also supported by a wire gauze on a ring above a burner. Position another clamp near the top of the ringstand and use this to hold the thermometer by the thick vacuum hose segment. Adjust the height of this clamp so that about 3 cm of the thermometer extends into the oil.

18. Use a rubber band at the height above the oil-immersion segment of the thermometer to secure to the thermometer both the capillary and the bulb melting-point tubes. The camphor in the capillary should be at the same height as the mercury reservoir of the thermometer. The bulb, containing the camphor–unknown mixture, should be against the bottom part of the thermometer mercury reservoir.

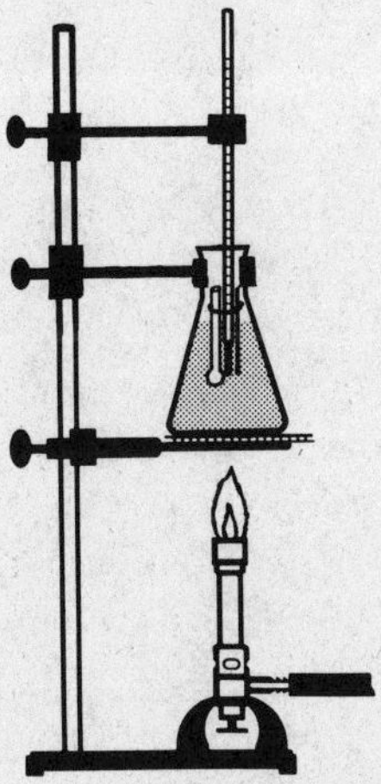

19. Replace the thermometer in its clamp and begin to heat the oil. Practice varying the air and gas controls of the burner to change the rate of heating. Also vary the location of the burner to find the best position for causing convection currents in the oil. You may heat rather rapidly to a temperature around 160°C, but above 170°C the temperature must not rise at a rate of more than 1°C per 30 sec. Watch only the *capillary* melting point tube at this stage. Record the temperature at which liquid first appears in this tube and the temperature at which the last of the camphor has melted. The temperature at which the last of the solid disappears is recorded as the melting "point" of pure camphor. If your technique is good, the temperature range should be small.

After you have determined the melting point of the pure camphor, heat the oil to about 180°C until the camphor–unknown mixture has entirely melted to form a clear liquid. Turn off the burner and let the oil cool to around 120°C. If you are to make a second determination, begin again at procedure 1 while the oil bath is cooling.

20. The camphor–unknown mixture should have solidified by the time the oil temperature dropped to 120°C. (If not, consult your instructor.) Reheat the oil, at a rate of 5–10°C per minute, watching the camphor–unknown mixture in the bulb carefully. Record the temperature at which liquid first appears and the temperature at which the last cloudiness due to camphor crystals disappears. These two temperatures provide you with the *approximate* melting range of the mixture. Turn off the burner and allow the oil to cool until the mixture has again solidified.

If you are going to make a second determination with a new mixture, you can continue your preparations while the oil is cooling.

When the mixture in the bulb has solidified, reheat the oil at a rate of 5–10°C per minute to just about 10°C below your "approximate" melting range. Then reduce the heating rate to 1°C per minute and determine an *accurate melting point* of the mixture, defined as the temperature at which the last trace of solid disappears.

If time permits, complete a second determination, using fresh samples of camphor and of the same "unknown."

When you have finished your work, allow the oil to cool to room temperature while you clean up your work area, then pour the cool oil into the storage container provided.

CALCULATIONS AND REPORT

Be sure all data are recorded accurately in your permanent notebook. Using the melting points of the pure camphor and the camphor–unknown mixture, calculate the approximate molecular weight of the unknown compound. The molal melting-point constant for camphor is −37.7°C kg (camphor) $mole^{-1}$ (solute).

Using your notebook records and calculations, complete the Report Form. From comparison of the experimental molecular weight with those calculated from the molecular formulas of the "sample compounds," suggest possibilities for the identity of your unknown.

GRADE	INSTRUCTOR'S INITIALS

Report Form

FOR INVESTIGATION 14: **Molecular Weights of Some Pharmaceutical Chemicals**

NAME: ______________________ ID NO.: ______________________

LAB SECTION ______________________ INSTRUCTOR ______________________

Sample Pharmaceuticals

Determine molecular formulas and molecular weights.

Include Others Provided by Your Instructor

Compound	Molecular formula	Molecular weight	Compound	Molecular formula	Molecular weight
caffeine			scopolamine		
sulfanilamide			acetylsalicyclic acid		
saccharin			camphor		
chlorpheniramine					
benzocaine					
phenolphthalein					
N, N-diethyl-*meta*-toluamide					

	First determination	Second determination (optional)
Melting point of pure camphor	______ °C	______ °C
Melting point of camphor/unknown mixture	______ °C	______ °C
Composition of the mixture		
Mass of camphor	______ mg	______ mg
Mass of "unknown"	______ mg	______ mg

Approximate molecular weight of the unknown compound: ______________

Show calculations on reverse side.

Possible identity of the unknown (from list above)

Unknown Number__________

DATE ______________ SIGNATURE ______________________

Investigation 15

TRUTH IN LABELING

INTRODUCTION Recent "consumer-protection" legislation has seemed to suggest that we will at last be able to read the labels of drugs, beverages, packaged foods, and other products to determine not only what chemicals we are consuming, but also how much. Misleading or unsupported advertisements and "short weights" can result in federal corrective action. How effective these improved standards will be and how much new knowledge the public will derive from more precise product labels remain to be seen.

Will the average citizen know more about a drug when the label reads, "Each tablet contains 120 mg acetaminophen" than he did before? The U.S. citizen is still relatively unfamiliar with the metric system, so "120 mg" may not tell him very much about the quantity of the drug. He probably doesn't know what acetaminophen is anyway. (Not very many trained chemists would recognize it from its "trade name," although, hopefully, most would know how to look it up.) Of even greater importance, will the label reader know what a chemical ingredient *does*? Further information on the label will, of course, inform him that acetaminophen is an aspirin substitute and will reassure him that the drug is "unlikely to cause the stomach upset" and "not likely to cause a reaction" that might result from aspirin. But what about an overdose or long-term effects of extensive or prolonged use? That information probably won't be on the label, possibly because it isn't fully known or (as has been suggested*) because some information is "better supressed." This is not to suggest, by any means, that this particular drug is dangerous. But we should recognize that every year we are consuming more and more chemicals whose long-range effects are not known to us, and the "truth in labeling" laws have not done much about that problem. It has been reported† that some 2000 chemicals are permitted as food additives alone by the Food and Drug Administration, with little information on long-term effects.

The specification of chemical ingredients on a label, then, provides us with some protection, but with less information than we might like to have. In fairness to the pharmaceutical and food technology industries, as well as to the Food and Drug Administration, we should recognize that we still have far greater protection from improper or impure drugs and food additives than most other peoples on earth. That is not to say that we would not benefit from a greater concern for human welfare on the part of both industry and government.

What about weights on labels? Consider our "120 mg" tablet of acetaminophen, costing about 12 cents. Strictly

*Van Allen, Gene M. and Judith. 1972. *Food Pollution: The Violation of Our Inner Ecology*. New York: Holt, Rinehart, and Winston.
†The National Observer, March 4, 1972.

speaking, "120 mg" could mean a weight "rounded up" from as low as 115 mg to a weight "rounded down" from 124 mg. Suppose a drug company deliberately set out to make a little extra profit by packaging 115-mg tablets while, quite properly, reporting the weight as "120 mg." The consumer who purchased 300 tablets a year (not an unusual number for a "normal" family) would have paid $1.50 for drugs he only thought he had purchased. That is not very much extra for the consumer, but multiply that by 30-million consumers and it makes a tidy extra profit for the company—all strictly legal, if a label weight is acceptable, "rounded" according to established rules.

In this experiment we will investigate some product weights to see how closely they conform to label specifications. We will use this as an exercise in becoming more proficient in the use of the analytical balance, as well as in seeing how various ways of determining weights compare in reliability.

Before starting this investigation, you should be thoroughly familiar with the concepts of significant figures, *accuracy*, and *precision* (Appendix C), and with the use of exponential notation. In addition, you should have studied *carefully* the procedures required for using an analytical balance of the type assigned to you (Appendix A or class handout, or both).

References

1. O'Connor, Rod. 1977. *Fundamentals of Chemistry*. New York: Harper & Row (Mathematical Skills).
2. Kermode, G. O. "Food Additives." *Scientific American* (March 1972).

SAFETY *Do not taste any of the products* used in this experiment. They may have been contaminated by storage in chemical stock areas. If you are working in a laboratory where chemical reagents are set out, wear approved eye protection.

OJBECTIVES

1. Be able to use proper weighing techniques for obtaining reliable results.
2. Be able to demonstrate proper use and care of a precision analytical balance.
3. Be able to determine average values, range, and average deviation for sets of mass determinations.
4. Be able to express exponential results using significant figures only.

PROCEDURES

Good Balance Techniques

1. Do not lean on the balance table.
2. Turn the mass dials only when the balance is arrested or on partial release.
3. Never add or remove an item unless the balance pan has first been arrested. It is proper, however, to add small increments of chemicals to achieve a certain mass, even when the arrest knob is in the released position.
4. Keep the balance doors closed except when adding or removing the item to be weighed.

5. If the balance pan swings when the pan control knob is turned to "full release," move the knob back and forth gently between the release and the arrest positions until the pan has stabilized.
6. Read the division line in the light window with your eyes at the height of the indicator gap to avoid parallax error.
7. Leave the balance spotlessly clean, with the doors closed and the pan arrested.
8. If you have any difficulties, consult your instructor. Do not attempt to "fix" a balance.

Secure five small packages of a product, the mass of which is given on the label. Enter the name of the product and the claimed mass in your notebook. Zero your analytical balance.

A. Weigh one of your objects five times "directly" in the following manner:

1. Be sure the pan is arrested. Check the balance's zero point and adjust it, if necessary. Then place the object on the balance pan.
2. Determine and record the mass to the nearest 0.1 mg.
3. Arrest the pan and return the *fractional* mass dial to the 00 reading. Do not touch the other mass dials.
4. Turn the pan control knob to the full release position.
5. Cover the numerical readings on the fractional, mass scale with a small piece of paper so you can read the indicator gap, but not the numbers in the fractional mass window. Turn the fractional mass dial until a line is properly centered in the indicator gap. Then read and record the mass.
6. Repeat steps 3, 4, and 5 until you have a total of five masses for the same item.
7. Check that the zero position of the balance has not changed.
8. With the balance pan arrested remove the package from the balance pan and mark it by bending one of its corners for identification. This package will be used again. Record which balance you have used.

Calculate the average mass and the average deviation of your results.

B. Weigh the remaining four packages, each only once, on the same balance.

Calculate the average mass of the packages, the range, and the average deviation. Use the average mass of the package employed in procedure A as a fifth mass.

C. Repeat procedure A with the same package originally used, but now employ a different balance. Record your results for later comparison.

D. Repeat procedure B with the same packages originally used, employing the balance you used in procedure C.

E. Next you will weigh a package employing the "difference" weighing method. Return to your original balance and weigh a clean watch glass. Record its mass.

Determine the mass of the watch glass only once. Leave the watch glass on the pan and place the package used in procedure A on the watch glass. Repeat procedure A for the package–watch glass combination. How does your "package mass data" set compare with that of procedure A?

F. Repeat procedure E, using the balance employed in procedure C.

G. Carefully open the package used in procedure A and empty the contents completely into a waste container, being careful not to lose any of the package material. Now place the package material on the pan of the balance used in procedure A and determine the mass of the empty package only once.

H. Calculate the masses of the package contents (the net masses) as the differences between the masses of the full packages (procedure B) and the mass of the packaging material (procedure G). Assume that the masses of the packaging materials from different packages are the same.

I. Weigh one of the packages used in procedure B five times on a triple beam balance. In order to avoid biased results, move the 1-g mass behind a sheet of paper that prevents you from seeing the mass. When the balance is again in equilibrium, remove the paper, read, and record the mass to the nearest 0.01 g. Move the 1-g mass and repeat the above procedure. Calculate the average, the range, and the average deviation of your results.

J. Obtain copies of the results of procedures H and G (averages and average deviations only) from three other students, who have used the same balance and same product as you.

REPORT Using your notebook records and the information obtained from three other students (procedure J) complete the Report Form. *Do not* write the collected data into the Report Form *during the laboratory period.* You may copy the Report Form tables into your notebook to organize your data collection. The Report Form must be completed using these collected data. A copy of all the required calculations must be submitted with the Report Form. The discussion questions must be answered on separate pages and should be submitted when the report is due.

GRADE / INSTRUCTOR'S INITIALS

Report Form

FOR INVESTIGATION 15:

Truth in Labeling

NAME: ______________________ ID NO.: ______________________

LAB SECTION ______________________ INSTRUCTOR ______________________

Product name: ______________________ Label mass: ______________________

Enter your results below. Calculations must be submitted on separate pages.

Procedure	Trial or package 1	Trial or package 2	Trial or package 3	Trial or package 4	Trial or package 5	Average	Average deviation	Range
A. Balance No.								
B. Balance No.						+	+	+
C. Balance No.								
D. Balance No.								
E. Package & watch glass								
Watch glass							*	*
Package								
F. Package & watch glass								
Watch glass							*	*
Package								
H. Package mass†								
Empty package‡							*	*
Product mass						+	+	+
I. Triple beam balance								

+Use significant figures only.

*Enter here the average deviation and range you would expect on the basis of your results obtained in procedure A (see "discussion" Questions 1 and 2).

†Taken from procedure B.

‡Procedure G.

G. Mass of packaging material ______________ g

J. Data from procedures H and G obtained by three other students using the same balance as you. (It is your responsibility to secure these data.)

Student's name	Average–average deviation† (procedure H)	Packaging material (procedure G)
*		

*Your own name.

†Express these averages and their deviations using significant figures only.

(OVER)

Average the "Averages (procedure H)" and "Packaging material (procedure G)" and enter the new averages below.

Average package content ______________ g ± ______________ g*

Average packaging material mass

______________ g± ______________ g*

DISCUSSION QUESTIONS

Answer the following questions on separate pages and submit these with your Report Form.

1. A coin has been weighed on the balance you have used in this experiment (procedure A). The average mass was found to be 25.6743 g. Estimate the likely average deviation.
2. Although you have weighed your watch glass and the packaging material (procedures E and G) only once, you should be able to estimate the average deviations. What are these values?
3. Compare your average result obtained in procedure A with that of procedure C.
4. Compare your average result obtained in procedure D with that of procedure B.
5. Compare your average result obtained in procedure A with that of procedure D.
6. Compare the average result (expressed in significant figures only) of procedure B with that of procedure C (expressed similarly).
7. With the help of the discussion on accuracy and precision (Appendix C) write the product net mass (procedure H) using significant figures only. Indicate the likely average deviation.
8. Compare the label net mass with the average product mass (procedure H). According to these results is the label mass "true?" On the average, does the manufacturer fill his packages just right? Does he overfill? Does he add less than claimed on the label?
9. Discuss the merits of determining a mass "by difference" (with the watch glass) over the direct weighing procedure (without a watch glass), when two different balances must be used. Base your discussion on the results of procedures A and C, D and B, and A and E. Why should the same balance be used during an experiment?
10. Compare the precision obtained with the triple beam balance (procedure I) with that obtained in procedure A.
11. Most package labels indicate "net weight" rather than "net mass." What is the difference, if any?

______________ ______________

DATE SIGNATURE

*Use significant figures only.

Investigation 16

CAFFEINE IN A CHLOROFORM–WATER SYSTEM

INTRODUCTION Caffeine is a cardiac, respiratory, and psychic stimulant drug found in tea, coffee, maté leaves, guarana paste, and cola nuts. The drug also has diuretic properties. Prolonged use of caffeine typically results in habituation. Caffeine is one of the least toxic drugs, although large overdoses may cause cardiac arythmia, tremors, or extreme nervousness. Recent evidence has suggested possible adverse long-range side effects with habitual use, but the research studies have not yet been conclusive.

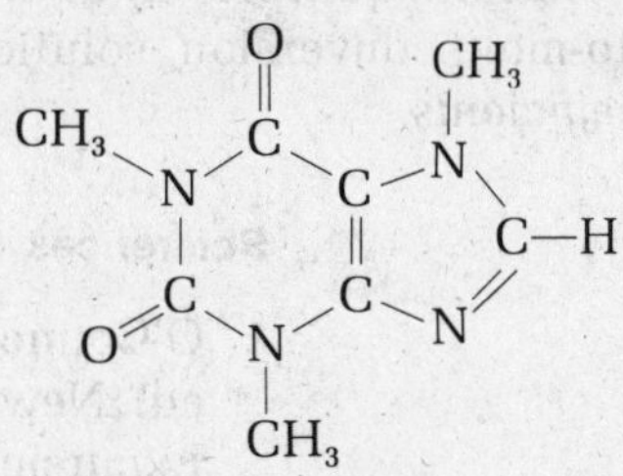

caffeine

Caffeine is prepared commercially from tea dust (1.5–3.5 percent caffeine by mass) or as a by-product of the production of "decaffeinated" coffee. Its stimulant properties and low toxicity make caffeine a commercially valuable nonprescription drug.

Caffeine has a water solubility of about 2.2 g per 100 ml at 25°C. Its solubility increases appreciably with temperature, reaching 67 g per 100 ml in boiling water. It is more soluble in chloroform, however, (18.2 g per 100 ml at 25°C) and advantage may be taken of this in extracting caffeine into chloroform from an aqueous preparation of coffee or tea. Chloroform and water are said to be immiscible, although they do have measurable mutual solubilities. A typical extractive separation would involve shaking an aqueous solution of caffeine with chloroform until an equilibrium distribution of caffeine is attained between the two liquid phases. The more dense chloroform solution would then be drained away, dried, and evaporated to obtain the dissolved caffeine. The equilibrium system for distribution of caffeine between water and chloroform may be represented by

$$\text{caffeine(aq)} \rightleftharpoons \text{caffeine}(CHCl_3)$$

for which we may write an expression for the equilibrium constant called the *distribution coefficient* (K_D):

$$K_D = \frac{C_{\text{caffeine (in } CHCl_3)}}{C_{\text{caffeine (in } H_2O)}}$$

(The concentrations are typically expressed as moles/liter, although any consistent mass/volume units could be used.)

The distribution coefficient may be approximated from the ratio of the solubilities of caffeine in chloroform and in water, respectively. However, this is not a very accurate approximation because the chloroform becomes saturated with water (and the water with chloroform) during the extraction and these mutual saturations change the solvent characteristics appreciably. The determination of optimum conditions for the commercial production of caffeine by chloroform extraction of aqueous solutions requires a more accurate value of the distribution coefficient, which must be determined experimentally.

We will be investigating a method for determining this distribution coefficient in this experiment.

Before starting this investigation, you should be familiar with the principles governing solubility, colloid formation, and liquid-liquid extraction (distribution equilibrium) and with calculations involving mass-to-mole conversion, solution *molarity,* and *distribution coefficients.*

References

1. O'Connor, Rod. 1977. *Fundamentals of Chemistry* (2nd ed). New York: Harper & Row (Units 13, 14, and 18, and Excursion 4).
2. Dreisbach, Dale. 1966. *Liquids and Solutions.* Boston: Houghton-Mifflin.

SAFETY Be very careful in working with chloroform. It is flammable and may be absorbed as a cumulative poison. Avoid inhaling chloroform vapors and if you spill chloroform on your skin or clothing, wash immediately and thoroughly with *soap* and water. Never shake a separatory funnel containing an organic solvent without venting it (carefully) periodically to relieve pressure from solvent vapors.

OBJECTIVES

1. Be able to perform properly and safely the techniques of liquid-liquid extraction and solvent evaporation.
2. Be able to make calculations involving mass, moles, molarity, and the use of a distribution coefficient.

PROCEDURES

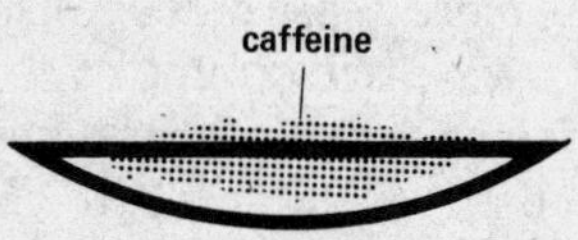

1. Weigh a clean, dry 9-cm watch glass to the nearest milligram. Remove the watch glass from the balance and transfer onto it between 1.1 and 1.2 g of caffeine. Reweigh and record the mass of caffeine to the nearest milligram.

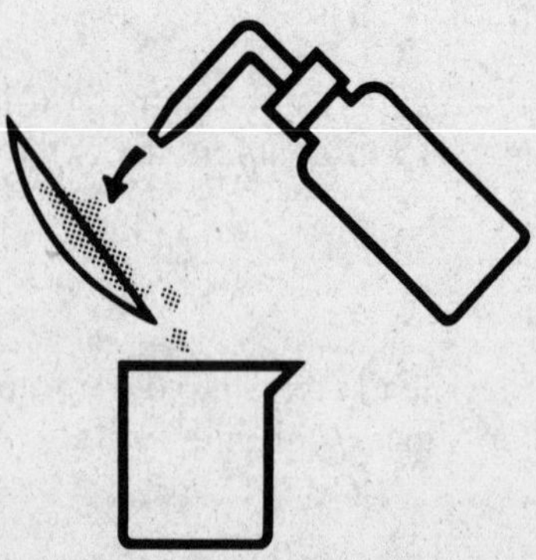

2. Use distilled water to rinse the caffeine completely from the watch glass into a 100-ml beaker. Add additional distilled water to fill the beaker about one-third full.

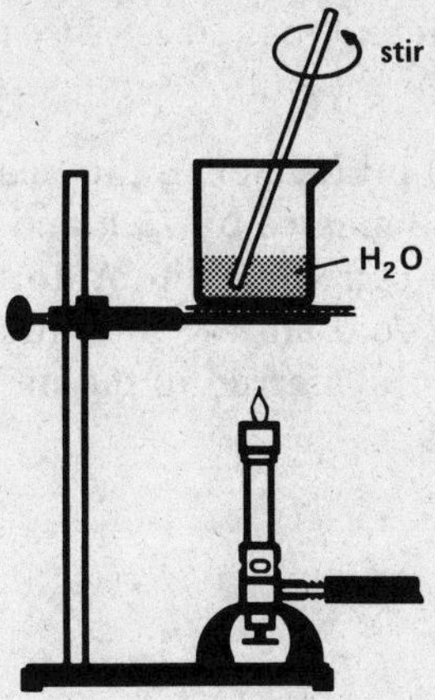

3. Warm the mixture over a low flame, with stirring, until all the caffeine has dissolved. *Do not boil the solution.* When all caffeine has dissolved, carefully set the beaker on a pad to cool and remove the stirring rod, rinsing any liquid from the rod into the beaker with distilled water.

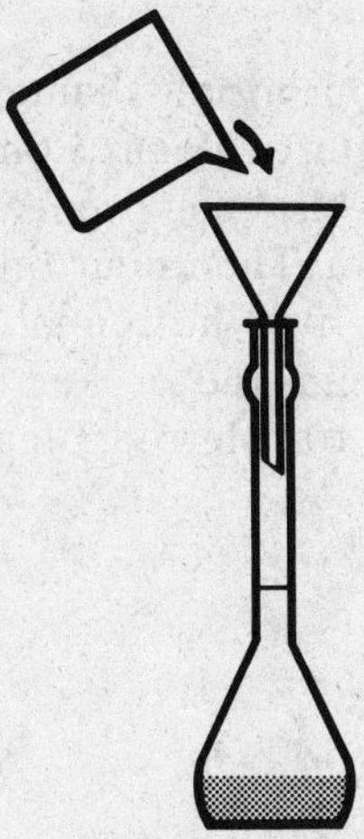

4. Add about 20 ml of distilled water to a clean 100-ml volumetric flask. Pour the liquid from the beaker through a funnel into the flask.

5. Rinse the beaker into the flask with five successive small portions of distilled water.

6. Rinse and remove the funnel. Be sure the contents of the flask have cooled to room temperature. Then fill the flask to the mark, stopper it securely, and mix the contents thoroughly by inverting and shaking the flask. (Hold the stopper in.)

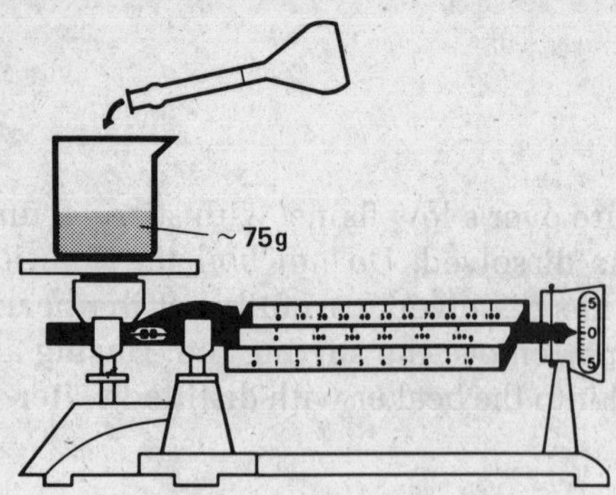

7. Weigh a clean, dry 250-ml beaker to the nearest 0.1 g. Then advance the balance masses by 75 g and pour into the beaker enough of the caffeine solution to return the balance to equilibrium. Avoid adding too much. Record the final mass of solution transferred to the nearest 0.1 g.

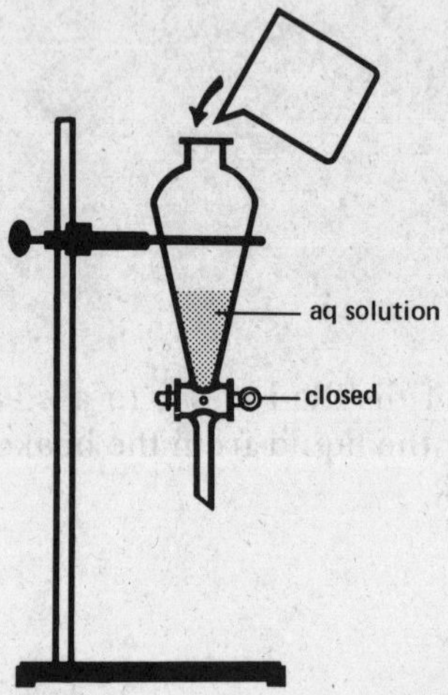

8. Attach a ring clamp, of appropriate diameter, to a ring-stand and use the ring to support a clean 250-ml separatory funnel. Be sure the stopcock has been properly lubricated and set in the closed position. Then pour the contents of the 250-ml beaker carefully into the funnel. Stopper the separatory funnel and reweigh the beaker to determine the mass of solution actually transferred. Clean the beaker and wipe it dry.

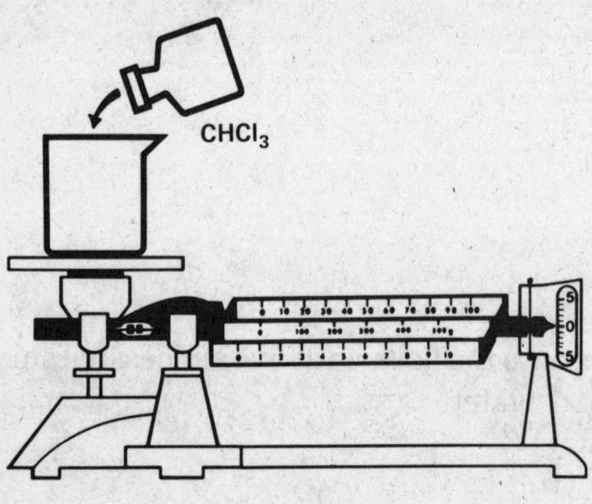

9. Ask your instructor to specify the amount of chloroform you are to use (38, 75, 110, or 150 g). Weigh the amount specified, using the method described in procedure 7.

Caution: Avoid inhaling the vapors.

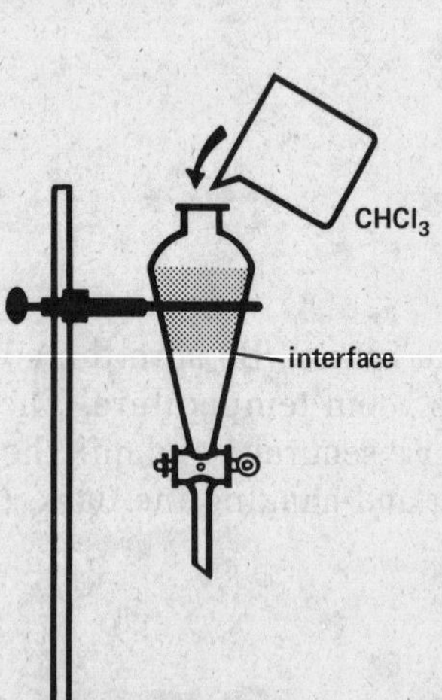

10. Transfer the chloroform carefully into the separatory funnel containing the aqueous solution and stopper the funnel. Reweigh the beaker quickly. (Chloroform is volatile.) Record the mass of chloroform actually transferred.

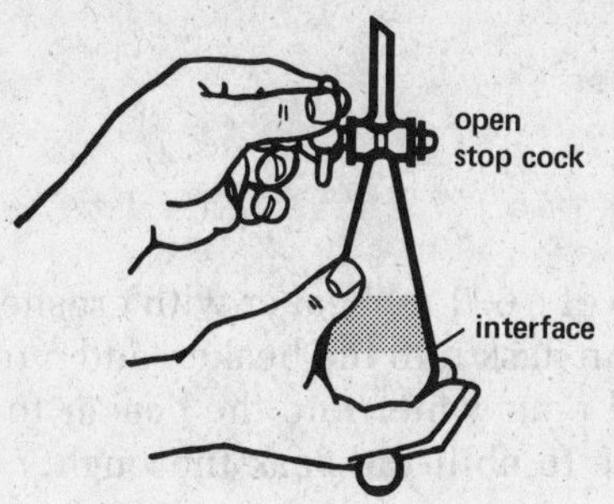

11. Holding the stopper securely, remove and invert the separatory funnel. Open the stopcock cautiously to vent solvent vapors that have built up pressure. Then close the stopcock.

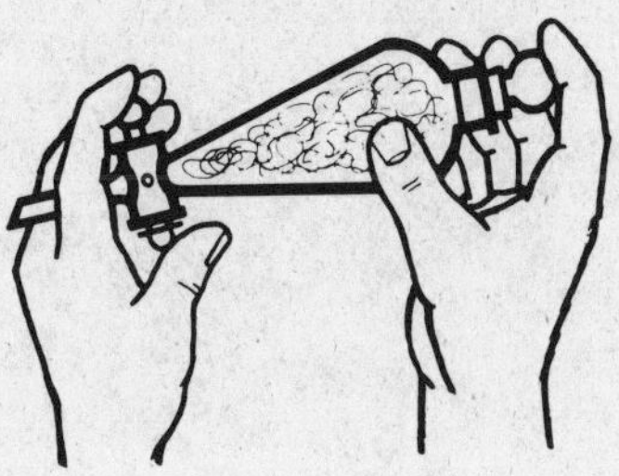

12. Hold *both* the stopper and stopcock securely and shake the separatory funnel *gently* to mix the liquids.

Caution: Vent periodically (procedure 11) to relieve pressure.

Note the proper way of holding the funnel. Mixing is best accomplished by rocking the funnel up and down from the horizontal position. Vigorous shaking is likely to produce an emulsion that is difficult to work with. After 5 min of gentle shaking, vent the funnel and (stopcock closed) return it to its ring support to allow the two phases to separate.

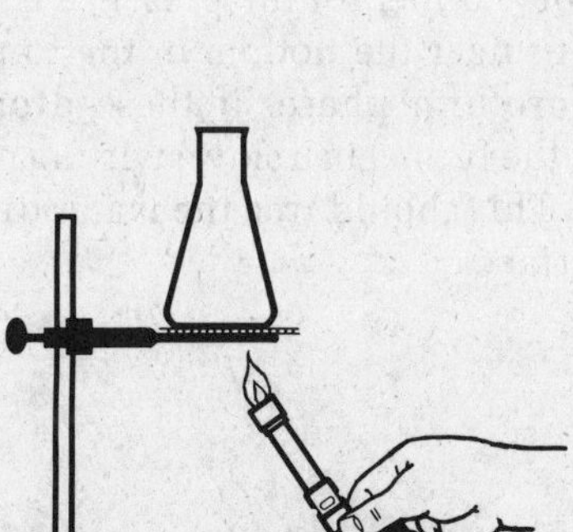

13. Clean a 125-ml Erlenmeyer flask by washing with detergent and rinsing, first with tap water and then with distilled water. Wipe the outside of the flask dry with a clean towel and shake as much of the water as possible out of the flask. Complete the drying by heating the flask with a "cool" flame to expel the last traces of moisture. Allow the dry flask to cool to room temperature, then weigh it to the nearest milligram.

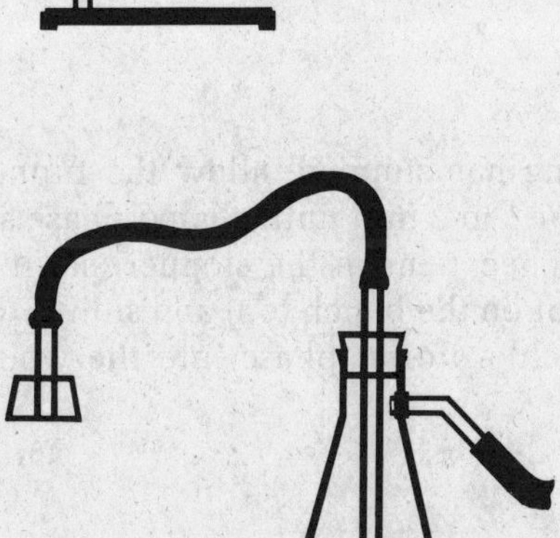

14. Lubricate and insert (carefully) a short glass tube into the smaller side of a No. 7 rubber stopper until the tube nears, but does *not protrude from,* the wider side of the stopper. Lubricate and insert a longer tube through a stopper so that one end of the tube protrudes above the stopper about 5 cm and the other end reaches nearly to the bottom when the stopper is placed in a suction flask. Connect the glass tubes by a 60-cm length of vacuum hose.

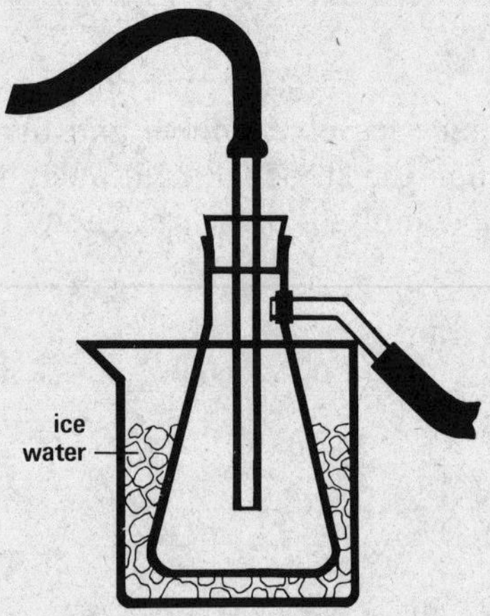

15. Cover the bottom of a 600-ml beaker with crushed ice. Then place the suction flask into the beaker, add more ice around the flask, and pour water into the beaker to form an ice–water mixture to chill the flask thoroughly.

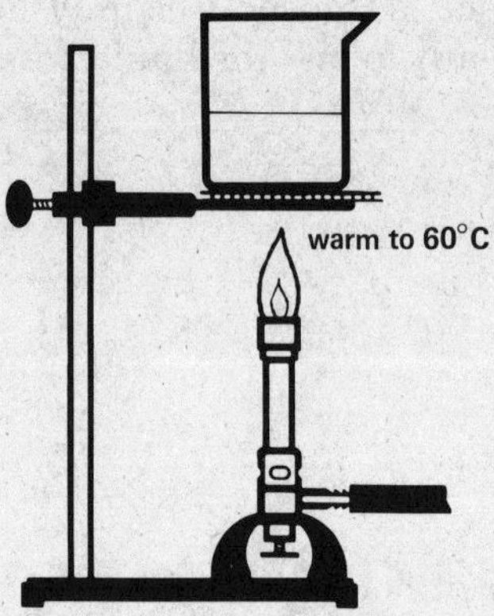

16. Warm about 250 ml of distilled water in a 400-ml beaker to around 60°C (a temperature at which you could still hold the beaker without much discomfort.)

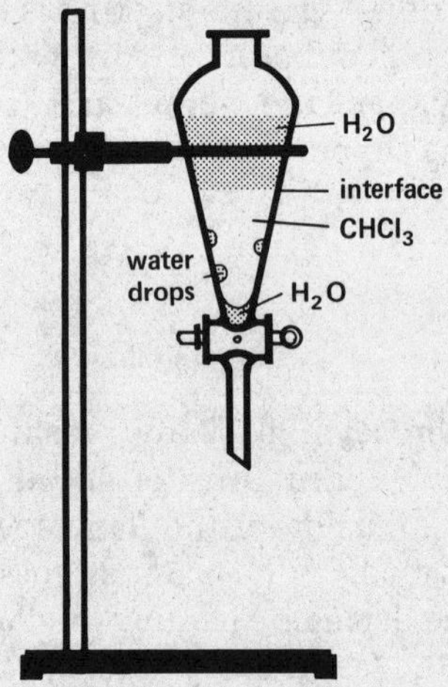

17. Check the separatory funnel for the presence of water droplets on the walls or near the bottom of the funnel in the region of the chloroform phase. If these drops are observed, gently move the funnel in a slow swirling motion to form a "whirlpool." This should free the water drops to rise to the liquid interface.

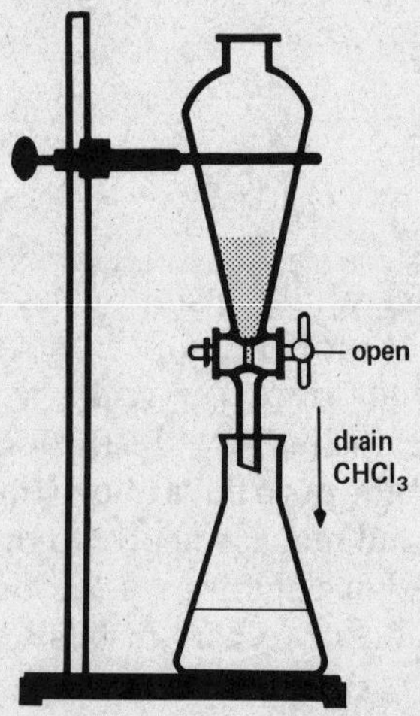

18. After the swirling has stopped, allow the funnel to stand undisturbed for 1 to 2 min until a good phase separation is observed. Then remove the stopper (set it in a clean watch glass, not on the bench top) and *slowly* drain most of the lower (chloroform) phase into the weighed Erlenmeyer flask.

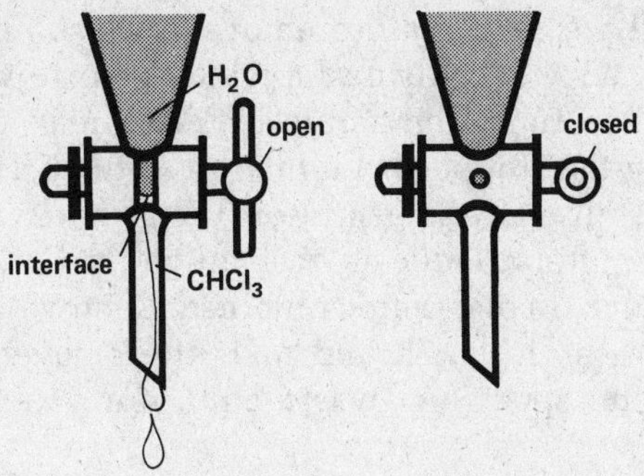

19. Adjust the stopcock to give a dropwise flow when only a small volume of chloroform remains in the funnel. Close the stopcock as soon as the water–chloroform interface reaches the lower end of the stopcock bore.

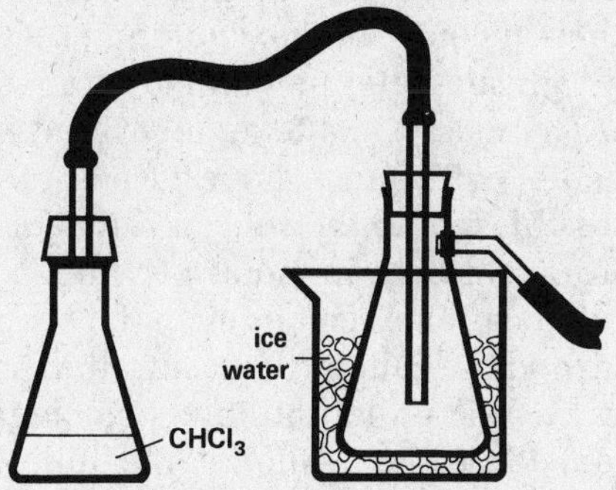

20. Connect the suction flask to the aspirator and turn it on. Press the inverted rubber stopper onto the top of the Erlenmeyer flask until a partial vacuum is established. (You will hear a change in the sound of the water aspirator.) Holding the stopper in place, swirl the flask gently and continuously as the chloroform begins to evaporate.

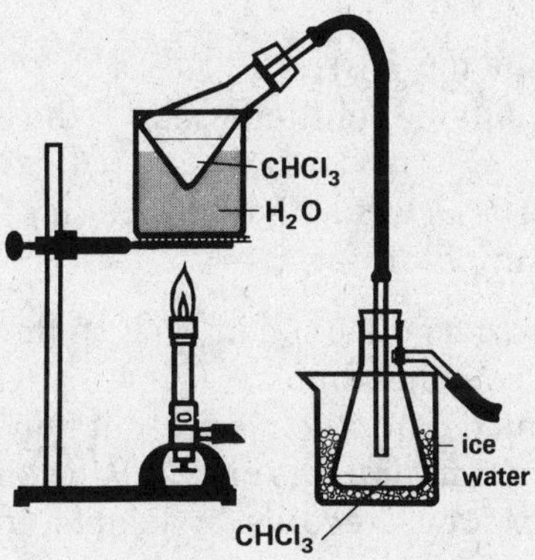

21. To speed up evaporation, immerse the bottom of the flask in the warm water bath, which is maintained at $\sim$60°C over a *low* flame.

Caution: Keep the hose away from the flame!

Continue holding the flask and shaking it gently to avoid "bumping" as the chloroform begins to boil. The chloroform vapors will recondense to liquid inside the chilled suction flask.

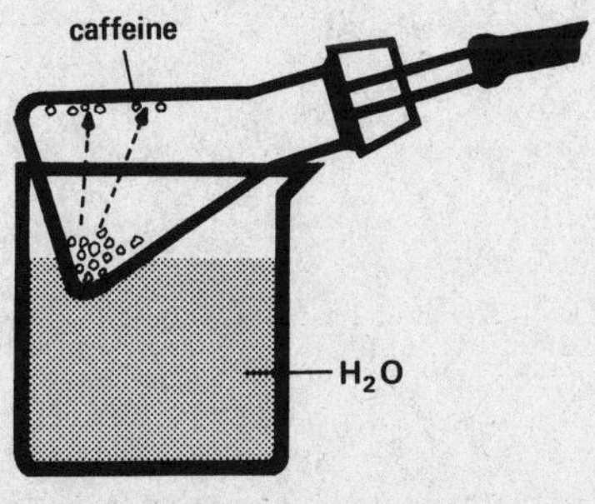

22. Near the end of the evaporation, you may note some splattering. Holding the flask at an angle will avoid splattering onto the stopper. Continue heating and gentle shaking until all the chloroform has evaporated, leaving a residue of caffeine. Turn off the burner, but leave the flask in the warm water for about 5 min after the last liquid chloroform has disappeared.

23. When only a dry solid remains, remove the flask from the water bath and wipe the outside dry with a clean towel. Let the flask cool to room temperature, then break the vacuum by sliding the stopper from the top of the Erlenmeyer flask. *Then* turn off the aspirator. Weigh the flask containing the dry caffeine to the nearest milligram. Calculate and record the mass of caffeine obtained. Remove the suction flask from the ice bath and pour the recovered chloroform into the specified "waste chloroform" container.

CALCULATIONS AND REPORT

From the mass of chloroform employed and its density (1.48 g ml^{-1}), calculate the volume of chloroform used. From the mass of the aqueous caffeine solution used and its density (1.0 g ml^{-1}), calculate the volume of aqueous solution employed. The mass of caffeine recovered from the chloroform together with the volume of chloroform used will make it possible to find the concentration of caffeine in the chloroform solution in moles per liter. Next, calculate the amount of caffeine remaining in the aqueous phase at equilibrium and the caffeine concentration in the water in moles per liter. The distribution coefficient K_D can be derived from these data. Submit the following information to your instructor or directly to those students who used approximately the same amount of chloroform as you.

Grams of caffeine in 100 ml H_2O.
Grams of aqueous caffeine solution used for the extraction.
Grams of chloroform used for the extraction.
Grams of caffeine in chloroform.

Your instructor will post a collection of such student data for you to copy into your notebook.

Using these student data, which were obtained with volumes of chloroform similar to the one used in your own experiment and your own records, calculate average values of the equilibrium concentrations of caffeine in chloroform and in water [in units of mole (caffeine) per liter (solvent)]. Then calculate the distribution coefficient for caffeine in a chloroform–water system.

Use your notebook records and calculations to complete the Report Form.

GRADE	
	INSTRUCTOR'S INITIALS

Report Form

FOR INVESTIGATION 16:
Caffeine in a Chloroform—Water System

NAME: ______________________ ID NO.: ______________________

LAB SECTION ______________________ INSTRUCTOR ______________________

Your Own Experiment

Mass of caffeine (procedure 1) ____________ g

Mass of aqueous solution transferred (procedure 7) ____________ g

Volume of aqueous solution transferred (procedure 7) ____________ ml

Concentration of aqueous caffeine solution ____________ moles caffeine per liter of solution

Mass of chloroform used (procedure 9) ____________ g

Volume of chloroform used (procedure 9) ____________ ml

Mass of caffeine extracted into chloroform (procedure 23) ____________ g

Concentration of caffeine in chloroform ____________ moles per liter

Distribution coefficient, K_D ____________

Other Students' and Your Own Experiments

(Note: Average only the equilibrium concentrations of caffeine and use these average values to obtain an average K_D.)

Number of experiments used ____________

Data obtained from (list names)

Average equilibrium concentration of caffeine in chloroform

____________ ± ____________ moles liter^{-1}

Average equilibrium concentration of caffeine in water

____________ ± ____________ moles liter^{-1}

Average distribution coefficient: $K_D =$ ____________ ± ____________

Attach separate pages showing the details of all calculations. Consult the Appendix if you are not familiar with the method of calculating averages.

(OVER)

Additional Calculations

Attach a separate page with details.

Using the distribution coefficient calculated from the collected student data (reverse side), determine for a system using chloroform to extract 100 ml of an aqueous solution containing 1 g of caffeine:

a. How many extractions of 100 ml each of $CHCl_3$ would be required to remove at least 98 percent of the caffeine from the aqueous solution?

b. Which would remove the larger fraction of caffeine from the aqueous solution, a single extraction with 100 ml of chloroform or two successive extractions with 50 ml portions of chloroform?

DATE ____________ SIGNATURE ____________

Investigation 17

SOME COMMERCIAL ANTACIDS

INTRODUCTION Foods may be classified roughly into three categories: carbohydrates, fats, and proteins. Each of these types of materials undergoes a complex and extensive series of reactions, referred to as *digestion* and *metabolism,* within the body. Most of these reactions are catalyzed by very specific enzymes, without which the reactions would proceed too slowly to sustain life.

The catalytic enzymes are, in many cases, so specific in their action that seemingly minor differences in the molecular structures of two similar compounds will permit only one of them to be "attacked" by the enzyme. Both starches and celluloses, for example, are large polymers made up of connected glucose molecules. For the polymer to be useful as food, it must be broken down in the body to simpler units, eventually glucose. The most important difference between starches and celluloses is the geometry of the so-called glycoside link which connects the glucose units. Humans contain enzymes capable of catalyzing the hydrolytic cleavage of the glycoside linkages in starches, but not of those in celluloses. As a result, starches can provide us with foods that furnish the glucose necessary as an energy source, but celluloses—containing the same glucose—pass through the body virtually intact.

The various enzymes involved in the digestion and metabolism of foods can function properly as catalysts only within certain pH ranges. The digestion of starches, for example, begins in the mouth, where the material is mixed with saliva containing the enzyme α-amylase. Active under the alkaline conditions of saliva, this enzyme begins to break the large starch molecules into smaller ones. However, as soon as the food is swallowed, the α-amylase system is deactivated by the acid in the stomach. Further carbohydrate digestion must then await the transfer of the food from the stomach to the small intestine, where *pancreatic* α-amylase—again in an alkaline medium—continues the hydrolytic reactions begun in the mouth.

The enzyme systems that initiate the digestion of proteins, on the other hand, require acidic conditions. The digestion of these materials, then, does not begin, to any significant extent, until the foods enter the stomach. Protein foods could be swallowed intact were it not for the important function of "chewing," which physically macerates the material into a form more easily digested.

As food enters the stomach, an intricate biochemical mechanism is triggered, one result of which is the secretion of dilute hydrochloric acid. This acid has several functions. Acting on chunks of the food, the acid helps to swell and loosen the fibrous portions for more effective contact with the aqueous medium of the stomach. To a slight extent, the acid itself catalyzes the hydrolysis of various linkages in proteins (the peptide, or amide, bond), carbohydrates (the glycoside link), and fats (ester links). This function is minor compared to the role of HCl in the deactivation of salivary α-amylase and the activation of proteolytic ("protein-breaking") enzymes. The enzymes that attack food proteins are secreted initially in their inactive forms, probably to protect tissue protein. Activation by acid in the stomach is essential to the function of these enzymes in beginning the digestion of protein foods.

For various reasons, not all of which are yet understood, the stomach may secrete an excess of hydrochloric acid. Such a condition, known as gastric hyperacidity, may cause an acute discomfort, familiar to most of us as "indigestion." This condition is often induced by highly spiced or fat-rich foods, or simply by packing too much food into the stomach. Mild cases of gastric hyperacidity may be treated by taking any of a number of commercial "antacids" (substances acting to neutralize excess hydrochloric acid in the stomach). Since an excess of alkali must be avoided if we are not to upset the delicate pH requirements of the digestive enzyme systems, such antacids must be taken only in controlled amounts. Some of the commercial preparations are designed to minimize this problem by providing components that help to "buffer" the stomach within an acidic pH region.

In this experiment, we will investigate some of the more common commercial antacids to determine how much hydrochloric acid they can neutralize.

Before starting this investigation, you should be thoroughly familiar with the fundamental aspects of acid-base chemistry, including pH notation, and the types of calculations involved in solution stoichiometry employing *molar* concentrations. In addition, you must perform (or set up for cases requiring experimental data) the calculations indicated in the Procedures section of this investigation.

References

1. O'Connor, Rod. 1977. *Fundamentals of Chemistry*. New York: Harper & Row (Units 13, 19, and 34).
2. O'Connor, Rod and Charles Mickey. 1974. *Solving Problems in Chemistry*. New York: Harper & Row (Unit 6).

SAFETY Do not, *under any circumstances*, taste or swallow any of the antacids used for this experiment. Storage in the chemical stockroom may have resulted in contamination. Wear approved eye protection and use proper care in working with acids and alkalies.

OBJECTIVES

1. Be able to mix and dilute chemical reagents properly.
2. Be able to perform titrations accurately, including the use of an endpoint indicator.
3. Be able to make appropriate calculations dealing with solution concentrations, solution volumes, and solute amounts, including the use of these quantities in stoichiometry.

PROCEDURES

Obtain two burets which, when filled with distilled water, will drain without leaving water droplets on the interior walls of the buret. If necessary, clean the burets as described in Appendix A.

Calculate the volume of concentrated (12 *M*) hydrochloric acid required for preparation of 400 ml of 0.1 *M* HCl and the mass of sodium hydroxide required for preparation of 400 ml of 0.1 *M* NaOH. Have your calculations checked by your instructor before proceeding further.

A. Preparation of Solutions

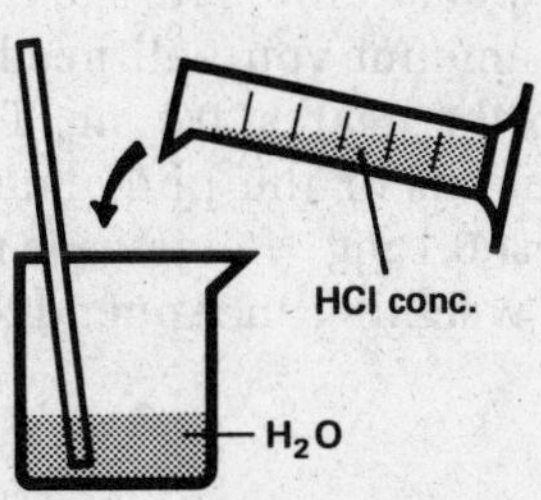

1. Pour 200 ml of distilled water into a 600-ml beaker and add, slowly with stirring, the calculated volume of 12 *M* HCl. Then add sufficient distilled water to bring the total volume to ~400 ml. Stir the solution carefully, then cover the beaker with a paper towel and label the beaker "0.1 *M* HCl."

Caution: Sodium hydroxide is very corrosive. Avoid contact with the pellets. If you spill any, clean them up with a damp paper towel and dispose of them as directed by your instructor. In case of contact, wash *immediately and thoroughly* with water.

2. Obtain 20 NaOH pellets (no more) on a clean dry watch glass. Reseal the stock bottle.

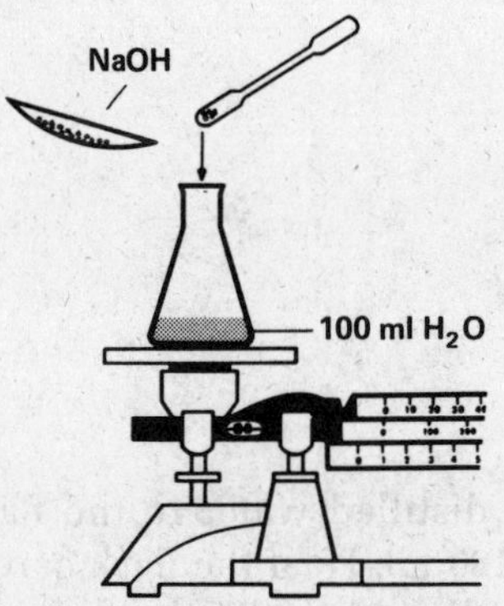

Weigh a 500-ml Erlenmeyer flask containing 100 ml of distilled water to the nearest 0.1 g. Advance the balance masses by the amount calculated for the NaOH needed. Add pellets, one by one, to the flask (using your spatula) until the balance returns to equilibrium. Dispose of excess pellets as directed by your instructor. Remove the flask from the balance and swirl it carefully until all the sodium hydroxide has dissolved. Then add sufficient distilled water to the flask to bring the total volume to ~400 ml. Mix the contents thoroughly by careful swirling, stopper the flask, and label it "0.1 *M* NaOH."

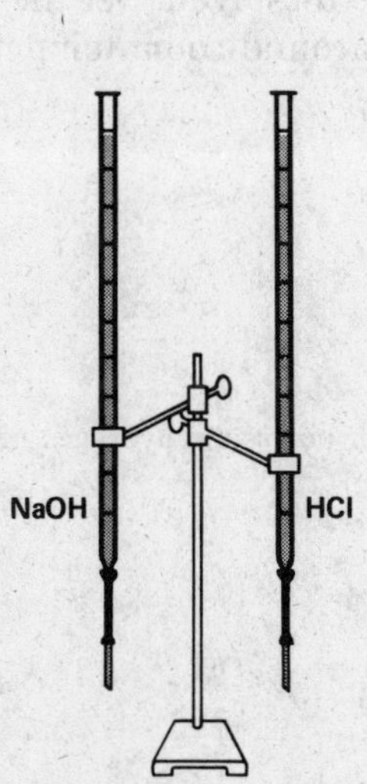

3. Rinse and fill (Appendix A) one buret with your 0.1 *M* HCl and the other with your 0.1 *M* NaOH. Drain enough solution from each buret to expel air from the buret tips and to drop the liquid levels to just below the zero marks.

B. Standardization of the Acid and the Base

Calculate the mass of potassium hydrogen phthalate* that would be neutralized by 30 ml of 0.1 *M* NaOH. Have your calculation checked by your instructor before proceeding further.

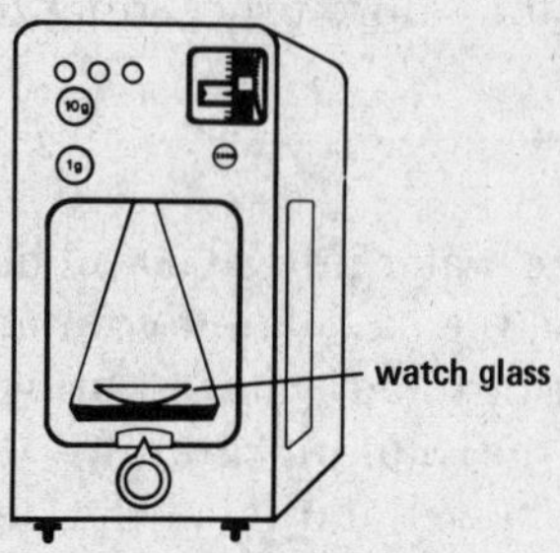

1. Obtain from your instructor, on a clean piece of aluminum foil, a sample of potassium hydrogen phthalate slightly in excess of the amount you will need. Weigh a clean, dry watch glass to the nearest 0.1 mg. Transfer to this the approximate amount of the phthalate required and reweigh to the nearest 0.1 mg. (Review, as necessary, the section on precision weighing in Appendix A.)

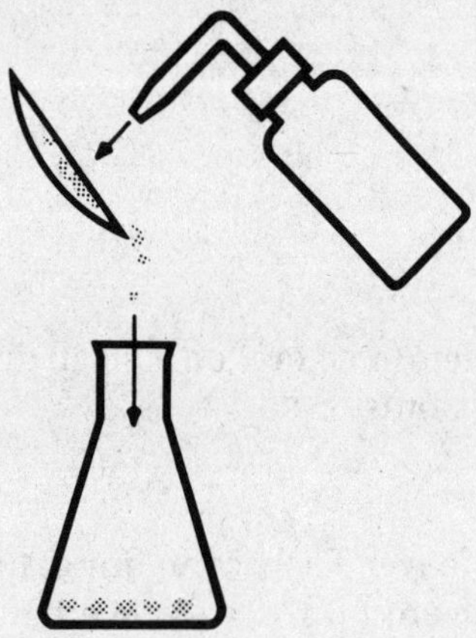

2. Carefully remove the watch glass from the balance and transfer the weighed phthalate *completely* into a clean 250-ml Erlenmeyer flask, rinsing the watch glass contents into the flask with a stream of distilled water.

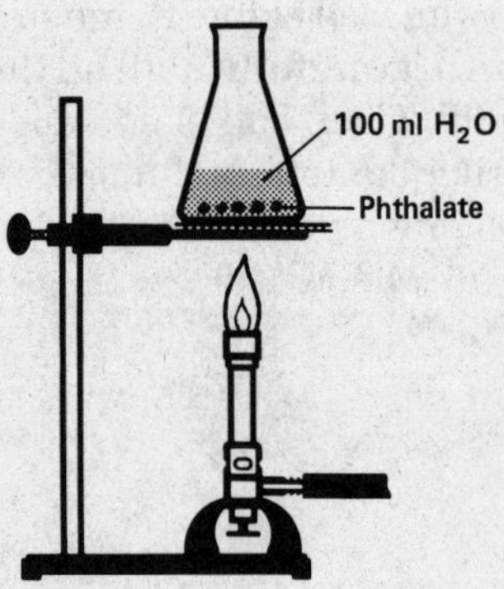

3. Add enough additional distilled water to the flask to make a total volume of ~100 ml. Heat the flask carefully until all the phthalate has dissolved. Set the flask on an insulating pad to cool to around room temperature.

*Note: The structural formula for potassium hydrogen phthalate is

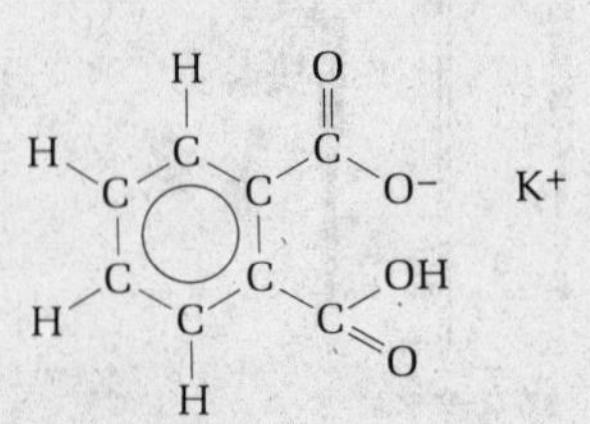

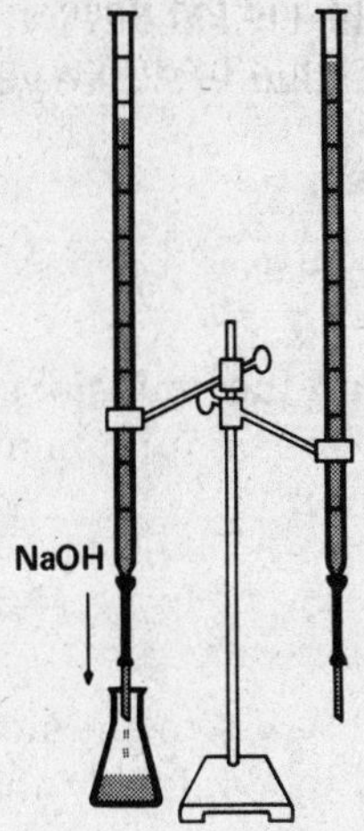

4. Add three drops of phenolphthalein indicator solution to the flask and set it on a sheet of white paper below the "NaOH" buret. Read and record the volume level in this buret, estimating to the nearest 0.01 ml. Titrate the phthalate solution to a permanent pale pink endpoint. Record the final buret reading. Clean the flask and rinse it with distilled water. Repeat the standardization if time allows.

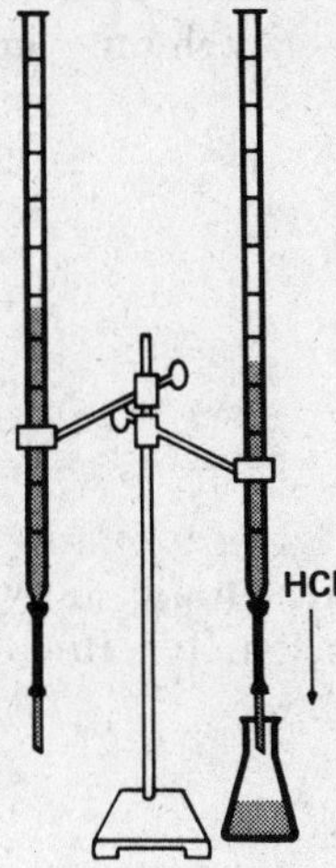

5. Record the reading of the "HCl" buret, then drain about 30 ml of this solution into the clean 250-ml Erlenmeyer flask. Record the final buret reading. Add about 70 ml of distilled water and three drops of the bromophenol blue* solution to the flask and swirl to mix the contents.

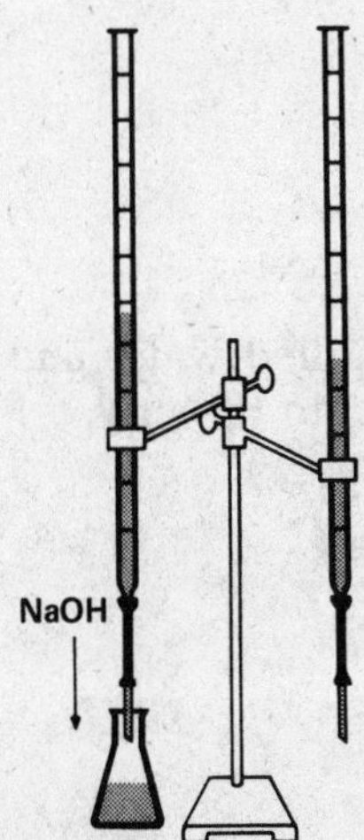

6. Refill the "NaOH" buret and adjust the liquid level to just below the zero mark. Record the buret reading, then titrate the HCl solution to a "pure" blue endpoint. Record the final buret reading. Clean the flask and rinse it with distilled water. Refill both burets and adjust the liquid levels to just below the zero marks. Repeat this standardization if time permits.

*Bromophenol blue is yellow in solutions of pH 3.0 or below and blue in solutions of pH 4.6 or above. At the desired endpoint of this titration the indicator color should be the first "pure" blue without any greenish tint. Familiarize yourself with the correct endpoint color by titrating 30 ml of 0.01 *M* HCl (prepared by mixing 3 ml of 0.1 *M* HCl with 27 ml of distilled water) with 0.1 *M* NaOH. When you reach the "basic" endpoint (blue), titrate to the "acidic" endpoint (yellow) with 0.1 *M* HCl. Go back and forth between these two endpoints until you are certain you can recognize the correct endpoint colors.

Calculate the molarities of the HCl and NaOH solutions in the burets to three significant figures. Have your instructor check these calculations.

C. Determination of an "Antacid's" Neutralizing Ability

Obtain from your instructor the antacid sample to be tested. Ask for suggestions on sample amount and HCl volume for sample dissolution.

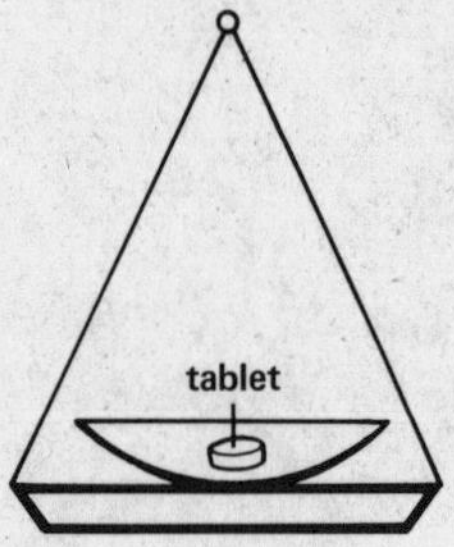

1. Weigh a clean dry watch glass to the nearest 0.1 mg. Then determine the mass of one antacid tablet. Ask your instructor how much antacid sample you should use in your titration. Weigh out this amount, breaking a tablet into pieces if necessary. Transfer the weighed sample completely into a clean 250-ml Erlenmeyer flask.

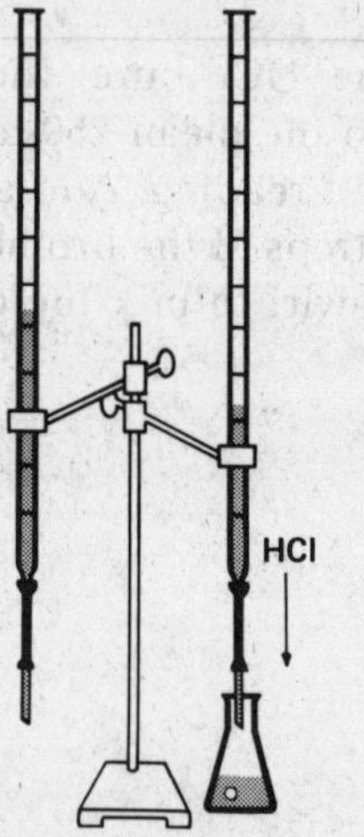

2. Using a buret, add to your sample the volume of HCl solution suggested by your instructor. Record the volume added to the nearest 0.01 ml.

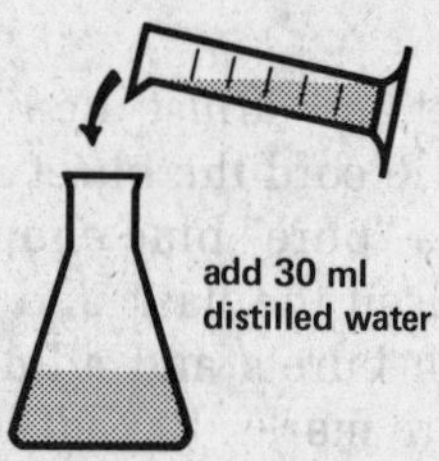

3. Add about 30 ml of distilled water and five drops of bromophenol blue solution to the flask.

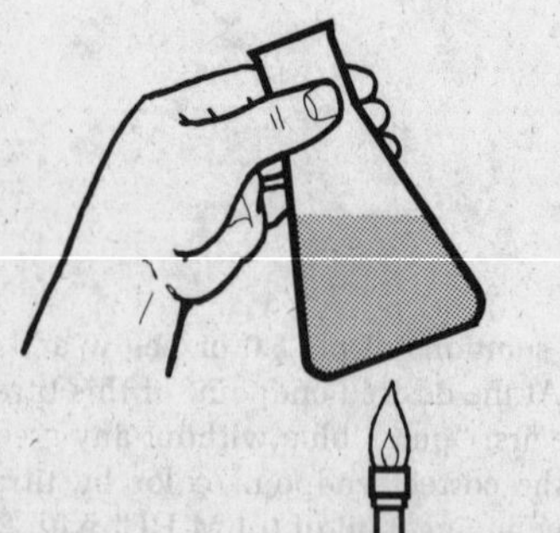

4. Warm the flask over a low flame, while swirling, until the antacid sample has disintegrated. (Do *not boil.*) Not all the material will dissolve (some inert "binders" and "fillers" are usually present), but you must obtain as complete a dissolution as possible. If the solution turns blue, add enough HCl from the buret to eliminate the blue color, plus 5 ml excess. Record the total volume of HCl added.

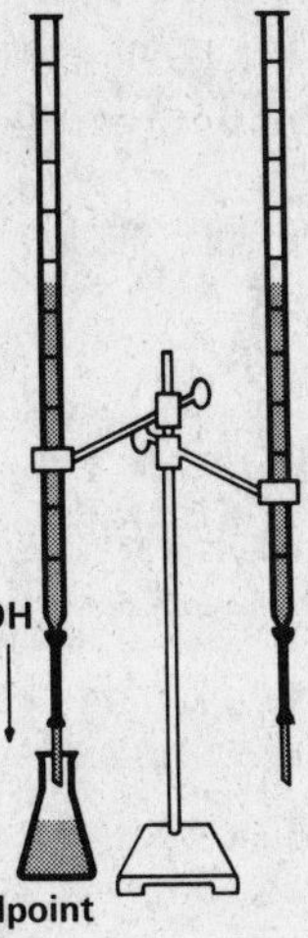

5. Allow the solution to cool to around room temperature, then titrate it with the sodium hydroxide solution, recording initial and final volumes accurately. You have reached the desired endpoint when a blue color without greenish tint persists for at least 30 sec.

CALCULATIONS AND REPORT

Using the molarities found for your HCl and NaOH solutions, the volume of HCl solution added to your antacid sample, and the volume of NaOH solution required to neutralize the acid not consumed by the antacid sample, calculate the number of moles of HCl consumed by your sample. From this result, the mass of your sample, and the mass of a single tablet of the antacid, calculate the number of moles of HCl that would be consumed by 1 g of the antacid preparation.

Report this result, and associated data, to your instructor. Near the end of the laboratory period, your instructor will post a collection of student reports. Copy this set of data into your permanent notebook. Use the data for the particular antacid you were given to calculate the average value for "moles of HCl per gram of antacid" and the average deviation of these results. Also, if so instructed, calculate the average value for "moles of HCl per gram of antacid" for other antacids used.

Use your notebook records and calculations to complete the Report Form.

GRADE / INSTRUCTOR'S INITIALS

Report For

FOR INVESTIGATION 17: Some Commercial Antacids

NAME: ____________ ID NO.: ____________

LAB SECTION ____________ INSTRUCTOR ____________

Solution Preparation
Show calculations on attached pages.

Volume of 12 *M* HCl required for 400 ml of 0.1 *M* HCl ____________ ml

Mass of NaOH required for 400 ml of 0.1 *M* NaOH ____________ g

Solution standardization
Show calculations on attached pages.

Phthalate sample

Sample & watch glass ____________ g

Watch glass ____________ g

Sample ____________ g

HCl solution sample

Final volume HCl ____________ ml

Initial volume HCl ____________ ml

Volume sample ____________ ml

Titration of phthalate

Final volume NaOH ____________ ml

Initial volume NaOH ____________ ml

Volume NaOH consumed ____________ ml

Titration of HCl

Final volume NaOH ____________ ml

Initial volume NaOH ____________ ml

Volume NaOH consumed ____________ ml

Concentration of NaOH solution ____________ *M*

Concentration of HCl solution ____________ *M*

Your Antacid Analysis
Show calculations on attached pages.

Name of antacid used ____________

One antacid tablet

Tablet & watch glass ____________ g

Watch glass ____________ g

Tablet ____________ g

Antacid sample used

Sample & watch glass ____________ g

Watch glass ____________ g

Sample ____________ g

HCl used

Final volume HCl ____________ ml

Initial volume HCl ____________ ml

Volume HCl added ____________ ml

Moles of HCl added ____________

Titration of excess HCl

Final volume NaOH ____________ ml

Initial volume NaOH ____________ ml

Volume NaOH consumed ____________ ml

Moles of NaOH consumed ____________

(OVER)

Moles of HCl consumed by antacid sample ______________

Moles of HCl consumed by one gram of the antacid formulation ______________

Average the values for "moles of HCl per gram of antacid." (Use the data of other students who analyzed the same antacid as you, and your own results.) List all the values for "moles of HCl per gram of antacid" and mass of one tablet and calculate the averages and average deviations (see Appendix C).

Average moles of HCl per gram of antacid ______ ± ______ moles

Average mass of one tablet ______ ± ______ g

Additional Problem

The hydrochloric acid concentration in a particular hyperacidic stomach was 0.03 M. If the volume of liquid in the stomach is 300 ml, what mass of the particular antacid tablet you worked with in this experiment would have to be taken in order to change the contents of the stomach to a "more normal" pH of 3.5?

______________ DATE ______________ SIGNATURE

Investigation 18

DETERMINATION OF AN IONIZATION CONSTANT

INTRODUCTION When we talk about acid strength, we are referring to the extent to which the acid is ionized in solution. This concept is often confused with considerations of corrosive activity, unique chemical properties, or concentration. Most of us have seen television commercials for a "household ammonia" product with the user exclaiming, "This smells *strong*." The odor "strength" in this case, of course, is a function of the concentration of the solution. (Ammonia is classified as a *weak* base, at any concentration, because the equilibrium system consists principally of aqueous ammonia, with only small relative amounts of the NH_4^+ and OH^- ions from reaction of water with NH_3). Many people think of hydrofluoric acid as "strong" because it will etch glass. Actually, HF is a weak acid ($K_a = 6.9 \times 10^{-4}$). The reaction of HF with glass results from the formation of the strong Si—F bonds, in SiF_4 or $[SiF_6]^{2-}$, not from the "acidity" of the solution. *Aqua regia*, a mixture of nitric and hydrochloric acids, is often considered as "super strong" because this is one of the few reagents that will react with gold or platinum. In fact, both HCl and HNO_3 are strong in terms of percentage ionization, but not to the extent of $HClO_4$ (which does not react with gold or platinum). Aqua regia reacts with gold or platinum, not as just a proton source, but by the combined effect of oxidation by nitrate ion and coordination by chloride ion. The "driving force" for the reaction is principally the formation of complexes such as $[AuCl_4]^-$ or $[PtCl_6]^{2-}$.

Acid strength, then, in proper chemical usage, simply refers to the degree of ionization of the acid in solution. The organic acids, either phenols or carboxylic acids, are typically weak. For example,

phenols

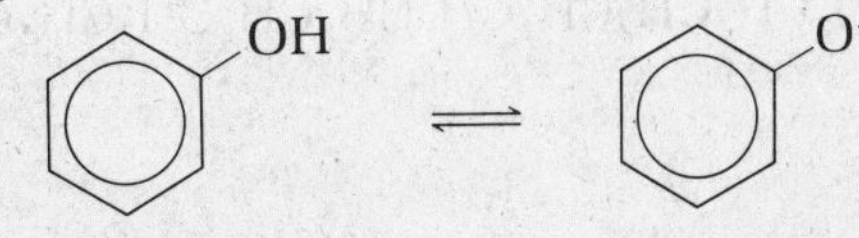

$+ H^+$ $(K_a = 1.1 \times 10^{-10})$

(phenol)

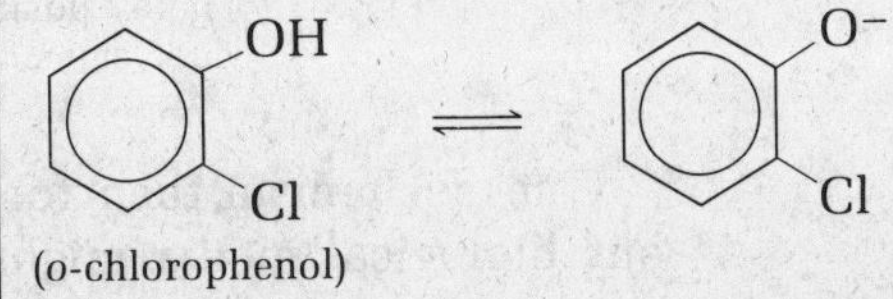

$+ H^+$ $(K_a = 7.7 \times 10^{-9})$

(o-chlorophenol)

carboxylic acids

$$HC(=O)—OH \rightleftharpoons HC(=O)—O^- + H^+ \quad (K_a = 2.0 \times 10^{-4})$$

(formic acid)

$$CH_3CH_2CH_2C(=O)—OH \rightleftharpoons CH_3CH_2CH_2C(=O)—O^- + H^+ \quad (K_a = 1.5 \times 10^{-5})$$

(butyric acid)

The fact that carboxylic acids are weak (i.e., only slightly ionized) has implications in biological systems. Fatty acids from the hydrolysis of glyceryl esters of the oils and fats in food (Investigations 6 and 10) are only very slightly soluble in water. If we are to metabolize fatty acids, they must be converted to some form suitable for transport through the aqueous systems of the blood and other biological fluids. Although the carboxyl group readily forms hydrogen bonds with water, this factor alone is insufficient to overcome the nonpolar characteristics of the long carbon chain. For example,

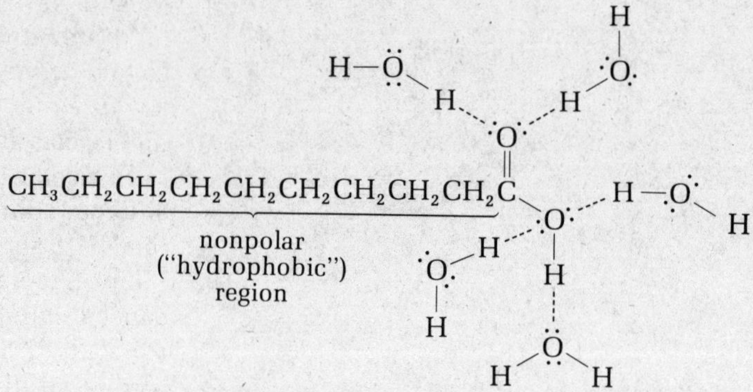

(an undissociated fatty acid: water insoluble)

If the fatty acid were ionized, the attraction of polar water molecules to the negatively charged region of the carboxylate anion would be significantly greater than the simple hydrogen-bonding attractions of the electrically neutral carboxyl group. Salts of fatty acids are soaps (Investigation 6) and these readily form stable dispersions in water:

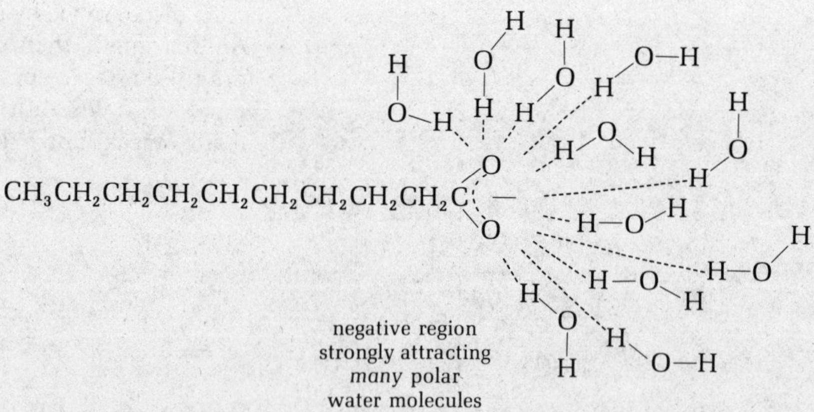

One mechanism, then, for dispersing fatty acids in aqueous biological media might involve the reaction of the weakly ionized acid with a base stronger than water, for example, hydroxide ion, to produce the anion of the fatty acid to a much greater extent than would the "normal" ionization of the acid alone (with water as the only proton acceptor). This mechanism is not utilized to any significant extent because of the relatively low hydroxide-ion concentrations of most biological fluids. However, the formation of fatty acid anions *is* involved in the rapid transport system of the blood. Salts of fatty acids with the "soluble"

serum albumin (proteins) are transported through the blood to provide delivery of fatty acids for metabolic processes serving as energy sources. The salt formation results from the action of basic nitrogen centers in the albumin as proton acceptors.

The degree of ionization of a weak acid is also related to the function of that acid, and its anion, in a conjugate-pair buffer system (Investigation 19). We may represent the equilibrium system of an aqueous solution of a carboxylic acid, for example, as

$$RCO_2H \rightleftharpoons RCO_2^- + H^+$$

(R represents the carbon chain)

for which

$$K_a = \frac{C_{RCO_2^-} \times C_{H^+}}{C_{RCO_2H}}$$

(ionization constant)

We may rearrange this expression to give

$$C_{H^+} = K_a \times \frac{C_{RCO_2H}}{C_{RCO_2^-}}$$

Thus the hydrogen-ion concentration (which may also be expressed as the pH) of this system depends only on the value of the ionization constant and the *ratio* of concentrations of the acid and its anion. This assumes, of course, that the ionization of the acid is the only significant source of H^+. This assumption is valid only for those cases in which the concentrations and K_a values are such that H^+ from the ionization ("autoprotolysis") of water is negligible.

We should be able to determine the ionization constant for an acid by measuring the pH of a solution for which the equilibrium concentration of undissociated acid is known under conditions such that $C_{H^+} = C_{RCO_2^-}$. Alternatively, K_a could be found from a measurement of equilibrium pH and of the equilibrium concentrations of both the undissociated acid and its anion. The problem in either of these cases is the difficulty of measuring *equilibrium* concentrations, since the analytical method employed must, of necessity, be one that does not disturb the equilibrium. Any chemical reaction, as in a titration, for example, *will* disturb the equilibrium system. We could, of course, use a titration to determine the total amount of acid in a solution. If either the hydrogen-ion or the anion (RCO_2^-) concentration could then be measured under *equilibrium* conditions, appropriate calculations could be employed for the K_a determination. Experimentally, the hydrogen-ion concentration of a solution *can* be measured by a pH meter under conditions that, for all practical purposes, correspond to those of equilibrium.

EXAMPLE

An aqueous solution of propionic acid ($CH_3CH_2CO_2H$) was found to have a pH of 2.70. A 25.0-ml aliquot of this solution was titrated to the equivalence point by 38.6 ml of 0.185 *M* NaOH. What is the ionization constant for propionic acid?

SOLUTION

Since the propionic acid is monoprotic, the equivalence point of the titration corresponds to an addition of the number of moles (or millimoles) of OH^- equal to the total number of moles (or millimoles) of the acid *available*. The "label concentration"* (which specifies the *total available* solute, rather than the equilibrium concentration) may be calculated as

$$\frac{0.185 \text{ mmol (NaOH)}}{\text{ml (NaOH)}} \times \frac{38.6 \text{ ml (NaOH)}}{25.0 \text{ ml } (CH_3CH_2CO_2H)} \times \frac{1 \text{ mmol } (CH_3CH_2CO_2H)}{1 \text{ mmol (NaOH)}} = 0.286\ M$$

Then, from the equilibrium pH

$$C_{H^+} = 2.0 \times 10^{-3}\ M$$

and, assuming the equilibrium system to be adequately represented by

$$CH_3CH_2CO_2H \rightleftharpoons CH_3CH_2CO_2^- + H^+$$

the equilibrium concentrations are

$$C_{CH_3CH_2CO_2^-} = C_{H^+} = 2.0 \times 10^{-3}\ M$$

$$C_{CH_3CH_2CO_2H} = 0.286 - 0.002 = 0.284\ M$$

The ionization constant is then

$$K_a = \frac{C_{H^+} \times C_{CH_3CH_2CO_2^-}}{C_{CH_3CH_2CO_2H}} = \frac{(2.0 \times 10^{-3})^2}{0.284} = 1.4 \times 10^{-5}$$

Experimentally, it is difficult to find the "label concentration" (formality) of a weak acid *accurately* by a conventional titration. As the reaction progresses, the combination of acid and anion acts as a buffer, so that the pH change is gradual over a wide concentration range. At the equivalence point, the solution (of the salt of the weak acid) is alkaline and added increments of NaOH will produce large pH changes. Although such a titration can yield a reasonable *approximation* of the "label concentration" (as in Investigation 32), it is often less accurate than we would like. In addition, the pH at the equivalence point of the titration is a function of the ionization constant:

$$RCO_2^- + H_2O \rightleftharpoons RCO_2H + OH^-$$

$$K_{b(RCO_2^-)} = \frac{K_w}{K_{a(RCO_2H)}}$$

If the ionization constant is not known, it is difficult to select the best indicator for the titration.

A better method for accurately determining the information needed for a K_a calculation utilizes a graphical treatment of pH versus concentration data. Such *titration curves* (Figure 18.1) indicate conditions under which the data most appropriate for accurate calculation may be selected. In this Investigation we will use this method, employing a pH meter to monitor the pH change during an acid-base titration.

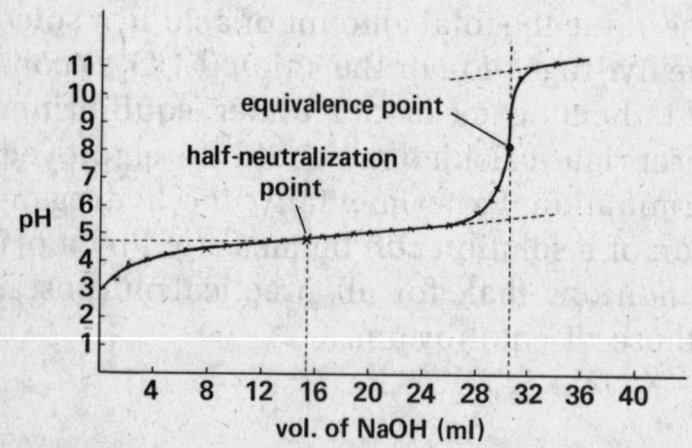

FIGURE 18.1
Titration curve for acetic acid. (The equivalence point corresponds to the inflection point of the curve in the nonbuffered region.)

*This is sometimes referred to as the *formality* (F) of the solution.

Distilled water saturated with air is slightly acidic, due to the CO_2/HCO_3^- equilibrium system resulting from dissolved CO_2 from the air:

$$CO_2(aq) + H_2O(l) \rightleftharpoons HCO_3^-(aq) + H^+(aq)$$

To avoid the complications from this system, the distilled water used may be boiled to expel dissolved CO_2. If this is not done, then appropriate compensation must be made for the CO_2/HCO_3^- system.

Before starting this investigation, you should be familiar with the concepts and calculations involved in solution preparation, pH of weak acids and bases, and acid-base titration. You should be familiar also with the construction and interpretation of titration curves. In addition, prior to coming to the laboratory, you must have studied the information provided by your instructor concerning the proper use of a pH meter.

References

1. O'Connor, Rod. 1977. *Fundamentals of Chemistry* (2nd ed.). New York: Harper & Row (Units 19–21 and Excursion 2).
2. Drago, Russell and N. A. Matwiyoff. 1968. *Acids and Bases*. Lexington, Mass.: Raytheon.

SAFETY Wear approved eye protection at all times and exercise standard precautions in working with reagents.

OBJECTIVES

1. Be able to use a pH meter properly.
2. Be able to perform an acid-base titration, using a pH meter to monitor the course of reaction.
3. Be able to prepare and interpret titration curves accurately.
4. Be able to use data from titration curves for the calculation of ionization constants.

Before starting experimental work, check with your instructor to be sure that you are fully aware of the proper techniques for using the type of pH meter available to you, including:

a. How to protect the meter from electrical damage.
b. How to protect the electrodes from physical, chemical, or electrical damage.
c. How to rinse and dry the electrodes, and the occasions on which rinsing or drying (or both) is necessary.
d. How to use the meter's "temperature setting," if appropriate.
e. How to calibrate the pH meter.
f. How to use the pH meter for pH measurement.
g. How the meter is to be left when you have finished using it.
h. How to record meter use and any meter malfunction, if such records are maintained.

In addition, be sure you are familiar with the proper operation of a magnetic stirrer for use in conjunction with the pH meter.

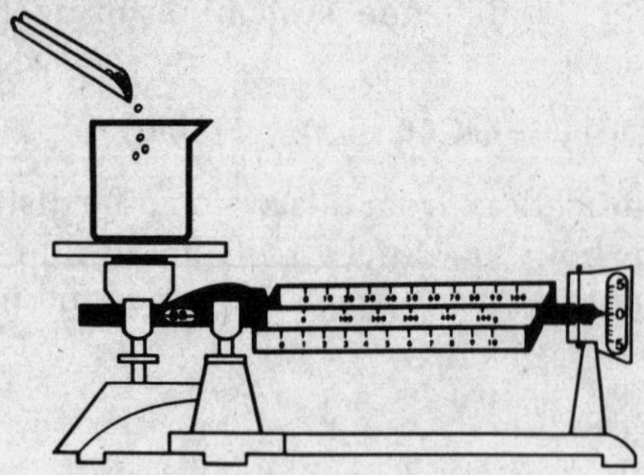

1. Calculate the mass of NaOH required for preparation of 200 ml of 0.1 *M* NaOH. Weigh out this amount of NaOH pellets (*corrosive*) into a preweighed 400-ml beaker, using a triple beam balance. Remove the beaker from the balance and, using your volumetric flask as a measuring device, pour 200 ml of "CO_2-free" distilled water* into the beaker. Stir or swirl the mixture until a homogeneous solution is obtained. Label the beaker, cover it with a paper towel, and set it aside for later use.

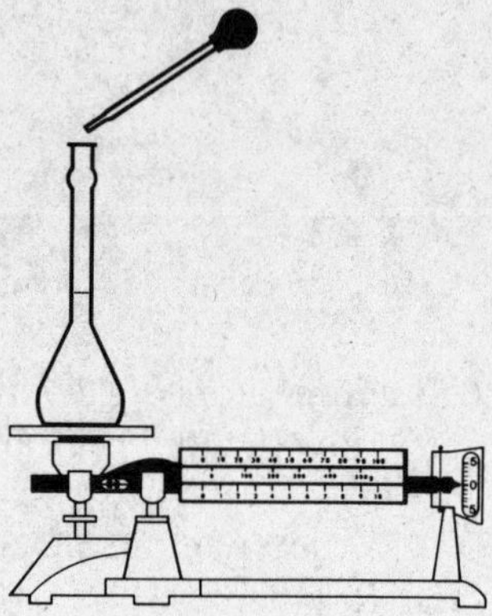

2. Calculate the mass of CH_3CO_2H required for preparation of 100 ml of 0.10 *M* CH_3CO_2H. Weigh a clean (but not necessarily dry) 100-ml volumetric flask on a triple beam balance. Using a clean medicine dropper, add the calculated mass of acetic acid to the flask. Remove the flask from the balance and add "CO_2-free" distilled water to the 100-ml mark. Stopper the flask and invert several times to mix the contents.

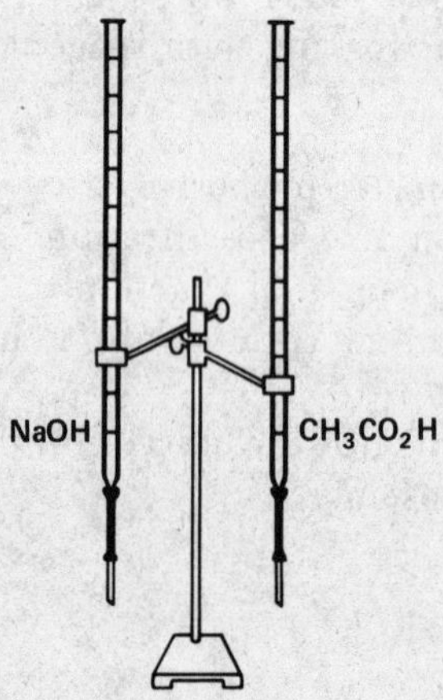

3. Obtain and, if necessary, clean (Appendix A) two 50-ml burets. Mount these securely on a ringstand. Rinse and fill the left buret with the 0.1 *M* NaOH solution and the other buret with the 0.10 *M* CH_3CO_2H solution. Label the burets as to their respective contents. Drain enough liquid from each buret into a "waste" flask to expel air from the buret tip and to drop the liquid level below the zero mark.

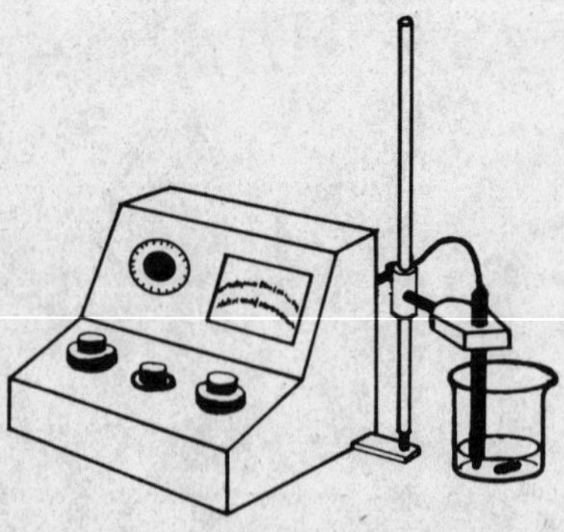

4. After your instructor has checked your understanding of the proper use of a pH meter and has assigned a meter for your use, obtain a standard buffer solution and use this to standardize the assigned pH meter. Record all procedures used in your notebook.

*If "CO_2-free" water is not used, ask your instructor how to compensate for dissolved CO_2.

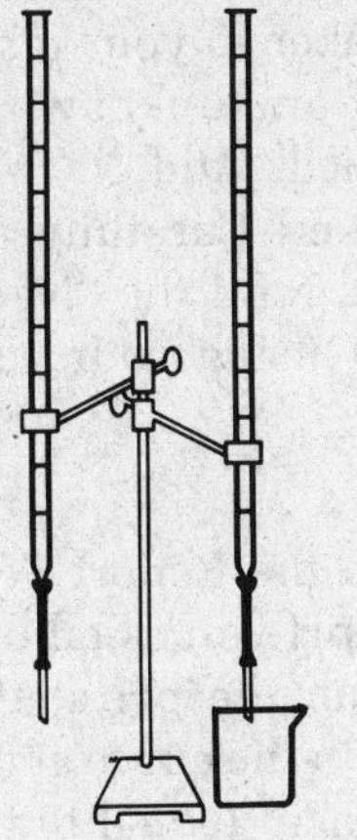

5. Measure 30.0 ml of your 0.10 *M* acetic acid from the buret into a clean 250-ml beaker. Add two drops of phenolphthalein indicator and 40 ml of "CO_2-free" distilled water. *Carefully* add a magnetic stirring bar to the solution.

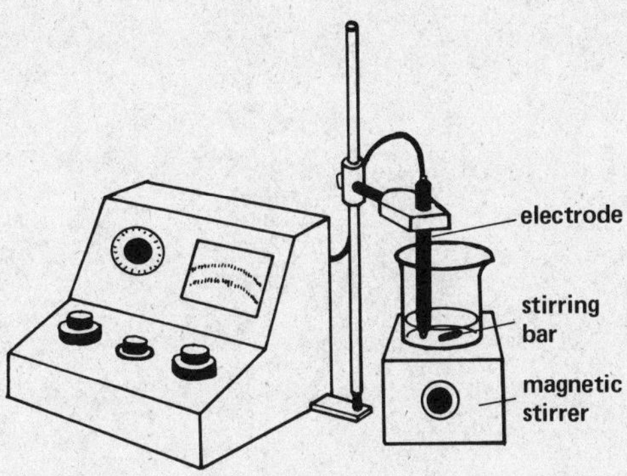

6. Check that the switch is off on the magnetic stirrer, then set the beaker of acetic acid solution on the stirrer. With the speed control at a "slow" setting, turn on the stirrer and check that the stirring bar turns properly in the center of the beaker. Turn the stirrer off and *carefully* insert the pH electrode into the solution so that it is immersed at least 1 cm, with the bottom of the electrode at least 0.5 cm above the stirring bar. If the solution is too shallow for proper electrode immersion, add just enough more "CO_2-free" distilled water to allow for proper electrode immersion. Using the pH meter properly, measure the pH of the solution. Record this value, as the "initial pH," in your notebook.

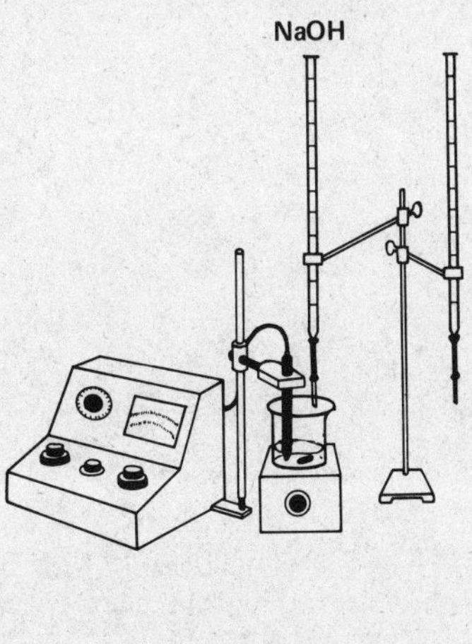

7. Arrange your burets to permit addition of the 0.1 *M* NaOH solution into the reaction beaker. With the stirrer on, begin addition of the NaOH solution. You will need to use the following technique, recording the information indicated:

a. Add NaOH solution in 2.0 ml increments, recording each volume to the nearest 0.01 ml and the pH after each addition, until a pH of 2 units greater than the "initial pH" is obtained.
b. Add NaOH solution in 0.5 ml increments, recording each volume to the nearest 0.01 ml and the pH after each addition, until the phenolphthalein indicator changes to a permanent pale pink. Identify the pH at this point as the "endpoint pH."
c. Continue addition of 0.5 ml increments, recording volumes and pH readings as before, until the pH is at least 10 and further increments of NaOH solution each produce less than a 0.05 change in pH. (Do not exceed pH 12.)

Then turn off the stirrer, remove the reaction beaker, rinse the electrode, and immerse the electrode in the "storage solution" as directed by your instructor.

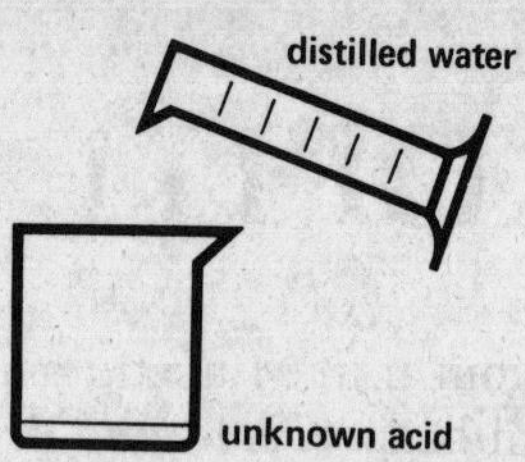

8. Take a clean 250-ml beaker to your instructor and obtain a specified volume of an "unknown acid" solution. Record the volume obtained. Add 50 ml of "CO_2-free" distilled water to the beaker. Carefully add a magnetic stirring bar to the solution, refill the "NaOH" buret, and repeat procedures 6 and 7 using your "unknown acid."

CALCULATIONS AND REPORT

Prepare titration curves on the Report Form for each of your titrations by plotting pH versus ml of NaOH added. Determine for each solution the pH at the equivalence point (the inflection point* in the curve at the nonbuffered region) and the pK_a of the acid (equal to the pH at "half-neutralization," i.e., at the inflection point in the buffered region).

Use your notebook records and titration curves to complete the remainder of the Report Form.

*Consult your instructor for a suitable method for determining the inflection point.

GRADE / INSTRUCTOR'S INITIALS

Report Form

FOR INVESTIGATION 18:
Determination of an Ionization Constant

NAME: ______________________ ID NO.: ______________________

LAB SECTION ______________________ INSTRUCTOR ______________________

Use the back of this page for construction of titration curves.

Solution Description

NaOH	CH_3CO_2H	"Unknown"
NaOH + beaker ________ g	CH_3CO_2H + flask ________ g	Number (or code) ________
Beaker ________ g	flask ________ g	Volume obtained ________ ml
NaOH ________ g	CH_3CO_2H ________ g	
Solution volume ________ ml	Solution volume ________ ml	
$\sim C_{NaOH}$ ________ M	$\sim C_{CH_3CO_2H}$ ________ M	

Titration and pH Data

	CH_3CO_2H	"Unknown"
Initial pH		
pH at phenolphthalein endpoint		
pH at equivalence point (from curve)		
Volume (ml) NaOH to reach equivalence point		
pK_a (equal to pH at "half-neutralization")		
K_a		

Questions

Attach separate pages with answers.

1. Why is the pH of the endpoint not necessarily the same as the pH of the equivalence point?
2. Why was it unnecessary to know the reactant concentrations to a high degree of accuracy?
3. What was the approximate "label concentration" (formality) of your "unknown" acid? How did you determine this?

______________ DATE ______________ SIGNATURE

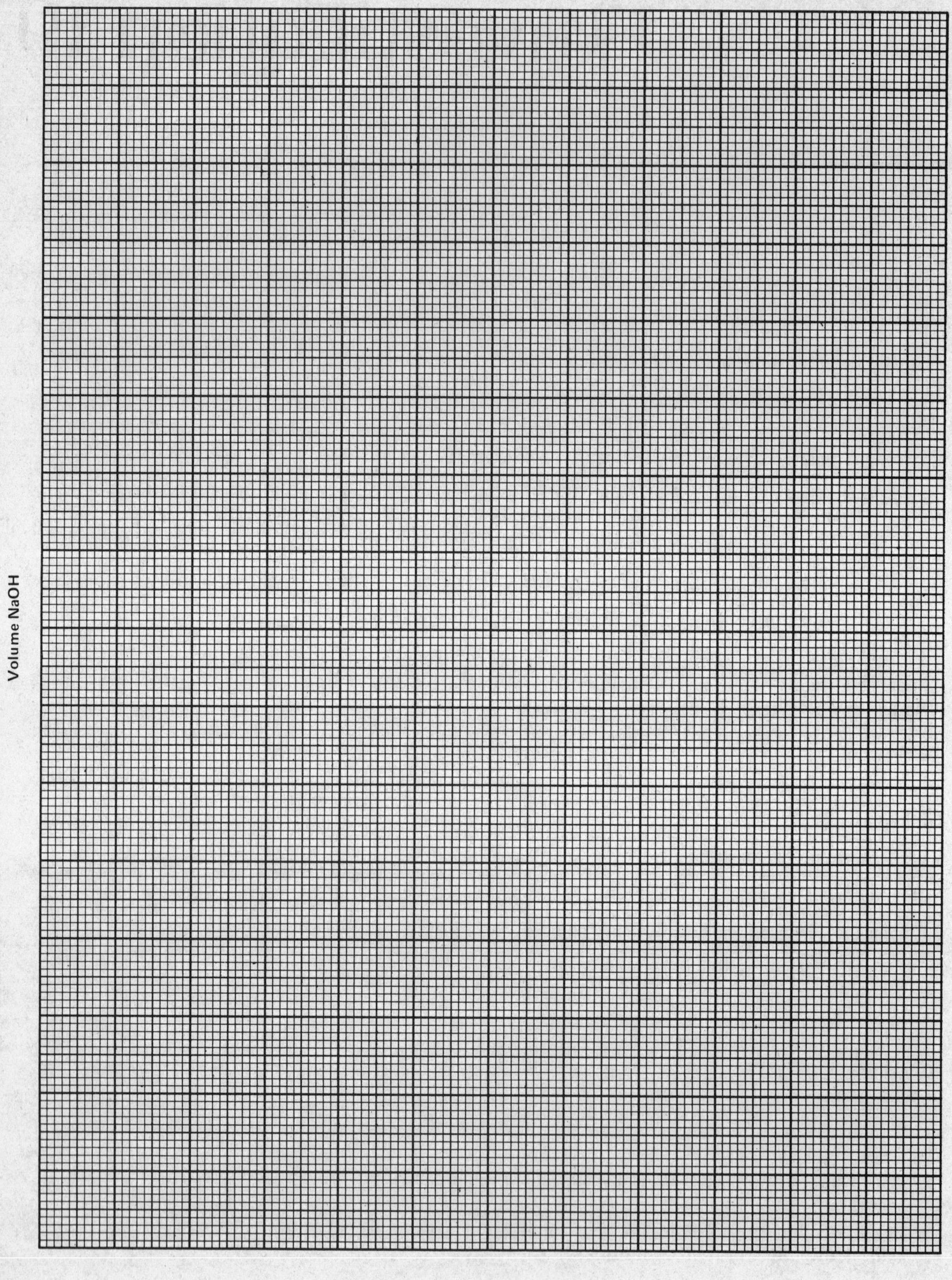
Volume NaOH
pH

Investigation 19

BUFFER SYSTEMS

INTRODUCTION The complex chemical systems of living organisms are often so pH dependent that a relatively small pH change may be fatal to the organism. Human blood, for example, normally has a pH of 7.4 (± 0.1). Should the blood vary by as little as 0.5 pH units, death usually results. If one drop (0.05 ml) of 6 *M* HCl is added to a liter of pure distilled water, the pH changes by 3.6 pH units (from 7.0 to 3.4). If the blood had no more resistance than distilled water to pH change, a little acid entering the blood stream would quickly produce a fatal reaction. Fortunately, the blood *is* protected against pH change. Addition of the drop of 6 *M* HCl to a liter of blood, for example, causes a pH change of less than 0.05 units. Human blood and many other important biological systems are protected against serious pH change by *buffers.*

Buffer solutions have two special characteristics: They maintain a fairly constant pH when diluted and they contain species capable of neutralizing both added bases and acids. Most biological buffers are fairly complex. That of the blood, for example, contains dissolved CO_2, H_2CO_3, HCO_3^-, CO_3^{2-}, H_3PO_4, $H_2PO_4^-$, HPO_4^{2-}, and PO_4^{3-}. Only a few of these species are present in significant concentrations; the rest play no appreciable role in the buffer action. Even so, the blood buffer system is difficult to treat by a simple mathematical analysis. Let's consider, instead, a less complex system that still illustrates the important characteristics of buffers.

The simplest buffer systems are those containing approximately equimolar amounts of both members of a monoprotic conjugate pair, such as NH_3/NH_4^+ or $CH_3CO_2H/CH_3CO_2^-$. We can represent the equilibrium system of an acetic acid-acetate solution by

$$CH_3CO_2H \rightleftharpoons CH_3CO_2^- + H^+$$

for which the ionization constant is

$$K_a = \frac{C_{H^+} \times C_{CH_3CO_2^-}}{C_{CH_3CO_2H}}$$

(K_a for acetic acid is 1.8×10^{-5}.)

If we solve this expression for the hydrogen-ion concentration

$$C_{H^+} = K_a \times \frac{C_{CH_3CO_2H}}{C_{CH_3CO_2^-}}$$

we can see why the buffer maintains a fairly constant pH when diluted. Since the same solution contains both acetic acid and acetate ion, both concentrations are affected equally by dilution. The hydrogen-ion concentration (and, hence, the pH) is a function only of the value of K_a and the *ratio* of the concentrations of the two buffer components.* A buffer of this type also illustrates why added acid or base can be consumed without a large pH change (within certain limits). Added acid is neutralized by the acetate in the buffer:

$$CH_3CO_2^- + H^+ \longrightarrow CO_3CO_2H$$

whereas the acetic acid component can react with added base:

$$CH_3CO_2H + OH^- \longrightarrow CH_3CO_2^-$$

As long as significant amounts of *both* buffer components remain unconsumed, the pH of the solution changes very little.

This is an important restriction. Although the pH of a buffer is a function of component concentration *ratio,* the amount of strong acid or strong base that a buffer can absorb without a large change in pH (that is, the *buffer capacity*) depends on actual amounts of buffer components available.

In this experiment we will investigate the buffer capacities of some simple systems. The concept of *buffer capacity* is important in understanding why the buffers of biological systems do not provide an unlimited protection against added acid or base.

The *buffer capacity* of a solution is defined as the number of moles of added H^+ (or OH^-) that will change the pH of 1 liter of the buffer by 1.0 pH unit. Note that the *buffer capacity* of a solution may be different for added acid (H^+) or added base (OH^-). For a conjugate-pair buffer containing different numbers of moles of buffer components, the amount of H^+ that will deplete the basic component of the buffer will not be the same as the amount of OH^- for depletion of the acidic component.

Before starting this investigation, you should be familiar with the concepts of conjugate-pair buffer activity and of buffer capacity, as well as with the calculations of solution concentration, dilution, pH, and buffer pH change. In addition, before coming to the laboratory, you must perform the calculations indicated in the Procedures section of this investigation.

References

1. O'Connor, Rod. 1977. *Fundamentals of Chemistry* (2nd ed.) New York: Harper & Row (Units 13 and 20.)
2. O'Connor, Rod and Charles Mickey. 1974. *Solving Problems in Chemistry.* New York: Harper & Row (Unit 11).

*To illustrate one of the limitations on this simplistic approach, calculate the pH of

a. A solution that is 0.1 *M* in CH_3CO_2H and 0.1 *M* in $CH_3CO_2^-$.
b. A solution that is 0.01 *M* in CH_3CO_2H and 0.01 *M* in $CH_3CO_2^-$.
c. A solution that is 10^{-15} *M* in CH_3CO_2H and 10^{-15} *M* in $CH_3CO_2^-$.

Does it seem reasonable to you that solution (c) might have a pH close to 7, rather than the pH calculated? If so, why?

SAFETY Wear approved eye protection and exercise proper care in working with acids and bases. In case of skin contact with reagents, wash quickly and *thoroughly* with water.

OBJECTIVES

1. Be able to prepare solutions accurately.
2. Be able to perform titrations properly.
3. Be able to make appropriate calculations dealing with a conjugate-pair buffer system.

PROCEDURES

In part A of this experiment you will be working in teams of six, as assigned by your instructor. Remember to observe carefully and keep records, even when you are not personally working on some part of the experiment.

A. Solution Preparation

Calculate the mass of sodium hydroxide required for preparation of 600 ml of 0.1 *M* NaOH. Check this calculation with the five other team members. The team will prepare and then equally divide this solution.

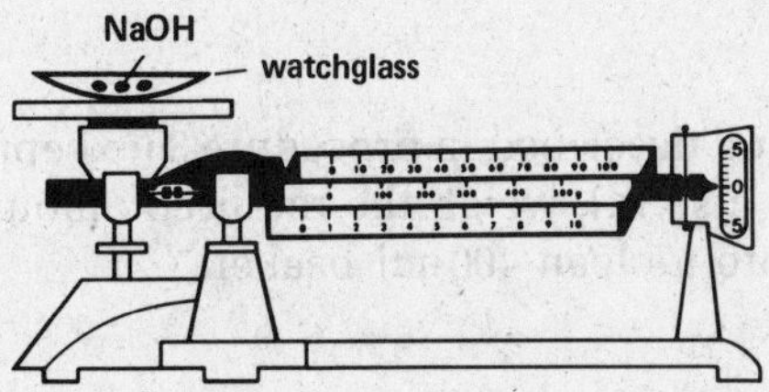

1. Weigh a watch glass to the nearest 0.1 g. Then weigh the amount of sodium hydroxide calculated.

Caution: Sodium hydroxide is a very caustic chemical.

2. Using a 250-ml graduated cylinder, measure 200 ml of distilled water into a 600-ml beaker. Add the weighed sodium hydroxide to this, with stirring. Then pour in an additional 400 ml of distilled water, stirring until all NaOH has dissolved and a homogeneous solution has been obtained. Each team member is to obtain 100 ml of this solution in a clean 250-ml beaker. Cover and label the beaker.

Calculate the mass of "glacial" acetic acid required for preparation of 400 ml of 0.10 *M* acetic acid. Check this answer with your teammates. The team will prepare and equally divide this solution.

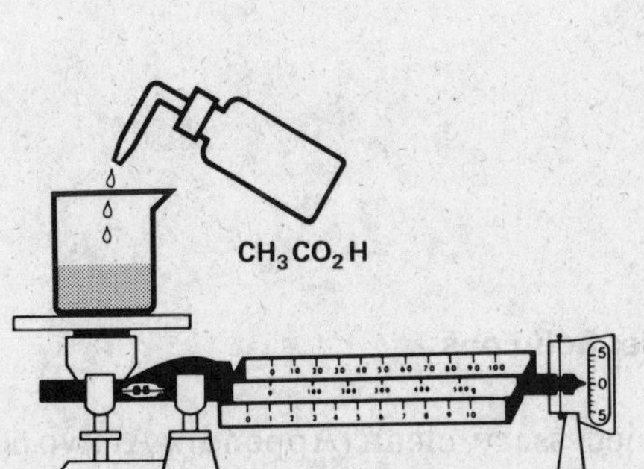

3. Weigh a clean 400-ml beaker to the nearest 0.1 g. Add 100 ml of distilled water and reweigh. Then advance the balance masses by the amount calculated for the acetic acid needed. Carefully add *just enough* acetic acid from a squeeze bottle to return the balance to equilibrium. If too much is added, you will have to start over.

Caution: Glacial acetic acid is a corrosive reagent.

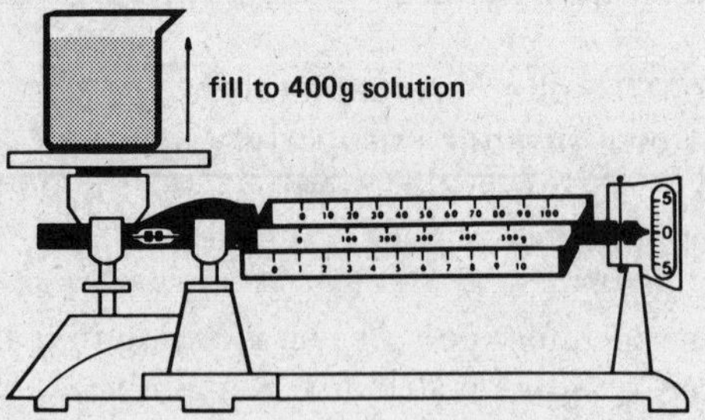

4. Then advance the balance masses by an amount 400 g heavier than the *empty* beaker weight. Carefully add *just enough* distilled water to return the balance to equilibrium. Add final increments dropwise. Remove the beaker and stir its contents until a homogeneous solution is formed. Each team member is to obtain 65 ml of this solution in a clean 100-ml beaker. Cover and label the beaker.

Calculate the mass of sodium acetate required for preparation of 400 ml of a 0.10 *M* solution and check your answer with your teammates. The team will prepare and equally divide this solution.

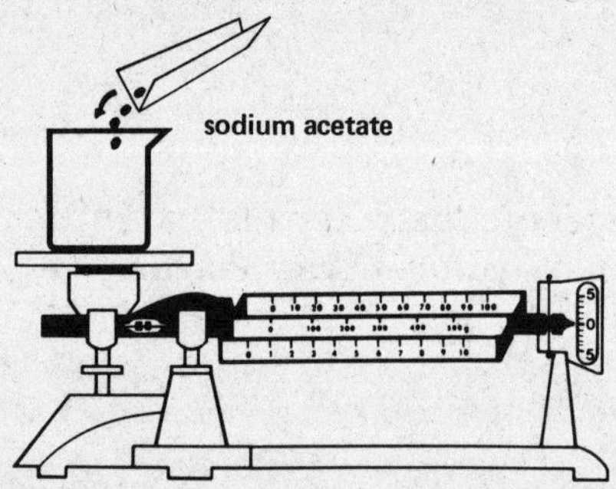

5. Using the method described in procedure 3 (except that sodium acetate is a solid), weigh the required amount of sodium acetate into a clean 400-ml beaker.

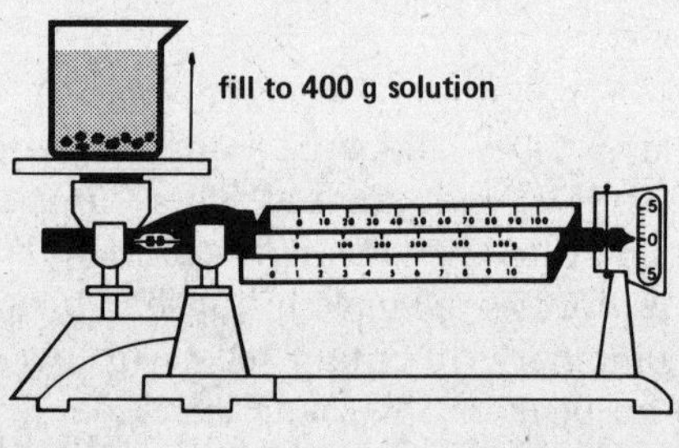

6. Using the method described in procedure 4, weigh sufficient distilled water to provide 400 g of total solution. Remove the beaker and stir until a homogeneous solution is formed. Each team member is to obtain 65 ml of this solution in a clean 400-ml beaker. Cover and label the beaker.

Note: If your solutions have been prepared properly, the concentrations of acetic acid and of sodium acetate will be known with sufficient accuracy. Since sodium hydroxide pellets absorb both CO_2 and moisture from the air rather rapidly, the composition of this solution is not very accurately determined from solute mass and solution volume. Your instructor may wish to have you standardize your sodium hydroxide solution against potassium acid phthalate, as described in the experiment on Some Commercial Antacids (Investigation 17).

B. Titration of Buffer Solutions

1. Obtain and, if necessary, clean (Appendix A) two 50-ml burets. Mount these securely on a ringstand and rinse and fill the left buret with the 0.1 *M* NaOH solution. Label this buret "NaOH." Then pair up with one of your neighbors.

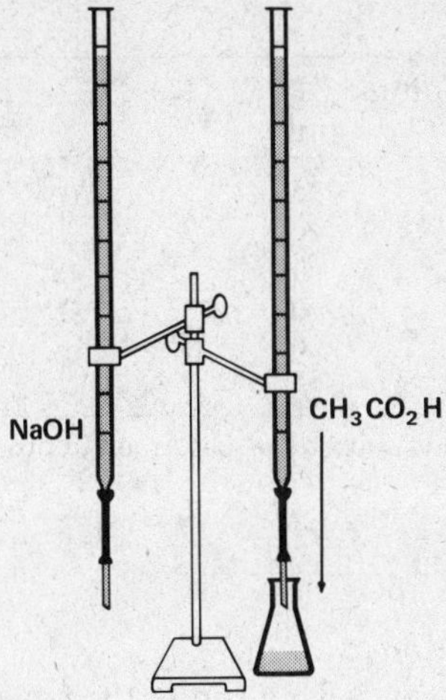

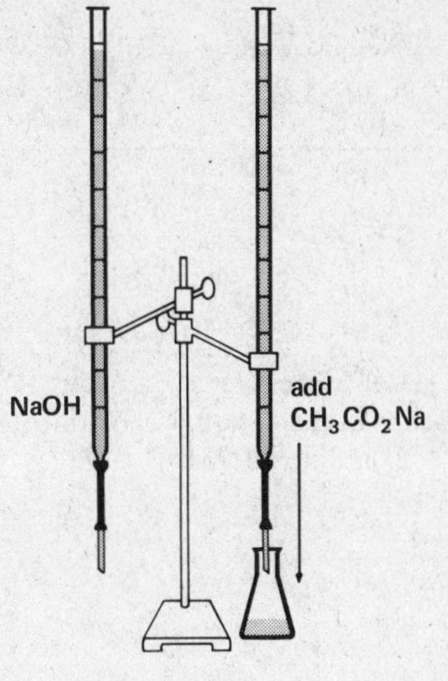

One of you will rinse and fill his second buret with the 0.10 *M* acetic acid and the other will rinse and fill buret with the 0.10 *M* sodium acetate solution. Label the burets properly. Drain just enough liquid into a waste flask from both your burets to expel air from the buret tips and to drop the liquid levels just below the zero marks.

2. Each of you will prepare the solutions* specified in Table 19.1 and will titrate each of these to a pale pink endpoint, recording titrant volumes accurately. Use your 125-ml Erlenmeyer flask as the titration vessel. (Review the titration procedures described in Investigation 17.)

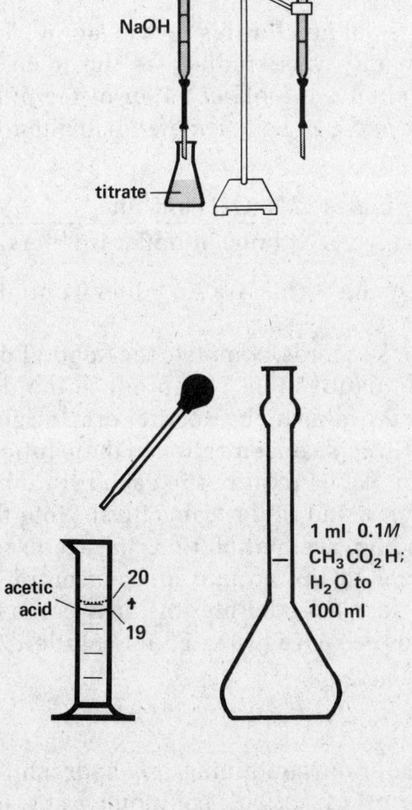

3. In order to prepare solution 4 proceed as follows: Measure 1.0 ml of 0.1 *M* acetic acid from the buret into a clean (but not necessarily dry) 100-ml volumetric flask. Fill the flask with distilled water to the 100-ml mark and mix the solution thoroughly. Add distilled water to your 25-ml graduated cylinder until the water meniscus just reaches the 19-ml mark. Then add the dilute acetic acid solution by means of a medicine dropper from your volumetric flask to the graduated cylinder until the meniscus of the solution just reaches the 20-ml mark. Stir the solution in the cylinder with a glass rod. Pour all the liquid into the Erlenmeyer flask and then add 20 ml of distilled water to bring the total volume to 40 ml. Titrate this solution with 0.1 *M* sodium hydroxide to the phenolphthalein endpoint. Calculate the initial pH of the dilute acetic solution you prepared.

*If a pH meter is available, ask your instructor to teach you how to use it to measure the pH of the solutions prepared *before* beginning the titrations.

Table 19.1

Solution No.	Volume H_2O (ml)	Volume CH_3CO_2H (ml)	Volume CH_3CO_2Na (ml)	Phenolphthalein solution
1	10	15	15	2 drops
2	20	10	10	2 drops
3	30	5	5	2 drops
4	39	1*	0	2 drops
5	40	0	0	2 drops
6	30	0	10	2 drops

*Add 1 ml of the dilute acetic acid solution *prepared from* the 0.1 *M* acetic acid solution as described in procedure B3. The acetic acid used for solutions 1, 2, and 3 is 0.1 *M*.

Solutions to be Titrated

Use a 25-ml graduated cylinder to measure the distilled water required.

C. pH Change

If you used a pH meter in procedure B2, use it again to measure the pH of the solutions at the titration endpoint.

If no pH meter is available to you, assume that the endpoint occurs at pH 8.7. Calculate the pH values for each of the solutions numbered 1, 2, 3, 4, and 6 before any addition of sodium hydroxide solution. Assume that solution 5 (distilled water) had an initial pH of about 6 because of some dissolved carbon dioxide:

$$(CO_2 + H_2O \rightleftharpoons HCO_3^- + H^+)$$

Enter the results of these calculations along with the corresponding NaOH titrant volumes in your notebook. Include the data from standardization of the sodium hydroxide if you did that part of the experiment.

CALCULATIONS AND REPORT

The buffer capacity is defined for this investigation (since only the "base" capacity was studied) as the moles of NaOH necessary to change the pH of 1 liter of the buffer solution by one unit and can be calculated using the following equation:

$$\text{buffer capacity} = \frac{\text{moles of NaOH consumed}}{\text{pH change} \times \text{volume of buffer (in liters)}}$$

In each of the experiments 40 ml (4×10^{-2} liters) of buffer solution were employed.

Using your notebook records, complete the Report Form and perform the calculations that will produce the data required to complete the table in the Report Form (original pHs of solutions, acetic acid concentration in the solutions, pH changes, moles of NaOH required in each titration to reach the endpoint pHs, and buffer capacities). Note that in the acetic acid–sodium acetate solutions the acetic acid concentration is equal to the acetate ion concentration before any base has been added. Enter the answers to the following problem in the space provided for solution 7 in the Report Form table.

PROBLEM

Forty milliliters of a solution containing only hydrochloric acid and water had a pH of 2.0. This solution was titrated with 0.1 *M* sodium hydroxide solution to the phenolphthalein endpoint. Calculate the "buffer capacity" of this hydrochloric acid solution. (Assume negligible volume change.)

Plot the buffer capacities of solutions 1–4 versus the initial concentrations of acetic acid.

Submit on a separate page your comments about your results and the graph. Use the graph to determine the buffer solution needed in the reaction described below.

A reaction mixture of 1 liter volume and initial pH 4.6 will generate during the reaction 1×10^{-2} moles of hydroxide ions. How many moles each of acetic acid and acetate ions must be present in this reaction mixture in order to prevent its pH value from rising above pH 5.6?

A more accurate determination *of the buffer capacity* would require plotting a titration curve from data obtained by following the course of the titration with a pH meter, and then finding the slope of the curve in the region of effective buffer action. You may wish to consult your instructor if you are interested in how this is done.

GRADE	INSTRUCTOR'S INITIALS

Report Form

FOR INVESTIGATION 19:

Buffer Systems

NAME: ______________________ ID NO.: ______________________

LAB SECTION ______________________ INSTRUCTOR ______________________

Show calculations on attached pages.

Solution Preparation

Mass of NaOH required for 600 ml of 0.10 *M* NaOH ____________ g

Mass of CH_3CO_2H required for 400 ml of 0.10 *M* CH_3CO_2H ____________ g

Mass of CH_3CO_2Na required for 400 ml of 0.10 *M* CH_3CO_2Na ____________ g

Molarity of dilute CH_3CO_2H solution prepared in procedure B3 ____________ mole liter^{-1}

Titrations and Buffer Capacity

If you standardized the NaOH solution, record the molarity determined. Otherwise, use 0.10 *M*.

Concentration of sodium hydroxide titrant ____________ *M*

Volumes of solutions 1–6 used in titrations ____________ liter

pH at endpoint for solutions 1–7 ____________

Solution No.	Original pH	CH_3CO_2H Concentration	ΔpH	Volume NaOH (ml)	Moles NaOH	Buffer capacity
1						
2						
3						
4						
5						
6						
7						

Attach your comments, calculations and answer to the problem on separate pages.

______________ DATE ______________________ SIGNATURE

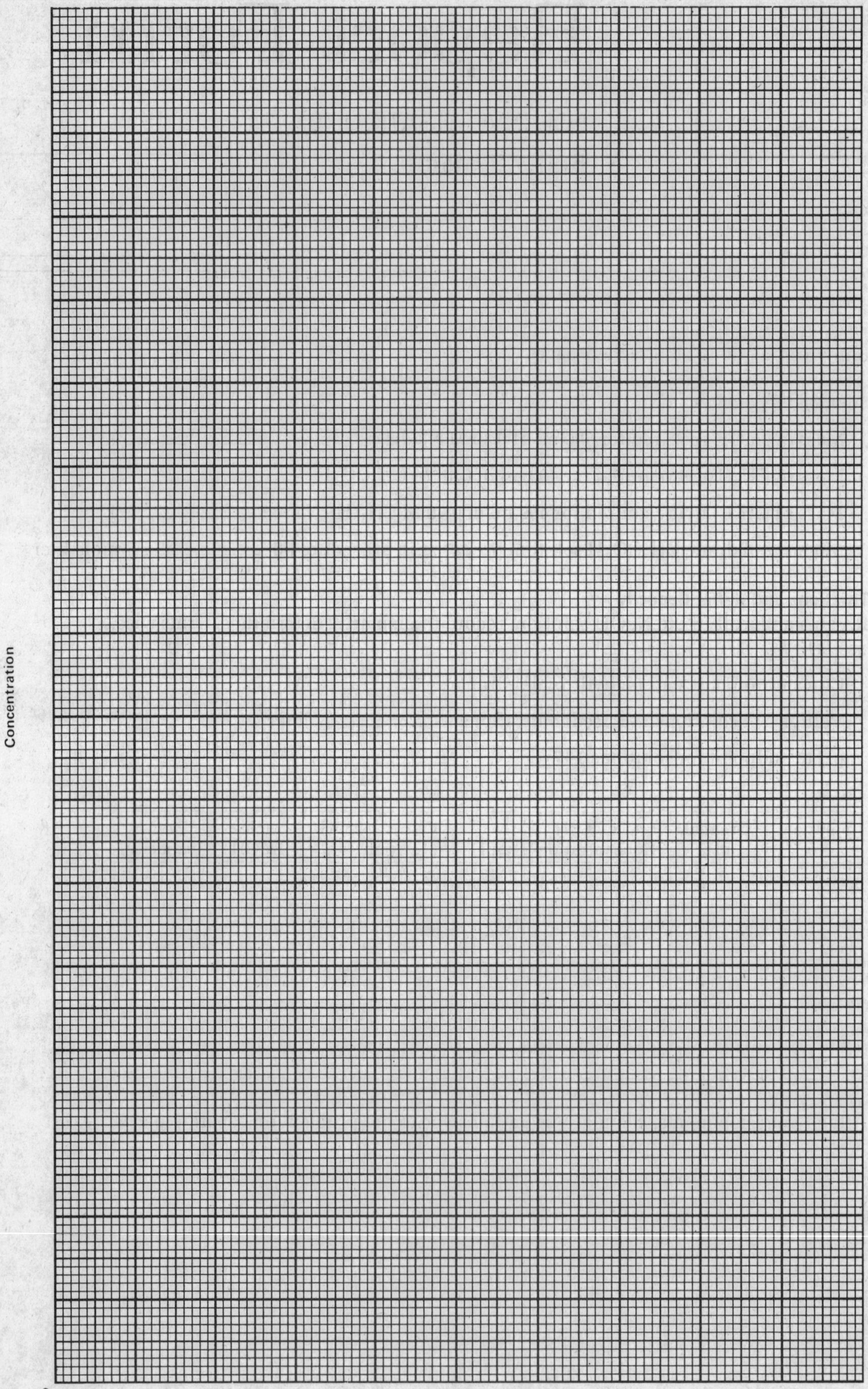
Concentration
0
Buffer capacity

Investigation 20

pH-CONTROLLED SEPARATIONS

INTRODUCTION Many of our more important metals occur in nature as sulfide ores. Among these are *galena* (PbS), *zinc blende* (ZnS), *cinnabar* (HgS), *argentite* (Ag_2S), *stibnite* (Sb_2S_3), and *chalcocite* (Cu_2S). Some of these have found uses in the native forms. Galena was employed during the early twentieth century for the "crystal" in "crystal radios" and zinc sulfide is a valuable white pigment for paints. Most sulfide ores, however, are used as sources of the metals. Industrial smelters employing "sulfide roasting" processes were major sources of air pollution by SO_2 until the recent clean air standards imposed stricter control on smelter emissions.

Soluble salts are leached from the soil, over a period of time, as rainwater or groundwater percolates through the soil and rock deposits. The existence of large deposits of sulfide ores suggests, then, that these metal sulfides must be relatively insoluble in water. Such is, indeed, the case as can be seen from the data in Table 20.1.

Table 20.1 Solubilities and solubility products for some common metal sulfides (at 20°–25°C)

Compound	Solubility (g per 100 ml H_2O)	K_{sp}
MnS	2.3×10^{-7}	7.0×10^{-16}
FeS	5.6×10^{-9}	4.0×10^{-19}
NiS	5.0×10^{-10}	3.0×10^{-21}
CoS	2.0×10^{-10}	5.0×10^{-22}
ZnS	3.9×10^{-11}	1.6×10^{-23}
CdS	1.4×10^{-13}	1.0×10^{-28}
PbS	2.0×10^{-13}	7.0×10^{-29}
Ag_2S	2.8×10^{-16}	5.5×10^{-51}
CuS	8.5×10^{-18}	8.0×10^{-37}
HgS	8.5×10^{-26}	1.6×10^{-54}

It is often feasible to take advantage of the different solubilities of metal sulfides, and when these differences are sufficiently great, to effect a separation of two or more metal ions. Since the sulfides of the alkali metals (Group IA) and the alkaline earth metals (Group IIA) are water soluble, ions of these metals remain in solution during precipitations of insoluble metal sulfides. Until recent years, sulfide precipitations were commonly employed for the separation and analysis of metal ions and the unpleasant odor of hydrogen sulfide was characteristic of industrial and academic analytical laboratories. In modern laboratories, instrumental techniques have largely replaced the classical wet chemical methods for routine metal ion analyses, but sulfide precipitations are still employed in special cases and these precipitations provide an excellent illustration of the principles of pH-controlled separations.

In an aqueous solution of hydrogen sulfide we can identify the following interactive equilibria:

$$H_2S(aq) \rightleftharpoons H^+(aq) + HS^-(aq) \qquad (K_{a_1} = 1.0 \times 10^{-7})$$
$$HS^-(aq) \rightleftharpoons H^+(aq) + S^{2-}(aq) \qquad (K_{a_2} = 1.3 \times 10^{-13})$$
$$H_2O(l) \rightleftharpoons H^+(aq) + OH^-(aq) \qquad (K_w = 1.0 \times 10^{-14})$$

It should be apparent, as suggested by Le Chatelier's principle, that changing the hydrogen-ion concentration of an aqueous solution of hydrogen sulfide will change the relative concentrations of H_2S, HS^-, and S^{2-} and the hydroxide-ion concentration

$$C_{OH^-} = \frac{1.0 \times 10^{-14}}{C_{H^+}}.$$

Three mathematical relationships are useful in understanding the influence of hydrogen-ion concentration (i.e., the pH influence) on the composition of the aqueous H_2S equilibrium system:

$$K_{a_1} = 1.0 \times 10^{-7} = \frac{C_{H^+} \times C_{HS^-}}{C_{H_2S}}$$

$$K_{a_2} = 1.3 \times 10^{-13} = \frac{C_{H^+} \times C_{s^{2-}}}{C_{HS^-}}$$

$$K_{a(net)} = K_{a_1} \times K_{a_2} = 1.3 \times 10^{-20}$$
$$= \frac{C_{H^+} \times C_{HS^-}}{C_{H_2S}} \times \frac{C_{H^+} \times C_{S^{2-}}}{C_{HS^-}} = \frac{C_{H^+}{}^2 \times C_{S^{2-}}}{C_{H_2S}}$$

It would appear that we could employ the $K_{a(net)}$ expression to find the sulfide-ion concentration at any pH for which the H_2S concentration is known. Such is, of course, the case, but the problem lies in "knowing the H_2S concentration." The only way we "know" the concentration of any solute is by measurement. The concentrations that must be employed in equilibrium calculations are *equilibrium concentrations*. These are not simple to measure for the aqueous H_2S equilibrium system. *Only under certain conditions* is the *equilibrium* H_2S concentration reasonably approximated by the "label concentration" (i.e., in terms of how the solution was prepared). If we are interested in the pH control of sulfide ion concentration for application to selective precipitation of metal sulfides, we must analyze the aqueous H_2S system a bit more closely.

The solubility of hydrogen sulfide gas in water, around room temperature, is limited to about 0.10 mole H_2S liter^{-1} of water. To simplify our considerations, let's assume that we will be working with solutions prepared by adding 0.10 mole of H_2S gas to sufficient water to form 1 liter of solution. Then, at equilibrium, a total of 0.10 mole of *all sulfide-containing species* will be present per liter of solution, that is,

$$C_{H_2S} + C_{HS} + C_{S^{2-}} = 0.10\ M$$

In order to select the most appropriate mathematical formulation, we need to know the *principal* solute species as a function of pH. From the K_{a_2} expression:

$$C_{S^{2-}} = \left[\frac{1.3 \times 10^{-13}}{C_{H^+}}\right] C_{HS^-}$$

at pH 12($C_{H^+} = 10^{-12}$), the concentration of sulfide ion is only 13 percent of that for HS^-, and at higher H^+ concentrations, the percentage of S^{2-} decreases. Thus we may safely conclude that S^{2-} *is not a principal solute species at* pH < 12. At pH > 13, S^{2-} *is a principal solute species.*

From the K_{a_1} expression

$$C_{HS^-} = \left[\frac{1.0 \times 10^{-7}}{C_{H^+}}\right] C_{H_2S}$$

at pH 7, $C_{HS^-} = C_{H_2S}$. That is, in particular solution under consideration

$$C_{H_2S} = C_{HS} = 0.05\ M$$

and *at* pH ≤ 6 *only* H_2S *is a principal solute species.* In the pH range 8–12, HS^- *is the principal solute species.*

These considerations are summarized in Table 20.2, along with conclusions as to the most appropriate mathematical relationship between H^+ and S^{2-} in each pH range.

Hydrogen sulfide gas has a most disagreeable odor, reminiscent of rotten eggs, and the gas is quite poisonous. The H_2S gas itself is now rarely employed for sulfide precipitations, having been replaced by the use of thioacetamide solutions. Thioacetamide hydrolyzes slowly at room temperature, but rapidly when the solution is heated, to generate aqueous H_2S:

$$\underset{\text{thioacetamide}}{CH_3\overset{\overset{\large S}{\|}}{C}NH_2(aq)} + 2H_2O(l) \longrightarrow \underset{\text{acetate ion}}{CH_3\overset{\overset{\large O}{\|}}{C}O^-(aq)} + \underset{\text{ammonium ion}}{NH_4^+(aq)} + H_2S(aq)$$

In this investigation we will use aqueous thioacetamide to generate H_2S for the pH-controlled separation of some common metal ions by selective precipitation of metal sulfides.

Table 20.2 The pH dependence of solute species in a 0.10 *M* aqueous hydrogen sulfide equilibrium system

pH	Principal solute species	To find $C_{S^{2-}}$, given C_{H^+}	To find C_{H^+}, given $C_{S^{2-}}$
≥14	$S^{2-}(\sim 0.10\ M)$	assume $C_{S^{2-}} = 0.10\ M$	use as $C_{H^+} < 10^{-14}\ M$ ($C_{OH^-} \geq 1.0\ M$)
14 ↓ 12	S^{2-} and HS^- ($C_{S^{2-}} + C_{HS^-} \approx 0.10\ M$) $C_{S^{2-}}$ range: $\sim 0.09\ M \longrightarrow \sim 0.01\ M$	$1.3 \times 10^{-13} = \dfrac{C_{H^+} \times C_{S^{2-}}}{(0.10 - C_{S^{2-}})}$	$C_{H^+} = (1.3 \times 10^{-13}) \left[\dfrac{(0.10 - C_{S^{2-}})}{C_{S^{2-}}}\right]$
12 ↓ 8	$HS^-(\sim 0.10\ M)$ $C_{S^{2-}}$ range: $\sim 0.01\ M \longrightarrow 10^{-6}\ M$	$C_{S^{2-}} \approx \dfrac{1.3 \times 10^{-14}}{C_{H^+}}$	$C_{H^+} \approx \dfrac{1.3 \times 10^{-14}}{C_{S^{2-}}}$
8 ↓ 6	HS^- and H_2S ($C_{HS^-} + C_{H_2S} \approx 0.10\ M$) $C_{S^{2-}}$ range: $\sim 10^{-6}\ M \longrightarrow \sim 10^{-9}\ M$	first find C_{HS^-} from $1.0 \times 10^{-7} = \dfrac{C_{H^+} \times C_{HS^-}}{0.10 - C_{HS^-}}$ then find $C_{S^{2-}}$ from $1.3 \times 10^{-13} = \dfrac{C_{H^+} \times C_{S^{2-}}}{C_{HS^-}}$	first find C_{HS^-} in terms of C_{H^+} from $1.3 \times 10^{-13} = \dfrac{C_{H^+} \times C_{HS^{2-}}}{C_{HS^-}}$ then substitute in $1.0 \times 10^{-7} = \dfrac{C_{H^+} \times C_{HS^-}}{0.10 - C_{HS^-}}$
≤6	$H_2S(\sim 0.10\ M)$ $C_{S^{2-}} < 10^{-9}\ M$	$C_{S^{2-}} \approx \dfrac{1.3 \times 10^{-21}}{C_{H^+}{}^2}$	$C_{H^+} \approx \sqrt{\dfrac{1.3 \times 10^{-21}}{C_{S^{2-}}}}$

To select conditions for pH-controlled separations, we must know the minimum concentration of S^{2-} that will induce a desired precipitation and the maximum permissible S^{2-} concentration that will avoid an unwanted precipitation. The key information for these decisions includes the concentration of the metal ion involved and the solubility product of the corresponding metal sulfide. Consider, for example, a solution 1.0×10^{-8} *M* in Sn^{2+}. The solubility product for SnS is 1.1×10^{-26}. If a precipitate of SnS is formed, the resulting saturated solution would be represented by

$$SnS(s) \rightleftharpoons Sn^{2+}(aq) + S^{2-}(aq)$$

for which

$$K_{sp} = C_{Sn^{2+}} \times C_{S^{2-}}$$

and

$$C_{S^{2-}} = \frac{K_{sp}}{C_{Sn^{2+}}}$$

The sulfide-ion concentration for a saturated solution in which

$$C_{Sn^{2+}} = 1.0 \times 10^{-8}$$

is then given by

$$C_{S^{2-}} = \frac{1.0 \times 10^{-26}}{1.0 \times 10^{-8}} = 1.1 \times 10^{-18}\ M$$

To induce precipitation, some of the Sn^{2+} ion must be removed from solution, requiring that $C_{S^{2-}}$ exceed 1.1×10^{-18} *M*, so the minimum sulfide ion concentration is "just greater" than 1.1×10^{-18} *M*. By similar reasoning

$$C_{S^{2-}} = 1.1 \times 10^{-18}\ M$$

is the *maximum* sulfide-ion concentration permissible if precipitation is to be avoided.

The appropriate pH conditions may then be determined from information in Table 20.2, assuming a 0.10 *M* H_2S system.

For $C_{S^{2-}} < 10^{-9}$ *M*, we are in the range pH < 6 (Table 20.2), for which

$$C_{H^+} = \sqrt{\frac{1.3 \times 10^{-21}}{C_{S^{2-}}}}$$

To induce precipitation, C_{H^+} must be as follows:

$$C_{H^+} < \sqrt{\frac{1.3 \times 10^{-21}}{1.1 \times 10^{-18}}} \qquad (<3.4 \times 10^{-2}\ M)$$

That is, the pH ($-\log C_{H^+}$) must be >1.47.

To prevent precipitation, the C_{H^+} must be high enough to reduce $C_{S^{2-}}$ to 1.1×10^{-18} or less, that is, pH ≤ 1.47.

In this investigation we will employ calculations similar to those described in the preceding discussion to determine the feasibility of certain pH-controlled separations. An actual separation will then be effected and the results of this separation will form the basis for analysis of an "unknown" solution.

Before starting this investigation, you should be thoroughly familiar with the principles and calculations involved in pH-controlled sulfide precipitations. You should trace through the derivations of the information presented in Table 20.2. In addition, prior to coming to the laboratory, you must complete two sets of sample calculations of the type required in the Procedures section of this investigation. One set should be for Ag_2S and one set for any other sulfide listed in Table 20.1. (Your instructor may wish to suggest one.)

References

1. O'Connor, Rod. 1977. *Fundamentals of Chemistry* (2nd ed.). New York: Harper & Row (Units 19–21 and Excursion 2).
2. Moeller, Therald and Rod O'Connor. 1972. *Ions in Aqueous Systems.* New York: McGraw-Hill.

OBJECTIVES

1. Be able to make appropriate calculations concerning pH-controlled sulfide precipitations.
2. Be able to devise and test an analytical scheme for the separation of metal ions on the basis of differing solubilities of the metal sulfides.
3. Be able to analyze an "unknown" solution containing one or more of the metal ions for which the analytical scheme was devised.

PROCEDURES

Your instructor will assign to you a pair of metal ions from those forming the sulfides listed in Table 20.1. From the information in Tables 20.1 and 20.2 and the following facts and assumptions, devise an experimental procedure for the sequential precipitation of the metal sulfides. Have your procedure checked by your instructor.

The solution you will be working with contains 10 mg ml^{-1} of each metal ion, before any other solutions are added.

The precipitations are to be performed by mixing 5 drops of the metal ion solution, 35 drops of a thioacetamide solution, and 40 drops of aqueous acid (or other pH-control solution).

One drop from a typical medicine dropper has a volume of 0.05 ml. You may assume that the concentration of thioacetamide used is such as to produce, *in the total* mixture, a solution approximating that of a 0.10 *M* aqueous H_2S equilibrium system.

For the required separation scheme, you must include calculations, for both the metal ions assigned, of all the steps described in part 1(a–f) of the Procedures section of this investigation.

After your instructor has checked your sample calculations and your proposed sequential precipitation scheme, obtain the two assigned metal-ion solutions from the instructor in separate *clean* test tubes. Label the tubes properly and use these solutions to test your analysis scheme.

1. Before beginning experimental work, you will need to make the following calculations, *in your notebook,* for the specific metal ion solutions assigned to you:

a. For each solution, the minimum $C_{S^{2-}}$ required to initiate precipitation of the metal sulfide from a solution prepared by diluting 5 drops of the original metal ion solution (10 mg of metal ion per milliliter) to a total volume of 80 drops.

b. For each *diluted* solution [as described in (a)], the *minimum* $C_{S^{2-}}$ necessary to achieve "complete precipitation" of the metal sulfide, defined as having reduced the metal ion concentration to one-thousandth of its original value.

c. For each diluted solution, the maximum C_{H^+} *realistically obtainable* from HCl of concentration $\leq 6\ M$ that will provide a sufficient $C_{S^{2-}}$ for complete precipitation of the metal sulfide [as calculated for (b)].

d. For each diluted solution, the minimum C_{H^+} that will provide a $C_{S^{2-}}$ low enough to *avoid* precipitation [i.e., a $C_{S^{2-}}$ just less than that required to initiate precipitation as calculated for (a)].

(Note: There is no way to prevent precipitation of the metal sulfide from a 0.10 *M* aqueous H_2S system by pH control if this calculation indicates a C_{H^+} not "realistically obtainable.")

e. For each diluted solution, the total number of millimoles of H^+ required to provide the C_{H^+} calculated for (c) and (d), in a total solution volume of 4.0 ml (80 drops), *assuming**:

If the required C_{H^+} is $\leq 10^{-6}\ M$, no additional H^+ is needed.

If the required C_{H^+} is $10^{-4}\ M$ to $10^{-6}\ M$, about 0.3 mmol of additional H^+ should be added.

If the required C_{H^+} is $10^{-3}\ M$ to $10^{-4}\ M$, about 0.5 mmol of additional H^+ should be added.

If the required C_{H^+} is $10^{-3}\ M$, 0.5 mmol of H^+ *plus* the number of millimoles needed to change the C_{H^+} from $10^{-3}\ M$ to the desired value should be added.

f. The number of drops (assuming 0.05 ml per drop) of a "stock acid" that will provide approximately the number of millimoles of H^+ calculated for each case in (e). The "stock acids" available are 12 *M* HCl, 6 *M* HCl, and 0.3 *M* HCl and no more than 40 drops of a "stock acid" may be used.

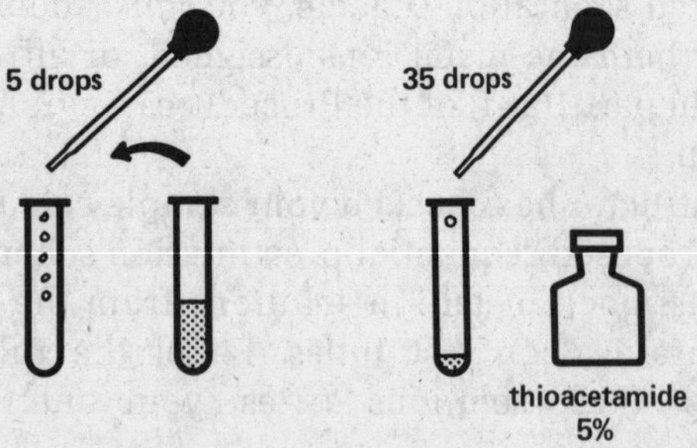

2. Now, using clean medicine droppers, transfer 5 drops of each of your metal ion solutions into separate small test tubes. Add to each of these solutions 35 drops of 5 percent thioacetamide solution and mix by shaking.

*These assumptions are suitable approximations for the effect of the ammonium acetate buffer resulting from hydrolysis of the thioacetamide used.

3. To each test tube add the number of drops of the appropriate "stock acid" solution calculated [procedure 1(f)] to produce a C_{H^+} appropriate for "complete precipitation" of the particular metal sulfide. Then add distilled water, dropwise, until the total volume of added water and "stock acid" is 40 drops (for each test tube). Mix by shaking.

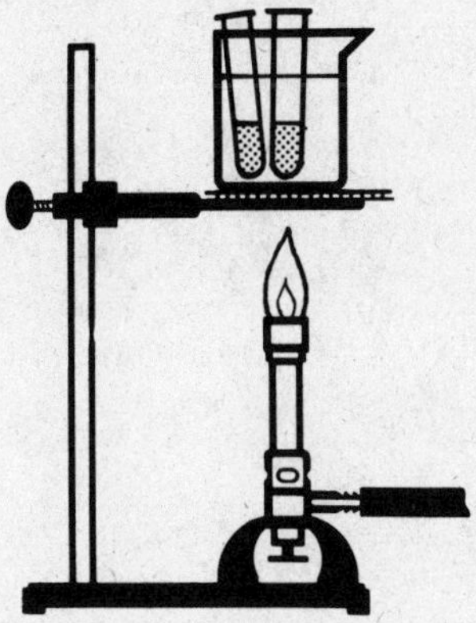

4. Place your test tubes into a 400-ml beaker about one-third filled with water and heat the water to a gentle boil. Continue heating for 5 min. If a precipitate does not appear in one of the test tubes, remove this test tube and add to it 5 drops of 5 *M* ammonia. (If the ammonia is required, record this in your notebook.) Return this test tube to the hot-water bath and continue heating for another 5 min. Note and record the colors of the precipitates in both tubes.

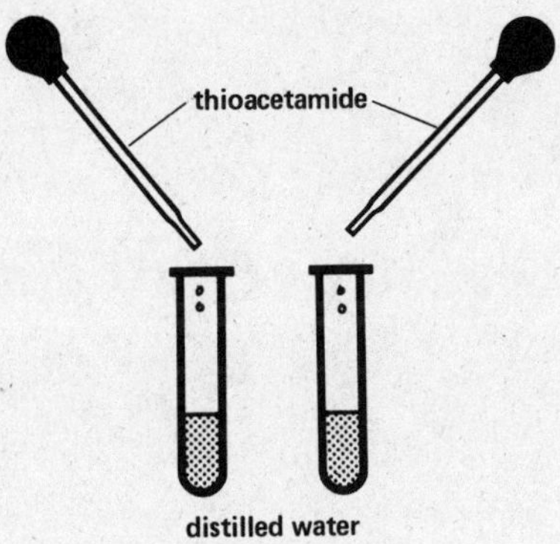

5. Centrifuge both test tubes, decant the clear solutions from the precipitates into two clean test tubes. Dilute the decanted solutions to twice their volumes with distilled water, add 5 drops of the thioacetamide solution, mix, and heat for another 5 min. If more precipitates form, record this observation in your notebook and suggest an explanation of why the initial precipitation was not "complete."

6. In a clean test tube, mix 5 drops each of your two metal-ion solutions. Add 30 drops of thioactamide solution and adjust the C_{H^+} to that appropriate for "complete" precipitation of the *less soluble sulfide without initiating precipitation of the more soluble sulfide*. Place the test tube into the water bath and heat it for 5 min. Centrifuge and then decant the solution from the precipitate into a clean test tube.

7. Assume that the decanted solution still has a total volume of 80 drops and the C_{H^+} used for procedure 6. Add the number of drops of 15 *M* ammonia calculated to be *twice* that required to consume the H^+ present. Mix and heat the solution in the water bath for 5 min. Centrifuge and test for "complete precipitation" of the more soluble sulfide, as described in procedure 5.

8. Take a clean test tube to your instructor and obtain an "unknown solution" which may contain either or both of the metal ions you have previously studied, or which may be just distilled water. Use your tested separation scheme to determine which, if any, of the metal ions are present in your "unknown."

CALCULATIONS AND REPORT Use your notebook records to complete the Report Form. Attach separate pages containing neatly written calculations as required for procedures 1(a–f) and 7.

GRADE	INSTRUCTOR'S INITIALS

Report Form

FOR INVESTIGATION 20:
pH-Controlled Separations

NAME: ______________________ ID NO.: ______________________

LAB SECTION ______________________ INSTRUCTOR ______________________

Show calculations on attached pages.

Assigned metal ions ______________________ and ______________________

	Less soluble sulfide ________	More soluble sulfide ________
a. Minimum $C_{S^{2-}}$ to initiate precipitation		
b. Minimum $C_{S^{2-}}$ for "complete precipitation"		
c. Maximum "realistic" C_{H^+} for "complete precipitation"		
d. Minimum C_{H^+} to *avoid* precipitation (indicate if "not realistically obtainable")		
e. Millimoles H^+ for "complete precipitation"		
Millimoles H^+ to *avoid* precipitation		
f. Drops of "stock acid" (specify) for complete precipitation		
Drops of "stock acid" (specify) to avoid precipitation		
Color of precipitate		

Drops of 15 *M* NH_3 required for procedure 7: ______________ drops

On the back of this page, briefly describe the steps taken and the observations made in testing your analysis scheme and in analyzing your "unknown solution."

The "unknown solution" contained ______________________

______________ DATE ______________________ SIGNATURE

Investigation 21

ANALYSIS BY SULFATE PRECIPITATIONS

INTRODUCTION Most metal ions form water soluble sulfates. The only common metal ions forming insoluble sulfates are barium, calcium, lead, strontium, silver, and mercury(I).* Barium sulfate is used in x-ray examination of the gastrointestinal tract because the heavy barium ion is particularly effective in scattering x-rays and the low solubility of $BaSO_4$ ensures a Ba^{2+} concentration well below toxic limits. Lead sulfate is formed during the discharge of the automobile battery. The low solubility of $PbSO_4$ is important since the automobile battery is designed for a repetitive discharge-recharge sequence and the adhering of a $PbSO_4$ deposit on the electrode surfaces is convenient for the cell reaction during recharge. The insoluble calcium sulfate occurs in nature as the dihydrate, $CaSO_4 \cdot 2H_2O$, *gypsum*, which is used in the production of wallboard (Sheetrock). When powdered gypsum is heated to around 100°C a partial dehydration occurs, forming *plaster of paris*, $(CaSO_4)_2 \cdot H_2O$. When plaster of paris is mixed with water, the tiny crystals dissolve and recrystallize as long needlelike crystals of $CaSO_4 \cdot 2H_2O$. As these crystals grow together, an intricate network forms and the plaster is said to "set." Strontium sulfate is found naturally as the mineral *celestite*, often in the form of beautiful crystals. The mineral has no significant value, since neither strontium metal nor most of its salts have found commercial applications. Strontium nitrate, mixed with sulfur and carbon, is used in flares and fireworks because of the red color imparted by strontium to a flame, but this salt is more readily prepared from *strontianite* ($SrCO_3$) than from *celestite*. Neither Ag_2SO_4 nor Hg_2SO_4 has any significant commercial uses.

Since there are so few insoluble sulfates, sulfate ion can be employed to separate Ca^{2+}, Sr^{2+}, Ba^{2+}, and Pb^{2+} from most other metal ions in aqueous solution. Of the other common metal ions, only Ag^+ and Hg_2^{2+} pose possible problems in moderately dilute solutions. As we shall see in Investigation 22, Ag^+ and Hg_2^{2+} may be removed by precipitation of their chlorides if their presence might interfere with a desired sulfate precipitation. Since calcium sulfate is significantly more soluble than $SrSO_4$, $BaSO_4$, or $PbSO_4$ (Table 21.1), it is possible to avoid the precipitation of $CaSO_4$ by a proper control of sulfate concentration.

Sulfuric acid ionizes essentially 100% in aqueous solution in terms of loss of one proton, but the resulting bisulfate ion is a weak acid:

$$HSO_4^- \rightleftharpoons H^+ + SO_4^{2-} \qquad K_a = 1.3 \times 10^{-2}$$

*Under appropriate conditions of pH and sulfate concentrations, mercury(II) ion forms an insoluble "basic sulfate," formulated as $HgSO_4 \cdot HgO$.

Table 21.1 Sparingly soluble sulfates

Compound	Approximate water solubility (20°–25°C) (g per 100 ml water)	(mole liter^{-1})	K_{sp} (20°–25°C)
Ag_2SO_4	0.78	2.5×10^{-2}	6.4×10^{-5}
$CaSO_4$	0.068	5.0×10^{-3}	2.5×10^{-5}
Hg_2SO_4	0.058	1.18×10^{-3}	1.4×10^{-6}
$SrSO_4$	0.0105	5.7×10^{-4}	3.2×10^{-7}
$PbSO_4$	0.0034	1.14×10^{-4}	1.3×10^{-8}
$BaSO_4$	0.00021	8.9×10^{-6}	7.9×10^{-11}

Since the ionization constant for bisulfate ion is relatively large, we may safely assume that essentially all of the sulfate from sulfuric acid is "available" in dilute solutions. That is, unlike the case of H_2S (Investigation 20), we cannot realistically control sulfate concentration in dilute solutions by pH-control. Thus either sulfuric acid or some soluble sulfate salt, such as Na_2SO_4, could be employed as a sulfate source. There is, however, one consideration favoring the choice of sulfuric acid. Caution must be used in selecting sodium sulfate if ions are present that form very insoluble hydroxides since the hydrolysis of sulfate might furnish sufficient hydroxide ion to precipitate some metal hydroxides:

$$SO_4^{2-} + H_2O \rightleftharpoons HSO_4^- + OH^- \qquad K_b = 7.7 \times 10^{-13}$$

To determine the conditions appropriate to the sequential precipitation of metal sulfates, it is necessary to calculate conditions under which the number of millimoles of sulfate added will stoichiometrically precipitate the less soluble sulfate without leaving an excess sulfate concentration high enough to initiate precipitation of the more soluble sulfate. For our purposes, we shall define "complete precipitation" as having reduced the metal ion concentration to one-thousandth (10^{-3}) of its original value. The pH-dependence of the sulfate-to-bisulfate ratio need not be considered unless some of the metal ions involved form insoluble hydroxides (Table 21.2), in which case the pH must be kept low enough to avoid hydroxide precipitation. Because of the problems of hydroxide precipitation, dilute sulfuric acid is usually employed for sulfate precipitation for solutions containing metal ions forming highly insoluble hydroxides.

Table 21.2 Solubility products for some common metal hydroxides*

Compound	K_{sp} (at 20°–25°C)	Compound	K_{sp} (at 20°–25°C)
$Ca(OH)_2$	4.0×10^{-6}	$Zn(OH)_2$	4.5×10^{-17}
$Mg(OH)_2$	1.2×10^{-11}	$Cu(OH)_2$	1.8×10^{-19}
$Mn(OH)_2$	2.0×10^{-13}	$Hg(OH)_2$	3.2×10^{-26}
$Cd(OH)_2$	2.0×10^{-14}	$Sn(OH)_2$	3.3×10^{-27}
$Pb(OH)_2$	4.1×10^{-15}	$Cr(OH)_3$	7.0×10^{-13}
$Fe(OH)_2$	1.8×10^{-15}	$Al(OH)_3$	5.0×10^{-33}
$Co(OH)_2$	2.0×10^{-16}	$Fe(OH)_3$	6.0×10^{-38}
$Ni(OH)_2$	1.6×10^{-16}		

*Both $Ba(OH)_2$ and $Sr(OH)_2$ are soluble in water. Both Ag^+ and Hg_2^{2+} may form insoluble precipitates of the oxides, Ag_2O and Hg_2O, in alkaline solutions.

EXAMPLE

To illustrate the calculations involved in planning a sulfate precipitation, let's consider a 0.005 *M* solution of Pb^{2+} in terms of:

a. What is the minimum $C_{SO_4^{2-}}$ that will initiate precipitation of $PbSO_4$?
b. What is the maximum $C_{SO_4^{2-}}$ allowable if precipitation of $PbSO_4$ is to be avoided?
c. What number of drops of 0.06 *M* H_2SO_4 will provide enough sulfate for "complete precipitation" of $PbSO_4$ from 5 drops of the Pb^{2+} solution? (Assume 1 drop = 0.05 ml.)
d. What is the *excess* $C_{SO_4^{2-}}$ resulting from addition of enough distilled water to the mixture from (c) to give a total volume of 40 drops?
e. What is the pH of the solution from (d), assuming complete ionization of the sulfuric acid used?
f. What is the maximum pH allowable if precipitation of $Pb(OH)_2$ is to be avoided?
g. What do these calculations suggest as a reasonable procedure for either avoiding precipitation of $PbSO_4$ or achieving "complete precipitation" of $PbSO_4$, without formation of $Pb(OH)_2$?

SOLUTION

(a, b) If any $PbSO_4$ precipitates, the resulting saturated solution is represented by

$$PbSO_4(s) \rightleftharpoons Pb^{2+}(aq) + SO_4^{2-}(aq)$$

for which (Table 21.1)

$$K_{sp} = 1.3 \times 10^{-8} = C_{Pb^{2+}} \times C_{SO_4^{2-}}$$

As precipitation just begins from a 0.005 *M* Pb^{2+} solution

$$C_{SO_4^{2-}} = \frac{1.3 \times 10^{-8}}{5 \times 10^{-3}} = 2.6 \times 10^{-6}\ M$$

and, since precipitation would remove some Pb^{2+} from solution:

a. The *minimum* $C_{SO_4^{2-}}$ to initiate precipitation must be "just greater than 2.6×10^{-6} *M*."
b. The *maximum* $C_{SO_4^{2-}}$ to avoid precipitation is 2.6×10^{-6} *M*.
c. To determine the *amount* of sulfate needed involves considerations of stoichiometry, based on the net equation

$$Pb^{2+}(aq) + SO_4^{2-}(aq) \longrightarrow PbSO_4(s)$$

The amount of Pb^{2+} consumed is given by

$$\underbrace{\left[5\ \text{drops} \times \frac{0.05\ \text{ml}}{1\ \text{drop}} \times \frac{0.005\ \text{mmol}}{\text{ml}}\right]}_{\text{amount of } Pb^{2+} \text{ originally present}}$$

$$- \underbrace{\left[5\ \text{drops} \times \frac{0.05\ \text{ml}}{1\ \text{drop}} \times \frac{5 \times 10^{-6}\ \text{mmol}}{\text{ml}}\right]}_{\text{amount of } Pb^{2+} \text{ left after complete precipitation}}$$

$$\approx 1.25 \times 10^{-3}\ \text{millimole}$$

and the volume of 0.06 M H_2SO_4 needed is

$$1.25 \times 10^{-3}\ \text{mmol} \times \frac{1\ \text{ml}}{0.06\ \text{mmol}} \times \frac{1\ \text{drop}}{0.05\ \text{ml}}$$
$$= \sim 0.4\ \text{drop}$$

(Realistically, we can't use "part of a drop," so 1 drop of 0.06 M H_2SO_4 would be used.)

d. To find the excess $C_{SO_4^{2-}}$, we must assume addition of 1 drop of 0.06 M H_2SO_4 and removal of $\sim 1.25 \times 10^{-3}$ mmol of SO_4^{2-} by the $PbSO_4$ precipitation. Then, the *concentration* of SO_4^{2-} remaining is

$$\frac{[(1\ \text{drop} \times 0.05\ \text{ml drop}^{-1} \times 0.06\ \text{mmol ml}^{-1}) - (1.25 \times 10^{-3}\ \text{mmol})]}{(40\ \text{drops} \times 0.05\ \text{ml drop}^{-1})}$$

$$C_{SO_4^{2-}}(\text{excess}) = \sim 9 \times 10^{-4}\ M$$

e. Assuming complete ionization of the H_2SO_4 added (1 drop of 0.06 M),

$$C_{H^+} = \frac{0.06\ \text{mmol}\ (H_2SO_4)}{\text{ml}} \times \frac{2\ \text{mmol}\ (H^+)}{1\ \text{mmol}\ (H_2SO_4)} \times \underset{\substack{\uparrow \\ \text{dilution factor}}}{\frac{1}{40}} = 0.003\ M$$

This corresponds to pH 2.5.

f. If any $Pb(OH)_2$ precipitates, the resulting saturated solution is represented by

$$Pb(OH)_2(s) \rightleftharpoons Pb^{2+}(aq) + 2\ OH^-(aq)$$

for which (Table 21.2)

$$K_{sp} = 4.1 \times 10^{-15} = C_{Pb^{2+}} \times C_{OH^-}^2$$

$$\text{and } C_{OH^-} = \sqrt{\frac{4.1 \times 10^{-15}}{0.005 \times \underset{\substack{\uparrow \\ \text{dilution factor}}}{(5/40)}}} = \sim 2.6 \times 10^{-6}\ M$$

We conclude, therefore, that $Pb(OH)_2$ will precipitate if the C_{OH^-} exceeds 2.6×10^{-6} M, that is, if the pH exceeds ~8.4.

g. From calculations (a) through (f), we may conclude:

1. There is no way to avoid precipitation of $PbSO_4$ by adding the minimum practical volume (1 drop) of 0.06 M H_2SO_4 to 5 drops of 0.005 M Pb^{2+} solution.
2. A single drop of 0.06 M H_2SO_4 is sufficient for "complete precipitation" of $PbSO_4$ from 5 drops of 0.005 M Pb^{2+} solution.
3. The excess sulfate concentration in the mixture considered is about 9×10^{-4} M.
4. Any solution of pH < 7 will avoid precipitation of $Pb(OH)_2$ with the concentration of Pb^{2+} involved.

In this *Investigation* we will devise and test an analytical scheme based on the selective precipitation of metal sulfates.

Before starting this investigation, you should be thoroughly familiar with the principles and calculations involved in solubility equilibria. In addition, prior to coming to the laboratory, you must complete two sets of calculations of the type required in the Procedures section of this investigation. One set should be for Ca^{2+} and the other set for one of the ions Pb^{2+}, Ba^{2+}, Sr^{2+}, Ag^{+}, or $Hg_2{}^{2+}$. For Ag^{+} or $Hg_2{}^{2+}$, you should assume that the pH must be maintained below 7 to avoid precipitation of the oxide. (Your instructor may wish to suggest the particular ion you should consider.)

References

1. O'Connor, Rod. 1977. *Fundamentals of Chemistry* (2nd ed.). New York: Harper & Row (Units 18–21 and Excursion 2).
2. Moeller, Therald, and Rod O'Connor. 1972. *Ions in Aqueous Systems*. New York: McGraw-Hill.

SAFETY Wear approved eye protection at all times and exercise standard precautions in working with chemicals. Be particularly careful with sulfuric acid.

OBJECTIVES

1. Be able to make appropriate calculations concerning sulfate and hydroxide precipitations.
2. Be able to devise and test an analytical scheme based on the selective precipitation of metal sulfates.
3. Be able to analyze an "unknown" solution containing one or more of the ions studied in this investigation and, at the option of your instructor, an ion studied in Investigation 20.

PROCEDURES

Your instructor will assign to you a pair of metal ions from those forming the sulfates listed in Table 21.1. From the information in Tables 21.1 and 21.2 and the following facts and assumptions, devise an experimental procedure for the selective precipitation of the metal sulfates. Have your procedure checked by your instructor.

The solution you will be working with contains 10 mg ml^{-1} of the metal ion, before any other solutions are added.

The precipitations are to be performed by mixing 5 drops of the metal ion solution with sufficient other solutions and distilled water to form a final total volume of 40 drops.

One drop from a typical medicine dropper has a volume of 0.05 ml.

For the required analysis scheme, you must include calculations, for both the metal ions assigned, of all the steps described in procedure 1(a–h).

After your instructor has checked your sample calculations and proposed analysis scheme, obtain the two assigned metal ion solutions in separate *clean* test tubes. Label the test tubes properly and use these solutions to test your analysis scheme.

1. Before beginning experimental work, you will need to make the following calculations *in your notebook*, for the specific metal ions assigned to you:

a. For each solution, the *minimum* $C_{SO_4{}^{2-}}$ required to initiate precipitation of the metal sulfate from a solution prepared by diluting 5 drops of the original metal ion solution (10 mg of metal ion per ml) to a total volume of 40 drops.

b. The number of millimoles of H_2SO_4 needed for "complete precipitation" of the *less soluble* sulfate.

c. The *minimum* number of drops of 0.06 *M* H_2SO_4 (assuming 0.05 ml $drop^{-1}$) required to provide the number of millimoles of H_2SO_4 calculated for (b).

d. The *minimum* number of drops of 0.6 *M* H_2SO_4 required to provide the number of millimoles of H_2SO_4 calculated for (b).

e. The *excess* $C_{SO_4{}^{2-}}$ in the mixture resulting from calculation (c), after addition of enough distilled water to make a total volume of 40 drops.

f. The *excess* $C_{SO_4{}^{2-}}$ in the mixture resulting from calculation (d), after addition of enough distilled water to make a total volume of 40 drops.

g. The maximum pH allowable to avoid precipitation of metal oxides or hydroxides, and the pH of the final total solutions in (e) and (f).

h. Appropriate conditions for precipitating the less soluble sulfate *without* initiating precipitation of the more soluble sulfate.

Note: If calculations (a) through (f) suggest that this is not practical with the H_2SO_4 solutions involved, consult your instructor.

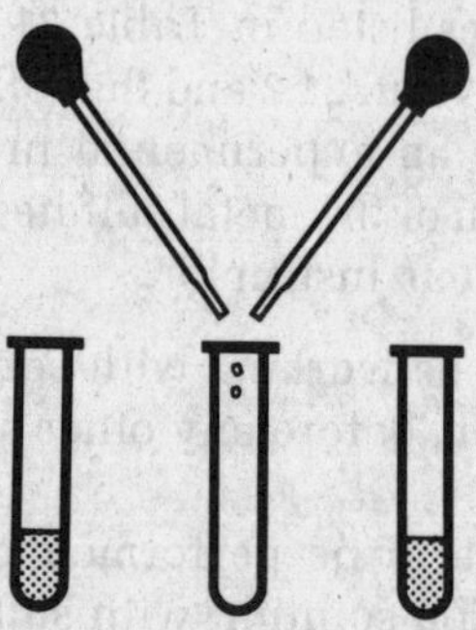

2. Now mix 5 drops of each of your metal ion solutions in a clean test tube. Calculate the number of drops (no more than 30) of the *appropriate* "stock H_2SO_4" (0.06 *M* or 0.6 *M*) that will "completely precipitate" the *less soluble* metal sulfate *without* initiating precipitation of the *more soluble* sulfate or of any metal hydroxides. Then determine the number of drops of distilled water to make a total mixture volume of 40 drops (metal ion solutions + "stock" reagent + water).

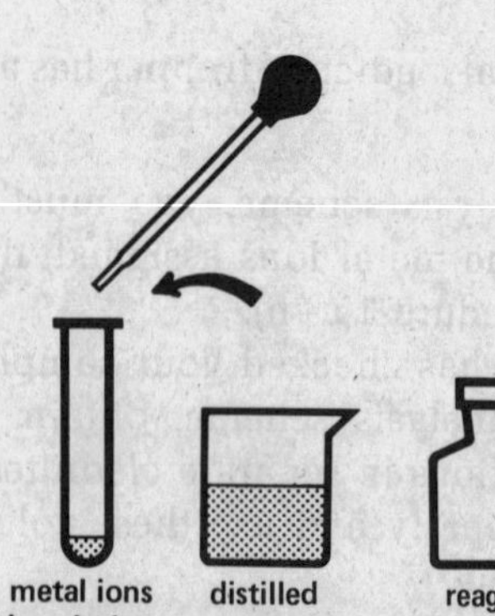

3. Add to the test tube containing the metal ion solution the number of drops of distilled water calculated in procedure 2. *Then* add the calculated number of drops of the appropriate "stock" H_2SO_4. Mix well by carefully shaking the test tube.

4. Centrifuge the test tube, decant the clear solution from above the precipitate into a clean test tube. Record in your notebook the appearance of the precipitate. Calculate the number of drops of either 6 *M* or 0.6 *M* H_2SO_4 solution that will "completely precipitate" the *more soluble* metal sulfate. (*Note:* You must consider that each drop of reagent added will dilute the metal ion solution. It should be apparent that dilution is minimized by employing the maximum suitable concentration of H_2SO_4.)

5. Add the calculated number of drops of the appropriate H_2SO_4 solution, *one drop at a time,* to the clear solution obtained in procedure 4. Mix by careful shaking after each drop. Record your observations. If a slight precipitate appears at a sulfate concentration well below that calculated for initial precipitation of the more soluble sulfate, this should suggest to you that the prior precipitation of the less soluble sulfate was "incomplete." Should this occur, consult your instructor.

6. Take a clean test tube to your instructor and obtain an "unknown." This "unknown" may contain:

a. Only distilled water.
b. Only one of the ions you have just studied.
c. Both of the two ions you have just studied.*
d. One of the two ions studied in this investigation and one of the ions you studied in Investigation 20. (Your instructor will inform you if option (d) is included.)

7. Using the analysis scheme tested for sulfate precipitations and—if option (d) is possible for your "unknown"—your notebook records from Investigation 20, determine the composition of your "unknown."

CALCULATIONS AND REPORT Use your notebook records to complete the Report Form. Attach separate pages containing neatly written calculations as required for Procedures 1 (a through h), 2, and 4.

*If your calculations and experimental work were performed properly, you should have discovered that either Ca^{2+} or Ag^+ can be separated from the other ions studied, but not completely from each other, and that no sulfate separation of the remaining ions is feasible. This information, plus the results of Investigation 20, should give you useful clues to an approach to a two-ion unknown.

GRADE
INSTRUCTOR'S INITIALS

Report Form

FOR INVESTIGATION 21:

Analysis by Sulfate Precipitations

NAME: ______________________ ID NO.: ______________________

LAB SECTION ______________________ INSTRUCTOR ______________________

Show calculations on attached pages.

Assigned metal ions ______________ and ______________

	Less soluble sulfate ______	More soluble sulfate ______
Minimum $C_{SO_4^{2-}}$ to initiate precipitation		
Number of mmols H_2SO_4 for "complete precipitation"		
Maximum pH to avoid precipitation of hydroxide or oxide		
Drops of "stock" H_2SO_4 (specify) *used* for "complete precipitation"		
Maximum drops of "stock" H_2SO_4 (specify) allowable to avoid precipitation		

On the back of this page, briefly describe the steps taken and the observations made in testing your analysis scheme and in analyzing your "unknown."

The "unknown" contained ______________

______________ DATE ______________________ SIGNATURE

Investigation 22

ANALYSIS BY CHLORIDE PRECIPITATIONS

INTRODUCTION Very few metal ions form insoluble chlorides. With the exceptions of AgCl, Hg_2Cl_2, and $PbCl_2$, all common* metal chlorides are water soluble. These three compounds occur as the natural minerals *cerargyrite* (AgCl), *calomel* (Hg_2Cl_2), and *cotunnite* ($PbCl_2$). Calomel has been used in medicine as a stimulant for bile secretion. Its low solubility keeps the concentration of Hg_2^{2+} well below toxic limits, although some reports have been made of adverse reactions to calomel therapy.

Chloride precipitations present few complications compared to those involving sulfides (Investigation 20) or sulfates (Investigation 21). Since hydrochloric acid is a strong acid, chloride ion is not significantly hydrolyzed and the Cl^- concentration is, for all practical purposes, pH independent. As with most salts, the solubilities of metal chlorides increase with temperature (Table 22.1) and lead chloride is sufficiently soluble in hot water to permit its separation from the less soluble AgCl and Hg_2Cl_2 by selectively dissolving $PbCl_2$ from a mixed chloride precipitate with hot water.

Table 22.1 Solubility products for common metal chlorides.

Compound	K_{sp} (20°–25°C)	K_{sp} (90°–100°C)
$PbCl_2$	1.6×10^{-5}	$\sim 6 \times 10^{-3}$
AgCl	2.4×10^{-10}	$\sim 2 \times 10^{-8}$
Hg_2Cl_2	1.2×10^{-18}	$\sim 10^{-11}$

There are some complications that are usually ignored in considerations of solubility product calculations for the metal chlorides. There is evidence suggesting, for example, that there are several different species present in solutions containing Pb^{2+} and Cl^- ions. The $[PbCl]^+$, $[Pb(OH)]^+$, $[PbCl_3]^-$, and $[PbCl_4]^{2-}$ ions have all been identified and it is probable that some molecular $PbCl_2$ occurs in aqueous solution. The unique diatomic ion Hg_2^{2+} also appears to form species such as $[Hg_2(OH)]^+$ and $[Hg_2Cl]^+$ and the $[AgCl_2]^-$ ion has definitely been identified. Although "intermediate" species such as $[PbCl]^+$ and $[Hg_2Cl]^+$ can be neglected in most simple calculations, there is a practical problem associated with such species as $[AgCl_2]^-$, $[PbCl_3]^-$, and $[PbCl_4]^{2-}$. The solid Hg_2Cl_2 does not react to any significant extent with excess Cl^-. The stability of the *complex ions* of silver and lead in aqueous solution means that the solid chloride salts may be dissolved by excess Cl^-. We can estimate the magnitude of the effect of excess Cl^- on AgCl or $PbCl_2$ precipitates from the appropriate equilibrium constants:

*A few of the "less common" metals form insoluble chlorides. For example, TlCl is only sparingly soluble in water.

$$AgCl(s) + Cl^-(aq) \rightleftharpoons [AgCl_2]^-(aq)$$

$$K_{25^\circ} = 2 \times 10^{-4} = \frac{C_{[AgCl_2]^-}}{C_{Cl^-}}$$

from which

$$C_{[AgCl_2]^-} = (2 \times 10^{-4})C_{Cl^-}$$

$$PbCl_2(s) + Cl^-(aq) \rightleftharpoons [PbCl_3]^-(aq)$$

$$K_{25^\circ} = 4 \times 10^{-4} = \frac{C_{[PbCl_3]^-}}{C_{Cl^-}}$$

from which

$$C_{[PbCl_3]^-} = (4 \times 10^{-4})C_{Cl^-}$$

These considerations apply, of course, only to the equilibria in systems containing some undissolved AgCl or $PbCl_2$, respectively, but they do indicate clearly that the loss of solid by formation of a soluble complex ion increases with increasing chloride concentration. As a result, we must keep in mind the fact that adding more of a "precipitating ion" can, indeed, result in *less* precipitate.

In this investigation we shall study the formation of metal chloride precipitates and the possible separation of $PbCl_2$ from AgCl and Hg_2Cl_2 by taking advantage of the relatively high solubility of $PbCl_2$ in hot water. We will need to consider the problem of complex-ion formation and the need to maintain a pH low enough to avoid hydioxide precipitations when these are possible (as discussed in Investigation 21).

Before starting this investigation, you should be familiar with the principles and calculations involved in solubility, acid-base, and complex-ion equilibrium systems. In addition, prior to coming to the laboratory, you must complete the calculations required in the Procedures section of the investigation for all three metal ions.

References

1. O'Connor, Rod. 1977. *Fundamentals of Chemistry* (2nd ed.). New York: Harper & Row (Units 18–21 and 27 and Excursion 2).
2. Moeller, Therald and Rod O'Connor. 1972. *Ions in Aqueous Systems.* New York: McGraw-Hill.

SAFETY Wear approved eye protection at all times. Avoid contact with any chemical reagents (Ag^+ and Pb^{2+} ions are toxic; $Hg_2{}^{2+}$ and Ag^+ cause skin damage). In case of contact with chemicals, wash immediately and thoroughly with water.

OBJECTIVES

1. Be able to make appropriate calculations concerning chloride and hydroxide precipitations and the formation of complex chlorides.
2. Be able to devise and test an analytical scheme based on the differing solubilities of $PbCl_2$, AgCl, and Hg_2Cl_2.
3. Be able to analyze an "unknown" solution containing one or more of the metal ions studied in this investigation and, at the option of your instructor, an ion studied in either Investigation 20 or 21.

PROCEDURES

Your instructor will assign to you Pb^{2+} and either Ag^+ or Hg_2^{2+} to study. From the information in Table 22.1, the equilibrium constants given for $[AgCl_2]^-$ and $[PbCl_3]^-$, and the following facts and assumptions, devise an experimental procedure for the precipitation and separation of $PbCl_2$ and the chloride of the other metal ion assigned. Have your procedure checked by your instructor.

The solution you will be working with contains 10 mg ml^{-1} of the metal ion, before any other solutions are added.

The precipitations are to be performed by mixing 5 drops of the metal-ion solution with sufficient other solutions and distilled water to form a final total volume of 40 drops.

One drop from a typical medicine dropper has a volume of 0.05 ml.

For the required analysis scheme, you will need all of the calculations indicated for procedure 1(a–f). (These should have been completed prior to your coming to the laboratory.)

After your instructor has checked your calculations and your proposed analysis scheme, obtain the two assigned metal-ion solutions in separate *clean* test tubes. Label the test tubes properly and use these solutions to test your analysis scheme.

1. Before beginning experimental work, you will need to have the following calculations completed *in your notebook,* and checked by your instructor, for *each* of the ions Pb^{2+}, Ag^+, and Hg_2^{2+}.

a. For each solution, the minimum C_{Cl^-} required to initiate precipitation of the metal chloride from a solution prepared by diluting 5 drops of the original metal-ion solution (10 mg of metal ion per milliliter) to a total volume of 40 drops.

b. For each *diluted* solution [as described in (a)], the number of millimoles of Cl^- necessary to achieve "complete precipitation" of the metal chloride, defined as having reduced the metal ion concentration to 10^{-3} of its original value.

c. For each diluted solution, the maximum pH allowable if precipitation of the metal hydroxide or oxide is to be avoided. [Assume pH 7 maximum for Ag^+ or Hg_2^{2+} and see Investigation 21 for information on $Pb(OH)_2$.]

d. The maximum "excess" C_{Cl^-} acceptable if the concentration of the complex anion ($[PbCl_3]^-$ for lead, $[AgCl_2]^-$ for silver, none for mercury) in equilibrium with the solid metal chloride is not to exceed one-thousandth of the original metal-ion concentration of the *diluted* solution [as described in (a)].

e. The minimum volume (in drops or milliliters) of hot water (90°–100°C) required to dissolve all the metal chloride formed by "complete precipitation" [as described in (b)].

f. The number of drops of "stock HCl solution" that will, when diluted to a total volume of 40 drops, provide *approximately* each of the Cl^- concentrations or amounts calculated for (a), (b), and (d). The "stock HCl solutions" available are 12 *M*, 3 *M*, and 0.3 *M*.

g. The pH corresponding to each HCl concentration at dilutions calculated for (f).

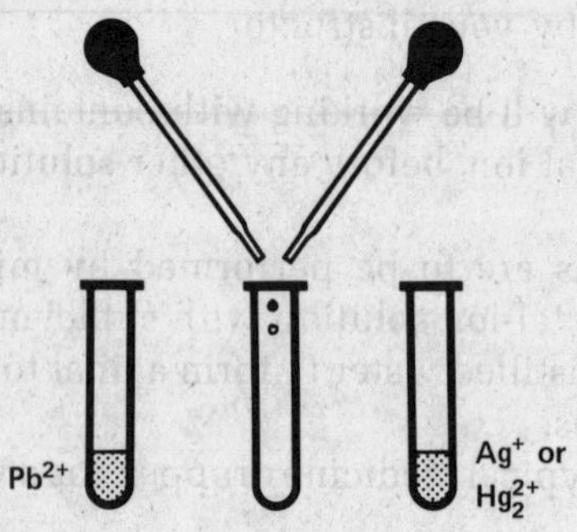

2. Now mix 5 drops of each of your metal-ion solutions in a clean test tube. Calculate the number of drops (no more than 30) of the *appropriate* "stock HCl solution" that will "completely precipitate" both metal chlorides without significant formation of $[PbCl_3]^-$ or $[AgCl_2]^-$ [i.e., limited as described in procedure 1(d)] and that will avoid any hydroxide precipitation. Then determine the number of drops of distilled water to make a total mixture volume of 40 drops (metal-ion solutions + "stock HCl" + water).

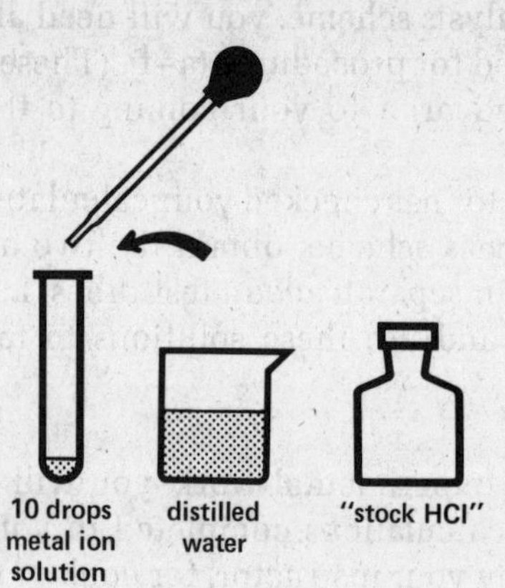

3. Add to the test tube containing the metal-ion solution the number of drops of distilled water calculated in procedure 2. Then add the calculated number of drops of the "stock HCl solution." Mix well by carefully shaking the test tube.

4. Centrifuge the test tube and decant the clear solution from above the precipitate into a clean test tube. Add to this solution 1 drop of 3 *M* HCl. (If additional precipitate forms, shake this mixture carefully and pour it back into the test tube containing the initial precipitate. Centrifuge again, decant, and discard the clear solution from above the precipitate.)

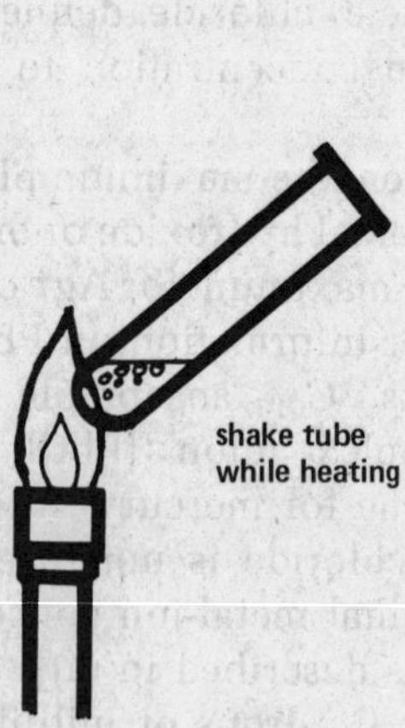

5. To the precipitate in the test tube add just the number of drops of distilled water calculated as sufficient to dissolve, at 90°–100°C, all of the precipitated lead chloride without dissolving any significant fraction of AgCl or Hg_2Cl_2. Shake the test tube carefully to disperse the precipitate and then heat the test tube carefully until the water boils gently; continue to shake the tube to avoid superheating.

6. Centrifuge the *hot* test tube and decant the clear solution from above the remaining precipitate into a clean test tube. Record any observations. Allow the test tube containing the hot solution to cool in a cold bath (cold water in a beaker or, if available, ice and water in a beaker) and see if any $PbCl_2$ precipitates. If no precipitate forms, discuss with your instructor a possible method of detecting Pb^{2+} in this solution based on the studies made in Investigations 20 and 21. Record any tests and observations made.

7. Take a clean test tube to your instructor and obtain an "unknown." This "unknown" may contain:

a. Only distilled water.
b. Only one of the two ions you have just studied.
c. Both of the ions you have just studied.
d. One of the two ions studied in this investigation *and* one of the ions you studied in either Investigation 20 or 21. [Your instructor will inform you if option (d) is included.]

8. Using the analysis scheme tested for chloride separations and (if option (d) is possible for your "unknown") your notebook records from Investigations 20 and 21, determine the composition of your "unknown."

CALCULATIONS AND REPORT Use your notebook records to complete the Report Form. Attach separate pages containing neatly written calculations as required for procedures 1 (a–g), 2, and 5.

GRADE	
	INSTRUCTOR'S INITIALS

Report Form

FOR INVESTIGATION 22:

Analysis by Chloride Precipitation

NAME: ______________________ ID NO.: ______________________

LAB SECTION ______________________ INSTRUCTOR ______________________

Show calculations on attached pages.

Assigned metal ions: $Pb^{2+}$and ____________

	Less soluble chloride ____________	$PbCl_2$
a. Minimum C_{Cl^-} to initiate precipitation		
b. Number of millimoles of Cl^- for "complete precipitation"		
c. Maximum pH to avoid oxide or hydroxide precipitation		
d. Maximum C_{Cl^-} to avoid loss by complex ion formation		
e. Minimum volume of hot water to dissolve all metal chloride (indicate units)		
f. Drops of "stock HCl" (specify) used for complete precipitation		
g. pH of total mixture after addition of "stock HCl"		

On the back of this page, briefly describe the steps taken and the observations made in testing your analysis scheme and in analyzing your "unknown."

The "unknown" contained ______________________

______________ DATE ______________________ SIGNATURE

Investigation 23

DEVELOPING AN ANALYTICAL SCHEME

You will be assigned two, three, or four metal ions from among those studied in Investigations 20, 21, and 22. Based on your calculations and notebook records from these investigations, design an analysis scheme suitable for separating each of these ions from the assigned mixture. Have your scheme checked by your instructor, then obtain samples of solutions of the assigned ions, mix five drops of each, and test your analysis scheme. Record *in your notebook* all procedures used and observations made.

When you are sure that you have a workable and tested scheme, obtain an "unknown" from your instructor. This "unknown" may contain any combination of the ions assigned, all of these ions, or only distilled water. Use your tested analysis scheme to determine the composition of your "unknown," recording *in your notebook* all procedures and observations.

Use your notebook record to complete the Report Form.

GRADE

INSTRUCTOR'S INITIALS

Report Form

FOR INVESTIGATION 23:
Developing an Analytical Scheme

NAME: ______________________ ID NO.: ______________________

LAB SECTION ______________________ INSTRUCTOR ______________________

Metal ions assigned ________ ________ ________ ________

Assigned ion(s) forming insoluble sulfides ________

Assigned ion(s) forming insoluble sulfates ________

Assigned ion(s) forming insoluble chlorides ________

"Unknown" number ________

The "unknown" contained ________

Outline of analysis scheme and summary of observations made during analysis of "unknown":

(Continue on back of this page, as necessary)

________ DATE ________ SIGNATURE

Investigation 24

PREPARATION OF A METAL FROM ITS ORE

INTRODUCTION There may come a time, perhaps sooner than we think, when the study of many of the metal ores will be a part of a class in the history of chemistry, and the only place you will see a sample of high-grade ore will be in a museum.

We have already seen the combined effects of the depletion of U.S. ore deposits and the loss of many foreign sources. The prices of metals have soared and we have been forced to adapt our metallurgical technology to the more costly processing of low-grade ores and "recyclable" scrap metal. Nearly half of our present lead and copper, a fourth of our steel, and almost 20 percent of our aluminum, nickel, and zinc production is from scrap metal.

The recycling of metals is, of course, sound policy from an ecological point of view. It is much better to use scrap metal as a raw material than to allow it to accumulate in unsightly dumps while continuing to strip our natural mineral resources. It was not, in most cases, the desire to "clean up America" that prompted the large-scale recovery of scrap metals. As long as high-grade ores were abundant, their processing was considerably less expensive than that of most scrap-recycling processes. At first glance, that seems surprising. Surely it should be easier to recover copper from scrap metal than to mine and process an ore which, at best, is much less than 80 percent copper. Such is not the case, however, for three reasons. First, most of our metals industry has been designed for refining ores. Quite different processes are necessary for scrap metal treatment, so the extensive modification of existing facilities or the construction of new ones represents a large capital investment. This could not be justified on an economic basis until the costs of high-grade ores had risen to the point where scrap recycling was a profitable competition. The second reason for the high cost of scrap metal processing has to do with the location of the competitive raw materials. Ore deposits, to be profitable, have to occur within a small geographic area so that mining and shipping costs are minimized, but scrap metal is accumulated at every city dump, farm gully, and roadside. The cost of accumulating and transporting scrap to central locations for recycling is enormous. There is still a third reason. Metal ores contain the metal in the form of one of its salts (e.g., Cu_2S, $PbCO_3$). The isolation of the crude salt from "gangue" materials, such as clay or silica, is usually a fairly simple physical process, easily designed for mass production and low energy requirements. The conversion of the salt to the metal is also relatively simple. Scrap metals, on the other hand, normally contain *alloys* of two or more metals. The recovery of a single metal from an alloy is, in most cases, considerably more complicated than its production from a salt. In addition, on any large scale, scrap metal contains a variety of alloys. The sorting of scrap to select single alloys or even collections of quite similar alloys could be, on any reasonable production scale, prohibitively slow and expensive.

Even if we can eventually solve most of the problems connected with scrap recycling, we are not going to reach (as some have hopefully suggested) a steady state ecology in which all the metals we need are recovered from those we discard. No industrial process is 100 percent efficient and, in addition, common sense tells us that we are not going to collect and transport every empty can or bent nail to a recycling plant. Unless our demand for metals suddenly starts to decrease (and it is still increasing every year), we must continue to rely, to some extent, on natural ores. Recycling of metal wastes can reduce the rate of ore depletion. These ores are not inexhaustible and someday a generation "smarter" than ours may have to learn to get along without them. In the meantime, and for the foreseeable future, the refining of ores will continue to play an essential role in our technology.

We will concentrate in this experiment on a single metal—lead. We are going to produce some lead from one of its natural ores and also prepare a valuable lead compound, the commercial pigment *chrome yellow*. This experiment will illustrate several aspects of metals technology: the chemistry of ore refining, the reason for the large energy demand by the metals industry, and the economic advantages of producing both a free metal (for direct use or for alloy production) and metal compounds. In addition, as your nose will inform you during the experiment, we will discover why some smelting operations are associated inherently with the emission of undesirable waste products.

Before starting this investigation, you should be familiar with the concepts of oxidation and reduction, including the balancing of "redox" equations, and with calculations involving stoichiometry, solution concentrations, and ideal gas laws.

References

1. O'Connor, Rod. 1977. *Fundamentals of Chemistry* (2nd ed.). New York: Harper & Row (Units 22 and 25, and Excursion 3).
2. Readings from Scientific American. 1973. *Chemistry in the Environment.* San Francisco: Freeman (Units 15, 26, 27, 35, and 36).

SAFETY There are several significant hazards in this experiment. Careful work is essential.

Wear approved eye protection *at all times* and be especially careful in handling the reagents used. In case of contact with chemicals, wash *immediately* and *thoroughly* with water. Avoid inhaling gaseous products generated. Remember that lead compounds are cumulative poisons. Use proper care in working with hot equipment and chemicals.

OBJECTIVES

1. Be able to work properly and safely with hazardous reagents.

2. Be able to use proper techniques for a high-temperature process.
3. Be able to balance equations for both oxidation-reduction and metathetical reactions, and to make appropriate calculations based on the stoichiometry of such reactions.

PROCEDURES

One of the purposes of this experiment is to illustrate the economic advantage to a metals industry of producing both a free metal and commercially valuable metal compounds. To simulate this situation, you will act as both parts of the plant operation. While one part is refining the ore, the other part is using metal already produced to prepare a compound. This is efficient, both for you and the real plant, only if the two processes go on at the same time. You can do this best if you follow the suggested work sequence:

1. Begin heating the lead sulfide (procedures A1–A3).
2. Dissolve a lead sample in nitric acid (procedures B1–B3).
3. Calculate the amount of $K_2Cr_2O_7$ needed (procedure B4).
4. Prepare the dichromate solution (procedures B4–B5).
5. Filter the $Pb(NO_3)_2$ solution and prepare $PbCrO_4$ (procedures B6–B7).
6. Filter and dry the $PbCrO_4$ (procedure B8).
7. Reduce lead oxide with carbon (procedures A4–A8).
8. Isolate the lead (procedures A9–A10).
9. Weigh the lead and the lead chromate (procedures A11 and B9).
10. Hand in both products to your instructor (procedures A11 and B9).

A. Preparation of Lead from Galena Ore (PbS)

1. Obtain from your instructor a sample of galena ore weighing between 6 and 10 g. If massive pieces are distributed, select a cube measuring between 0.5 and 1.0 cm on the edge and weigh it to the nearest 0.05 g. If it weighs more than 10 g, chip off a piece and reweigh. Return any fragments to your instructor. Describe the appearance of your sample, in detail, in your permanent notebook.

2. Place the weighed galena sample in a clean, dry mortar and rap it sharply (but carefully) with the pestle. Record observations of the fracture patterns of galena. Then smash and grind the ore into a *very fine* powder. Transfer the powder, as completely as possible, into a clean, dry crucible.

If galena samples are not available, omit procedures 1 and 2. Obtain from your instructor—in a clean, dry, preweighed crucible—between 6 and 10 g of powdered PbS. Determine and record the weight of this sample to the nearest 0.05 g.

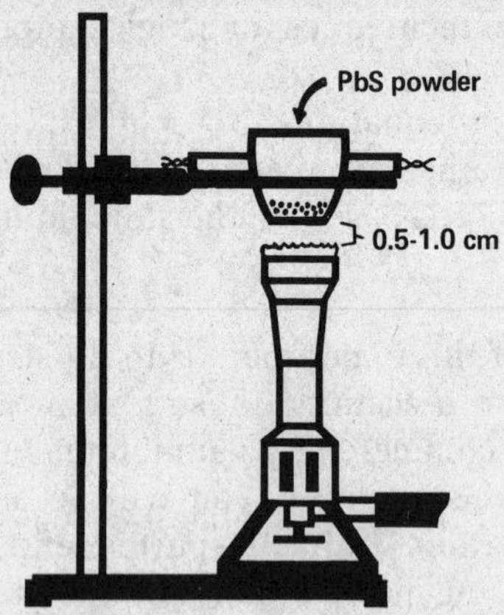

3. (If space is available, work at the fume hood.) Set your crucible in a clay triangle supported on a ring clamp. Position the clamp so that the bottom of the crucible is 0.5 to 1.0 cm above the top of an *unlighted* Fisher burner. (An ordinary laboratory burner will not give a hot enough flame.) Light the burner and *gradually* adjust from a cool to a very hot flame. Heat the crucible until it glows red. The contents will melt and SO_2 gas will be evolved.

Caution: Do not inhale this gas.

Continue heating until the contents of the crucible form a red-hot melt that no longer evolves SO_2. (You can see the mist formed when SO_2 contacts moist air if you look across the crucible toward a dark background.) (Heating time is about 1 hr.)

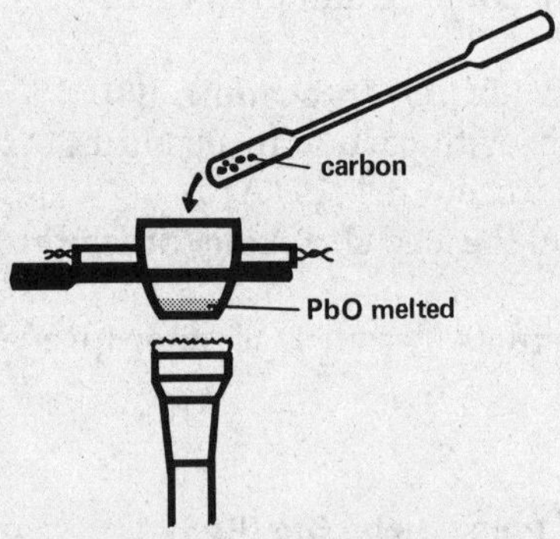

4. Carefully use a spatula load of carbon to cover the surface of the molten lead oxide in the crucible.

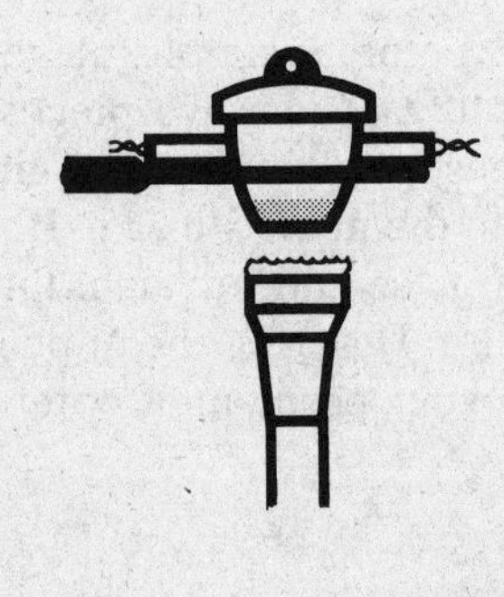

5. Using crucible tongs, carefully place the cover on the crucible. Heat the covered crucible to a red glow for 5 to 6 min. Then turn off and remove the burner.

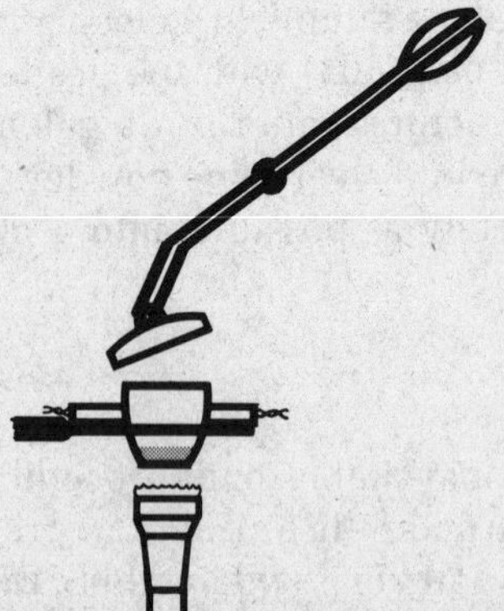

6. Lift the cover off the crucible, carefully, with crucible tongs and set it on an asbestos pad (*not* the bench top) to cool.

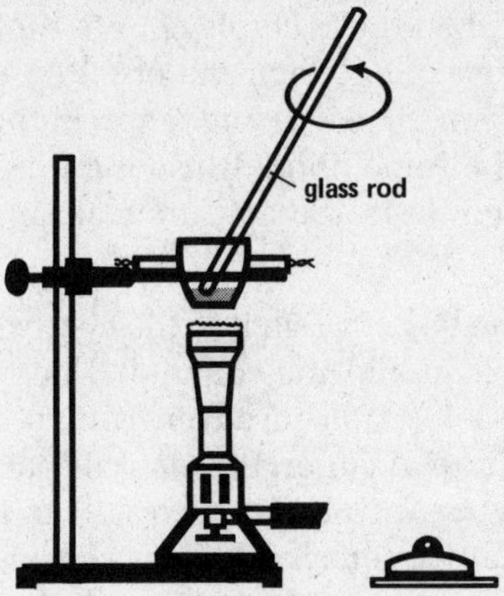

7. With a 6-in. glass rod, *carefully* stir the contents of the crucible.

Caution: Do not press against the bottom of the crucible.

Record any observations of lead globules. Add a second spatula load of carbon black, remove the glass rod, cover the crucible, and reheat the crucible to redness for 5 to 6 min.

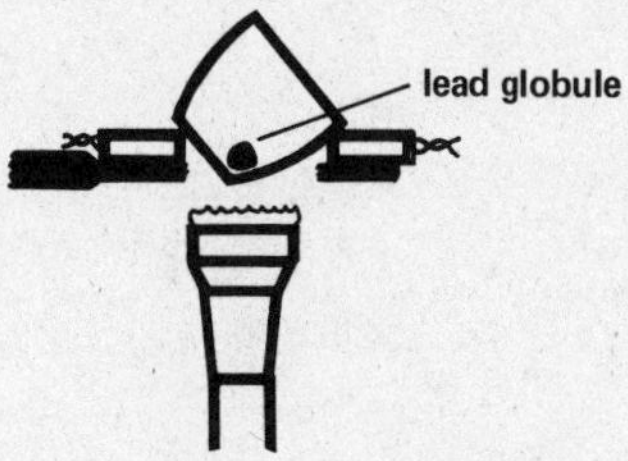

8. Use tongs to remove the cover and set it on the asbestos pad. Stir the crucible contents *carefully* to force small globules of lead to collect into a single large globule. Remove the stirring rod and, with tongs, tilt the crucible at an angle in the clay triangle. Heat the bottom and wall of the crucible to redness. (While this is heating, set a 400-ml beaker containing 300 ml of tap water near the ring-stand.)

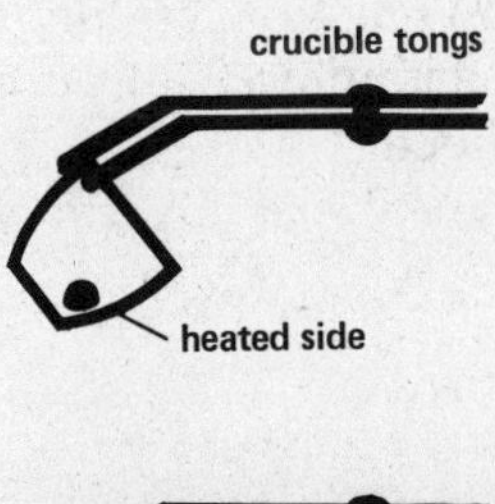

9. When the bottom side wall of the crucible is very hot, grip the *opposite* side with tongs and quickly, but *carefully*. . . .

10. . . . . tilt the crucible over the beaker just enough to let the molten lead slide into the water.

Caution: Do not allow the hot crucible to touch the beaker or the water. Watch out for splattering.

Inspect the inside of the crucible. If lead residues remain, repeat procedures 8–10.

Turn off the burner and let the crucible cool to room temperature on the clay triangle. When the crucible is *completely cool, carefully* immerse it (using tongs) in the nitric acid bath located in the fume hood. Rinse the tongs thoroughly. Leave the crucible in the acid to soak.

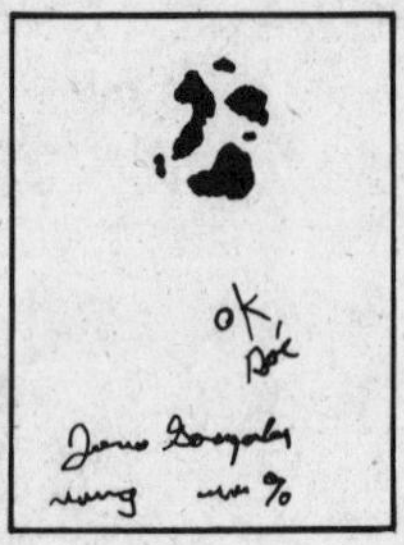

11. Remove and dry the lead. Weigh it to the nearest 0.05 g. Record the weight of lead and the calculated percentage yield (based on the mass of PbS used) in your notebook and also on a small square of paper. Present this paper and your lead sample to your instructor, who will initial the paper and tell you whether you should keep the lead sample or add it to the collection jar. Save the initialed paper to attach to your Report Form.

B. Preparation of Chrome Yellow from Lead

Since it will take you about 2 hr to prepare the lead from galena ore (Part A), you will use a lead sample provided by your instructor (and probably prepared by another student) for this part of the experiment. Obtain a sample of lead weighing about 2 g. Weigh a clean 100-ml beaker to the nearest 0.05 g, place your lead sample in the beaker, and reweigh. Record the mass of lead used in your notebook.

(Note: The reaction of lead with nitric acid *must* be performed in a fume hood.)

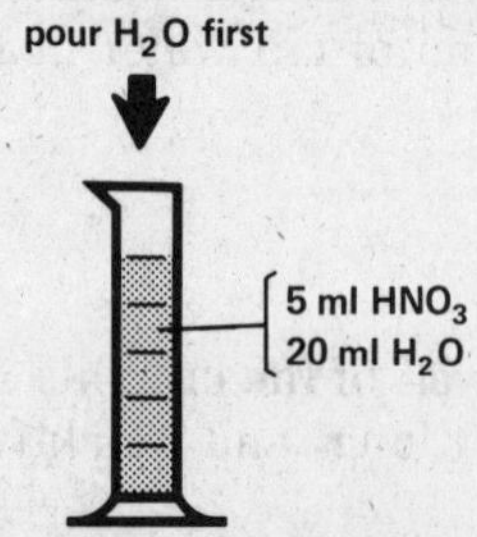

1. Pour 20 ml of distilled water into a 25-ml graduated cylinder. *Carefully* add, while stirring, sufficient concentrated nitric acid to make 25 ml of total solution. Pour this solution into the beaker containing the lead and set this mixture inside the fume hood.

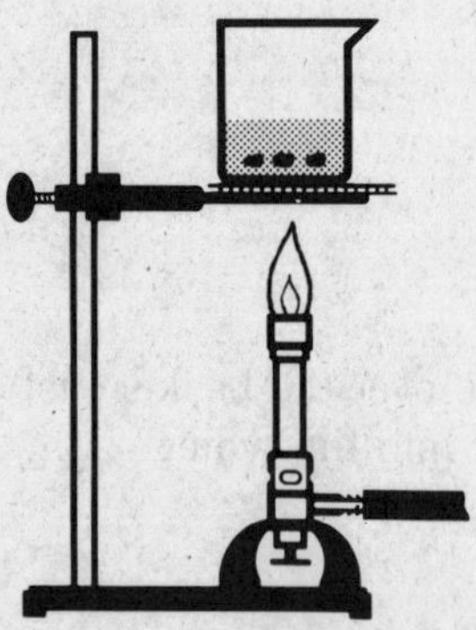

2. Place the beaker on a wire gauze supported on a ring clamp above a Bunsen burner. Heat the mixture with a *low* flame just enough to maintain a gentle boiling for about 20 min, by which time the lead should have all dissolved. Record in your notebook a description of any gaseous products evolved.

Caution: Do not inhale these gases.

If your lead sample was prepared by another student, it probably contained some unreacted carbon. This will not dissolve in the acid but will be suspended in the solution or float to the top. Lead will remain as spherical particles in the bottom of the beaker until consumed by the acid.

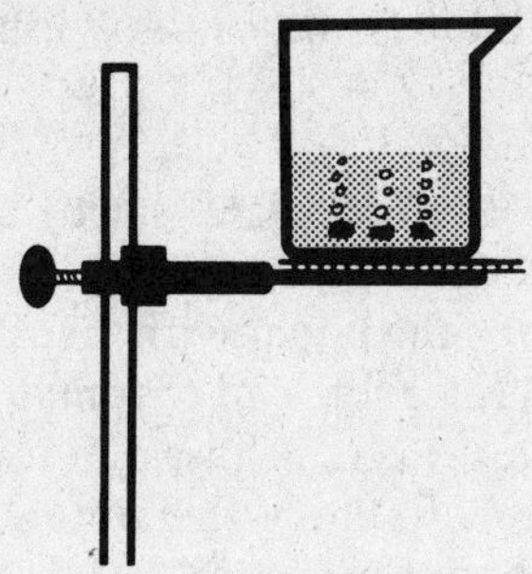

3. Check your mixture after about 10 min of boiling. Remove the flame and let the solution cool until boiling stops. Check to see if gas bubbles are still forming on the lead globules. If not, add 10 drops of concentrated nitric acid. When bubbles are observed, begin heating again until all the lead has been consumed. (If the liquid level drops below half the original level, add enough distilled water to restore the original volume.) When the last of the lead has disappeared, turn off the burner and allow the solution to cool.

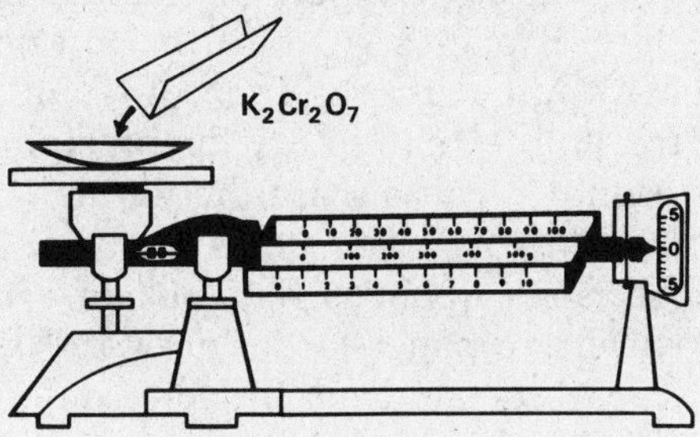

4. Using the mass of the lead sample consumed by the nitric acid, calculate the mass of potassium dichromate required for complete conversion of the lead nitrate to chrome yellow ($PbCrO_4$). Weigh out the calculated amount of $K_2Cr_2O_7$, on a watch glass, to the nearest 0.05 g.

5. Rinse the dichromate with distilled water through a funnel into a clean 125-ml Erlenmeyer flask. Add enough water to make a total volume of about 70 ml. Swirl (and warm gently, if necessary) to dissolve all the dichromate.

6. Filter your lead nitrate solution through a coarse filter (to remove carbon particles) into a clean 250-ml beaker. Rinse the lead nitrate beaker with 10 ml of distilled water and pour this through the filter.

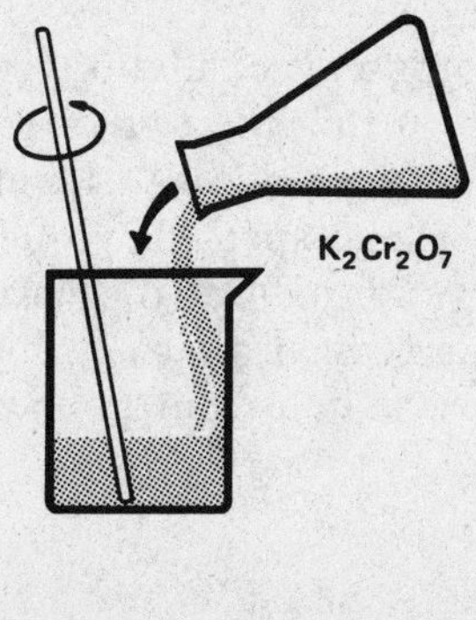

7. Add the dichromate solution dropwise, while stirring, to the filtered solution of lead nitrate. (If no precipitate forms, the lead nitrate solution is too acidic. Add dilute sodium hydroxide, dropwise with stirring, until a yellow precipitate begins to appear. Do not add excess NaOH.) Continue adding the dichromate solution, with stirring. Occasionally stop stirring, let the precipitate settle a bit, and pour a little of the dichromate solution down the wall of the beaker. When added dichromate no longer forms a precipitate where the dichromate solution enters the liquid in the beaker, you may assume that precipitation of $PbCrO_4$ is complete.

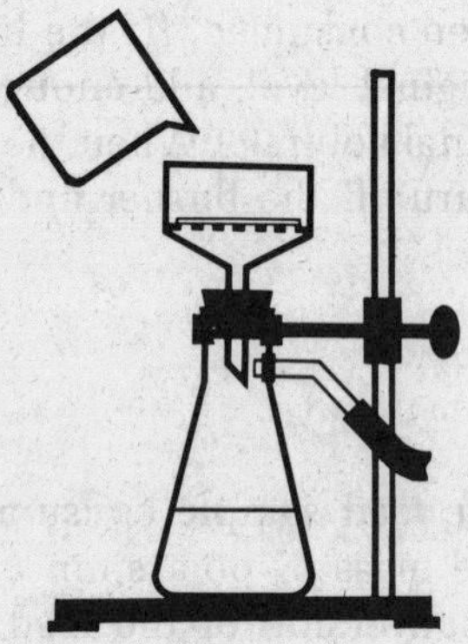

8. While the lead chromate precipitate is settling, assemble a vacuum filtration setup (Appendix A). Place a filter paper in the funnel, moisten the paper with distilled water, turn on the aspirator, and filter the lead chromate mixture. [If the filtrate is quite cloudy, pull the hose off the aspirator, then turn off the aspirator. Remove the funnel (keeping the filter paper in place), pour the suspension back into the beaker, and allow it to settle for a few minutes. Then filter the mixture again through the original filter paper. Repeat this procedure, if necessary, until the filtrate is only slightly cloudy.] Allow the aspirator to pull air through the filter for 10 to 15 min to dry the precipitate. Then shut off the aspirator and discard the filtrate.

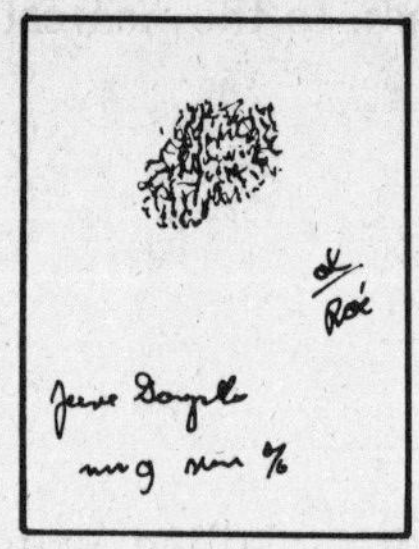

9. Cut out a paper rectangle about 10 cm × 12 cm, crimp the corners, and weigh the paper to the nearest 0.05 g. Scrape the dried lead chromate from the filter onto the weighed paper and reweigh. Label the paper with your name, the product mass, and the calculated percentage yield (based on the mass of lead used). Hand in your *chrome yellow* sample on this paper to your instructor, who will show you where to put the product. Obtain your instructor's initials on the emptied paper and save this to attach to your Report Form.

CALCULATIONS AND REPORT

Enter in your notebook balanced net equations for all reactions performed. (Assume that the reaction of lead with nitric acid produces lead nitrate, water, and NO_2 gas.)

From the percentage yield for your conversion of galena to lead, calculate:

a. The number of metric tons* of galena that would be required for the production of 5.0 metric tons of lead by this process.

b. The number of metric tons of sulfur dioxide gas that would be released during the production of 5.0 metric tons of lead, assuming that one mole of SO_2 is formed for each mole of *galena used.*

From the percentage yield for your lead chromate preparation, calculate:

a. The number of kilograms of lead required to produce 25 kg of $PbCrO_4$.

b. The volume of concentrated (15.7 *M*) nitric acid required to consume the amount of lead required.

c. The volume of NO_2 gas (at 25°C and 750 torr) generated by reaction of this amount of lead with nitric acid.

Use your notebook records and calculations to complete the Report Form. Attach to this the two slips of paper initialed by your instructor (procedures A11 and B9). Considering the necessity for using a special high-temperature burner (procedure A3), write a brief discussion of the energy requirements for a typical ore-smelting industry. Attach a copy of this discussion to your Report Form.

*1 metric ton = 1000 kg.

GRADE	
	INSTRUCTOR'S INITIALS

Report Form

FOR INVESTIGATION 24:

Preparation of a Metal from Its Ore

NAME: ______________________ ID NO.: ______________________

LAB SECTION ______________________ INSTRUCTOR ______________________

Show all calculations on reverse side.

Experimental Data

Mass of galena used ________ g   Mass of lead obtained ________ g

yield ________ %

Mass of lead used ________ g   Mass of crome yellow obtained ________ g

Mass of $K_2Cr_2O_7$ used ________ g   yield ________ %

Observations

Attach a separate page describing your observations of:

a. Fracturing of galena (if performed).

b. Appearance of lead while forming.

c. Properties of SO_2 evolved.

d. Properties of NO_2 evolved.

Equations (Show balanced net equations.)

$PbS + \text{air}$

$PbO + C$

$Pb + HNO_3$

$Pb(NO_3)_2 + K_2Cr_2O_7$

Calculations (Based on your experimental yields)

For production of 5.0 metric tons of lead, ____________ metric tons of galena are required and ____________ metric tons of SO_2 are released.

For production of 25 kg of *chrome yellow,* ____________ kg of lead and ____________ liters of concentrated nitric acid are required, while ____________ liters of NO_2 gas (at 25°C and 750 torr) are generated.

Attach the initialed slips from procedures A11 and B9.

Attach also the "energy discussion."

________________ DATE   ________________ SIGNATURE

Investigation 25

ANALYSIS OF IRON IN PYRITES

INTRODUCTION Modern society cannot survive without the continued production of metals. That is obvious. It should now be equally obvious that modern society cannot long survive the present technology for producing metals or the present philosophy that production must steadily be increased. The following examples indicate two of the most common methods for procuring metals from natural ores. All of these procedures irreversibly consume energy and mineral resources and all contribute, in one way or another, to the deterioration of the environment. There are better ways—often more expensive. The ultimate solution must involve changes in technology and, more important, basic changes in philosophy. We cannot continue to measure basic human values in terms of "gross national product."

The basic key to the smelting of sulfide ores (major sources of copper, silver, zinc, cadmium, mercury, vanadium, and molybdenum) has traditionally* been the fact that the sulfides, when roasted in air, evolve SO_2, leaving behind either the crude metals or their oxides.

Before the roasting stage, the ores must go through some preliminary operations. The exact steps, of course, vary with the particular ore used. Generally, the preliminary workup includes various physical operations such as crushing and washing. Crude separations may also be made to isolate the particular salt desired in a more "concentrated" form.

During the roasting stages, oxygen is consumed and sulfur dioxide is evolved. The gas, along with particulate matter, is the principal source of smelter pollution from operations using sulfide ores. Much attention has been focused on ways of reducing smelter pollution, and progress is being made. The result of the initial roasting may be either the metal or its oxide. The remaining steps vary considerably, depending on the metal involved and on the procedures deemed most economical by the particular smelter.

EXAMPLE 1: COPPER FROM A LOW-GRADE COPPER ORE

Step 1: Concentration by the flotation method

The crushed ore is treated with an oil-water emulsion. The sulfides are wetted by the oil and the "soils" by the water. Detergent is added, and the mixture is agitated vigorously. The oily sulfides are carried off by the foam, and the "soils" settle out.

Step 2: Roasting in air

The sulfide concentrates (mainly Cu_2S and FeS, with traces of other sulfides) are roasted in a blast of air; SO_2 is evolved, along with colloidal smokes of various volatile

*Some of the new processes under development do not lead to formation of SO_2.

oxides, such as As_2O_3. The product contains mainly copper(II) oxide and oxides and sulfides of iron. (FeS is much more resistant to oxygen than is Cu_2S.)

Step 3: Smelting to form "blister copper"

The roast products are mixed with sand and coke and heated in a blast of hot air. Under these conditions the iron is converted to a molten silicate ("slag"), which is drained away. The coke is oxidized to carbon monoxide, and the CuO is reduced to Cu_2S (obtaining the sulfur from the iron sulfide). After removal of the slag, the Cu_2S is partially converted by hot air to Cu_2O. The air is then shut off, and the Cu_2O is permitted to react with residual Cu_2S, forming SO_2 and molten copper:

$$Cu_2S + 2Cu_2O \longrightarrow 6Cu + SO_2$$

The SO_2 is moderately soluble in the liquid copper. As the metal cools, SO_2 escapes, leaving the metal with a blistered appearance.

Step 4: Electrolytic refining

The cast plates of crude* blister copper are used as the anodes in an electrolytic cell. Thin plates of pure copper are used as cathodes, and the electrolyte contains a copper(II) salt. Electrolytic refining typically produces copper containing less than 0.1 percent impurity.

EXAMPLE 2: "PIG IRON" FROM HEMATITE ORE (THE BLAST FURNACE PROCESS)

Iron ore, coke, and limestone ($CaCO_3$) are mixed and added at the top of the blast furnace and the mixture, heated initially to about 250°C, is allowed to sift down through the furnace against a rising blast of hot air (~1500°C). The following reactions occur as the mixture falls through the hot blast:

$$
\begin{aligned}
CaCO_3(s) &\longrightarrow CaO(s) + CO_2(g) \\
\underset{\text{coke}}{C(s)} + CO_2(g) &\longrightarrow 2CO(g) \\
3Fe_2O_3(s) + CO(g) &\longrightarrow 2Fe_3O_4(s) + CO_2(g) \\
Fe_3O_4(s) + CO(g) &\longrightarrow 3FeO(s) + CO_2(g) \\
FeO(s) + CO(g) &\longrightarrow Fe(l) + CO_2(g) \\
[C(s) + CO_2(g) &\longrightarrow 2CO(g)] \\
xCaO(s) + (SiO_2)_x(s) &\longrightarrow \underset{\text{silicate ("slag")}}{xCaSiO_3(l)}
\end{aligned}
$$

The calcium silicate (slag) is less dense than the molten iron. The two liquid phases are separated near the bottom of the furnace by first draining off the slag when the phase boundary reaches the upper drain outlet and then draining off the liquid iron from the lower furnace outlet. The iron obtained from the blast furnace (about 90 percent purity) is called "pig iron" because the molds originally used to collect the molten iron for cooling were shaped like fat pigs.

Various purification and alloying steps are used to convert this crude iron into different types of steel.

*Typically contaminated by silver and gold from the original ore. These metals are recovered as valuable by-products from most copper smelters. Some lead, iron, nickel, and zinc are usually present as well.

The iron industries must maintain analytical laboratories for continuously monitoring the composition of ore samples, since variations in ore content require corresponding variations in both stoichiometric and economic parameters in the production of iron. Depending on such factors as the speed and accuracy required, the nature of the ore, and the budget allocated, these analyses may involve any of a number of techniques. Spectroscopic assays are rapid, but often less accurate than more "conventional" methods. Spectroscopic equipment is not inexpensive and several preliminary workup steps are often required to prepare a sample in the form suitable for spectroscopic assay. Gravimetric analyses (those in which the iron from an ore sample is removed and converted to a form suitable for weighing, usually as Fe_2O_3) are accurate, fairly inexpensive, but typically slow and tedious. A third alternative, volumetric analysis, uses a titration with some appropriate reagent (of known concentration) of the iron removed from a weighed ore sample. Since volumetric analysis is usually more rapid than gravimetric analysis and less expensive than spectroscopic assay, it is frequently employed in analytical laboratories for the routine assay of iron ores.

In this experiment we will employ a volumetric method for determining the iron content of a pyrite ore, containing mainly FeS_2.

Before starting this investigation, you should be familiar with the concepts of oxidation and reduction, including methods for balancing "redox" equations, and with stoichiometric calculations involving solution molarities and volumes. In addition you should be familiar with the proper techniques for handling and heating chemicals, for preparing solutions, and for performing titrations.

References

1. O'Connor, Rod. 1977. *Fundamentals of Chemistry* (2nd ed.). New York: Harper & Row (Units 13, 22, and 27 and Excursions 2 and 3).
2. O'Connor, Rod and Charles Mickey. 1974. *Solving Problems in Chemistry*. New York: Harper & Row (Units 2 and 6).

SAFETY Wear approved eye protection at all times. Be cautious with all reagents, but *especially* with the mixture of nitric and hydrochloric acids ("aqua regia"). Avoid contact of the permanganate with organic material and *never* mix $KMnO_4$ with concentrated H_2SO_4. (This mixture can explode violently.) In case of contact with *any* of the reagents, wash *immediately* and *thoroughly* with water.

OBJECTIVES

1. Be able to transfer, mix, and heat hazardous chemicals safely.
2. Be able to accurately prepare solutions of specified concentrations.
3. Be able to perform a titration accurately, using a slight excess of a colored titrant to detect the endpoint.

4. Be able to balance equations for oxidation-reduction reactions and to use these for stoichiometric calculations involving solution concentrations and volumes.

PROCEDURES

On some parts of this experiment you will work as part of a six-student team, as specified by your instructor. When your team has been assigned, meet together and decide on your individual duties (starting with procedure C). Then proceed with your own work until you and your teammates are ready for the cooperative parts of the experiment.

A. Preparation of the $KMnO_4$ Solution

Balance the equation

$$MnO_4^- + Fe^{2+} + H^+ \longrightarrow Mn^{2+} + Fe^{3+} + H_2O$$

On the basis of your result, calculate the mass of $KMnO_4$ required for preparation of 150 ml of 0.10 *N* $KMnO_4$ solution. Have your instructor check your results before proceeding further.

If the 0.1 *N* $KMnO_4$ solution is to be supplied to you, obtain 150 ml of the solution in a clean 250-ml Erlenmeyer flask. Stopper the flask.

If you must prepare the solution, weigh a clean, dry watch glass to the nearest 0.05 g. Advance the balance masses by the amount calculated for the $KMnO_4$ required and carefully add crystals of the salt until the balance returns to equilibrium. Transfer these crystals into a clean 250-ml Erlenmeyer flask and add 150 ml of distilled water, using part of the water to rinse any remaining crystals from the watch glass into the flask. Swirl the flask until a homogeneous solution is formed, then stopper the flask.

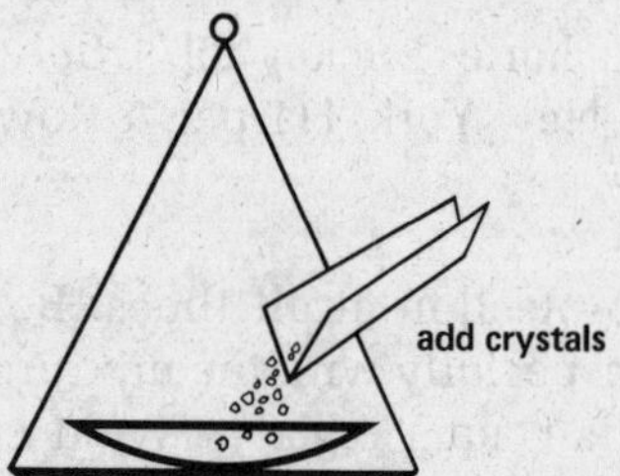

B. Conversion of a Pyrite Sample to $FeCl_3$

1. Take a 3-cm × 5-cm piece of aluminum foil to your instructor and obtain on it a pyrite sample of approximately the mass you will need. Transfer 0.9 to 1.0 g of this sample onto a *preweighed* watch glass and determine the sample mass to the nearest 0.1 mg.

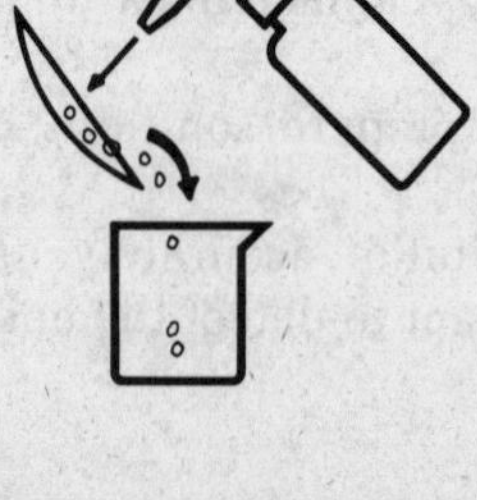

2. Transfer the solid completely into a clean 100-ml beaker, rinsing the watch glass with *not more than* 5 ml of distilled water. Set the beaker in the fume hood on a piece of paper labeled with your name.

3. In the fume hood, carefully pour 8 ml of concentrated nitric acid into a clean, dry graduated cylinder. Then *carefully* add concentrated hydrochloric acid until the total volume of the mixture is 24 ml. Slowly pour this into the beaker containing your pyrite sample while gently swirling the beaker. Set the beaker back on the labeled paper and *cautiously* rinse out the graduated cylinder.

Caution: This mixture, "aqua regia," is extremely corrosive.

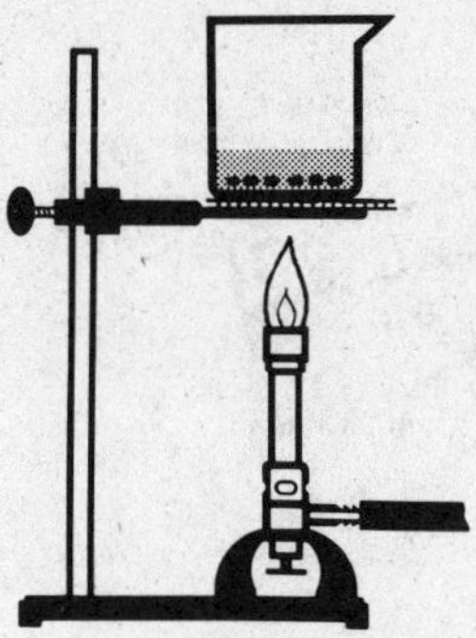

4. Set the beaker on a wire gauze supported by a ring clamp and heat with a low flame to a *gentle* boil. (Once the boiling rate is adjusted properly, you are ready to begin the "cooperative" part of this experiment, provided your other teammates are also ready.) Heating is continued until all the ore (except for some *white* silicates) has dissolved. When the solution volume has dropped to about 10 ml, the beaker must be covered with a watch glass to avoid splattering. At that point, each student must take over the work at the ore beaker, heating *carefully* to minimize splattering, until a dry, yellow residue (mostly $FeCl_3$) remains.

C. Preparation of Special Reagents

When all six team members have set up the ore-aqua regia mixture and adjusted the heating for slow, gentle boiling, Student 6 of the team will remain to watch all six beakers, notifying each teammate to return when the stage of covering the beaker with a watch glass is reached. In the meantime, the other members of the team will proceed as follows.

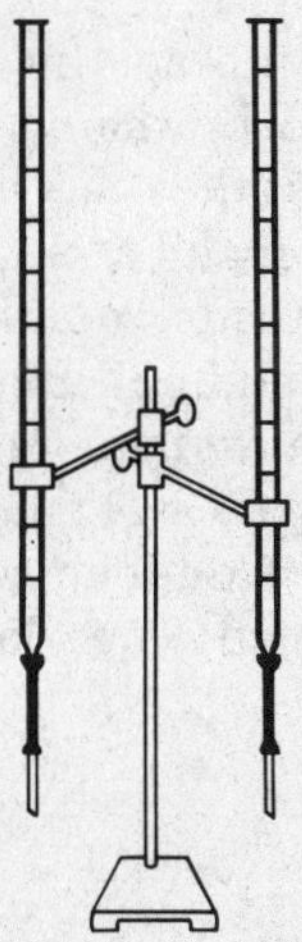

Student 1. Assemble two buret setups for each member of the team at their regular workbench areas. (Each student will clean his own burets, if necessary, as described in Appendix A.)

Student 2. Prepare a 5-percent $HgCl_2$ solution, to be shared by your teammates. Weigh a clean, dry watch glass to the nearest 0.1 g. Transfer onto this about 6 g of $HgCl_2$ (*poisonous*) and reweigh. Sample mass should be between 5.8 and 6.2 g. Fill a 25-ml graduated cylinder with distilled water and use this to rinse the $HgCl_2$ into a clean 250-ml Erlenmeyer flask. Add an additional 95 ml of distilled water and warm the flask, with swirling, until all solid has dissolved and the solution is homogeneous. Allow the solution to cool to room temperature, then label the flask "5% $HgCl_2$."

Student 3. Prepare a 1 *M* $SnCl_2$ solution to be shared by your teammates. Weigh a 125-ml Erlenmeyer flask and a powder funnel to the nearest 0.1 g. Advance the balance masses by 11.4 g and carefully add $SnCl_2 \cdot 2H_2O$ until the balance returns to equilibrium. Remove the flask from the balance and carry it to the fume hood. Use 20 ml of concentrated HCl (*corrosive*) to rinse any material adhering to the funnel into the flask. Heat the mixture cautiously, with swirling, until all solid has dissolved. Then add 30-ml of distilled water, and swirl until a homogeneous solution is formed. Take the solution back to your work area and label the flask "1 *M* $SnCl_2$."

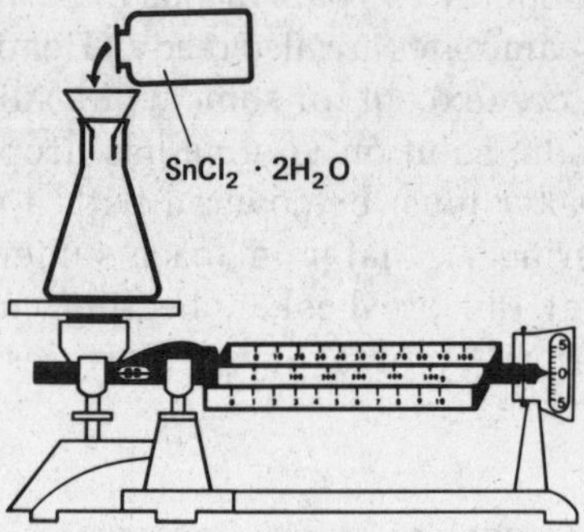

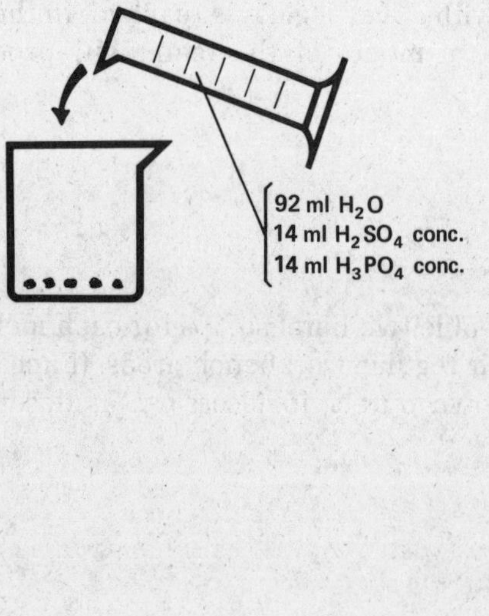

Students 4 and 5. Cooperate in preparing the Reinhardt-Zimmermann (R-Z) reagent to be shared by your teammates. Weigh a 250-ml beaker to the nearest 0.1 g. Advance the balance masses by 8.4 g and add $MnSO_4 \cdot 4H_2O$ to the beaker until the balance returns to equilibrium. Remove the beaker from the balance and add in sequence, with stirring, 92 ml of distilled water, then 14 ml of concentrated sulfuric acid (*corrosive*), then 14 ml of concentrated phosphoric acid (*corrosive*). Continue stirring until a homogeneous solution is formed. Then label the beaker "R-Z reagent."

All students. When notified by Student 6, return to the fume hood to complete the drying of the $FeCl_3$. When the product is dry, turn off the burner and allow the beaker to cool to room temperature. Then set the beaker on the paper labeled with your name at your regular working area.

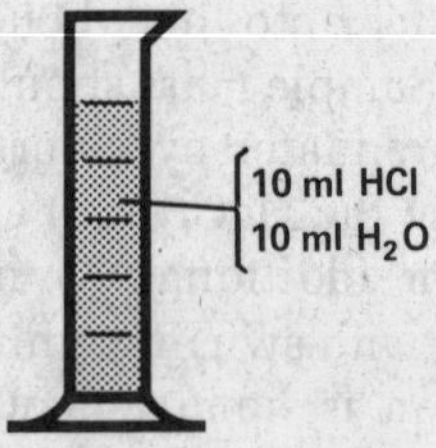

D. Preparation of the Fe(II) Solution

1. Prepare 20 ml of 6 *M* hydrochloric acid by careful addition of 10 ml of concentrated (12 *M*) HCl to 10 ml distilled water. (Add water, *then* acid, while stirring.)

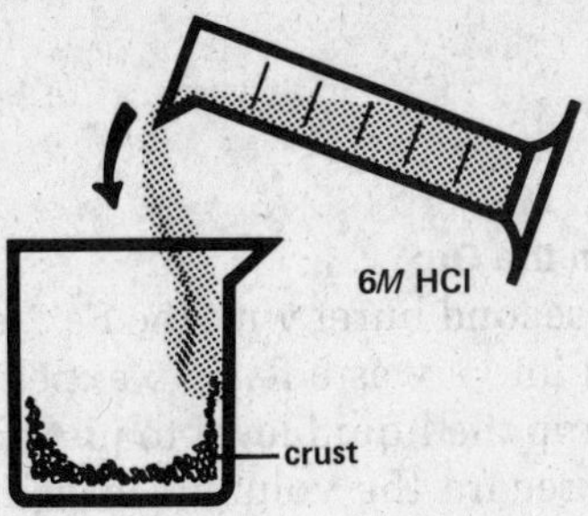

2. Use the 6 *M* HCl to rinse the underside of the watch glass into the beaker of $FeCl_3$ and to rinse any solid from the walls of the beaker down to the bottom. Pour any remaining acid into the beaker. Set the beaker on a wire gauze supported by a ring clamp above a burner. Warm the mixture, while stirring, until the solid has dissolved. Then heat the yellow (to brown) solution just until it *starts* to boil. Turn off the burner, leaving the beaker on the ring clamp.

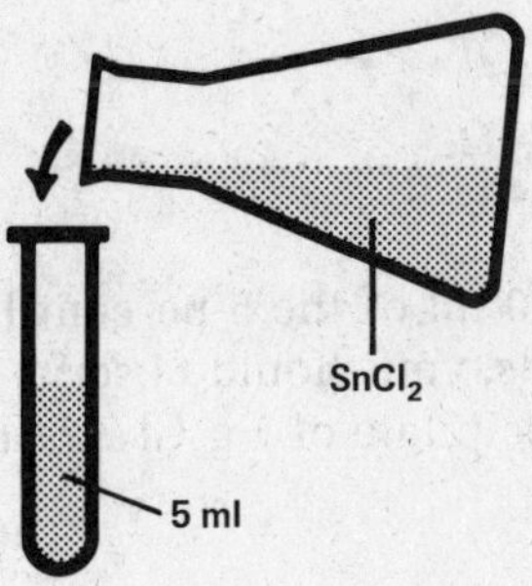

3. Obtain, in a clean test tube, 5 ml of the 1 *M* $SnCl_2$ solution prepared by one of your teammates. Using a medicine dropper, add this solution dropwise, while stirring, to the hot solution of $FeCl_3$ until the yellow color completely disappears. Allow the colorless solution (now containing Fe^{2+}) to cool to room temperature. Avoid adding *excess* $SnCl_2$.

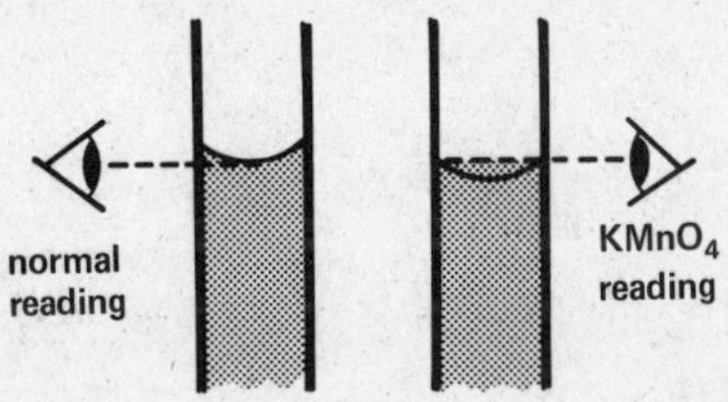

While the solution is cooling, check that your burets will drain properly without leaving droplets inside the buret. If not, clean them as described in Appendix A. Stir the $KMnO_4$ solution prepared at the beginning of the laboratory period thoroughly with a glass rod. Rinse one buret twice with 5-ml portions of the $KMnO_4$ solution, then fill it with that solution to above the mark. Drain just enough solution into a waste flask to expel air from the buret tip and to drop the solution level to just below the zero mark. Since the color of this solution obscures the bottom of the meniscus, all volume readings of the $KMnO_4$ solution must be made at the *top* of the solution level.

4. When the Fe^{2+} solution has cooled to room temperature, transfer it quantitatively into a clean 100-ml volumetric flask, using a funnel. Rinse the beaker several times with small portions of distilled water, also pouring these rinses through the funnel. Rinse the funnel and remove it, then fill the flask to the mark (using dropwise addition of final increments) with distilled water. Stopper the flask securely, invert it, and mix the contents thoroughly.

E. Titration of Fe^{2+} from the Ore

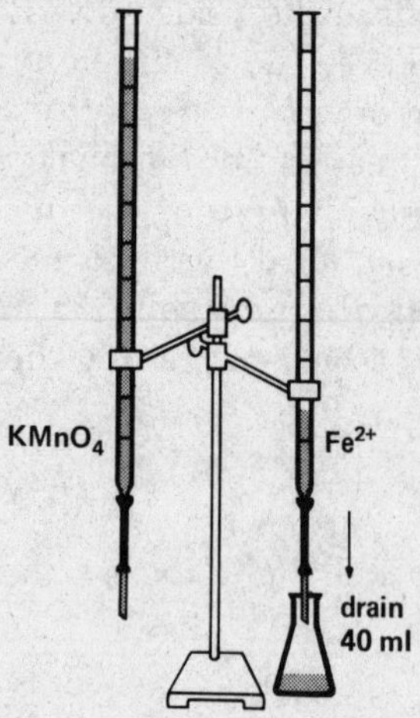

1. Rinse and fill your second buret with the Fe^{2+} solution. Drain out just enough into a waste flask to expel air from the buret tip and to drop the liquid level to just below the zero mark. Read and record the volumes in *both* burets. Then drain about 40 ml of the Fe^{2+} solution into a clean 500-ml Erlenmeyer flask. Record the final buret reading.

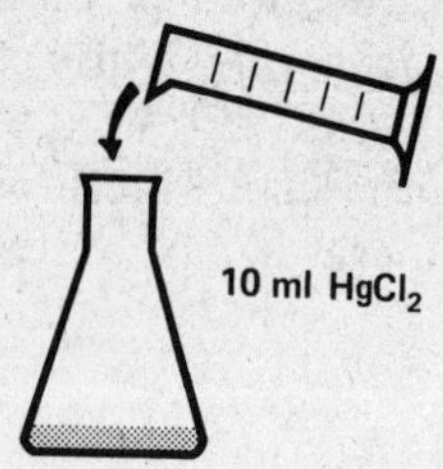

2. Pour into this flask 10 ml of the 5-percent $HgCl_2$ solution. After a few seconds, you should observe the formation of a fine white* precipitate of Hg_2Cl_2. After 2 min go on to procedure 3.

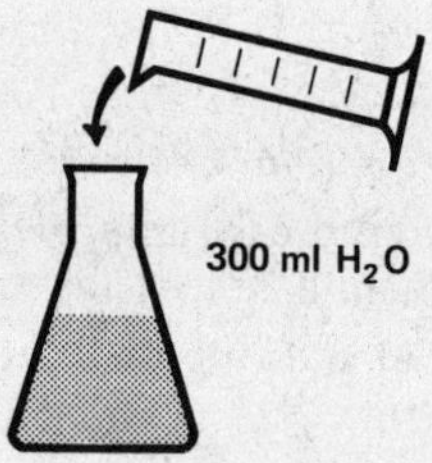

3. Add 300 ml of distilled water; swirl to mix the contents of the flask.

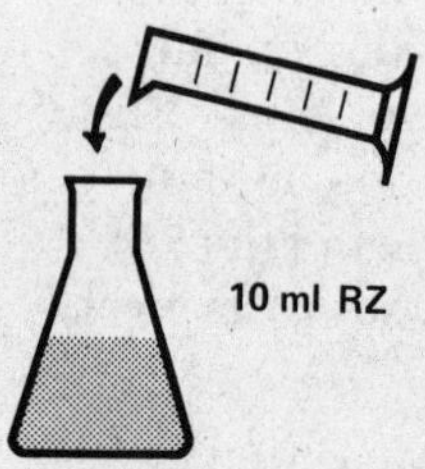

4. Then add 10 ml of the Reinhardt-Zimmermann (R-Z) reagent. Swirl to mix the contents of the flask.

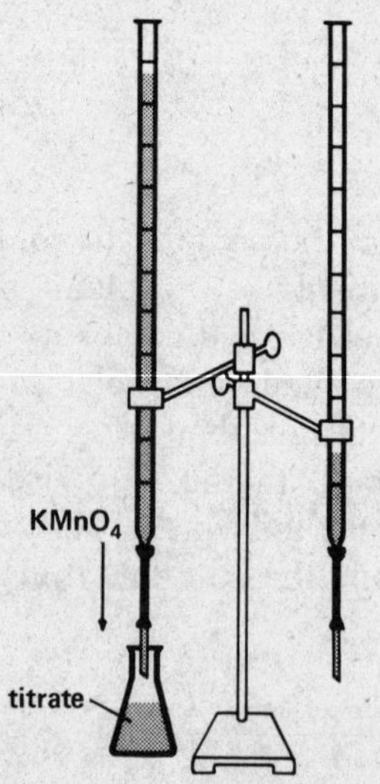

5. Titrate this mixture with the $KMnO_4$ solution. Initial addition can be fairly rapid, but you must slow to a dropwise addition near the endpoint. As the endpoint is approached, a faint pink color will form in the solution and disappear with continued swirling. The endpoint is reached when one drop of the permanganate solution produces a faint pink color that *persists* with continued swirling. Record the final $KMnO_4$ volume.

*If the precipitate is grey or black, you must have added too much $SnCl_2$ (step D3). In this case, your final results will probably be less accurate than desired.

Refill both burets and repeat procedures E1–E5, recording all necessary buret readings.

Calculate the amount of sodium oxalate required to consume 30 ml of 0.1 *N* $KMnO_4$ solution. Have your results checked by your instructor before proceeding.

F. Standardization of the $KMnO_4$ Solution

1. Weigh the required amount of sodium oxalate (*poison*) to the nearest 0.1 mg and transfer this completely into a clean 500-ml Erlenmeyer flask, using portions of a total 200-ml volume of distilled water. Add the remainder of this water to the flask.

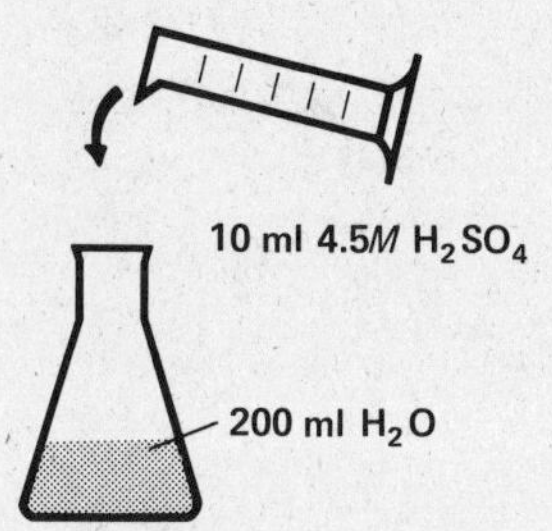

2. Prepare 10 ml of 4.5 *M* H_2SO_4 by adding 2.5 ml of the concentrated acid, *slowly, with stirring,* to 7.5 ml of distilled water in a 25-ml graduated cylinder. Pour this solution into the flask and heat the flask, with swirling, to about 70°C. (The flask will be almost too hot to hold. A folded paper towel around the flask neck will provide insulation.) Set the hot flask on a paper towel near the $KMnO_4$ buret.

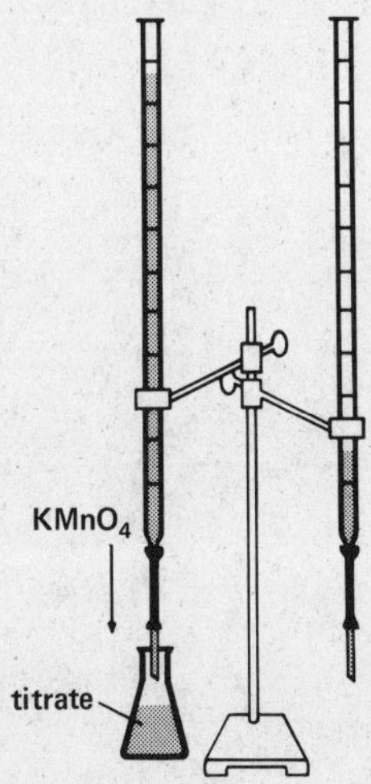

3. Refill the $KMnO_4$ buret and drain (into a waste flask) just enough solution to drop the level to just below the zero mark. Record the buret reading. Add 5 to 10 drops of the R-Z reagent to the flask containing the oxalate solution and titrate this solution while it is still hot (reheat if necessary) to a pale pink endpoint. Record the final buret reading.

If time permits, repeat procedures F1–F3.

CALCULATIONS AND REPORT Balance the equation

$$MnO_4^- + H_2C_2O_4 + H^+ \longrightarrow Mn^{2+} + CO_2 + H_2O$$

On the basis of the balanced equation and your data from procedure F, calculate the *normality* of the $KMnO_4$ solution to three significant figures.

On the basis of the balanced equation required for procedure A, the normality of the $KMnO_4$ solution, and your data from procedure E, calculate the *molarity* of the Fe^{2+} solution prepared.

Using the molarity of the Fe^{2+} solution, the volume of this solution, and the mass of your original pyrite sample, calculate the percentage (by mass) of iron in your ore sample.

Write balanced equations for all reactions performed in this experiment. From your calculations and notebook records, complete the Report Form.

GRADE
INSTRUCTOR'S INITIALS

Report Form

FOR INVESTIGATION 25:

Analysis of Iron in Pyrites

NAME: ______________________ ID NO.: ______________________

LAB SECTION ______________________ INSTRUCTOR ______________________

Analytical Data

Pyrite sample

Sample + watch glass	________ g
Watch glass	________ g
Pyrite	________ g

Sodium oxalate

	1st trial	(optional) 2nd trial
Sample + watch glass	________ g	________ g
Watch glass	________ g	________ g
$Na_2C_2O_4$	________ g	________ g

Ore titration (1st trial)

	Fe^{2+}	MnO_4^-
Final volume	________ ml	________ ml
Initial volume	________ ml	________ ml
Volume used	________ ml	________ ml

Ore titration (2nd trial)

	Fe^{2+}	MnO_4^-
Final volume	________ ml	________ ml
Initial volume	________ ml	________ ml
Volume used	________ ml	________ ml

Oxalate titration

	1st trial	(Optional) 2nd trial
Final MnO_4^- volume	________ ml	________ ml
Initial MnO_4^- volume	________ ml	________ ml
Volume MnO_4^- consumed	________ ml	________ ml

Discussion

Given that the conversion of MnO_4^- to Mn^{2+} is an autocatalytic process, what was the role of the R-Z reagent in this experiment?

(OVER)

Equations for Reactions Performed (Show balanced net equations.)

FeS_2 + aqua regia
Assume that the products include $[FeCl_4]^-$, NO_2, and SO_4^{2-}.

$FeCl_3 + SnCl_2$

$SnCl_2 + HgCl_2$

$Fe^{2+} + MnO_4^-$

$H_2C_2O_4 + MnO_4^-$

Calculations (Attach a separate page with details.)

Mass of $KMnO_4$ for 150 ml of 0.10 *N* solution ____________ g

Standardized concentration of $KMnO_4$ ____________ N

Concentration of Fe^{2+} solution ____________ M

Mass of iron in the pyrite sample ____________ g

Percentage of iron in the ore ____________ %

DATE ____________ SIGNATURE ____________

Investigation 26

TETRAAMMINECOPPER(II)

INTRODUCTION Copper is one of the essential metals for most living systems. Without an appropriate amount of copper in the diet, animals (including humans) may develop severe anemia—even though the diet contains more than sufficient iron for production of hemoglobin. The blood contains a protein, *ceruloplasmin,* which is complexed with Cu^{2+}. Ceruloplasmin appears to play an essential role in triggering the release from the liver of the iron needed for the production of hemoglobin.

A copper complex itself, *hemocyanin,* serves in the oxygen transport system of many of the invertebrates in the way that hemoglobin (an iron complex) does in vertebrates.

Copper, in the form of the Cu^{2+} ion, is also required for a number of essential enzyme systems, in all of which the Cu^{2+} ion is bound by coordination with electron-pair donor atoms of the enzyme protein. Among these enzymes are *tyrosinase* (which plays an important role in the chemistry of skin pigmentation), *cytochrome oxidase* (a very complicated enzyme involved in the generation of energy by oxidation of foods), *lysine oxidase* (necessary for maintaining the elasticity of the walls of the aorta), and *dopamine β-hydroxylase* (involved in the function of the adrenal glands).

Like many of the transition metal ions, Cu^{2+} forms strong covalent bonds with oxygen or nitrogen atoms in organic molecules. Coordination of Cu^{2+} with sulfur is also fairly common. An example of the formation of a copper(II) complex is that involving the amino acid *cysteine,* which may coordinate with Cu^{2+} ion in alternative ways:

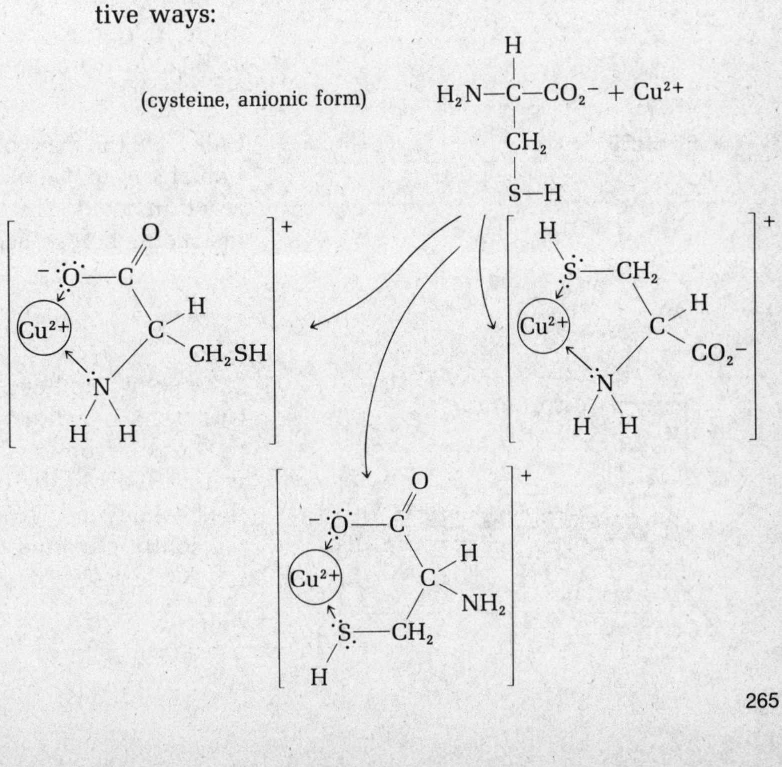

Many of the metalloproteins are very stable, suggesting that the metal ion is strongly held by the coordinate covalent bonds to oxygen, nitrogen, and sulfur. Metalloproteins containing copper range from relatively small molecules (such as the cytochrome oxidase cuproprotein of heart muscle, molecular weight ~25,000, one Cu^{2+} per molecule) to extremely large ones (such as the hemocyanin of the lobster, molecular weight ~780,000, 20 Cu^{2+} ions per molecule).

In this experiment we will prepare a much simpler complex, tetraamminecopper(II). The bonding in this complex is like that in the cuproproteins and it is the stability of the copper-nitrogen coordinate covalent bond that will permit us to isolate a stable crystalline salt of the complex cation.

Before starting this investigation, you should be familiar with the general concepts of complex-ion formation and stability and with stoichiometric calculations. In addition, before coming to the laboratory, you must complete the calculations (except for experimental data) indicated in the Procedures section of this investigation.

References

1. O'Connor, Rod. 1977. *Fundamentals of Chemistry* (2nd ed.). New York: Harper & Row (Units 13, 21, and 27).
2. Readings from *Scientific American*. 1973. *General Chemistry*. San Francisco: Freeman (Units 4, 30, and 31).

SAFETY Wear approved eye protection. A drop of concentrated ammonia in the eye can cause permanent damage. In case of contact with any reagents, wash immediately and thoroughly with water. Be cautious to keep the ethanol used away from flames.

OBJECTIVES
1. Be able to transfer solids without spillage.
2. Be able to prepare and mix reagent solutions properly.
3. Be able to make stoichiometric calculations.
4. Be able to prepare a crystalline coordination compound.

PROCEDURES

Calculate the mass of $CuSO_4 \cdot 5H_2O$ required for preparation of 3.1 g of $[Cu(NH_3)_4]SO_4 \cdot H_2O$, assuming a 90-percent reaction yield. Have your calculation checked by your instructor before beginning the experiment.

1. Without spillage, weigh the calculated amount of $CuSO_4 \cdot 5H_2O$ into a preweighed 100-ml beaker. Remove the beaker from the balance and pour in 15 ml of distilled water. Swirl to dissolve the salt. You may wish to warm the beaker gently to speed up dissolution. If you do, allow the solution to cool to room temperature before proceeding further.

Calculate the volume of concentrated (14.8 *M*) ammonia required to convert all of your $CuSO_4 \cdot 5H_2O$ sample to the tetraammine complex. Check your answer with your instructor before proceeding further. (You will use two times the calculated amount to improve the product yield.)

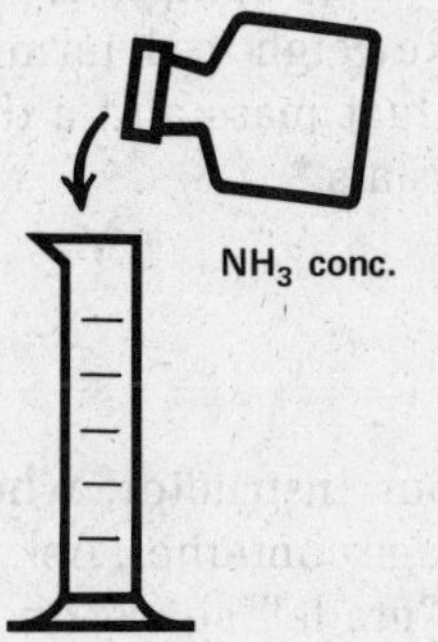

2. Take your 25-ml graduated cylinder and the beaker containing your $CuSO_4$ solution into the hood. Measure the required volume of concentrated ammonia into the graduated cylinder. Close the reagent bottle.

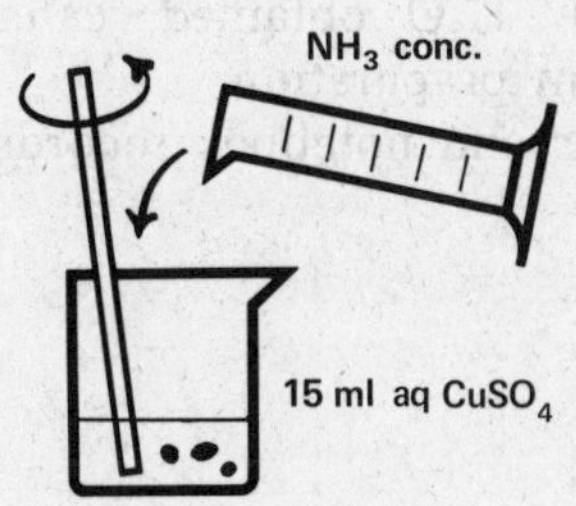

3. Slowly add the ammonia, with stirring, to the $CuSO_4$ solution. Continue stirring until all the initial precipitate of $Cu(OH)_2$, including any particles adhering to the beaker walls, has been dissolved.

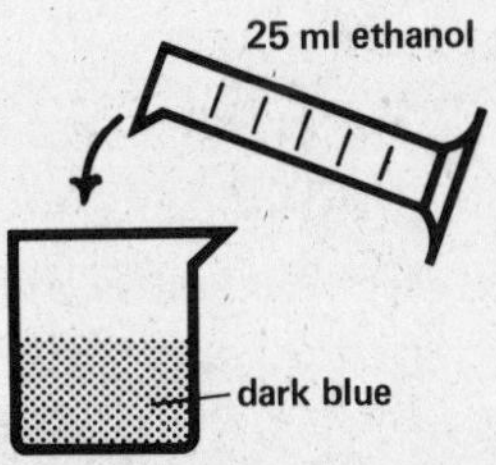

4. Very slowly, in small portions, pour 25 ml of ethanol into the deep blue solution. Stir the mixture thoroughly. Then set it aside for a few minutes to allow the crystals formed to settle. Prepare a "wash" solution by mixing 7 ml of ethanol and 5 ml of distilled water.

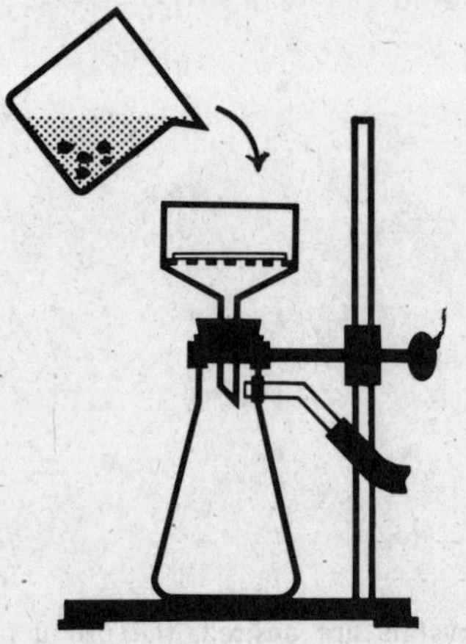

5. Assemble a vacuum filtration setup (Appendix A) and use it to isolate your crystals. Remember to moisten the filter paper and to decant most of the supernatant liquid through the funnel before transferring crystals from the beaker. Use four successive 3-ml portions of your ethanol-water solution to rinse the last of the crystals from the beaker into the funnel.

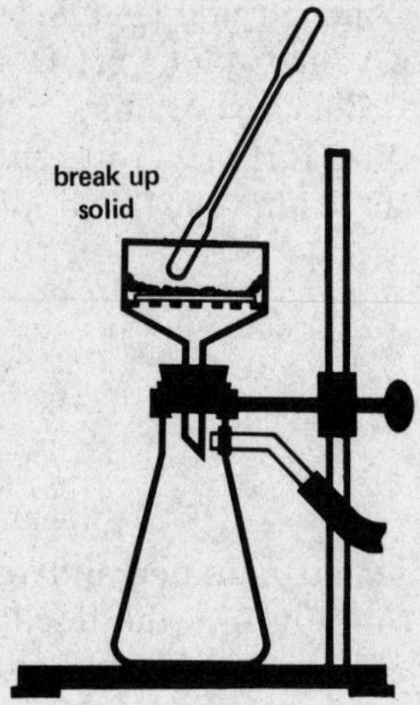

6. When all liquid has been pulled through the funnel, break up the blue cake on the filter with a spatula, being careful not to tear the filter paper. Allow the aspirator to pull air through the crystals to dry them for 5 to 10 min. Then pull the hose off the vacuum flask and turn off the aspirator. Transfer your crystals completely onto a clean, *preweighed* watch glass. Reweigh to determine the product yield. Record the product mass and a description of the appearance of the crystals.*

Show your product to your instructor, who will direct you to the appropriate storage container. Ask your instructor to mark the product "grade" space on your Report Form.

CALCULATIONS AND REPORT On the basis of the mass of $CuSO_4 \cdot 5H_2O$ used and the mass of $[Cu(NH_3)_4]SO_4 \cdot H_2O$ obtained, calculate the percentage yield for your preparation.

Use your calculations and notebook records to complete the Report Form.

*If larger, more regular crystals are desired, the product should be recrystallized from aqueous ethanol. Consult your instructor for details.

GRADE / INSTRUCTOR'S INITIALS

Report Form

FOR INVESTIGATION 26:

Tetraamminecopper(II)

NAME: ______________________ ID NO.: ______________________

LAB SECTION ______________________ INSTRUCTOR ______________________

Show all calculations on reverse side.

Mass of $CuSO_4 \cdot 5H_2O$ required (procedure 1) __________ g

Volume of 14.8 *M* NH_3 required (procedure 2)

(theoretical volume __________ ml) $\times$ 2 = __________ ml

Product yield

Product + watch glass __________ g

Watch glass __________ g

Product __________ g

Yield = __________ %

Product quality

Your description of product appearance

Instructor's evaluation:

☐ Reagent grade ☐ Tech grade ☐ Unsatisfactory

Instructor's initials: __________

__________ DATE __________ SIGNATURE

Investigation 27

A COBALT COMPLEX

INTRODUCTION Within recent years, we have become increasingly aware of the environmental hazards of mercury compounds. Although it has long been recognized that soluble mercury salts are toxic, the dangers of organomercury compounds (i.e., covalent compounds of mercury with carbon-containing groups) have attracted public concern only within the last few years. About 25 years ago, fishermen and their families around Japan's Minamata Bay were found to have a strange "disease." Even their household pets were affected. Investigations revealed significant brain damage, traced to high concentrations of mercury in seafood obtained from the bay. The mercury, ingested primarily in the form of methylmercury ion $[CH_3Hg]^+$ and dimethylmercury $[(CH_3)_2Hg]$, had originated from industrial wastes dumped into the bay. Since that time, a number of other cases of organomercury poisoning have been identified in various locations around the world and by 1970 there was widespread public alarm over mercury-contaminated seafoods. Although the most recent studies indicate that mercury concentrations in most food products are well below hazardous levels and suggest, indeed, that a trace concentration of mercury may be essential to most living organisms, the concern over organomercury poisoning still remains very real. The U.S. Food and Drug Administration has set a limit of 0.5 ppm of mercury in seafoods and regular testing indicates that, with rare exceptions, our seafood products are well within safe limits. There are still occasional reports of high levels of organomercury compounds in fish from mercury-contaminated waters. All this is interesting, if a bit frightening, but how does the problem of mercury-contaminated fish relate to the subject of this investigation?

It has been found that species such as dimethylmercury or methylmercury ion are produced from other, less hazardous, mercury compounds or from elemental mercury itself by certain bacteria present in most aquatic environments. A rapid growth of these bacteria is stimulated by increasing concentrations of mercury pollutants. The methylmercury compounds produced by these bacteria accumulate through the aquatic food chain in fish and appear to be particularly hazardous to humans because of their long biological half-lives and their apparent affinity for brain cells. The production of the methylmercury compounds by the responsible aquatic bacteria appears to be traced to a *cobalt complex* within the metabolic system of the bacteria.

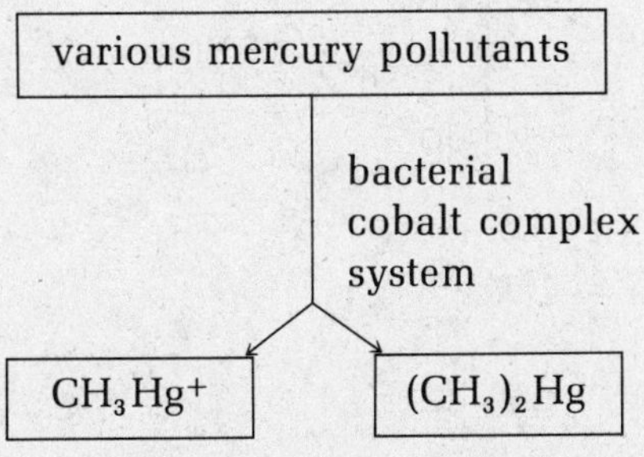

The mechanism of action of the bacterial cobalt complex system is not yet fully understood, but it appears to involve the transfer of a methyl group (CH_3) from methylcobalamin.

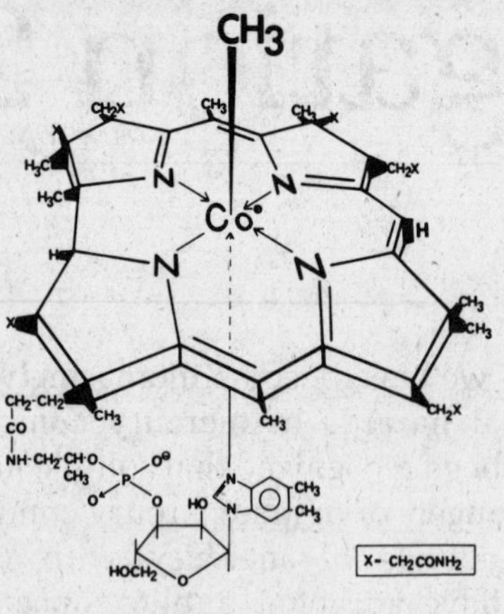

methylcobalamin

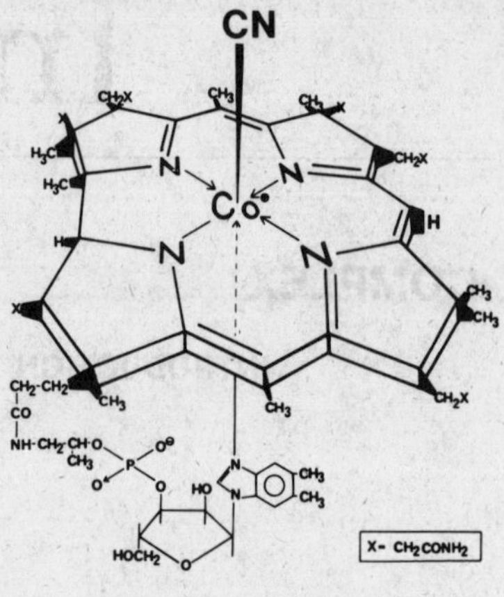

cyanocobalamin

Although some cobalt complexes obviously pose problems to man, others are extremely important. In 1926 it was found that the liver contains a chemical substance capable of curing pernicious anemia. Some 20 years later, this substance was identified as *vitamin* B_{12} and the complete structure of this valuable cobalt complex was determined in 1957.

The entire structure without the cyanide group is called *cobalamin* and a comparison of the structure of vitamin B_{12} (*cyanocobalamin*) with that of the bacterial complex (*methylcobalamin*) shows that these two complexes differ only in the nature of a single ligand group.

In this investigation we will prepare a different type of cobalt complex, but one that illustrates the coordination of a cobalt ion with a ligand group similar to one found in cobalamin. The IUPAC name of the complex to be prepared is chloro(pyridine)bis(dimethylglyoximato)cobalt (III) and the preparation of the complex may be represented (without balancing the equation) by

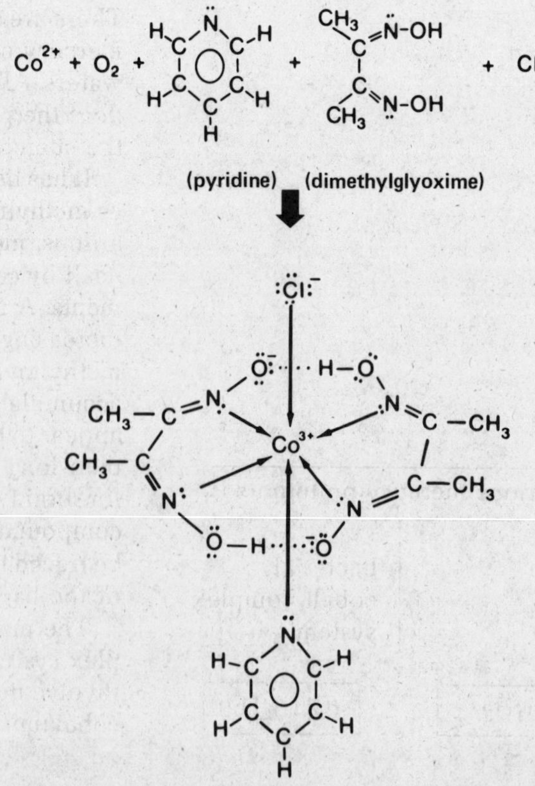

Before starting this investigation, you should be familiar with the general characteristics of coordination compounds and complex ions. In addition, *before coming to the laboratory,* you must write a balanced equation for the preparation of the cobalt complex and calculate the mass of $CoCl_2 \cdot 6H_2O$ required for preparation of 0.60 g of the complex, assuming a 70-percent yield, with $CoCl_2 \cdot 6H_2O$ as the limiting reagent.

In order to write this equation, you need to know that the products, in addition to the cobalt complex, are water and pyridinium chloride, C_5H_6NCl. You may assume that the oxidation of Co^{2+} to Co^{3+} by molecular oxygen (from the air) occurs in acidic solution, with the O_2 reduced to H_2O. You will probably find it simpler to balance the equation and to use the balanced equation for stoichiometric calculations if you employ *molecular* formulas for all species, for example, representing dimethylglyoxime by $C_4H_8N_2O_2$.

References

1. O'Connor, Rod. 1977. *Fundamentals of Chemistry* (2nd ed.). New York: Harper & Row (Unit 27).
2. Quagliano, J. V. and L. J. Vallarino. 1969. *Coordination Chemistry.* Indianapolis, Ind.: Heath-Raytheon.
3. Readings from *Scientific American* 1973. *Chemistry in the Environment.* San Francisco: Freeman (Unit 35).

Before starting work, have your instructor check your balanced equation and your calculation of the mass of $CoCl_2 \cdot 6H_2O$ required.

SAFETY Be very careful with open flames (or hotplates) to avoid igniting the ethanol used. In case of contact with any of the chemicals used, wash immediately and thoroughly with soap and water. Be careful in cutting, polishing, and inserting the glass tubing used for procedure 7.

OBJECTIVES
1. Be able to prepare and isolate a complex compound.
2. Be able to balance an equation and make stoichiometric calculations based on a reaction involving species having complicated structural formulas.

Procedures

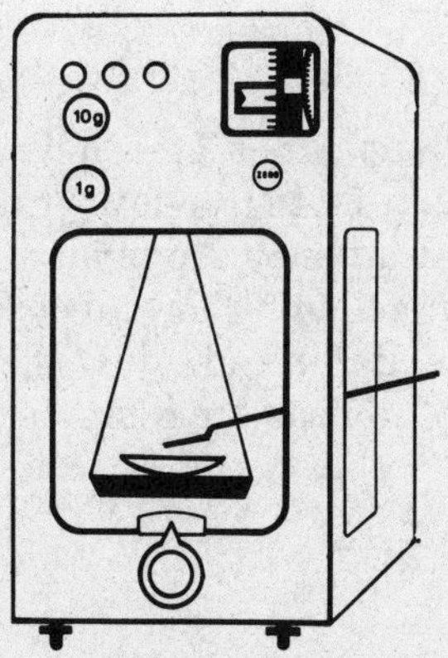

1. Weigh a clean, dry watch glass to the nearest 0.1 mg on the analytical balance. Add to the watch glass approximately the amount of $CoCl_2 \cdot 6H_2O$ calculated as needed for the preparation of the complex. Record the mass used (to ± 0.1 mg) in your notebook. Transfer the $CoCl_2 \cdot 6H_2O$ completely into a clean 100-ml beaker.

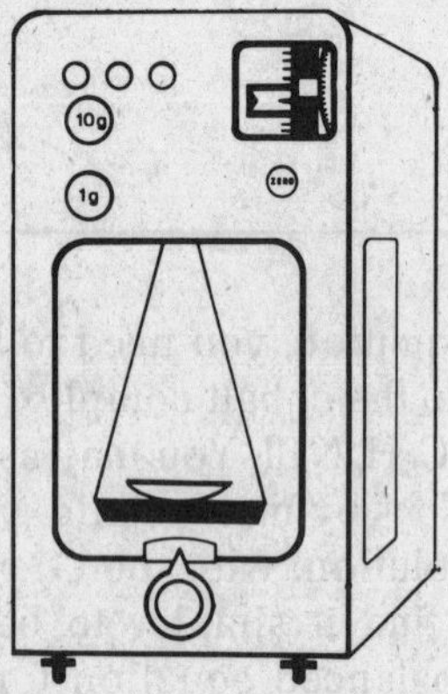

Using your balanced equation for the complex preparation, calculate the mass of dimethylglyoxime corresponding to 110 percent of that theoretically required to convert the amount of $CoCl_2 \cdot 6H_2O$ you are using completely to the complex. (This will afford a suitable excess of dimethylglyoxime.) Have your instructor check your calculation before proceeding further.

2. Using a clean, dry watch glass and the analytical balance, weigh out approximately the calculated mass of dimethylglyoxime. Record the mass used and transfer the compound completely into the beaker containing the $CoCl_2 \cdot 6H_2O$.

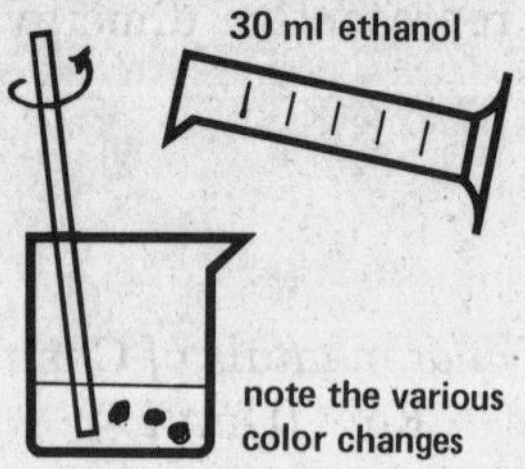

3. Slowly add 30 ml of 95 percent ethanol to the mixture of cobalt chloride and dimethyglyoxime. Record in your notebook the color changes observed during the initial stages of this procedure.

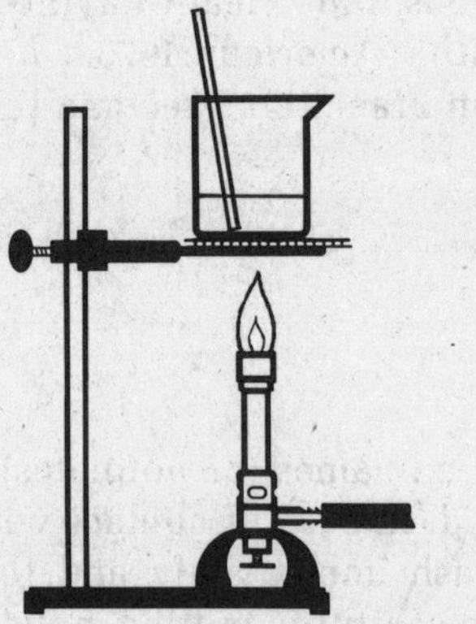

4. Place the 100-ml beaker containing the alcoholic mixture on a ringstand and bring to a very gentle boil. (*Exercise caution not to ignite the alcohol vapors.*) Some of the solid material in the mixture may not dissolve. Discontinue boiling after 5 min.

Warning: In case of fire, turn off the burner, remove the stirring rod with tongs, and smother the fire with an asbestos pad or damp paper towel.

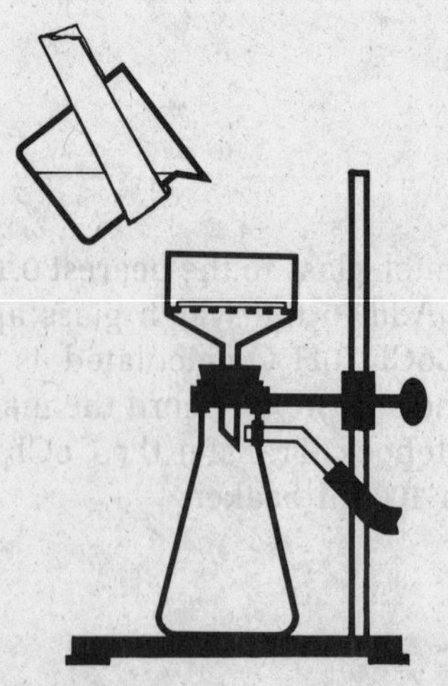

5. While the solution is still hot, filter it through a Buchner funnel into a 250-ml vacuum flask. (Review, if necessary, the section on vacuum filtration in Appendix A.) Rinse the beaker and Buchner funnel with 3 to 4 ml of 95 percent ethanol. Disconnect the aspirator and allow the solution in the vacuum flask to cool to near room temperature.

Using your balanced equation for the complex formation, calculate the mass of pyridine corresponding to 110 percent of that theoretically required to convert the amount of $CoCl_2 \cdot 6H_2O$ you are using completely to the complex. Assuming that one drop of liquid pyridine has a mass of 0.05 g, calculate the number of drops of pyridine corresponding to the calculated mass of pyridine needed. Have your instructor check these calculations before proceeding further.

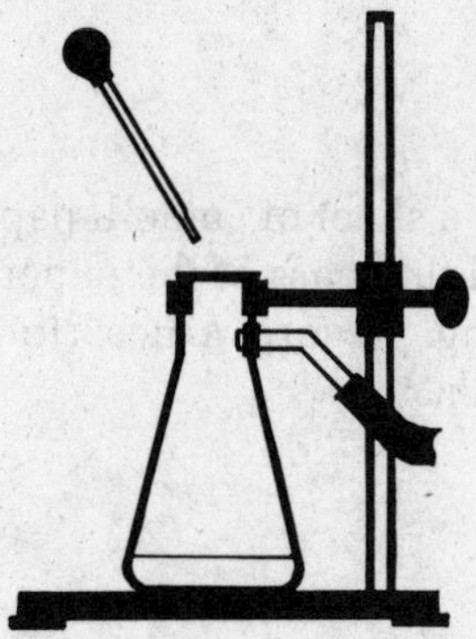

6. Take the vacuum flask containing the filtered ethanolic solution to the fume hood. *Keeping the pyridine container inside the fume hood,* add the calculated number of drops of pyridine. Swirl the flask gently to mix the contents. Record any observations of color change or other indications of reaction, if any.

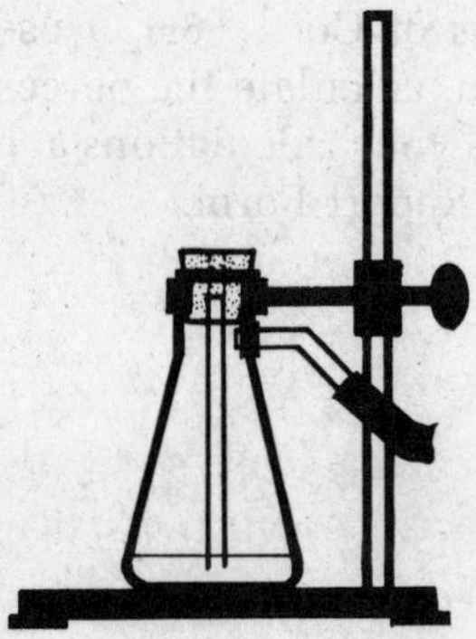

7. Obtain a one-hole rubber stopper that fits the top of the vacuum flask. Cut a piece of glass tubing just long enough to reach from about half-way into the stopper to about 1 cm above the bottom of the flask. Fire polish both ends of the tubing, allow the tubing to cool, and *carefully* insert one end of the tubing about half-way into the one-hole stopper (from the bottom of the stopper). Insert the stopper and tubing into the vacuum flask, checking that the tubing extends well below the surface of the liquid, but not against the bottom of the flask. Connect the aspirator and adjust the aspirator rate to draw a gentle stream of air bubbles through the solution. Continue bubbling air through the solution for 45 min. Record any observations of color change or crystal formation.

8. Disconnect the aspirator and transfer the reaction mixture completely into a clean 250-ml beaker. Use a small amount of distilled water from your squeeze bottle to wash the last crystals from the flask into the beaker.

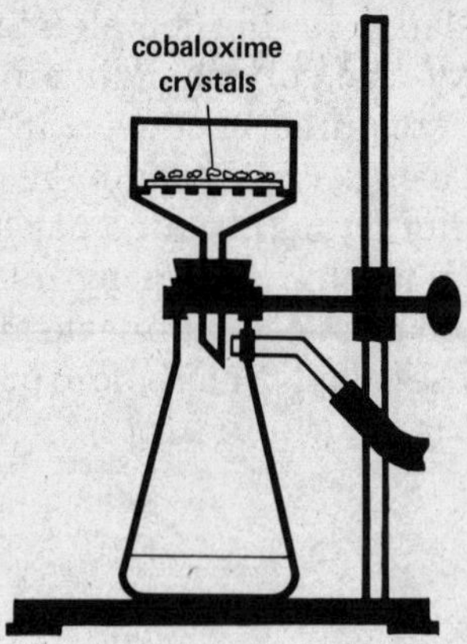

9. Filter the mixture using vacuum filtration. Wash the crystals with 10 ml of distilled water and then with 10 ml of 95 percent ethanol. By means of the aspirator, continue to draw air over the crystals for 5 min.

10. Weigh the crystals on a sheet of smooth paper. Record the mass of the paper and the mass of the paper plus crystals to the nearest 0.1 mg. Record a description of the appearance of the crystals.

Show your product to your instructor, who will direct you to the appropriate storage container. Ask your instructor to mark the product "grade" space on your Report Form.

CALCULATIONS AND REPORT On the basis of the mass of $CoCl_2 \cdot 6H_2O$ used and the mass of product obtained, calculate the percentage yield for your preparation. Use your calculations and notebook records to complete the Report Form.

GRADE / INSTRUCTOR'S INITIALS

Report Form

FOR INVESTIGATION 27:
A Cobalt Complex

NAME: ______________________ ID NO.: ______________________

LAB SECTION ______________________ INSTRUCTOR ______________________

Show all calculations on attached pages.

Cobalt(II) chloride hexahydrate ______________ g $mole^{-1}$

Dimethylglyoxime ______________ g $mole^{-1}$

Pyridine ______________ g $mole^{-1}$

Chloro(pyridine)bis(dimethylglyoximato)cobalt(III) ______________ g $mole^{-1}$

Reactants Used

$CoCl_2 \cdot 6H_2O$	Dimethylglyoxime	Pyridine
Watch glass + compound ______________ g	Watch glass + compound ______________ g	______________ drops, corresponding to approximately
Watch glass ______________ g	Watch glass ______________ g	______________ g
$CoCl_2 \cdot 6H_2O$ ______________ g	Dimethylglyoxime ______________ g	

Product

Paper + complex: ______________ g

Paper: ______________ g

Complex: ______________ g

Theoretical yield (based on $CoCl_2 \cdot 6H_2O$ used) ______________ g

Percentage yield ______________ %

Product quality ☐ good ☐ average ☐ poor ______________ Instructor's initials

Observations made

During procedure 3

During procedure 6

During procedure 7

(OVER)

Description of the product obtained

DATE

SIGNATURE

Investigation 28

CALCIUM IN NATURAL MATERIALS

INTRODUCTION Thirteen metals and 11 nonmetals are recognized as essential to life. Of the essential metals, calcium is the most abundant in man and other mammals (Table 28.1). The Ca^{2+} ion is required by some enzyme systems and calcium salts are the principal minerals of bones and teeth. Dietary calcium requirements are usually met easily. Milk and milk products are rich in calcium and even drinking water normally contains sufficient Ca^{2+} for adult needs.

Table 28.1 Essential metals for man and other mammals

Principal metals	Approximate concentration (percent, by mass) in the body	Trace metals	Approximate concentration (ppm) in the body
calcium	1.4	copper	4
sodium	0.3	tin	2
potassium	0.2	manganese	1
magnesium	0.04	molybdenum	0.2
iron	0.005	cobalt	0.04
zinc	0.002	vanadium	0.026
		chromium	(uncertain)

Calcium is the fifth most abundant element in the earth's crust, occurring in a wide variety of minerals such as *gypsum* ($CaSO_4 \cdot 2H_2O$), *calcite* ($CaCO_3$) *fluorite* (CaF_2), *perovskite* ($CaTiO_3$, a useful titanium ore), *scheelite* ($CaWO_4$, a tungsten ore), and the widely distributed *phosphate rock* [$Ca_3(PO_4)_2$]. Calcium is also a component of many of the more complex minerals. A number of these minerals are commercially important. Gypsum is used in making wallboard (Sheetrock) and plaster of paris. Phosphate rock is a valuable source of various phosphate fertilizers. Calcium carbonate, as *calcite, limestone,* or seashells can be decomposed by heating to form lime:

$$CaCO_3(s) \longrightarrow CaO(s) + CO_2(g)$$

Oyster shells are about 95 percent $CaCO_3$ and their abundance in seashore areas led to the development of an important seaside industry, the recovery of magnesium from seawater. Lime, produced by heating crushed oyster shells, was added to seawater to precipitate valuable magnesium hydroxide:

$$\underset{\text{(as oyster shell)}}{CaCO_3} \xrightarrow{\text{heat}} \underset{\text{(lime)}}{CaO} + CO_2$$

$$CaO(s) + \underbrace{Mg^{2+}(aq) + H_2O(l)}_{\text{(in seawater)}} \longrightarrow \underset{\text{(recovered)}}{Mg(OH)_2(s)} + Ca^{2+}(aq)$$

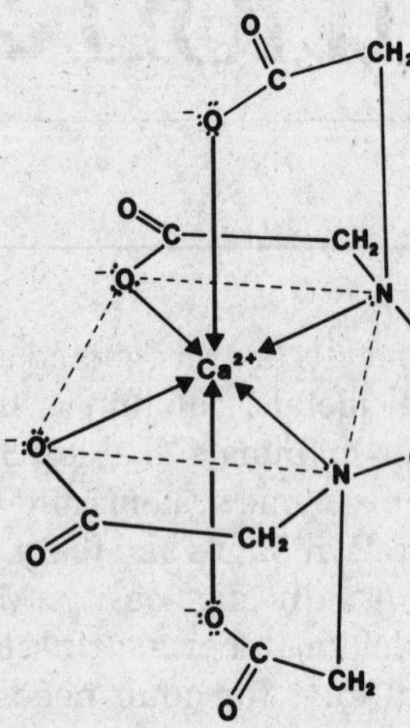

ethylenediaminetetraacetatocalcium(II)

Oyster shell chips are also sold as a food supplement for chickens, since the added dietary Ca^{2+} results in improved strength of eggshells.

One way of analyzing for Ca^{2+} in natural materials utilizes the strong affinity of Ca^{2+} for the ethylenediaminetetraacetate ion, $EDTA^{4-}$. The complex formed is quite stable. In this investigation, we will use $EDTA^{4-}$ (as the disodium dihydrogen salt) for the titration of Ca^{2+} obtained from oyster shell.

Before starting this investigation, you should be familiar with the general principles of coordination chemistry, solution concentrations, and stoichiometry. In addition, prior to coming to the laboratory, you should set up the calculations required in the Procedures section of this investigation.

References

1. O'Connor, Rod. 1977. *Fundamentals of Chemistry* (2nd ed.). New York: Harper & Row (Units 13, 21, 25, and 27 and Excursion 2).
2. O'Connor, Rod and Charles Mickey. 1974. *Solving Problems in Chemistry*. New York: Harper & Row (Units 6 and 12).
3. Ifft, James and John Hearst. 1974. Readings from *Scientific American. General Chemistry*. San Francisco: W. H. Freeman (Units 4, 11, 17, and 23).

SAFETY Wear approved eye protection at all times and exercise standard precautions in working with chemicals.

OBJECTIVES

1. Be able to make calculations dealing with weight-to-weight and solution stoichiometry.
2. Be able to perform a complexometric titration accurately.

Procedures

A. Solution Preparation

Calculate the mass of oyster shells required for the preparation of 100 ml of a 0.01 *M* Ca^{2+} solution, assuming the oyster shells to be 95 percent $CaCO_3$ with no other Ca^{2+} component. (If a different source of Ca^{2+} is to be used, your instructor will indicate appropriate modifications.) Have your calculation checked by your instructor before you begin experimental work.

1. Obtain a sample of crushed oyster shell about the size of a dime, or slightly smaller. Place the sample in a clean, dry mortar and use the pestle to grind the sample to a fine powder.

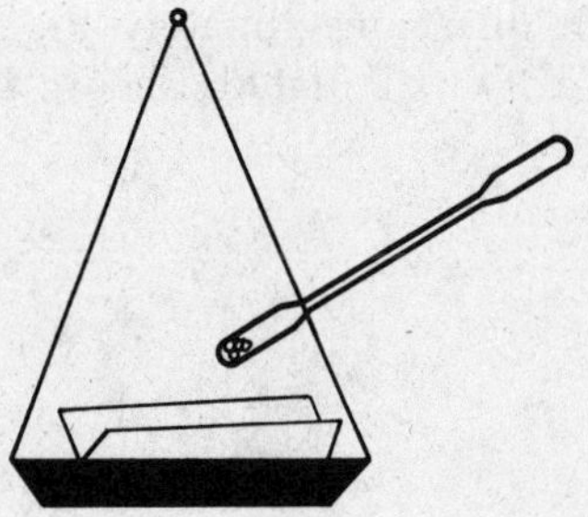

2. Use the analytical balance to weigh a creased piece of smooth paper, about 10 cm × 10 cm. Add to the paper approximately the mass of powdered oyster shell calculated as needed for the 0.01 *M* Ca^{2+} solution. Record the mass data in your notebook.

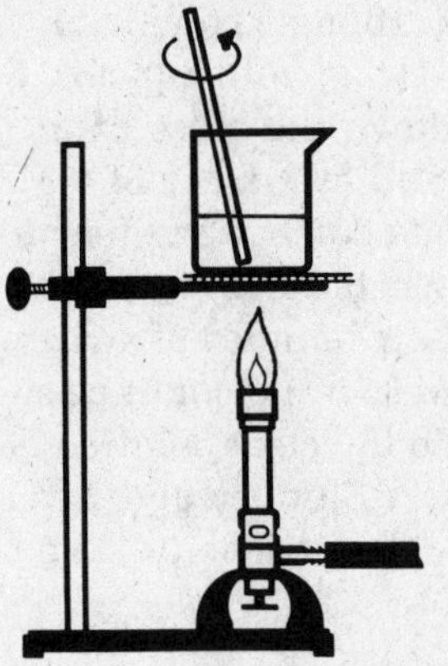

3. Transfer the weighed powder *completely* into a clean 100-ml beaker. Add 15 ml of distilled water and 30 drops of 6 *M* HCl. Warm the mixture *very gently* while stirring constantly. When all reaction has stopped and no Ca^{2+}-containing solid remains, discontinue heating.

4. Allow the mixture to cool to room temperature, then transfer it completely into a clean 100-ml volumetric flask, rinsing the beaker several times with small portions of distilled water and pouring these rinses into the volumetric flask. Drop two pellets of NaOH (Caution: *corrosive*) into the flask and add distilled water to the 100-ml mark. Stopper the flask and invert it several times to mix the contents.

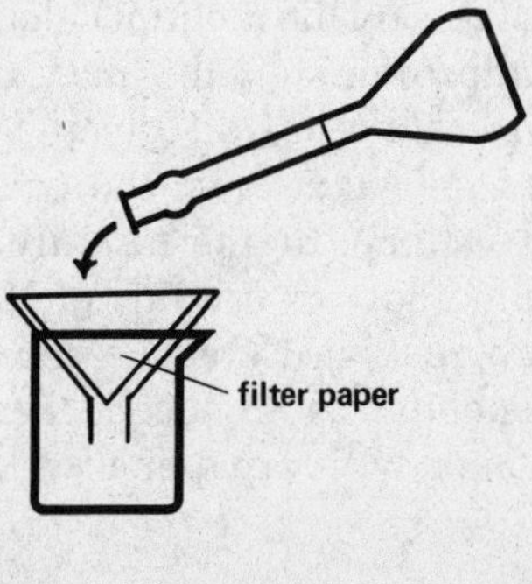

5. Fold an 11-cm-diameter Whatmann No. 1 filter paper into quarters. By opening one of the "pockets" formed, fit the filter paper into a powder funnel supported over a clean, dry 250-ml beaker. Filter the Ca^{2+} mixture from the volumetric flask into the beaker. Label the beaker as "Unknown Ca^{2+} solution," cover the beaker with a paper towel, and set it aside for later use. (Rinse the volumetric flask with distilled water, discarding the rinses.)

Calculate the mass of *pure* $CaCO_3$ required for preparation of 100 ml of 0.01 *M* Ca^{2+} solution.

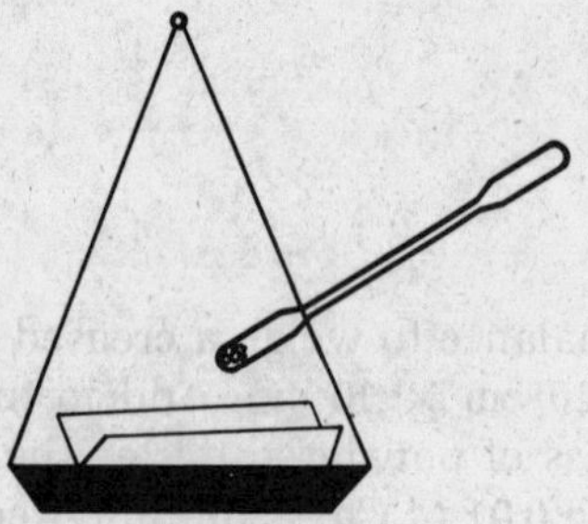

6. Using the analytical balance and a new 10-cm × 10-cm weighing paper, weigh out approximately the mass of pure (reagent grade) $CaCO_3$ calculated. Record the mass data in your notebook.

7. Transfer the calcium carbonate completely into a clean 100-ml volumetric flask. Using your squeeze bottle, flush the calcium carbonate down the sides of the flask. Add 10 ml of distilled water and five drops of 6 *M* HCl. Swirl the flask to ensure intimate contact and reaction between the calcium carbonate and the hydrochloric acid. If any particles remain after a few minutes of swirling, add two or three more drops of 6 *M* HCl and continue swirling until all solid has dissolved. To the clear solution add distilled water to the 100-ml mark. Stopper and invert the flask several times to ensure homogeneity. Label the flask "Standard Ca^{2+} solution."

Calculate the mass of disodium dihydrogen ethylenediaminetetraacetate dihydrate ($Na_2C_{10}H_{14}N_2O_8 \cdot 2H_2O$) required for the preparation of 200 ml of a 0.01 *M* solution. This solution will serve as the source of $EDTA^{4-}$ for your titration and will be referred to hereafter as the "EDTA solution."

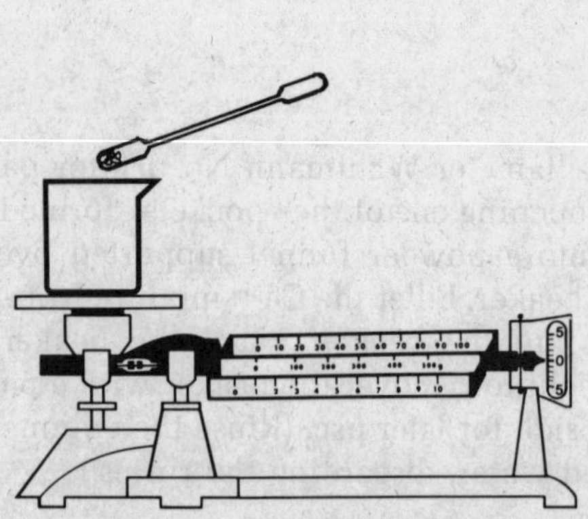

8. Weigh a clean, dry 500-ml beaker to the nearest 0.1 g. Advance the balance weights by the amount calculated for the EDTA salt and add approximately this mass of the salt into the beaker, *being careful not to add too much.* Record the mass data. Remove the beaker from the balance and add 200 ml of distilled water. Stir the mixture until all solute has dissolved, then add two crystals of $MgCl_2$ and two pellets of NaOH (*corrosive*). Continue stirring until the solution is homogeneous. Label the beaker "EDTA solution." Cover the beaker with a paper towel.

B. Standardization

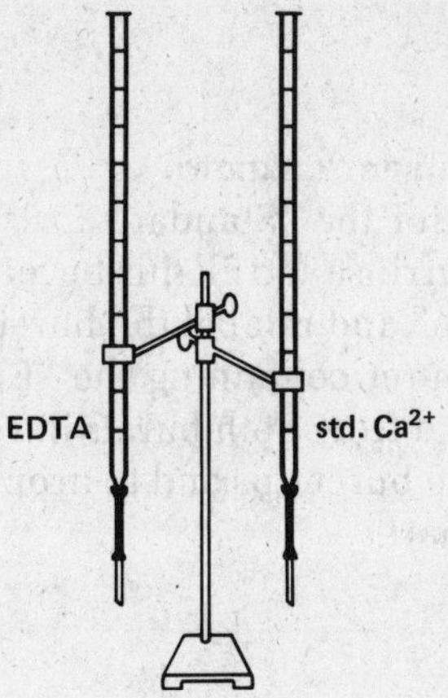

1. Obtain and, if necessary, clean (Appendix A) two 50-ml burets. Mount these securely on a ringstand. Rinse and fill the left buret with the 0.01 *M* "EDTA solution" and attach a proper label to the buret. Then rinse and fill the other buret with the "Standard Ca^{2+} solution" and attach a proper label to the buret. Drain just enough liquid from each buret into a "waste" flask to expel any air from the buret tips and to bring the liquid levels to just below the zero marks.

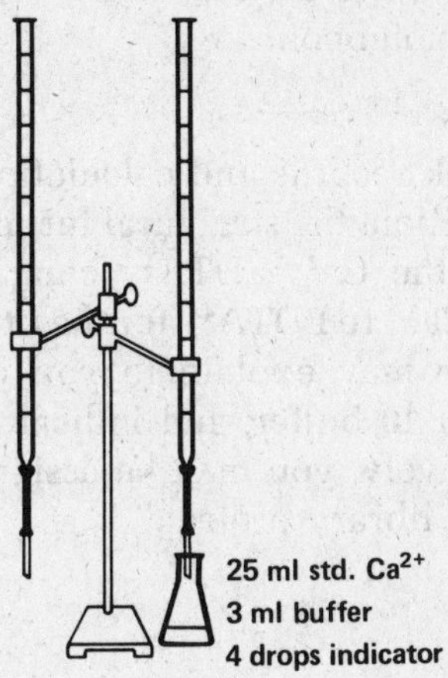

2. Record the buret reading, then drain about 25 ml of the "Standard Ca^{2+} solution" into a clean 125-ml Erlenmeyer flask. Record the final buret reading. Add to the flask 3 ml of a pH 10 buffer solution and two drops of Eriochrome Black T indicator.

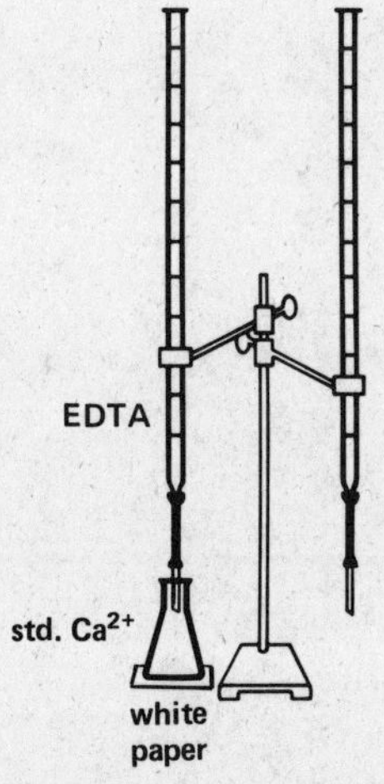

3. Set the flask containing the "Standard Ca^{2+} solution" on a sheet of white paper under the buret containing the "EDTA solution." Record the initial volume in this buret, then begin the titration, slowing to a dropwise addition near the endpoint and swirling the flask during the titration. *Near* the endpoint the indicator will change from red to purple. At the final endpoint, the indicator color will be a pure blue with no tint of red. Record the final buret reading.

4. Repeat procedures B2 and B3, but this time use a volume of "Standard Ca^{2+} solution" that should require 35 to 40 ml of the "EDTA solution." If the *ratio* of volumes of these two solutions for this second titration differs by more than 2 percent from that of the first titration, repeat the titration again with the remaining "Standard Ca^{2+} solution."

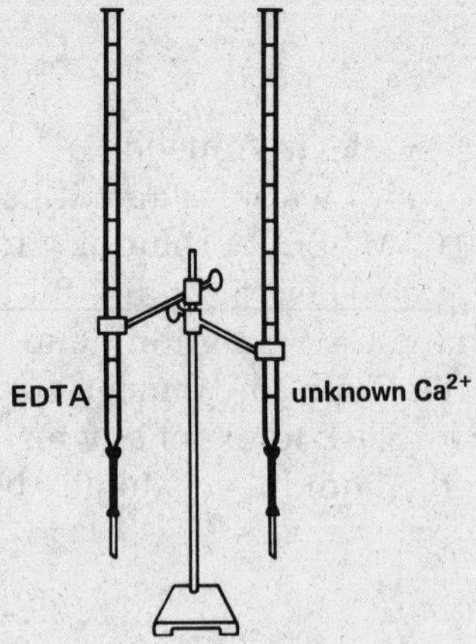

C. Determination of Ca^{2+} in an "unknown"

1. Rinse the buret used for the "Standard Ca^{2+} solution" with distilled water. Then rinse and fill this buret with your "Unknown Ca^{2+} solution" and relabel the buret properly. If necessary, refill the buret containing the "EDTA solution." Drain enough liquid from both burets into a "waste" flask to expel air from the buret tips and to drop the liquid levels below the zero marks.

2. Titrate the "Unknown Ca^{2+} solution," using the steps described in procedures B2, B3, and B4. Record all volume readings in your notebook.

CALCULATIONS AND REPORT

Use your notebook records and calculations to complete the Report Form. From the structural formula given in the Introduction for the Ca^{2+}-$EDTA^{4-}$ complex, determine the mole ratio of Ca^{2+} to $EDTA^{4-}$ for the titration reaction.

Your instructor may explain to you the role of the $MgCl_2$, NaOH, pH 10 buffer, and indicator used for this analysis. Alternatively, you may be assigned to find this information as a "library project."

GRADE	INSTRUCTOR'S INITIALS

Report Form

FOR INVESTIGATION 28:

Calcium in Natural Materials

NAME: ____________________ ID NO.: ____________________

LAB SECTION ____________________ INSTRUCTOR ____________________

Show calculations on attached pages.

Mass Data

Oyster shell
+ paper __________ g

Paper __________ g

(Powdered)
oyster shell __________ g

Pure $CaCO_3$
+ paper __________ g

Paper __________ g

Pure $CaCO_3$ __________ g

EDTA salt
+ beaker __________ g

Beaker __________ g

EDTA salt __________ g

TITRATION DATA

Standardization

	"Standard Ca^{2+}"			"EDTA"		
	Trial 1	Trial 2	Trial 3	Trial 1	Trial 2	Trial 3
final volume						
initial volume						
volume used						

Unknown

	"Unknown Ca^{2+}"			"EDTA"		
	Trial 1	Trial 2	Trial 3	Trial 1	Trial 2	Trial 3
final volume						
initial volume						
volume used						

Calculated Values

If any single value listed was not used in calculating an average, include a note explaining why the value was discarded.

(OVER)

Molarity of "EDTA Solution"

Trial 1 ______________________ *M*

Trial 2 ______________________ *M* Average ______________________ *M*

Trial 3 ______________________ *M* Average deviation ______________________

Molarity of "Unknown Ca^{2+} Solution"

Trial 1 ______________________ *M*

Trial 2 ______________________ *M* Average ______________________ *M*

Trial 3 ______________________ *M* Average deviation ______________________

Percentage of $CaCO_3$ in the Oyster Shell Sample

______________________ ± ____________ %

______________ ______________________

DATE SIGNATURE

Investigation 29

NITRATES IN NATURAL WATERS

INTRODUCTION Most natural waters contain detectable amounts of nitrate ion, some of which is present as part of the natural "nitrogen cycle" of the biosphere.* Man has added to this. The results have been noted in two major problem areas. Aquatic plant life, including algae, utilize nitrogen in varying forms. Nitrate is one such form and its increasing concentration, along with that of phosphates, is linked to the wide-spread eutrophication (nutrient enrichment) of lakes and streams, with the resulting "bloom" of algae and other water plants. In addition, nitrate poses a significant health problem to both humans and other animals. In the stomach, nitrate may be reduced to nitrite. This, in turn, may react with the iron of hemoglobin to form methemoglobin, which is incapable of acting as an oxygen carrier. Infants are particularly susceptible to this problem and a large number of "blue baby" deaths have been traced to nitrates in drinking water. As a result, the U.S. Public Health Service has set a limit of 45 ppm NO_3^- in drinking water for adults and a 10-ppm limit in drinking water for infants. Few water-treatment plants are capable of removing nitrate from drinking water supplies, so the increasing nitrate concentrations of natural waters represent very real public health problems in many localities.

There are a number of ways in which nitrates enter aquatic systems. In nature, atmospheric nitrogen is oxidized at high temperatures (typically produced in lightning discharges):

$$N_2 + O_2 \xrightarrow{1000°\text{–}2000°C} 2NO$$

The nitric oxide formed may be further oxidized to nitrogen dioxide which, in turn, may react with atmospheric moisture to produce nitric acid, eventually delivered to the surface of the earth in rain:

$$2NO + O_2 \longrightarrow 2NO_2$$
$$3NO_2 + H_2O \longrightarrow 2HNO_3 + NO$$

In addition to this process, various microorganisms convert nitrogen compounds, such as ammonia, to nitrate ion. Natural sources of nitrate have been supplemented to a significant extent by a number of processes associated with our technological society.

Agricultural fertilizers furnish enormous amounts of nitrate, either directly (e.g., as NH_4NO_3) or indirectly (e.g., from oxidation of ammonia or ammonium ion). Although the intent of fertilizer use is to increase soil nutrients needed for plant growth, a large percentage of chemical fertilizers is washed from the soil into neighboring water systems. Cations (e.g., NH_4^+ or K^+) from fertilizers are readily adsorbed by clays in the soil, but the nitrate anion

*Delwiche, C. C. "The Nitrogen Cycle," *Scientific American* (Sept. 1970).

is not. When you consider that chemical fertilizer use is expected to grow* from 63 million metric tons (1970) to 115 million metric tons (1980) over a 10-year period, it is not difficult to see why water pollution by agricultural runoff is an increasing problem. Nitrate will be a significant component in this runoff. In some localities, the drainage from feedlots and barnyards represents a larger source of nitrate (from decomposing animal wastes) than does the elution of nitrate fertilizers. In most cases, however, this is a fairly localized problem.

Agricultural fertilizers and feedlot drainage are primarily responsible for nitrate pollution in rural areas, but another source is equally significant in metropolitan regions. Although nitrogen is a relatively unreactive gas, it is oxidized at high temperatures (as in the natural process in lightning). Temperatures in this region occur in the internal combustion engine and in many of the fossil-fueled furnaces of industrial systems. Since nearly all combustion systems use air, rather than pure oxygen, for the burning of fuels, the oxidation of nitrogen from the air is an inherent part of such high-temperature combustions. We currently dump over 30 million tons of nitrogen oxides into the atmosphere from various combustion processes. About half of this comes from the emissions of fossil-fuel power plants and about 40 percent from internal combustion engines. New catalytic emission-control systems for automobiles are capable of reducing nitrogen oxides to molecular nitrogen, but it will be many years before these systems are in wide enough use to make a major impact. Some of the nitrogen oxides (N_2O, NO, NO_2) emitted into the atmosphere are consumed in the production of photochemical smog, but most appear within a few days as nitric acid, eventually entering the soil or water systems of the earth.

Nature has many ways of utilizing nitrate ion. Denitrifying bacteria, acting on NO_3^- or other nitrogen compounds, return molecular N_2 to the atmosphere. Plant forms of various types use nitrate as a nutrient, converting it to protein and other nitrogenous species. Although increasing nitrate concentrations stimulate some of the natural systems to remove nitrate at higher rates, in many areas we have exceeded the capabilities of these natural systems with the result that nitrate concentrations of natural waters have continued to increase.

In this experiment we will investigate ways of detecting nitrate ion in aqueous solutions and of estimating its concentration within certain limiting ranges.

Before starting this investigation, you should be familiar with the use of parts per million (ppm) and molar concentration units and with the chemical characteristics of nitrate ion. In addition, you must be thoroughly familiar with the safety precautions for working with hazardous chemicals. (Note that you are to bring to the laboratory a "natural" water sample for procedure C of this investigation.)

Chemical and Engineering News (July 5, 1971).

References

1. O'Connor, Rod. 1977. *Fundamentals of Chemistry.* (2nd ed.) New York: Harper & Row (Units 13 and 26).
2. Readings from *Scientific American.* 1973. *Chemistry in the Environment.* San Francisco: Freeman (Units 5, 9, 10, 26, and 27).

SAFETY Wear approved eye protection at all times. Be especially careful in using concentrated sulfuric acid. In case of contact with any chemical reagents, wash immediately and thoroughly with water.

(Caution: The naphthylamine used in this investigation is a very dangerous chemical and must never be allowed in contact with the skin. *Exercise extreme precautions* and in case of accidental contact wash the affected area *immediately and thoroughly* with soap and water and notify your instructor.)

OBJECTIVES

1. Be able to perform simple tests for nitrate proficiently and consistently.
2. Be able to determine the detection limits of simple nitrate tests.
3. Be able to determine the approximate concentration of nitrate in water samples.

PROCEDURES

A. Detection Limit of the "Brown Ring" Test

1. On the analytical balance, weigh between 0.95 and 1.05 g of potassium nitrate to the nearest 0.1 mg and transfer it completely into a clean 125-ml Erlenmeyer flask. Add 100 ml of distilled water, swirl to obtain a homogeneous solution, and label the flask "0.1 *M* NO_3^-."

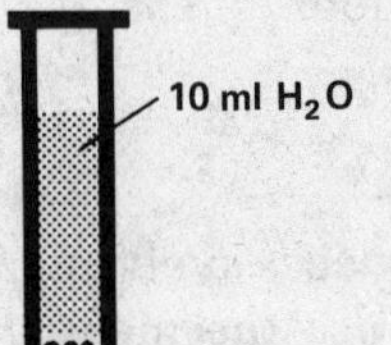

2. Fill a 12.5-cm test tube to a depth of 2 cm with crystals of $FeSO_4 \cdot 7H_2O$. Add 10 ml of distilled water, stopper the tube, and shake it vigorously until a homogeneous solution is formed. Set the tube in a test tube block.

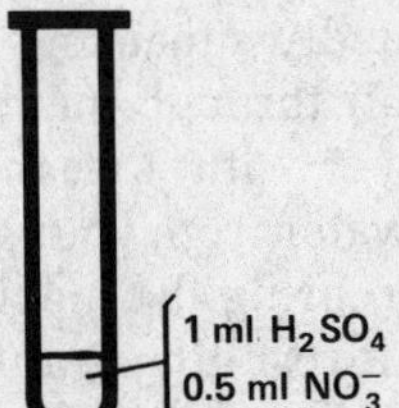

3. Place 0.5 ml (10 drops) of the 0.1 *M* NO_3^- solution in a small, clean test tube. Carefully add 1.0 ml of concentrated sulfuric acid from the reagent bottle and swirl or stir gently to mix the two liquids.

4. Tilt the test tube and, using a medicine dropper, very carefully *overlay* the dense KNO_3/H_2SO_4 solution with a shallow (~10 drop) layer of the $FeSO_4$ solution. The two liquids must not be allowed to mix.

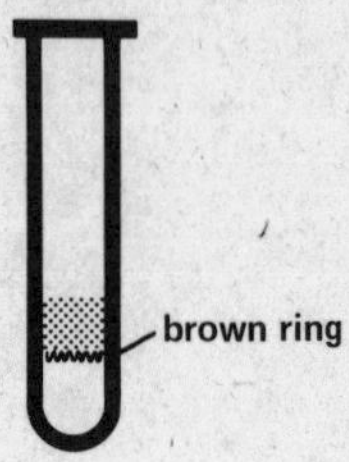

5. Carefully tilt the test tube back to a vertical position. You should observe a brown ring forming rather quickly at the interface of the two solutions. The brown color, due to the formation of the $[Fe(NO)]^{2+}$ complex ion, becomes fainter at lower concentrations of nitrate. At very low concentrations only a pale brown-to-violet ring is observed. (This test can only be used to detect nitrate in the absence of bromide and iodide ions, since both form colored products by reaction with H_2SO_4.) Repeat procedures A3–A5 as necessary until you can perform this test with facility.

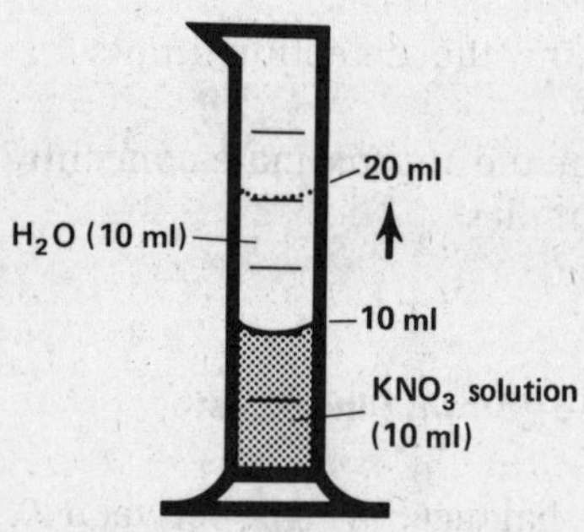

6. After you have developed proficiency with the "brown ring test," repeat the test on a series of successively diluted KNO_3 solutions to determine the lowest concentration at which you can detect NO_3^- by this test. To prepare the first solution, pour into a clean, dry 25-ml graduated cylinder a 10-ml volume of 0.1 *M* KNO_3 solution. Then add enough distilled water to bring the liquid meniscus just to the 20-ml mark. Mix the solution thoroughly by stirring with a long medicine dropper, then use the dropper to transfer 0.5 ml (10 drops) of this 0.05 *M* KNO_3 to a small test tube. Store this test tube in the block described in procedure A7.

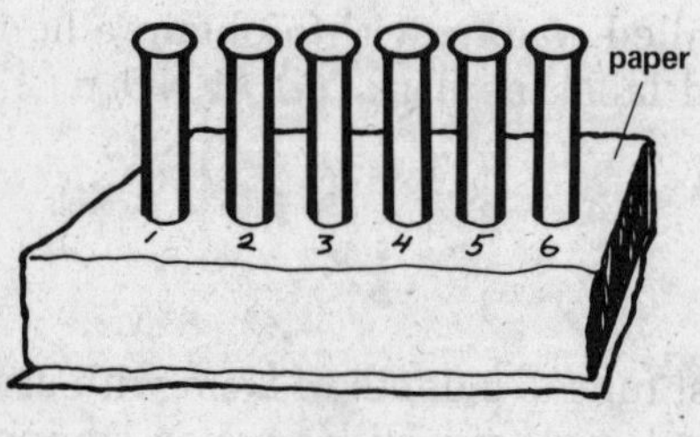

7. Wrap a test tube block in a paper towel. Push six small test tubes through the towel into the holes in the block. Number the test tube locations.

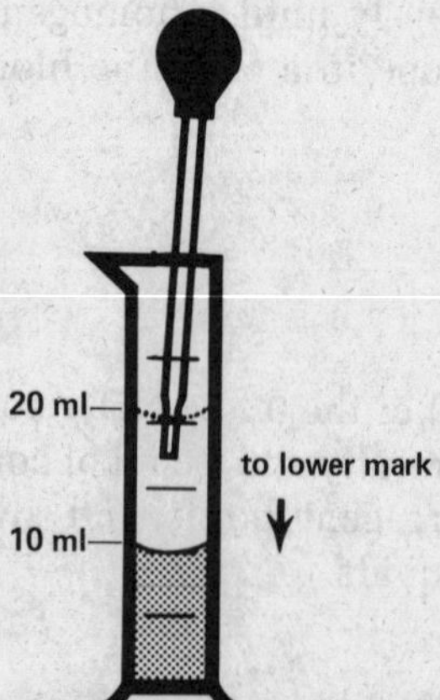

8. To prepare a series of successively more dilute solutions, proceed as follows: Use the medicine dropper to remove sufficient solution from the graduated cylinder (procedure A6) to drop the liquid meniscus to the 10-ml mark. Then rinse the medicine dropper inside and out several times with distilled water. Expel all the water from the dropper by repeatedly squeezing and releasing the dropper bulb to move air through the dropper. Wipe off the outside of the dropper with a clean paper towel. Then add enough distilled water from your squeeze bottle to the graduated cylinder to bring the meniscus back up to the 20-ml mark. Stir the resulting solution with the dropper until you are certain it is homogeneously mixed, then transfer 0.5 ml (10 drops) of this solution to the next

small test tube. Repeat this procedure until you have six samples of successively more dilute solutions in the six small test tubes.

In your notebook, prepare a table with headings as shown in the Report Form under "Brown ring test." Use the notebook table, *not* the Report Form, to keep a record of your successive dilutions and test results.

9. Perform the ring test on each one of your dilute solutions as described in procedures A3–A5. Keep the most dilute solution you have prepared in this series in your graduated cylinder until you are sure you have a solution too dilute to produce a brown ring. Report the lowest concentration at which a brown ring is still observable as the approximate "detection limit."

10. You probably will find that the brown ring test is not sensitive enough to detect nitrate ions at concentrations usually occurring in natural water samples. However, the brown ring test is very convenient to confirm qualitatively the presence of nitrate in minerals, fertilizers, and explosives. In a small test tube obtain from your instructor an unknown that might contain nitrate ion. If the unknown is a liquid, use 0.5 ml of it to perform the brown ring test. If your unknown is a solid, add 2 ml of distilled water and mix thoroughly. Since all nitrates are water soluble, you will be able to extract the nitrates from mixtures that contain other insoluble compounds. Record the result obtained with your unknown.

B. Detection Limit of Z–S–N Procedure*

1. Take a clean 25-ml graduated cylinder to your instructor and obtain 10 ml of a 50-ppm NO_3^- solution. Dilute and mix this thoroughly with an equal volume of distilled water. Also obtain, in a clean, dry test tube, a "reducing powder" consisting of equal parts by weight of zinc, sulfanilic acid, and naphthylamine hydrochloride (The Z–S–N reagent*).

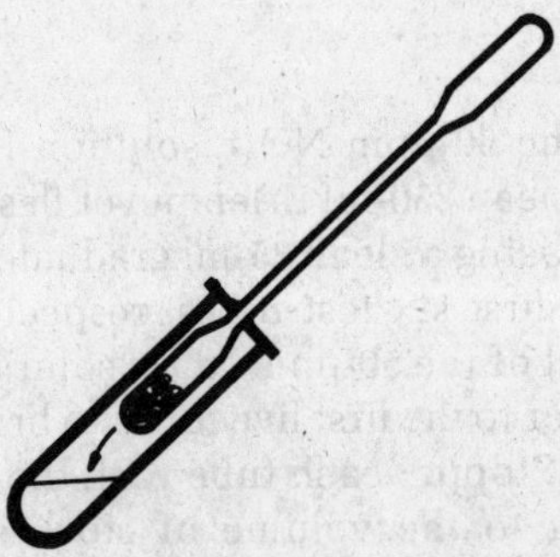

2. Pour about 1 ml of your diluted solution (now 25 ppm NO_3^-) into the first test tube. Add half a spatula full of the "reducing powder" to the tube, stopper it, and shake the tube for 1 min. A pink color should begin to develop. Set the tube into the test tube block. The pink color, of an intensity determined by the original nitrate concentration, will deepen slowly, reaching a maximum intensity in about 10 min.

Caution: This reagent is very dangerous. *Exercise extreme precautions* to avoid contact of this reagent with your skin. In case of accidental contact, wash *immediately* and *thoroughly* with soap and water and notify your instructor.

*Z–S–N represents zinc–sulfanilic acid–naphthylamine hydrochloride.

3. Dilute 10 ml of your 25-ppm NO_3^- solution with an equal volume of distilled water, as described in procedure A8. Repeat procedure B2 using 1 ml of this solution in the second test tube. Continue successive dilutions and testing until you reach the third dilution in a row that has not produced a pink color within the 1-min shaking period. (Since full color intensity requires 10 min, this should get you to the proper detection limit, which is defined as the concentration of the last solution giving a detectable pink color 10 min after mixing.)

In your notebook, prepare a table with headings as shown in the Report Form under "Z–S–N procedures." Use this notebook table to keep a record of your successive dilutions and test results.

C. Qualitative Tests for NO_3^- in Water Samples

Bring to the laboratory a sample of some "natural" water (e.g., from a lake, river, creek, or collected rainwater). Test this sample, and also the tap water and distilled water available in the laboratory, with the Z–S–N reagent by the method of procedure B2. Estimate the nitrate concentration of these samples by comparison with the results.

D. Semiquantitative Estimation of Nitrate Concentration

1. Line the bottoms of a 600-ml beaker and a 400-ml beaker with crumpled paper towels. In each beaker stand four 20-cm test tubes. Label solid No. 5 rubber stoppers for these tubes: 5, 10, 20, 30, 40, 45, X, and Y, respectively. (Note that the stoppers *must* be clean.)

2. Obtain 175 ml of the 50-ppm NO_3^- solution from the large stock bottle in a clean 250-ml Erlenmeyer flask. Label and stopper the flask. Using a clean 25-ml graduated cylinder, measure into the first six test tubes, respectively, 5, 10, 20, 30, 40, and 45 ml of the 50-ppm NO_3^- solution. Add sufficient distilled water to the first five tubes to bring each total volume to 45 ml. Stopper each tube with the stopper labeled to correspond to the volume of stock solution placed in the tube. Invert the stoppered tubes and shake to mix the contents thoroughly, then set the tubes back into the beakers. Take the tubes having the X and Y stoppers to your instructor, who will give you samples of two unknowns. Be sure to record the unknown code assigned to each tube. If you receive less than 45 ml of an unknown, add sufficient distilled water to make a total of 45 ml in each tube (i.e., to match the liquid level in the other tubes).

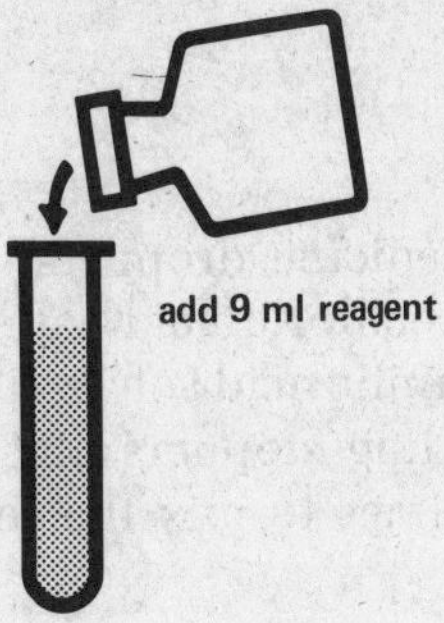

3. Clean and rinse thoroughly (with distilled water) five 12.5 cm test tubes. Label these, respectively: Z, S, N, Ac, and HCl. In the appropriately labeled tubes, obtain: Tube Ac, 9 ml of 2 *M* sodium acetate; Tube N, 9 ml of naphthylamine hydrochloride* solution (6 mg ml^{-1}); Tube S, 9 ml of sulfanilic acid solution (6 mg ml^{-1}); Tube Z, an amount of the zinc-sodium chloride powder equal in volume to 9 ml. Carefully mix 8 ml of distilled water and 2.0 ml of concentrated hydrochloric acid (pouring the acid *into* the water) and transfer this into the tube labeled HCl. Stand all five tubes in the test tube block.

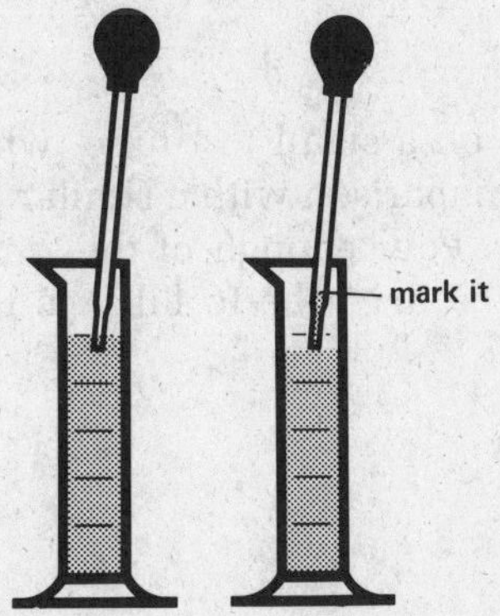

4. Calibrate a medicine dropper by using the dropper to withdraw enough water from a 25-ml graduated cylinder to drop the volume level from 25 to 24 ml. With a felt-tip pen or grease pencil, mark the 1.0-ml volume level in the dropper. Then allow the dropper to empty.

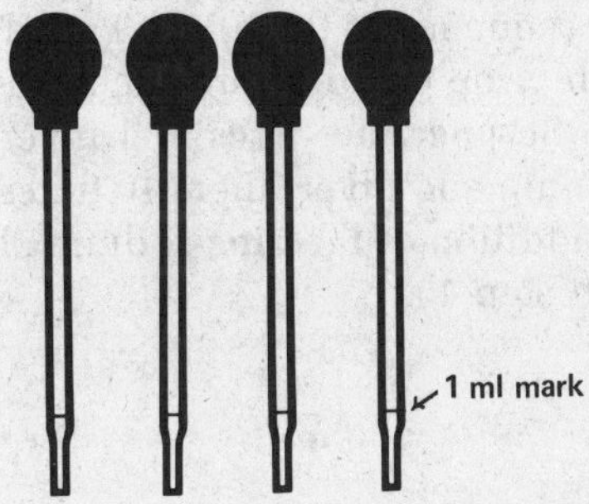

5. Using the marked dropper as a reference, mark three other medicine droppers at the same level.

The remaining procedures require that the contents of your eight 20-cm test tubes all be treated in the same way.

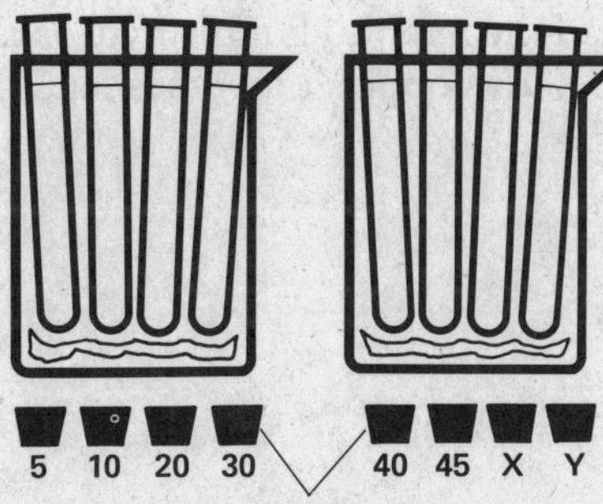

6. Remove the stoppers from the 20-cm tubes and set the stoppers in a row in front of the beakers containing the tubes so that the sequence of stoppers matches the sequence of the *corresponding* test tubes.

*Exercise *extreme precautions* with this reagent.

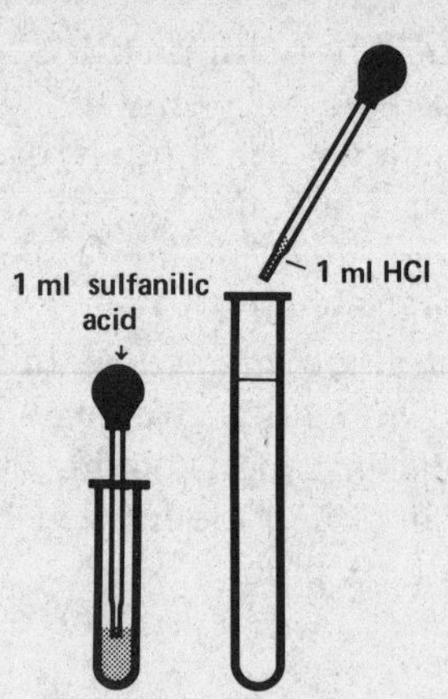

7. Using the calibrated medicine droppers, add to *each* test tube 1 ml of the hydrochloric acid solution (2.4 *M*), and then 1.0 ml of the sulfanilic acid solution. Stopper the tubes with the corresponding stoppers and invert each tube back and forth five times to mix the solutions.

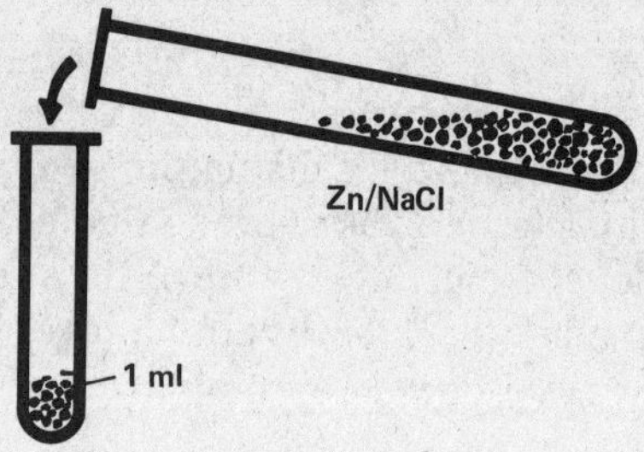

8. Mark the 1-ml level on a small test tube (which must be clean and *dry*) by comparison with a similar tube containing 1.0 ml of water. Pour enough of the zinc-sodium chloride powder into the dry tube to fill it to the 1.0-ml mark.

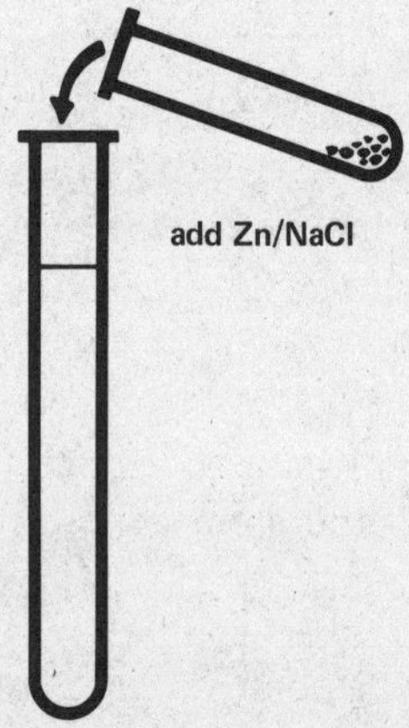

9. Remove the stoppers from the 20-cm tubes, again aligning these in the same sequence as the tubes. Transfer 1 ml of the zinc-sodium chloride powder into each tube. *Note the time of addition.* Stopper the tubes and invert them each 10 times. Wait 2 min and invert them 10 times again. Seven minutes after addition of the zinc-sodium chloride powder proceed with step 10.

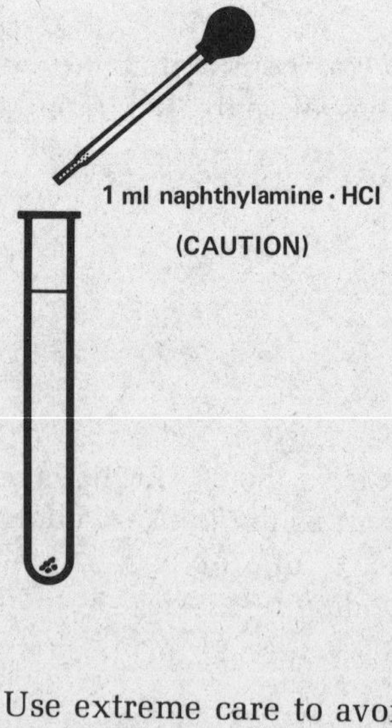

10. Unstopper the tubes, keeping the stoppers in proper sequence. Add 1.0 ml of the naphthylamine hydrochloride solution to each tube. Restopper the tubes and invert them each once.

Caution: Use extreme care to avoid contact of this mixture with the skin.

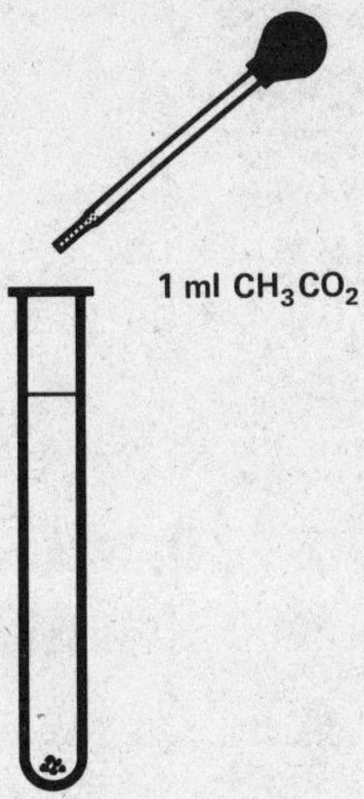

11. Unstopper the tubes, keeping the stoppers in proper sequence. Add 1.0 ml of the sodium acetate solution to each tube. Restopper the tubes and mix their contents by inverting ten times (*caution*).

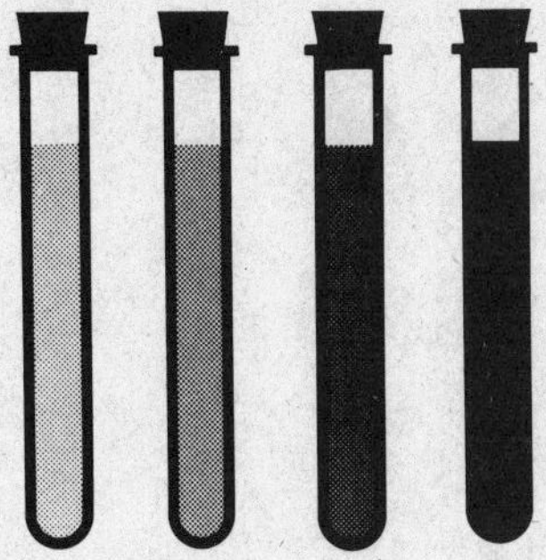

12. Five minutes after the addition of the sodium acetate solution, compare the color intensities of all eight tubes against a white background. Estimate the concentration of NO_3^- (in ppm) in your unknown solutions by comparison with the colors of the solutions of known nitrate concentrations.

13. Should a spectrophotometer be available, measure the absorbance of the solutions at 5200 Å against a reagent blank. Instructions on how to use the spectrophotometer available in your laboratory will be provided by your instructor. General operating procedures and instructions on how to prepare a calibration curve are provided in Investigation 35. When the color intensity is measured spectrophotometrically, the test tubes should be replaced by 50-ml volumetric flasks, which must be filled to the mark with distilled water after all the reagents have been added.

CALCULATIONS AND REPORT

Using your notebook records, complete the tables in the Report Form. The concentration unit ppm (parts per million) is defined as the milligrams of a solute species in 1000 g of solution. In dilute aqueous solutions, of density $\sim 1\ g\ ml^{-1}$, ppm is also given to a very good approximation as the milligrams of a solute species in 1 liter of solution. Make sure that you calculate the nitrate concentrations of the solutions prepared according to procedure D2 considering that the total volume is 50 ml. A more accurate experiment would use volumetric flasks instead of test tubes.

Complete the Report Form, submitting the calculations on a separate sheet. If you used a spectrophotometric method in this investigation, you should ask your instructor about modifications for the Report Form.

GRADE / INSTRUCTOR'S INITIALS

Report Form

FOR INVESTIGATION 29:

Nitrates in Natural Waters

NAME: ______________________ ID NO.: ______________________

LAB SECTION ______________________ INSTRUCTOR ______________________

"Brown Ring Test"

Mass of KNO_3 dissolved in 100 ml water ______________ g

Molarity of the KNO_3 solution ______________ moles liter^{-1}

Brown ring test unknown code ______________ nitrate: yes or no _____

Tube No.	KNO_3 solution used			Volume H_2O added (ml)	Concentration after mixing		Brown ring test
		NO_3^- conc.					
	Volume	(*M*)	(ppm)		Molarity	NO_3^- (ppm)	Observations
	original						

Approximate detection limit ______________ ppm (______________ *M*) [NO_3^-]

"Z–S–N Procedure"

Tube No.	Nitrate solution used		Volume H_2O (ml)	Conc. after mixing		Z–S–N Test observations
	volume (ml)	NO_3^- (ppm)		NO_3^- (ppm)	molarity	
1	original	25	0	25	4×10^{-4}	deep pink
2	10	25	10	12.5	2×10^{-4}	
3	10		10			
4	10		10			
5						

Approximate detection limit ______________ ppm (______________ *M*) [NO_3^-]

(OVER)

Nitrates in Water Samples (Procedure C)

"Natural water" (source ______________________________)
Results

Tap water
Results

Distilled water
Results

Approximate NO_3^- Concentration of the Unknown Solutions (X and Y)

Solution X

Solution Y

Which, if either, of these solutions or natural water samples contained a nitrate concentration above the allowed drinking water limit for infants? ____________

Which, if either, of these solutions contained a nitrate concentration above the allowed drinking water limit for adults? ______________________________

______________ DATE ______________________________ SIGNATURE

Investigation 30

MODELS FOR ORGANIC COMPOUNDS

INTRODUCTION Before man had a written history, he had learned to extract useful substances from sources provided by nature. Herbs provided flavors for foods or medicines for treating the sick. Bark extracts helped in tanning animal hides, and volatile chemicals in wood smoke helped in preserving his food. He could protect himself with arrows tipped in concentrated poisons from natural sources, celebrate his victories with alcoholic beverages, or seek temporary escape from the realities of his life with a variety of drugs. Often the arts of extraction and concentration of physiologically active compounds were carefully guarded secrets and provided their owner with wealth, prestige, and power in his society. He could choose to use his knowledge for the benefit of others or for purposes of destruction and personal gain.

Today man has progressed far beyond this primitive state. He need no longer rely primarily on natural sources for useful chemicals. He can isolate an active substance, purify it carefully, determine its complete structure, and, frequently, make it synthetically for wide-scale use much more economically than he could extract it from the natural source. In many cases he can prepare synthetic substitutes better than the original compound. He can choose to use his present knowledge for the benefit of others or for purposes of destruction and personal gain. Yes, man has progressed quite far. Synthetic antibiotics promise new eras of health; synthetic fabrics, synthetic pesticides, and synthetic fertilizers promise improved standards of living; but the major world powers have manufactured enough chemical warfare agents to kill every person on earth 30 times over. Man has progressed a long way. Or has he?

The arts of the chemist are no longer secret. A general understanding of the concepts involved and the techniques employed can be gained rather easily. Such knowledge should help in the wise determination of the uses for which these arts will be employed. In a free society the choices can no longer be made by the tribe's "wizard." In this day of public funding of research, the credit for developing new substances of potential benefit to man is shared to some extent by every citizen, and each must share in the blame for those new products that threaten the environment or that become new weapons in the arsenals of mass destruction.

Organic chemistry is primarily the study of covalent compounds of carbon. These compounds include a fantastic variety of substances ranging from the simple hydrocarbons used as fuels and lubricants to the complex nucleic acids of the genetic material.

Carbon forms more different chemical compounds than any other element.* This is attributed to the fact that carbon-carbon σ or π bonds are exceptionally stable, permitting the formation of long chains or rings of carbon atoms for a wide variety of compounds. In addition, carbon forms stable covalent bonds with hydrogen and with many other elements, the most important being oxygen, nitrogen, and the halogens.

Organic compounds may be categorized according to certain arrangements of atoms, the functional groups, which contribute characteristic properties to the molecules in which they occur. Such a classification is important in systematizing the study of the enormous number of organic compounds.

Since carbon normally forms four covalent bonds, there are many different ways in which the same atoms can be "connected." The existence of isomers, molecules having the same molecular formulas but different structures, is another reason for the immense number of different organic compounds.

In an attempt to produce more effective drugs with fewer adverse side effects, pharmaceutical chemists have synthesized thousands of compounds with structural variations designed to reveal those molecular characteristics essential for optimum biological activity. As an example, quinine has been largely replaced by synthetic antimalarial drugs developed over years of research into synthetic, structural, and biochemical areas. One of these new antimalarial drugs has the structure:

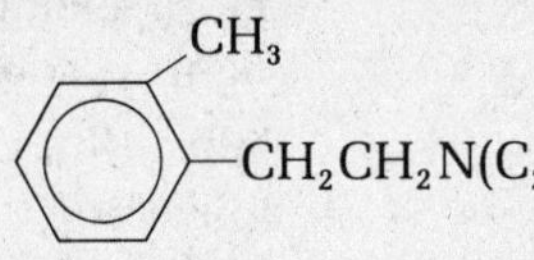

It is quite common to observe that seemingly minor variations in structure may have profound influences on biological activity. Neither of the following compounds is effective as an antimalarial drug, although they differ from the preceding substance only in the relative positions of the groups attached to the benzene ring:

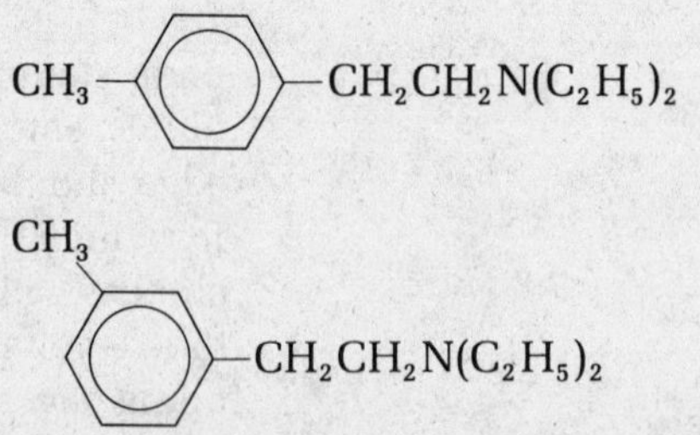

inactive

Isomeric compounds may vary widely in physical and chemical properties. Some, however, are essentially identical in these respects but show significant variations in their reactions within living organisms. The chemist must

*More than 2-million organic compounds are now known. This is at least 10 times the number of all other known compounds not containing carbon.

always be aware of isomeric differences and the problems they may pose in various aspects of his work. In no field are these problems of such importance as in the area of chemicals associated with living systems.

An organic compound with the molecular formula C_6H_8 might have several different structures, including, among others,

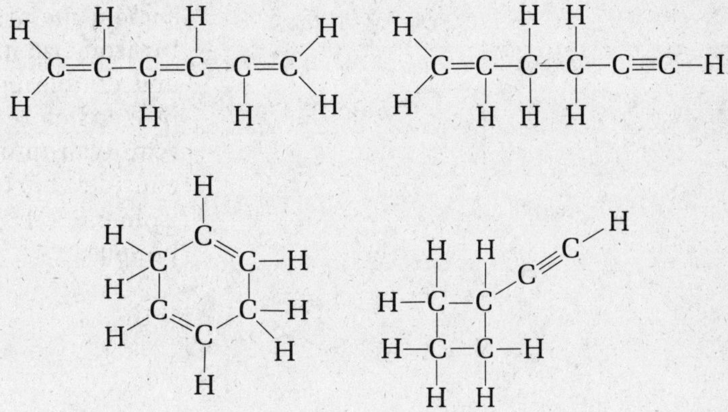

Substances like the preceding ones that have the same molecular formula but different structural formulas are called *isomers*. Carbon's ability to form stable σ or π bonds to other carbon atoms (as well as to atoms of many other elements) and to extend such bonding through long chains or rings of atoms accounts for the large number of isomeric compounds encountered. The determination of a complete (three-dimensional) structural formula for a compound is therefore a very complex operation, especially when there are many isomeric structures possible.

There are several types of isomerism. Positional isomerism is, perhaps, the simplest to describe by two-dimensional formulas. The isomeric structures shown below differ only in the position of the methyl (CH_3) group attached to the five carbon chain. It should be remembered that the most stable geometry around a carbon atom having four σ bonds is the tetrahedral pattern, corresponding to the sp^3 orbital description. Hence, structures such as those shown are only simplified two-dimensional representations.

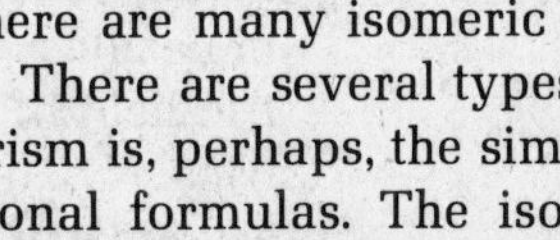

2-methylpentane

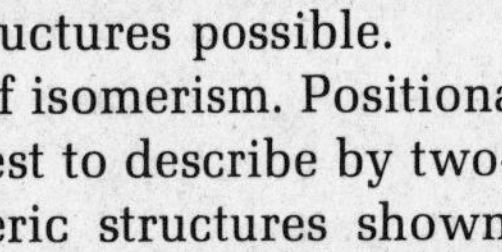

3-methylpentane

When you recall that free rotation is possible around σ bonds, it will be obvious that a substance such as 2-methylpentane could be described by several formulas like those below, which, however, are all equivalent.

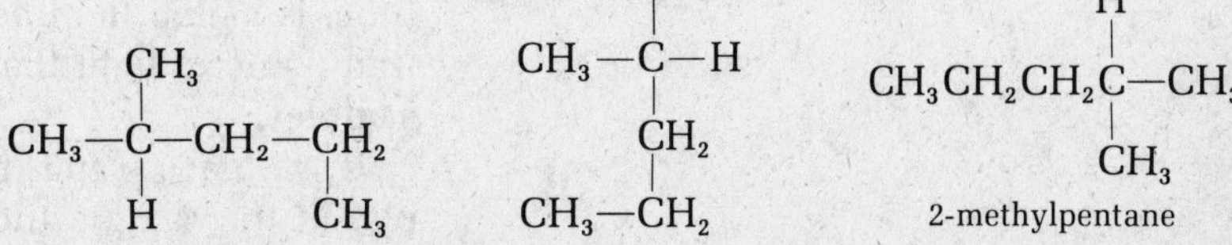

2-methylpentane

Certain structures that might appear at first glance to be simply different conformations (shapes) actually represent two different chemical compounds. Geometric isomers are species that differ, as in the case of conformational isomers, only in the relative spatial arrangement of atoms of otherwise equivalent groups. Geometric isomers, however, cannot be interconverted without breaking one or more covalent bonds. Such isomers may occur when some structural feature restricts the rotations necessary for isomeric interconversion. Two such features are rings and double bonds. In the former case, σ bonds must be broken before isomer conversion is possible, and the latter requires rupture of the π-bond system. In either situation considerable energy is needed and, as a result, geometric isomers are normally distinct and separable compounds.

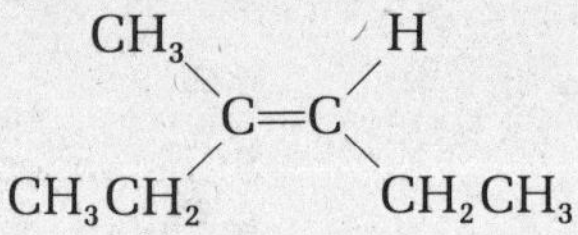

3-methyl-*cis*-3-hexene

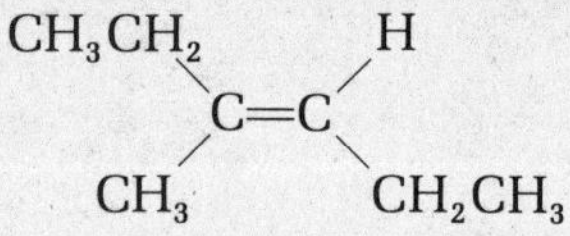

3-methyl-*trans*-3-hexene

Ordinary light is described as electromagnetic radiation, and we know that such radiation is capable of interacting with the electric field of electrons in atoms and molecules (e.g., as in absorption spectroscopy). The electric field of ordinary light is considered as being symmetrical about the line representing the light path, but asymmetry is introduced when such light is reflected from a flat surface or passed through certain materials such as Polaroid. This plane-polarized light is now represented as having its electric field described by vectors confined to a single plane.

If we view a light beam through a pair of identical Polaroid lenses from sunglasses and begin to turn one of the lenses, we notice that the light intensity begins to drop, reaching a minimum when the two lenses are in a position such that their longest dimensions are perpendicular (crossed). The plane-polarized light from the first lens is absorbed by the second lens unless the lenses are arranged in a parallel fashion. If we now insert a sample tube between the two lenses, we shall have prepared a simple instrument called a *polarimeter* with which we can observe the effects of plane-polarized light on a liquid or solution. One lens is called the *polarizer* (defining the polarization plane of the light entering the sample) and the other lens, placed at the "viewer end" of the sample tube, is called the *analyzer* (which defines the polarization plane of light that has been transmitted through the sample).

If we place water, methanol, or chloroform, for example, in the sample tube, the maximum light intensity is found with the same relative orientations of polarizer and analyzer as with the sample tube empty—even though the molecules of these compounds are unsymmetrical. However, if a saturated aqueous solution of table sugar (sucrose) is placed in the sample tube, the maximum light intensity is not observed until the analyzer has been turned to a new angle relative to the alignment of the polarizer. Apparently the sucrose molecules interact in some way with the plane-polarized light to rotate the plane of polarization. Substances capable of rotating the polarization plane of plane-polarized light are said to be "optically active."

It has been found that optical activity is associated only with substances whose molecules (or polyatomic ions) are not superimposable on their mirror images.

Optical isomers differ with respect to their interactions with plane-polarized light. The simplest optical isomers occur in organic molecules containing an asymmetric* carbon atom. Two optical isomers that are mirror images are called *enantiomers*. There are other types of optical isomers, but further discussion is beyond the scope of this text.

Conventional structural formulas may be used to represent optical isomers. Side groups are assumed to project up from the plane of the paper, while top and bottom groups project back behind the plane of the paper.

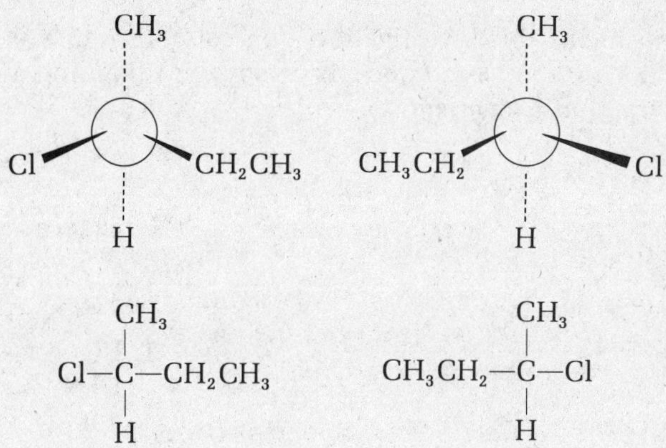

The molecular formula C_2H_6O may refer to either of two compounds:

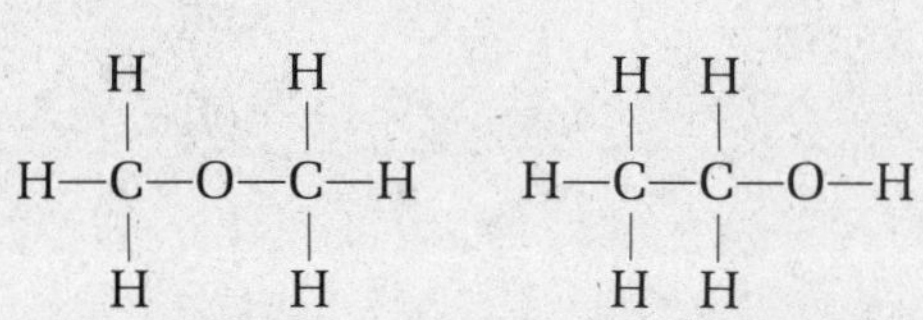

dimethyl ether   ethyl alcohol (ethanol)

The former is a gas at room temperature, sparingly soluble in water. The latter is a liquid (b.p., 78.5°C) and completely miscible with water. Even greater differences are observed in the chemical properties of these compounds. Ethanol reacts with carboxylic acids to form esters, with PCl_3 to form ethyl chloride, and may easily be "dehydrated" to form ethylene. Dimethyl ether undergoes

*An asymmetric carbon is one that is bonded to four different groups.

none of these reactions and, in fact, is inert to a variety of chemicals that would easily react with ethanol.

Such functional group isomers display the greatest differences in chemical and physical properties of all isomeric species. Other types of isomers may also differ to some extent in their physical and chemical properties. These differences are usually much less than those of functional group isomers. To some extent the differences in properties may be predicted on the basis of structural features. For example, variation in boiling points may often be estimated qualitatively on the basis of molecular arrangements related to polarity differences. Variations in physical properties among isomeric compounds can often be explained and predicted in terms of relatively simple considerations of intermolecular forces.

Mirror-image optical isomers (enantiomers) are identical in many respects. Their melting points, boiling points, solubilities, and most reactions in solution are the same. Such isomers differ in two important respects: They rotate the polarization plane of plane-polarized light in opposite directions (but by the same amount), and their biochemical properties are usually very different because of the stereoselective properties of the catalytic enzymes involved in the chemical functions of living systems.

Before starting this investigation, you should be thoroughly familiar with the basic concepts of covalent bonding, molecular geometry, and isomerism.

References

1. O'Connor, Rod. 1977. *Fundamentals of Chemistry*. (2nd ed.) New York: Harper & Row (Units 7, 8, 28, and 29).
2. Readings from *Scientific American*. 1973. *General Chemistry*. San Francisco: Freeman (Units 6, 7, 9, and 26).

SAFETY If you are working in a laboratory in which chemical reagents are set out, wear approved eye protection. (Consult your instructor if you have questions on the eye protection policy.)

OBJECTIVES

1. Be able to construct three-dimensional models of organic compounds and to draw representations of the models.
2. Be able to identify different types of isomerism.

PROCEDURES

You will be issued a stick-and-ball model kit. Be sure to disassemble your models and repack the kit to return it in good condition on completion of this investigation. For larger models, or for constructing isomer pairs, work with one or two of your neighbors.

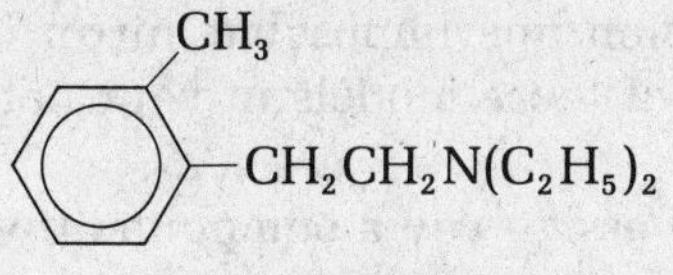

an active synthetic antimalarial
1-(2-N, N-diethylaminoethyl)-2-methyl benzene

Inactive isomers

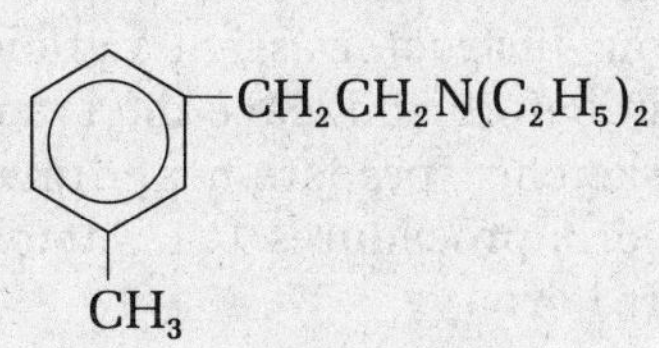

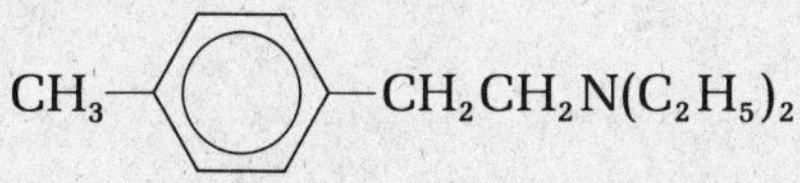

1. Construct models of the active synthetic antimalarial drug shown and of one of the inactive isomers. Draw three-dimensional representations of these models in your notebook. Identify the isomers by type (positional, functional group, geometric, or optical).

Odor	Receptor shape
camphorlike	hemispherical
musky	disk
floral	kite with tail
pepperminty	wedge
ethereal	rod

2. According to the stereochemical theory of odor,* the aroma of an organic compound is primarily due to its molecular shape and size (factors determining how the molecule could fit into available olfactory receptor sites). Construct models of the two geometric isomers of 1,4-dimethoxy-2-butene ($CH_3OCH_2CH{=}CHCH_2OCH_3$). Draw three-dimensional representations of these models and predict which isomer would be more likely to have a "pepperminty" odor, assuming the stereochemical theory is correct. (Identify the isomer as *cis* or *trans*. If you are unfamiliar with these terms, consult your instructor.)

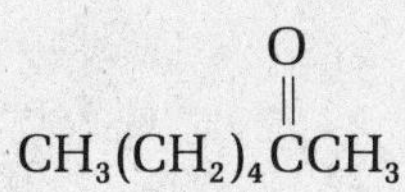

an ant alarm pheromone
2-heptanone

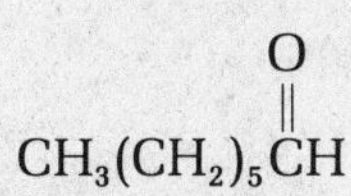

not "translated"
heptanal

3. Ants lay a trail of a volatile chemical for other ants to follow. Honeybees swarm after their queen by following a molecular "glide path beam," or mass to protect the hive when alerted by a different chemical substance. These and other *pheromones* (chemical communicating agents) suggest the fascinating possibility of "talking" with insects by means of a chemical vocabulary.† Construct models of the two functional group isomers shown. Draw representations of the models and identify each compound by its functional group "class."

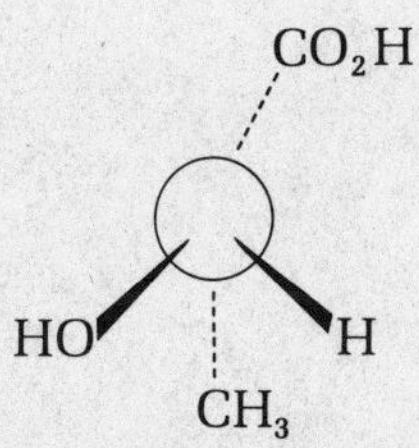

lactic acid
2-hydroxypropanoic acid
as isolated from muscle tissue

4. The typical laboratory synthesis of lactic acid, $CH_3CH(OH)CO_2H$, invariably yields a mixture of two forms which are nonsuperimposable mirror images. Lactic acid isolated from muscle tissue, however, is entirely one of these forms (shown at left), and the lactic acid isolated from yeast is 100 percent the other form. Construct models

*Amoore, John, James Johnston, Jr., and Martin Rubin. "The Stereochemical Theory of Odor," *Scientific American* (Feb. 1964).
†Wilson, Edward O. "Pheromones," *Scientific American* (May 1963).

of the two *optical isomers* of lactic acid and demonstrate that they are, indeed, nonsuperimposable mirror images. Draw representations of these models in your notebook.

5. Your instructor will assign you a compound (by structural formula) and ask you to construct and sketch a model of the compound. You will also construct a specified isomer of the compound. You will be given a time limit.

REPORT Complete your sketches for procedure 5 on the front of the Report Form and have your instructor assign a grade for this part of the experiment before you leave the laboratory. From your notebook sketches, prepare neat drawings of the models constructed in procedures 1–4. Attach these sketches to your Report Form.

GRADE	
	INSTRUCTOR'S INITIALS

Report Form

FOR INVESTIGATION 30:

Models for Organic Compounds

NAME: ______________________ ID NO.: ______________________

LAB SECTION ______________________ INSTRUCTOR ______________________

Compound assigned ______________________

Sketch of model

Isomer specification ______________________

Sketch of model

Grade for procedure 5 ☐

Instructor's initials ________

Isomer type (procedure 1) ______________________________

Isomer more likely to have a "pepperminty" odor (procedure 2)

☐ *cis*　　　☐ *trans*

Functional group class (procedure 3)

$CH_3(CH_2)_4\overset{\overset{\displaystyle O}{\|}}{C}CH_3$ ______________________________

$CH_3(CH_2)_5\overset{\overset{\displaystyle O}{\|}}{C}H$ ______________________________

[Attach neat drawings representing all models constructed in procedures 1–4.]

______________ DATE　　______________________________ SIGNATURE

Investigation 31

ORGANIC COMPOUND ANALYSIS

INTRODUCTION There are many reasons why we might wish to determine the identity of an organic compound. In natural product chemistry, for example, we might find some exotic herb that contains something of therapeutic value. If we could isolate from the complex mixture a single compound possessing the useful properties and find out what it is, we might discover a valuable new drug. Quinine, the first good antimalarial, was found in this way.

In the field of insect biochemistry, there are a host of interesting chemicals that have been identified: among them a sex attractant found useful in control of the gypsy moth population, a number of pheromones (chemicals used in communication among insects), and a variety of interesting compounds found in the venoms of stinging insects.

The identification of a biologically active component in a complex natural mixture is not a simple task. First, the compound must be isolated from the mixture in a pure form. Depending on the complexity of the mixture and the amounts available to work on, this may take a few days or several years of extensive research. Then structural information must be accumulated to permit an unambiguous assignment of chemical identity. If the compound of interest is one that is already known, this may involve relatively simple comparisons of properties to establish the identity. The determination of the structure of a new compound is much more difficult, especially if it is a relatively complex molecule. Isoamyl acetate, an alarm pheromone of the common honeybee, was isolated and identified by comparison with the known chemical in a fairly short time. On the other hand, many research teams all over the world have been investigating the complex venom system of the honeybee for more than 30 years and, although some 40 different compounds have been identified, the total composition of honeybee venom is not yet determined.

There are, of course, many reasons for identifying organic compounds in addition to those of the natural product chemist. The forensic chemist may be required to determine the nature of a confiscated drug to be used as evidence in a trial or to identify some exotic poison involved in a murder case. The environmental chemist may have to detect an undesirable organic pollutant in a water or air sample. The chemical oceanographer is interested in the identity of organic compounds in the marine environment and the "astrochemist" may use very special techniques for detecting organic compounds in the atmosphere of a far planet or even in the vast reaches of interstellar space.

The determination of the composition of a chemical mixture, or even the isolation and identification of a single component from a mixture, is a time-consuming task requiring special knowledge and skills. It is considerably easier to identify a single pure compound, especially if we have narrowed the possibilities to a fairly small list. That is our task in this experiment and we shall investigate only five classes of compounds: alcohols, aldehydes, amides, carboxylic acids, and ketones. The following are possibilities. Your instructor may add to or delete from this list.

Table 31.1 Possible "unknowns"

Class	Compound	m.p. (°C)	b.p. (°C)	Derivative* (m.p.)	Comments
alcohol	ethanol	(−117)	78	157	active ingredient of "alcoholic" beverages
	2-propanol	(−90)	82	153	commonly sold as "rubbing alcohol"
	methanol	(−94)	65	153	so-called "wood alcohol," very poisonous
	cyclohexanol	25	161	160	a common solvent for resins and insecticides
aldehyde	benzaldehyde	(−26)	179	237	almond flavoring agent
	salicylaldehyde	(−7)	197	252	used in perfumes, has a "bitter-almond" odor
	p-anisaldehyde	0	248	254	odor reminiscent of vanilla beans
	cinnamaldehyde	(−8)	252	253	cinnamon flavoring agent
	vanillin	80	285	269	vanilla flavoring agent
amide	acetamide	82	—	—	used as an additive to plastics and often to "denature" alcohol
	formamide	2	210	—	used as a softener for paper and animal glues
	phenylacetamide	157	—	—	used in manufacturing Penicillin G
	salicylamide	159	—	—	an aspirin substitute
carboxylic acid	acetic	16	118	—	acid component of "vinegar"
	butyric	(−19)	163	—	responsible for odor of rancid butter
	palmitic	63	390		found, as the glyceryl ester, in many vegetable oils and fats
	phenylacetic	76	265	—	used in production of some synthetic perfumes

Table 31.1 (Continued)

Class	Compound	m.p. (°C)	b.p. (°C)	Derivative* (m.p.)	Comments
ketone	acetone	(−95)	56	128	common solvent for varnishes, cements, etc.
	cyclohexanone	(−16)	156	162	valuable solvent and a reagent for preparing adipic acid for nylon synthesis
	2-butanone	(−86)	80	111	valuable solvent and used in preparing "smokeless" gunpowder
	acetophenone	20	202	237	used in perfumes to produce an orange-blossom aroma
	camphor	179	204	177	a moth repellant and a mild antiseptic

*The derivatives used will be 2,4-dinitrophenylhydrazones for aldehydes and ketones, and 3-nitrophthalates (half-esters) for alcohols.

Before starting this investigation, you should be familiar with the nomenclature, structural characteristics, and fundamental chemical properties of the classes of compounds included in this investigation. In addition, you should understand the general principles governing purification by recrystallization.

References

1. O'Connor, Rod. 1977. *Fundamentals of Chemistry* (2nd ed.). New York: Harper & Row (Units 28 and 29 and Excursion 4).
2. Owen, Terence C. 1969. *Characterization of Organic Compounds by Chemical Methods.* New York: Marcel Dekker.

SAFETY Wear approved eye protection and be very careful with open flames. (Most of the compounds used are highly flammable.) Under *no* circumstances should you taste any of the chemicals. (Although some are used as flavoring agents, many of these are toxic as the undiluted compound.) In case of chemical contact, wash thoroughly with *soap* and water.

OBJECTIVES

1. Be able to handle and heat organic chemicals safely.
2. Be able to use properly the techniques of crystallization, recrystallization, vacuum filtration, and melting-point determination.
3. Be able to use observations of the properties and chemical behavior of known compounds to devise analysis methods for the identification of similar unknown compounds.

PROCEDURES

In this experiment you will employ a number of techniques used in previous experiments. You should review the descriptions of such procedures carefully. You will need to keep careful notebook records.

You will be issued two "unknown" organic compounds, both of which are listed in Table 31.1 Possible "Unknowns" (as modified by your instructor). From preliminary investigations (procedures 2–5), you will determine the "class" of each unknown. After these determinations have been checked by your instructor, you will identify each of the compounds (procedures 6–11), by comparison of appropriate properties with those of known compounds.

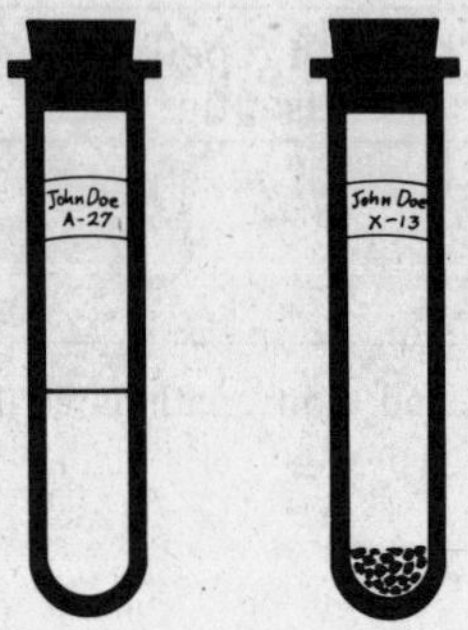

1. Obtain from your instructor approximately 2 g each of two different unknowns, in clean, dry test tubes labeled with a specified identification code and your name. Keep the tubes sealed except when removing a sample for use.

2. Examination of odors. Open one of the tubes and carefully waft the vapors from the tube toward your nose. Record the odor as precisely as possible (e.g., "A-27 smells like orange peels"). Repeat with the second unknown. Then sample the odors (*do not inhale vapors*) of some of the reference compounds provided, recording in your notebook the names of any compounds with odors similar to those of your unknowns.

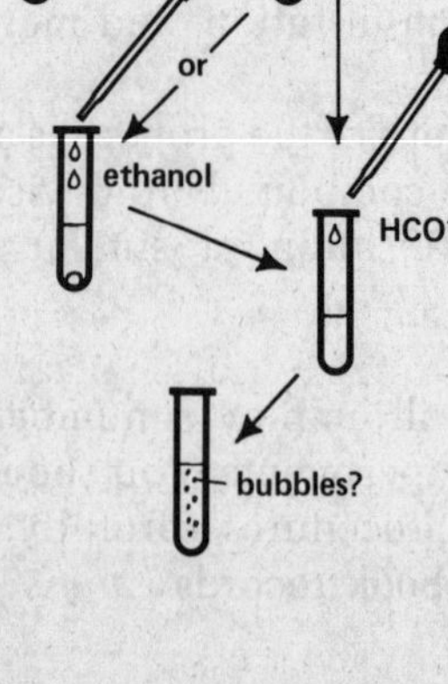

3. Test for Carboxylic Acids. Aqueous or aqueous-ethanolic solutions of carboxylic acids react with sodium bicarbonate to liberate carbon dioxide. The bubbles rising to the surface of the solution indicate that the compound tested is a carboxylic acid. Perform the test described below with acetic acid (a liquid miscible with water), palmitic acid (a solid, soluble in ethanol-water mixtures), and with your unknown. To a clean test tube add 2 drops of acetic acid, then 15 drops of distilled water. Swirl carefully to mix until the solution is homogeneous. (Other liquid carboxylic acids, e.g., butyric acid, are not very soluble in water and such a mixture will not become homogeneous on swirling. In this case add 15 drops of 95 percent ethanol and, if necessary, heat to speed up dissolution). Then add 15 drops of 5 percent sodium bicarbonate solution to the homogeneous mixture. Record all observations.

Repeat this procedure with a small amount (20 to 30 mg) of palmitic acid. Add 15 drops of 95 percent ethanol, heat until all the acid has dissolved, add 15 drops of water then 15 drops of 5 percent sodium bicarbonate. Record all observations.

Repeat the appropriate procedure with each of your unknowns, using 2 drops for liquid compounds and 20 to 30 mg for solids.

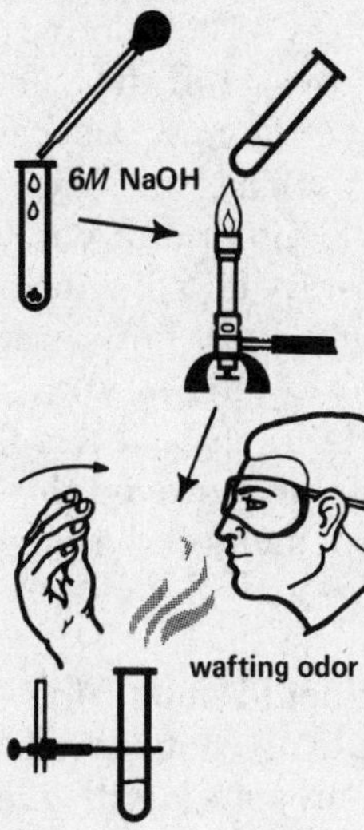

4. Test for Amides. (Primary amides react with aqueous NaOH to liberate ammonia, which is detected by odor.)

Place a small sample (about 50 to 100 mg) of acetamide in a clean test tube and add 15 drops of 6 *M* NaOH. Swirl carefully to mix the contents, then heat the tube carefully to a gentle boil for 2 to 3 min. Carefully sample the odor of the vapor to see if you detect ammonia. Record all observations. Repeat this procedure, substituting 2 drops of formamide for the acetamide. Then perform this test with your unknowns, using 20 to 30 mg for solids and 2 drops for liquids.

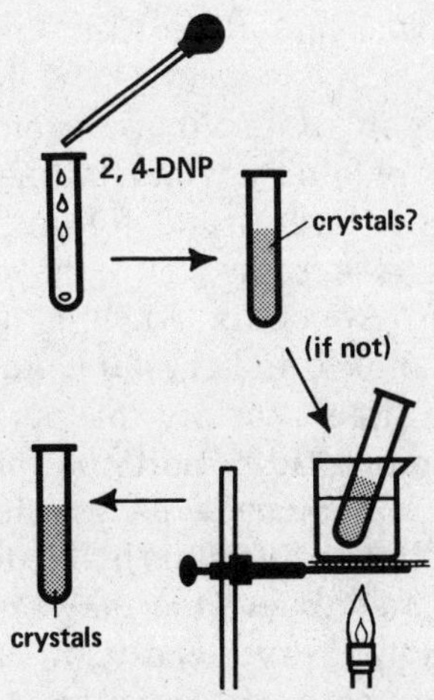

5. Test for Ketones and Aldehydes. (These compounds react with 2,4-dinitrophenylhydrazine to form crystalline 2,4-dinitrophenylhydrazones, for which melting points are determined.)

To 1 drop of cyclohexanone in a clean test tube, add 15 drops of the 2,4-dinitrophenylhydrazine (2,4-DNP) reagent. Swirl or stir to mix the contents, then set the tube aside for about 15 min. If no crystals have formed, heat the test tube for about 5 min in a beaker of water maintained at 60° to 70°C (no warmer), then allow the tube to stand undisturbed for 15 to 20 min. If you still find no crystals, scratch the inside of the test tube (within the solution) with a glass rod for 2 to 3 min and set the tube in a beaker of ice for another 15 min. Record all observations. Repeat this procedure with benzaldehyde and with your unknowns, using 1 drop for liquids and 20 to 30 mg for solids.

Study your notebook records carefully. From the results of your investigations (procedures 2–5), you should be able to determine the class of each of your unknowns. If not, consult your instructor for suggestions of additional tests. Have your instructor check your classifications before you proceed with further work.

6. Determination of Melting Point. If either of your unknowns is a solid, determine its melting point by the capillary method used for pure camphor in Investigation 14, "Molecular Weights of Some Pharmaceutical Chemicals." Repeat, for improved accuracy, with a second sample of the compound. (You may wish to practice with one of the reference compounds of known melting point.) Compare the melting point of your unknown with those tabulated for the possible "unknowns" in Table 31.1.

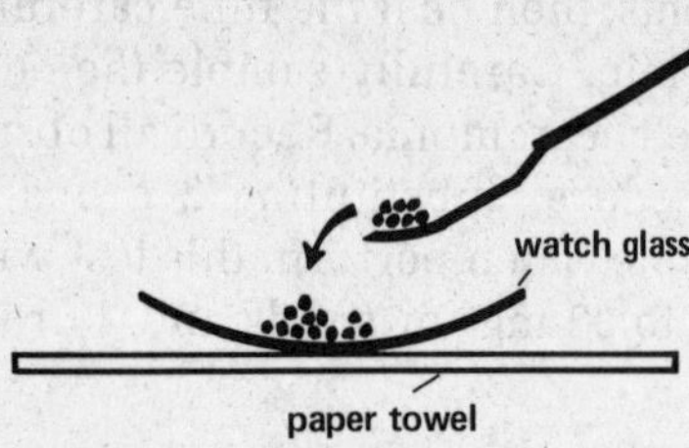

7. An accurately determined melting point should, if the compound is pure, enable you to identify it as one of the possibilities listed for that class of compounds. For verification, however, obtain from your instructor a tiny sample (20 to 30 mg) of the compound you believe to be identical with your unknown. On a clean, dry watch glass, mix this sample with an equal amount of your unknown, using the spatula blade to grind and stir the solids into a homogeneous mixture of fine (almost powdered) crystals. Determine the melting point of this mixture. Since the presence of a "foreign material" acts to lower the melting point of a substance, you will have proved the identity of your compound if this mixture melts at the same temperature as your pure unknown.

For liquid unknowns you could determine the boiling point. However, boiling points are not easy to determine accurately and they vary appreciably with the pressure. If you wish to try, consult your instructor for directions.

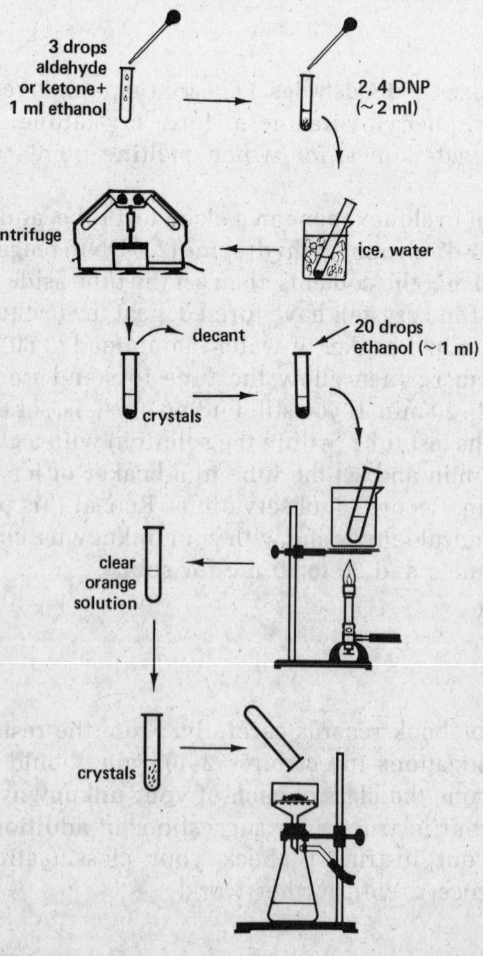

8. Preparation of 2,4-dinitrophenylhydrazones. If one of your unknowns is an aldehyde or ketone, place 3 drops of a liquid unknown or 50 mg of a solid unknown in a clean, dry small test tube. Add 20 drops ethanol and then 2 ml of the 2,4-dinitrophenylhydrazine reagent. Swirl the tube to mix the contents thoroughly. Refer to your notes for the conditions required for obtaining a crystalline product from this unknown (procedure 5) and follow the steps necessary. Once crystals have formed, chill the mixture in an ice bath. Centrifuge (Appendix A) and decant the liquid. Add 1 ml of ethanol to the crystals and heat the mixture in a water bath. Should the crystals not have dissolved after several minutes, add additional ethanol in small increments until a homogeneous solution is obtained. Cool the tube, isolate the crystals by the technique of vacuum filtration (Appendix A) and take the melting point of the dry compound. If crystals do not form in the cool solution even on scratching the inner test tube wall, reheat the solution and add warm distilled water, dropwise with stirring, until a precipitate begins to form. Warm the mixture in a water bath until the solution becomes clear. (If warming alone does not accomplish this, add warm ethanol, dropwise with stirring, until all the precipitate has dissolved.) Prepare a similar derivative from the compound suspected of being identical with your unknown and determine a "mixed melting point."

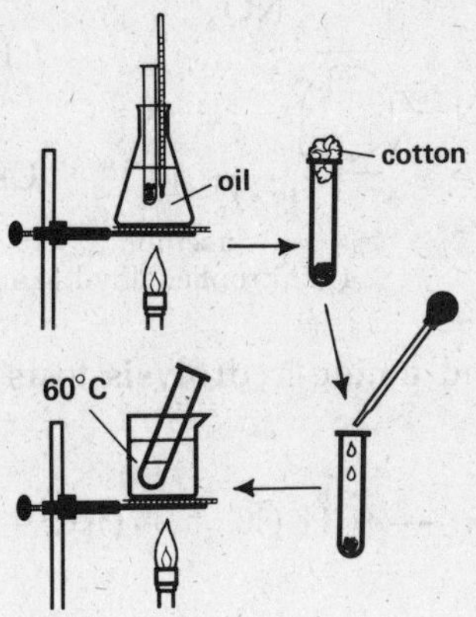

9. Test for Alcohols. (Alcohols react with 3-nitrophthalic anhydride to form crystalline esters, for which the melting point is characteristic of the "parent" alcohol.)

If one of your unknowns is an alcohol, obtain 0.2 to 0.3 g of 3-nitrophthalic anhydride in a clean, dry test tube. Heat this in an oil bath (*Caution:* fire hazard) to around 200°C for about 10 min. Remove the test tube carefully, allow it to cool, then wipe off the oil and stopper the tube loosely with a plug of cotton. When the test tube has cooled to room temperature, add 10 drops of your alcohol, stir, and heat the test tube in a 60°C water bath for about 10 min. Then set the test tube aside to cool, replacing the cotton plug. If crystals do not form on cooling, scratch the interior of the tube (within the liquid) with a glass rod to induce crystallization. When crystals have formed, isolate them by vacuum filtration and wash them on the filter with 5 to 10 ml of cold water. Recrystallize from ethanol-water (as in procedure 8) and determine the melting point and a mixed melting point (procedure 7) with the 3-nitrophthalate derivative of a known compound.

10. Formula Weight of a Carboxylic Acid. If one of your unknowns is an acid, weigh a clean, dry 125-ml Erlenmeyer flask, with stopper, to the nearest 1 mg. Then add about 0.2 g of the acid, restopper, and reweigh. Record the mass of acid used to the nearest 1 mg. Dissolve the acid in 15 ml of water (or 15 ml of ethanol if the acid is insoluble in water) and titrate this solution with *standardized* 0.1 *M* sodium hydroxide to a phenolphthalein endpoint. (For necessary techniques, see Investigation 32). Calculate the formula weight of the acid from your titration data and compare this with the formula weights of possible unknowns. (Check with your instructor to be sure that only monoprotic acids are possible unknowns.)

Study your notebook records carefully. You should now be able to identify your unknowns by name. If not, consult your instructor for suggestions. Be sure that you have sufficient evidence to defend your identifications.

REPORT Use your notebook records to complete the Report Form. The equations for the preparations of derivatives of alcohols, aldehydes, and ketones are illustrated by the following examples:

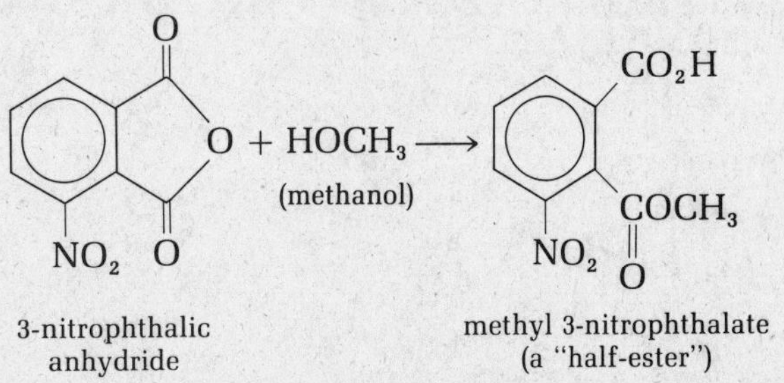

3-nitrophthalic anhydride

methyl 3-nitrophthalate (a "half-ester")

(b) 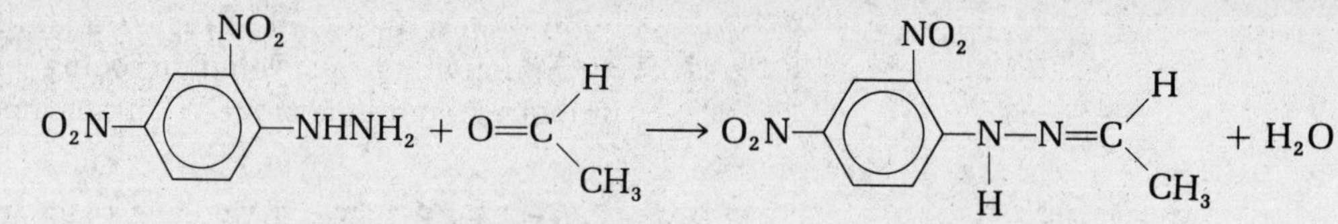

2,4-dinitrophenylhydrazine — acetaldehyde — acetaldehyde 2,4-dinitrophenylhydrazone

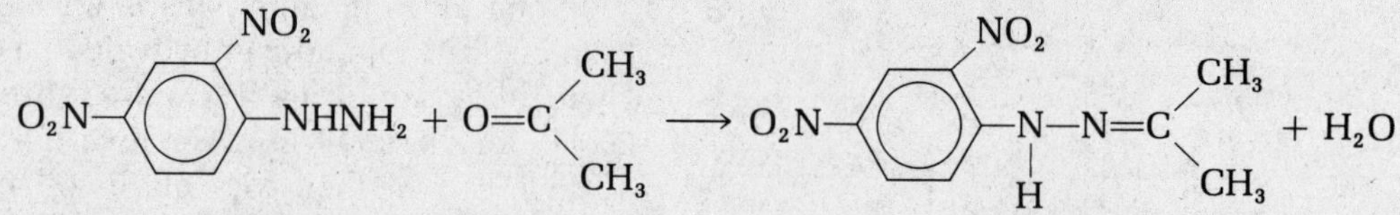

2,4-dinitrophenylhydrazine — acetone — acetone 2,4-dinitrophenylhydrazone

The bicarbonate and amide-hydrolysis tests are illustrated by

$$HCO_3^- + CH_3CO_2H \longrightarrow CH_3CO_2^- + H_2O + CO_2$$

$$OH^- + CH_3\overset{\displaystyle O}{\overset{\|}{C}}NH_2 \longrightarrow CH_3CO_2^- + NH_3$$

(Note: the general approach used in this experiment can be applied to the identification of any unknown organic compound if suitable comparison compounds are available, but the specific procedures would have to be supplemented and modified for compounds significantly different from those in Table 31.1, Possible "Unknowns." Modern instruments, such as the infrared spectrophotometer, are very useful in supplementing other analytical information.)

GRADE
INSTRUCTOR'S INITIALS

Report Form

FOR INVESTIGATION 31:

Organic Compound Analysis

NAME: ____________________ ID NO.: ____________________

LAB SECTION ____________________ INSTRUCTOR ____________________

Classification of Unknown

Properties or behavior	Unknown code ______	Unknown code ______
odor description		
odor similar to that of known		
indication of water solubility (procedure 3)		
behavior with HCO_3^- (procedure 3)		
behavior with 6 *M* NaOH (procedure 4)		
behavior with 2,4-DNP reagent (procedure 5)		

Classification		
Instructor's initials		

Which classes(class) of compounds liberate(s) CO_2 from HCO_3^-?

____

Which classes(class) of compounds liberate(s) NH_3 when boiled with 6 *M* NaOH?

____

Which classes(class) of compounds form(s) a precipitate when treated with the 2,4-dinitrophenylhydrazine reagent?

____

(OVER)

Specific Identification

List *all* information required for unambiguous identifications, including odor comparison, melting point, mixed melting point, or titration data and calculation.

Unknown code ____________________

____
Name of compound

Unknown code ____________________

____
Name of compound

Chemical behavior

Write balanced equations for *all* chemical reactions used in the identification of your unknowns, including reactions (if any) with HCO_3^-, NaOH, 2,4-dinitrophenylhydrazine, and 3-nitrophthalic anhydride. (Attach extra page, if necessary.)

"Unknown" name ____________________	"Unknown" name ____________________

____________________ DATE

____________________ SIGNATURE

Investigation 32

FORMULA WEIGHT OF A FATTY ACID

INTRODUCTION Fats and oils, esters of long-chain carboxylic acids, have been described in Investigations 6 and 10. A large number of different fatty acids have been isolated from these natural esters (Table 32.1) and many of these acids are commercially valuable. Before a particular reaction sequence can be planned, such as the conversion of a fatty acid to a long-chain alcohol (Investigation 10), the acid must be characterized. One important piece of information that can rather easily be obtained is the formula weight of the acid. This information is, of course, critical to any stoichiometric calculations involved in planning a reaction sequence.

Table 32.1 Some fatty acids from natural fats and oils

"Common" (trivial) name	Formula	(m.p., °C)	Obtained by hydrolysis of glyceryl esters found in:
butyric	$CH_3(CH_2)_2CO_2H$	−6	butter
caproic	$CH_3(CH_2)_4CO_2H$	−3	butter and goat fat
caprylic	$CH_3(CH_2)_6CO_2H$	16	butter, goat fat, and coconut oil
capric	$CH_3(CH_2)_8CO_2H$	31	butter, goat fat, and coconut oil
lauric	$CH_3(CH_2)_{10}CO_2H$	44	coconut oil and palm oil
myristic	$CH_3(CH_2)_{12}CO_2H$	54	butter, coconut oil, palm oil, beef tallow, and cod liver oil
palmitic	$CH_3(CH_2)_{14}CO_2H$	63	all of the above plus corn oil, cottonseed oil, lard, olive oil, peanut oil, and linseed oil
stearic	$CH_3(CH_2)_{16}CO_2H$	70	all of the above
oleic	$CH_3(CH_2)_7CH{=}CH(CH_2)_7CO_2H$ (*cis*)	16	all of the above
linoleic	$CH_3(CH_2)_4CH{=}CHCH_2CH{=}CH(CH_2)_7CO_2H$ (*cis*) (*cis*)	−5	all of the above, but to the largest extent in vegetable oils
linolenic	$CH_3CH_2CH{=}CHCH_2CH{=}CHCH_2CH{=}CH(CH_2)_7CO_2H$ (*cis*) (*cis*) (*cis*)	−11	soybean oil and linseed oil

Most of the acids obtained by the hydrolysis of fats and oils are *monoprotic,* that is each molecule can act as the donor of only a single proton. Such acids react on an equimolar basis with sodium hydroxide.

$$R{-}CO_2H + OH^- \longrightarrow R{-}CO_2^- + H_2O$$

(R represents the carbon chain.) We can use this reaction to determine the formula weight of any monoprotic acid. If we weigh the acid sample accurately and determine the concentration and volume of the base required to react completely with the weighed sample, we have all the necessary data for calculation of the formula weight of the acid. The product of the base concentration (in moles liter^{-1}) and the volume of base required (in liters) gives us the number of *moles* of base used. Since the reaction involves equimolar quantities of acid and base, we will now know the number of *moles* of the acid used. The ratio of mass of acid used to moles of acid used gives us the desired formula weight (as grams per mole). We could, if it is more convenient, work equally well with millimoles (mmol), defining the formula weight in units of milligrams per millimole (mg mmol^{-1}).

Example

A 0.8631-g sample of a fatty acid was completely neutralized by 31.96 ml of 0.1055 *M* sodium hydroxide. What was the formula weight of the acid?

Solution

Assuming equimolar reaction

$$\text{no. mmol acid} = \text{no. mmol base}$$
$$= \frac{0.1055 \text{ mmol}}{\text{ml}} \times \frac{31.96 \text{ ml}}{1} = 3.372 \text{ mmol}$$

$$\text{formula weight} = \frac{0.8631 \text{ g}}{3.372 \text{ mmol}} \times \frac{10^3 \text{ mg}}{\text{g}}$$
$$= \underline{256.0 \text{ (mg mmol}^{-1})}$$

In this investigation, we will use a sodium hydroxide titration to determine the formula weight of an "unknown" fatty acid from among those listed in Table 32.1. From a comparison of the experimentally determined formula weight with those calculated from the formulas of the possible "unknowns," we can make a statement of the possible identity of the "unknown."

Before starting this investigation, you should be familiar with the concepts of acid-base chemistry, the characteristics of carboxylic acids, and the calculations involved in solution preparation and solution stoichiometry. In addition, prior to coming to the laboratory, you must determine the formula weights of the fatty acids listed in Table 32.1 and set up the calculations required in the Procedures section of this investigation.

References

1. O'Connor, Rod. 1977. *Fundamentals of Chemistry* (2nd ed.). New York: Harper & Row (Units 13, 19, 28, and 29).
2. Goldwhite, Harold, 1972. *An Introduction to Organic Chemistry*. Columbus, Ohio: Merrill.

SAFETY Wear approved eye protection at all times and exercise standard precautions in working with chemicals, being particularly careful with NaOH. Use caution in heating ethanol to avoid a fire.

OBJECTIVES

1. Be able to make appropriate calculations involving solution preparation and solution stoichiometry.
2. Be able to accurately perform a titration of a fatty acid.

PROCEDURES

A. Preparation of NaOH Solution

Calculate the mass of sodium hydroxide needed to prepare 300 ml of 0.1 *M* NaOH solution. Check with your instructor before proceeding.

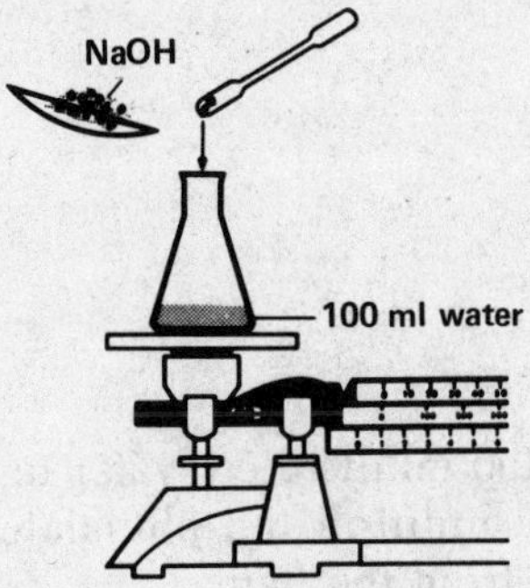

Caution: corrosive reagent.

1. Obtain 15 NaOH pellets on a clean, dry watch glass. Reseal the stock bottle. Weigh a 500-ml Erlenmeyer flask containing 100 ml distilled water to the nearest 0.1 g on a triple beam balance. Advance the balance masses by the amount calculated for the NaOH needed. With your spatula, add pellets one by one to the flask until the balance returns to equilibrium. Dispose of excess pellets as directed by your instructor. Remove the flask from the balance and swirl it carefully until all the sodium hydroxide has dissolved. Then add 200 ml of distilled water. Mix the contents thoroughly by carefully swirling, stopper the flask, and label it "0.1 *M* NaOH."

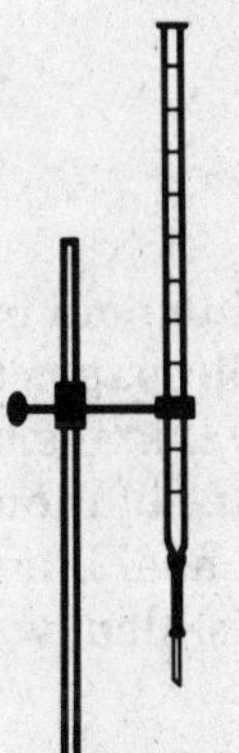

2. Rinse and fill one buret (Appendix A) with your 0.1 *M* NaOH. Drain ample solution from the buret to expel air from the buret tip and drop the liquid level to just below the zero mark.

B. Standardization of Base

Calculate the mass of potassium hydrogen phthalate, $C_8H_5O_4K$, that would be neutralized by 30 ml of 0.1 *M* NaOH. Check with your instructor before proceeding.

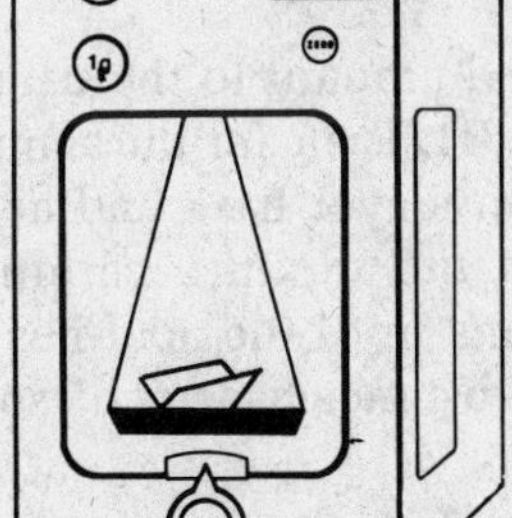

1. Obtain from your instructor on a clean piece of smooth paper, or aluminum foil, a sample of potassium hydrogen phthalate slightly in excess of the amount you will need. Weigh a clean creased piece of paper to the nearest 0.1 mg. Transfer to this the appropriate amount of phthalate required and reweigh to the nearest 0.1 mg.

2. Carefully remove the paper from the balance and transfer the weighed phthalate quantitatively into a clean 250-ml Erlenmeyer flask.

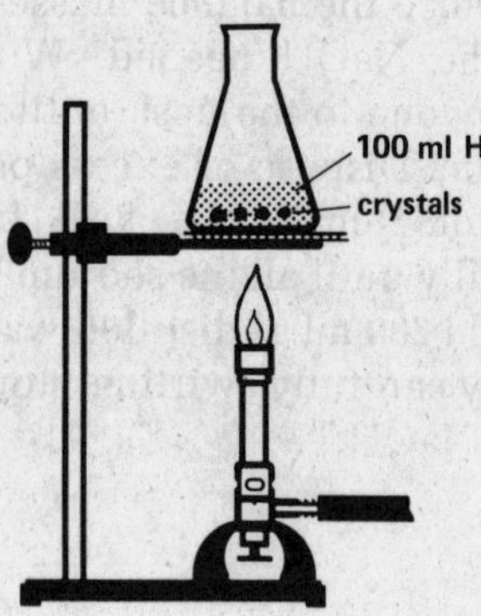

3. Add approximately 100 ml distilled water to the flask. Heat the flask carefully until all the phthalate has dissolved. Then remove it from the flame.

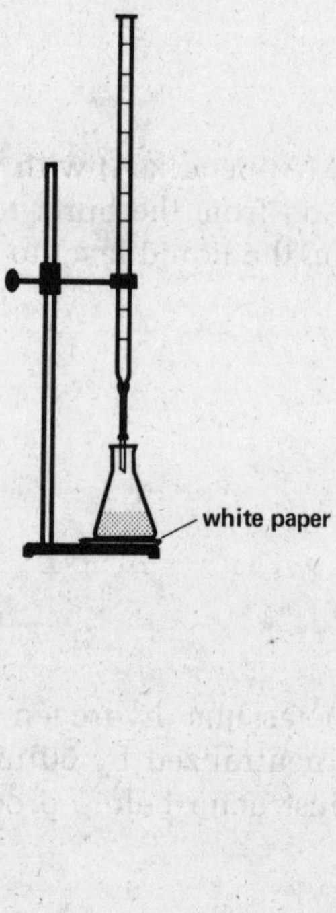

4. Add three drops of phenolphthalein indicator solution to the flask and set it on a white paper below the buret. Read and record the volume level in the buret to the nearest 0.01 ml. Titrate the phthalate solution to a permanent pale pink endpoint. Record the final buret reading. Clean the flask and rinse it with distilled water. Repeat the standardization if time allows.

C. Titration of an Unknown

1. From your instructor obtain an unknown, which should be placed into a clean, dry 100-ml beaker. This unknown could be the fatty acid prepared from soap or another solid carboxylic acid. Ask your instructor how much of your unknown you should use for a titration.

2. Weigh out the suggested amount to the nearest 0.1 mg as described in procedure B1. Transfer the sample quantitatively into a 250-ml Erlenmeyer flask and add 50 ml of 95 percent ethanol. Heat the mixture carefully (*flammable*) over a Bunsen flame until the acid has dissolved. Add more ethanol if the substance has not dissolved when the ethanol boils.

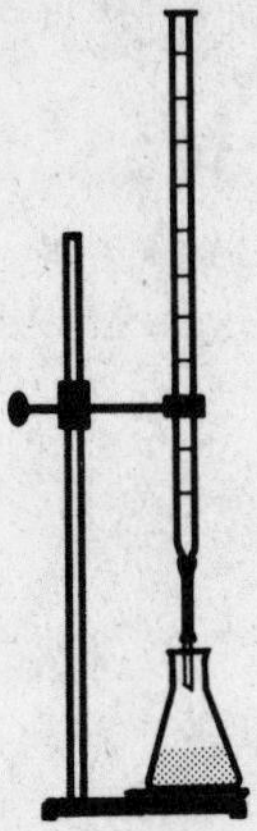

3. When you have obtained a homogeneous solution, add three drops of phenolphthalein and titrate with your NaOH solution to a pale pink endpoint. Record the initial and final buret readings. Calculate the formula weight of your unknown. Repeat the titration with another sample of the same unknown.

CALCULATIONS AND REPORT

Use your notebook and calculations to complete the Report Form. Calculate an average value for the formula weight of your "unknown" acid, based on the two titrations you performed. Then obtain from your instructor the results from a number of other students having the same "unknown" and calculate an average value for the formula weight using your data and the other student results. On the basis of the average formula weight, suggest one or more possible structures for your "unknown," assuming it to be one of the acids listed in Table 32.1. If more than one possibility is listed, suggest how these could be distinguished by further experimental work.

GRADE
INSTRUCTOR'S INITIALS

Report Form

FOR INVESTIGATION 32:
Formula Weight of a Fatty Acid

NAME: ____________________ ID NO.: ____________________

LAB SECTION ____________________ INSTRUCTOR ____________________

Unknown Number__________

Show all calculations on an attached page.

Solution Preparation and Standardization

Mass of NaOH calculated for 300 ml of 0.1 *M* solution __________ g

Mass of $C_8H_5O_4K$ calculated to neutralize 30 ml of 0.1 *M* NaOH __________ g

Phthalate standard	Trial 1	Trial 2
Sample + paper	______ g	______ g
Paper	______ g	______ g
Phthalate	______ g	______ g

Phthalate titration	Trial 1	Trial 2
Final volume NaOH	______ ml	______ ml
Initial volume NaOH	______ ml	______ ml
Volume NaOH consumed	______ ml	______ ml

Standardized concentration of NaOH solution __________ *M*

Formula Weight Determination

Fatty acid sample	Trial 1	Trial 2
Sample + paper	______ g	______ g
Paper	______ g	______ g
Fatty acid	______ g	______ g

Fatty acid titration	Trial 1	Trial 2
Final volume NaOH	______ ml	______ ml
Initial volume NaOH	______ ml	______ ml
Volume NaOH consumed	______ ml	______ ml

(OVER)

Formula Weight Calculation

Your results

Trial 1 ______________ g $mole^{-1}$

Trial 2 ______________ g $mole^{-1}$

Average ______________ g $mole^{-1}$ Average deviation ____________________

Group results

Your average value ____________________ g $mole^{-1}$

____________________ g $mole^{-1}$

____________________ g $mole^{-1}$

____________________ g $mole^{-1}$

____________________ g $mole^{-1}$

____________________ g $mole^{-1}$

Average ____________________ g $mole^{-1}$

Average deviation ____________________

The "unknown" acid studied was probably:

______________________________ (formula weight ______________)

Other possibilities, if any, are:

Suggestions for additional experiments, if needed:

_______________ DATE _____________________________ SIGNATURE

Investigation 33

A NATURAL FLAVORING AGENT

INTRODUCTION Although carboxylic acids generally have irritating or unpleasant odors and only a few of the alcohols have pleasant aromas, the products of reactions between alcohols and carboxylic acids usually have sweet, "fruity" odors. These reaction products, *esters*, are found as natural flavoring agents in many fruits (Table 33.1) and either single esters or mixtures of esters are used in a number of synthetic flavorings.

$$\underset{\substack{\text{butyric acid}\\\text{(odor of}\\\text{rancid butter)}}}{CH_3CH_2CH_2\overset{\overset{O}{\|}}{C}OH} + \underset{\substack{n\text{-butyl alcohol}\\\text{(pungent odor)}}}{HOCH_2CH_2CH_2CH_3} \rightleftharpoons \underset{\substack{n\text{-butyl butyrate}\\\text{(odor of pineapple)}}}{CH_3CH_2CH_2\overset{\overset{O}{\|}}{C}{-}OCH_2CH_2CH_2CH_3} + H_2O$$

Table 33.1 Some natural esters

Formula	Trivial name	IUPAC name	A flavoring agent and odor component of
$CH_3CH_2O\overset{\overset{O}{\|}}{C}H$	ethyl formate	ethyl methanoate	rum
$CH_3(CH_2)_3CH_2O\overset{\overset{O}{\|}}{C}CH_3$	*n*-amyl acetate	pentyl ethanoate	pears, bananas
$(CH_3)_2CHCH_2CH_2O\overset{\overset{O}{\|}}{C}CH_3$	isoamyl acetate	3-methylbutyl ethanoate	pears, bananas
$CH_3(CH_2)_6CH_2O\overset{\overset{O}{\|}}{C}CH_3$	*n*-octyl acetate	octyl ethanoate	oranges
$(CH_3)_2CHCH_2O\overset{\overset{O}{\|}}{C}CH_2CH_3$	isobutyl propionate	2-methylpropyl propanoate	rum
$CH_3O\overset{\overset{O}{\|}}{C}CH_2CH_2CH_3$	methyl butyrate	methyl butanoate	apples
$CH_3CH_2O\overset{\overset{O}{\|}}{C}CH_2CH_2CH_3$	ethyl butyrate	ethyl butanoate	pineapples
$CH_3(CH_2)_2CH_2O\overset{\overset{O}{\|}}{C}CH_2CH_2CH_3$	*n*-butyl butyrate	butyl butanoate	pineapples
$CH_3(CH_2)_3CH_2O\overset{\overset{O}{\|}}{C}CH_2CH_2CH_3$	*n*-amyl butyrate	pentyl butanoate	apricots
$(CH_3)_2CHCH_2CH_2O\overset{\overset{O}{\|}}{C}(CH_2)_3CH_3$	isoamyl valerate	3-methylbutyl pentanoate	apples
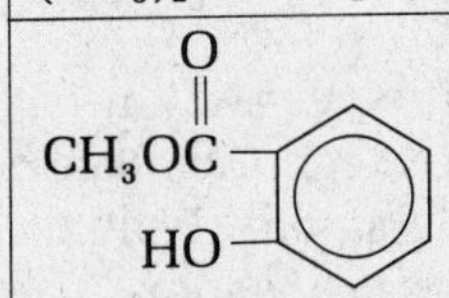	methyl salicylate	methyl 2-hydroxybenzoate	oil of wintergreen

Within living organisms, such as plants, esters are usually synthesized by a relatively complex stepwise process, typically employing enzymes as catalysts. The laboratory or industrial processes are considerably simpler, although often less efficient in terms of chemical and energy utilization. Several alternative synthetic routes are available. The *direct esterification* reaction between an alcohol and a carboxylic acid is a simple procedure, but yields are limited by the competitive reaction of ester hydrolysis, with maximum yields of 50 to 75 percent typical for the equilibrium system resulting from an equimolar mixture of alcohol and acid. For example,

$$CH_3CH_2OH + HO\overset{O}{\overset{\|}{C}}CH_3 \rightleftharpoons CH_3CH_2O\overset{O}{\overset{\|}{C}}CH_3 + H_2O \qquad K_{eq} = \sim 2 \text{ (at } 25^\circ\text{C)}$$

Both the ester formation (*esterification*) and hydrolysis reactions are slow, but both are catalyzed by H^+, and, as a result, equilibrium can be established rather quickly (typically within a few hours) in the presence of a small amount of a strong acid. As we would expect, from an understanding of Le Chatelier's principle, the equilibrium yield of an ester can be increased by use of an excess of either the alcohol or the carboxylic acid. If an ester is particularly valuable, an excess of alcohol or acid (whichever is less expensive) is usually employed. Alternatively, it may be possible to improve the yield by removing either the ester or water from the reaction mixture as it is formed. Depending on the relative boiling points of the mixture components, it is sometimes possible to carry out the reaction at a temperature at which either the ester or water is preferentially distilled out of the reaction mixture.

An alternative synthetic method utilizes the reaction of an acid chloride with an alcohol to form the ester. For example,

$$CH_3CH_2OH + Cl\overset{O}{\overset{\|}{C}}CH_3 \longrightarrow CH_3CH_2O\overset{O}{\overset{\|}{C}}CH_3 + HCl$$

This reaction is, for all practical purposes, irreversible. Since anhydrous conditions are used, no hydrolysis reaction is possible and yields are typically excellent (usually >90 percent). Acid chlorides are relatively expensive and economic factors dictate that this method is rarely employed for large-scale industrial processes.

It is interesting to note that the alcohol and carboxylic acids play rather "unexpected" roles in direct esterification. Although alcohols are typically no more acidic than water, it is the alcohol that acts to furnish the H^+, whereas the acid (typically a proton donor) furnishes the OH^-:

$$CH_3CH_2O\mathbf{H} + \mathbf{HO}{-}\overset{O}{\overset{\|}{C}}CH_3 \longrightarrow CH_3CH_2O\overset{O}{\overset{\|}{C}}CH_3 + \mathbf{H_2O}$$

By the use of "isotopic labeling" (with oxygen-18), it has been established that the oxygen of the alcohol molecule is retained in the ester. It should be understood, of course, that esterification is not just an acid-base reaction. The mechanism for the proton-catalyzed reaction is well established, as illustrated for the formation of ethyl acetate:

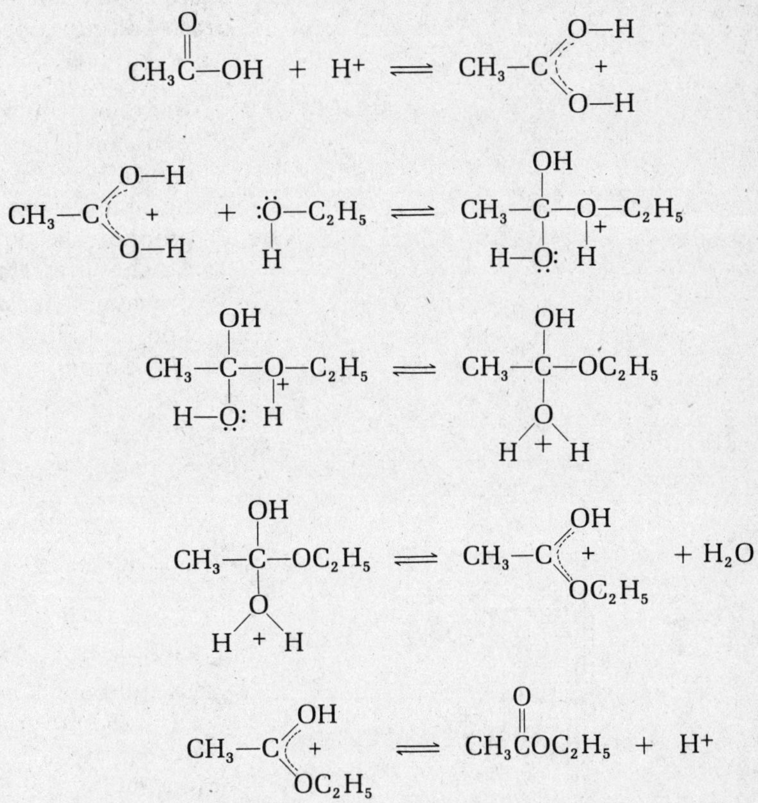

It is not necessary to have *protons* involved in the catalyzed reaction. Any *Lewis acid* (an electron-pair acceptor) can substitute for H^+ as a catalyst. The particular acid catalyst selected for any given esterification depends on a number of factors, such as solubilities, cost, and "efficiency" (as determined by comparisons of alternative procedures).

In this investigation, we will prepare an ester found in pears, *n*-amyl acetate (pentyl ethanoate), using zinc chloride as the Lewis acid catalyst.

Before starting this investigation, you should be familiar with the general characteristics of esters, alcohols, and carboxylic acids and with the considerations of equilibrium systems and stoichiometry applicable to esterification reactions. In addition, prior to coming to the laboratory, you must complete the calculations indicated in the Procedures section of this investigation.

References

1. O'Connor, Rod. 1977. *Fundamentals of Chemistry* (2nd ed.). New York: Harper & Row (Units 4, 17, 28, and 29 and Excursion 4).
2. O'Connor, Rod and Charles Mickey. 1974. *Solving Problems in Chemistry*. New York: Harper & Row (Units 4, 9 and 15).
3. Trahanovsky, Walter. 1971. *Functional Groups in Organic Chemistry*. Englewood Cliffs, N.J.: Prentice-Hall.

SAFETY Wear eye protection at all times and use all proper care to avoid ignition of the flammable vapors from the organic compounds used. Be particularly careful in use of ground-glass equipment. *Never* heat or *reheat* a liquid until after *fresh* boiling chips have been added to the *cool* liquid.

OBJECTIVES
1. Be able to make appropriate stoichiometric calculations, including considerations of limiting reagent, for a reaction of organic compounds.
2. Be able to properly assemble, use, and disassemble ground-glass equipment.
3. Be able to perform properly and safely the necessary procedures for an organic synthesis and product isolation (including the techniques of reflux, extraction, and distillation).

PROCEDURES

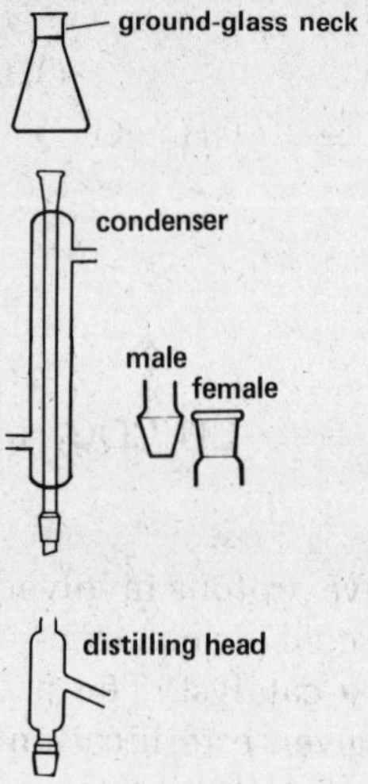

1. Obtain, clean, and dry the necessary items of ground-glass equipment, as directed by your instructor. You will need a 250-ml ground-glass-necked Erlenmeyer flask, a condenser with both "male" and "female" ground-glass joints, and a ground-glass distilling head.

Caution: Be *very* careful with this equipment. It is *expensive*.

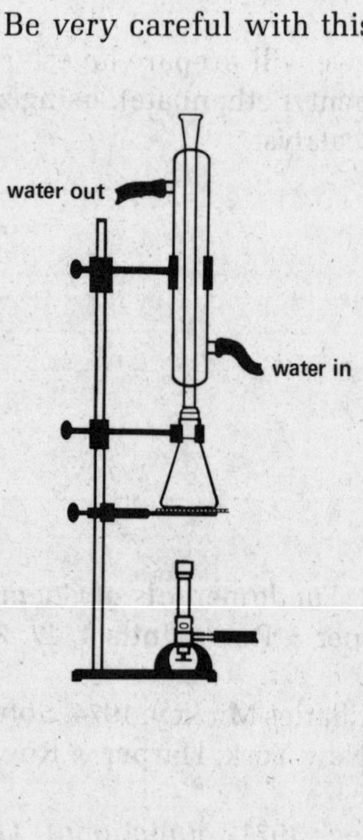

2. Moisten the tube connections on the condenser and attach the rubber condenser tubing snugly. Assemble the apparatus for a *reflux* setup, as shown, clamping the flask snugly in position with the flask supported by a wire gauze on a ring clamp. Do not clamp the condenser *tightly*. Attach the condenser clamp loosely to avoid a possible strain on the glass-to-flask connection. Have your instructor check the apparatus. Then remove the condenser and lay it in a safe place.

Calculate the *volumes* of acetic acid and 1-pentanol needed on the following basis:

a. You should use 0.10 mole of 1-pentanol. The density of this alcohol at room temperature is 0.814 g ml^{-1}.

b. You should use 0.17 mole of acetic acid for an improved equilibrium yield of ester (since this is the less expensive reagent). The density of acetic acid at room temperature is 1.05 g ml^{-1}.

3. Using a clean, dry graduated cylinder, measure the volume of acetic acid corresponding to 0.17 mole. *Carefully* test the odor of acetic acid. Describe this odor in your notebook.

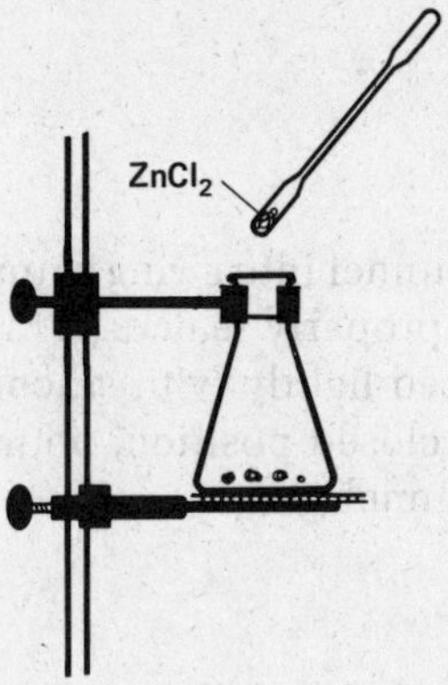

4. Place five spatula loads of anyhdrous zinc chloride into the dry ground-glass-necked Erlenmeyer flask. Then pour the measured acetic acid into this flask. Do not rinse the graduated cylinder.

Caution: Both these chemicals are corrosive.

Using your unrinsed graduated cylinder measure the volume of 1-pentanol corresponding to 0.10 mole, note and describe the odor of 1-pentanol *in its original container** and pour the measured alcohol into the flask. Add two or three fresh boiling chips.

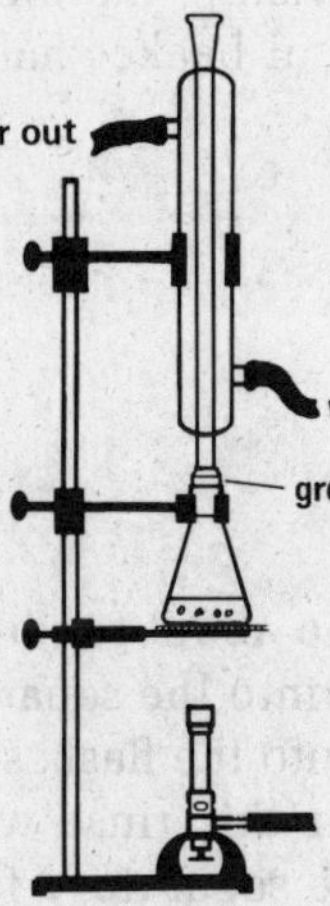

5. *Very* lightly grease the "male" joint of the condenser with a thin film of silicone grease. Insert the condenser into the flask and rotate the condenser to distribute the grease uniformly over the ground-glass joint. Secure the condenser with a clamp *just* tight enough to hold the condenser in place but *not* tight enough to place any strain on the apparatus. Connect the lower condenser tubing to a cold-water tap and place the end of the other condenser tubing (water outlet) in the sink or drain trough. Turn on the water *carefully* and adjust the flow rate so that a steady stream of water *runs* (not "gushes" or "trickles") out the drain tubing. Have your instructor check the apparatus.

*Since the graduated cylinder was previously filled with acetic acid, some of the odor of the acid will still be present.

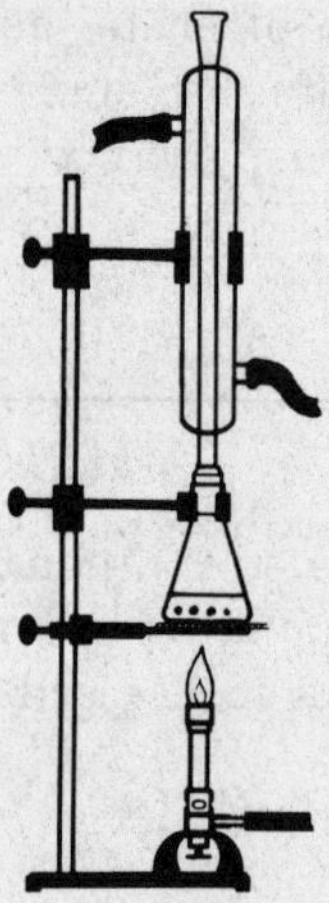

6. Light the burner and adjust the flame to achieve a slow rate of reflux (boiling of liquid with condensed vapor dripping back into the flask). Continue adjusting the flame until the reflux rate gives about two drops of condensate per second. Reflux the mixture at this rate for 20 min.

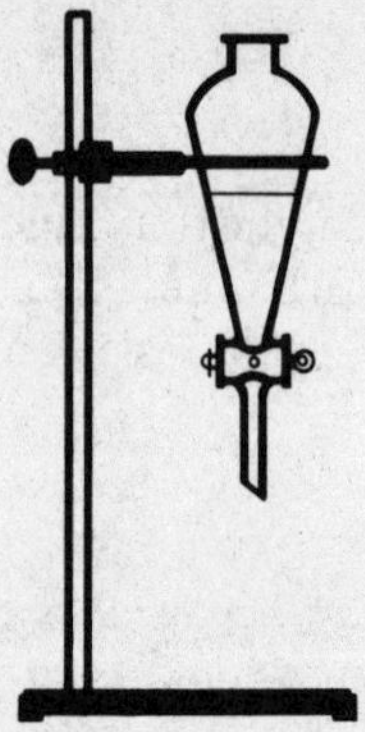

7. Set a clean separatory funnel into a ring clamp support. Check that the stopcock is properly seated. (If the stopcock is glass, it should be greased lightly with silicone grease.) With the stopcock in the closed position, pour 100 ml of distilled water into the funnel.

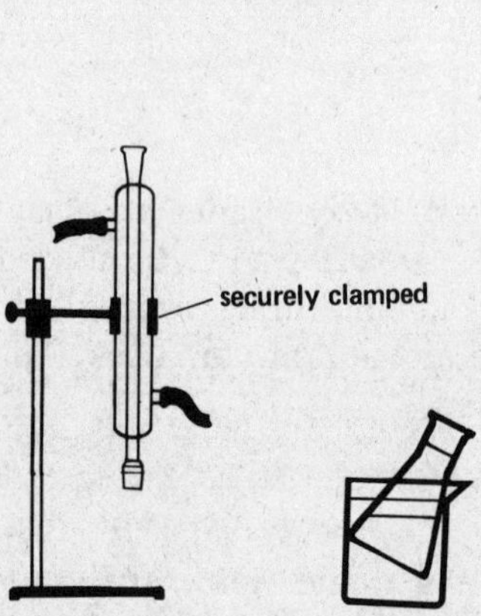

8. After the reaction mixture has refluxed for 20 min, turn off the burner. Continue running water through the condenser until the liquid in the flask has completely stopped boiling. Then turn off the water. Holding the flask neck securely (with a folded paper towel to protect your hand), loosen and remove the flask clamp. *While holding the flask neck,* tighten the condenser clamp so that it will support the condenser, then lower the ring clamp and turn it out of the way and, with a *gentle* twisting motion, remove the flask. Place the flask in a 600-ml beaker half full of cold tap water.

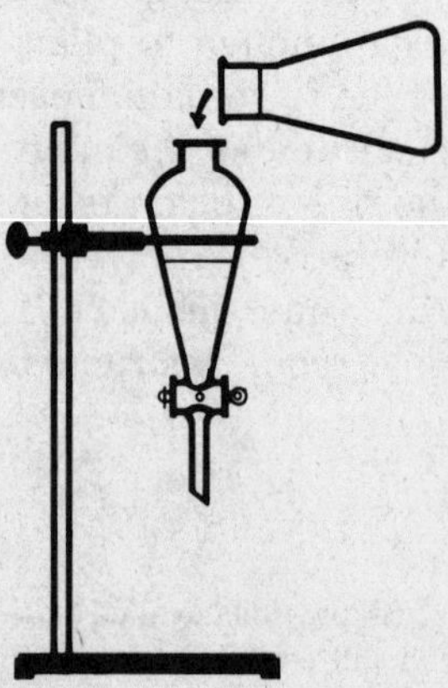

9. When the flask has cooled to about room temperature, pour the liquid from the flask into the separatory funnel. Pour 75 ml of distilled water into the flask, swirl the flask to rinse the interior, and pour this rinse water into the separatory funnel. Stopper the separatory funnel.

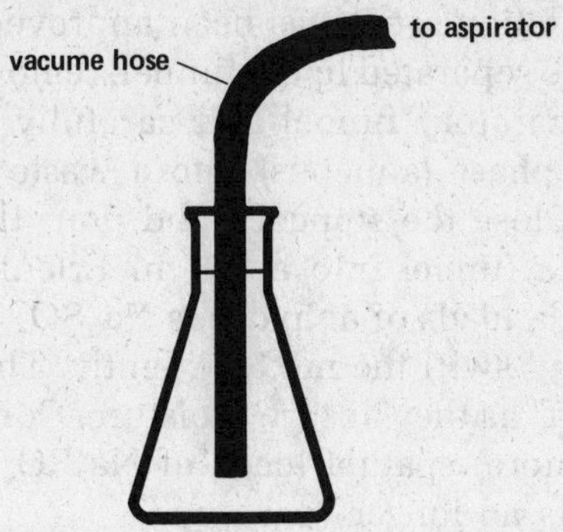

10. Wash the flask thoroughly with soap and water, then rinse it first with tap water and then with distilled water. Wipe the inside of the joint clean with a towel. Carefully shake the last of the rinse water from the flask, then rinse the flask with a small portion of acetone. Use an aspirator hose to pull air through the flask to evaporate all the acetone-water film.

Caution: Flammable vapors.

Review the proper technique for use of a separatory funnel, as described in Investigation 16.

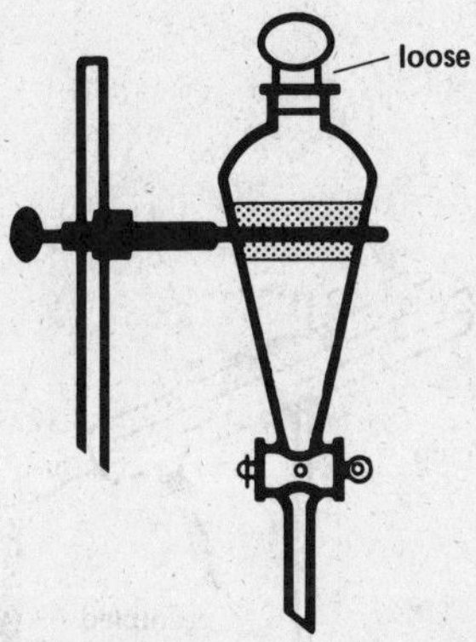

11. *Using proper techniques,* extract the reaction mixture thoroughly with the distilled water. Remember to hold the stopper *and* stopcock securely and to vent the funnel after its initial inversion and periodically during the shaking procedure. Then set the funnel back into the ring support, loosen (but do not remove) the stopper, and allow the funnel to stand until the two liquid phases have clearly separated.

12. While waiting for the separation, assemble a distillation apparatus as shown. All ground-glass connections should be greased lightly and the clamps should be arranged to hold the flask and condenser securely *but not to place any strain on the apparatus.* Before inserting the thermometer, grease the hole in the connecting tubing lightly. *Carefully* insert the thermometer, with a twisting motion, until the bulb is just opposite the side arm opening. Have your instructor check your setup.

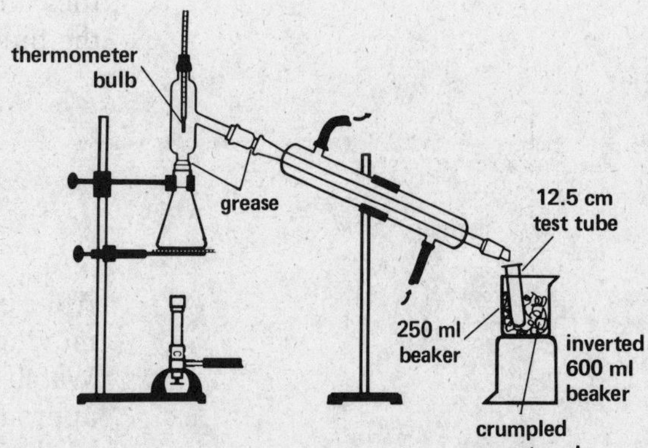

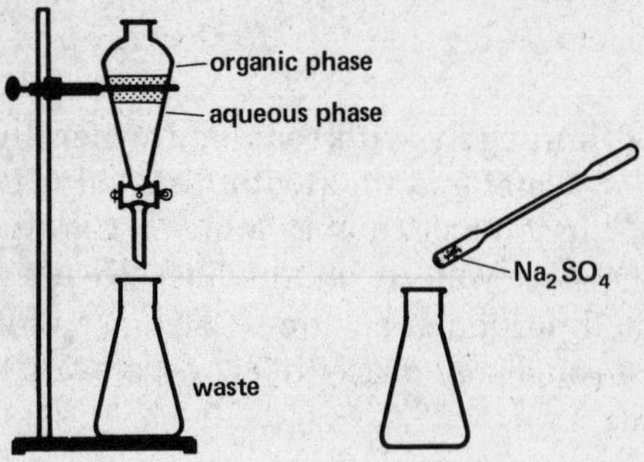

13. When your distillation setup has been approved and the liquid phases have separated in the funnel, remove the stopper from the separatory funnel and carefully drain *all* the lower liquid phase (aqueous) into a waste flask. Discard this phase. Close the stopcock and pour the organic phase from the funnel into a 125-ml Erlenmeyer flask. Add four spatula loads of anhydrous Na_2SO_4 to the liquid in this flask and swirl the mixture gently. The salt crystals will "cake up" as they absorb moisture. Continue adding (up to four more) spatula loads of Na_2SO_4 until additional increments no longer "cake up."

14. Carefully loosen the condenser clamp and disconnect the condenser and distilling head from the flask. Decant the organic phase from the Na_2SO_4 into the distilling flask through a funnel containing a small, loose cotton plug, being careful to keep the bulk of the Na_2SO_4 in its original flask. Add two or three fresh boiling chips to the distilling flask and reassemble the distilling apparatus. Adjust the proper water flow rate through the condenser. Your instructor will check your setup.

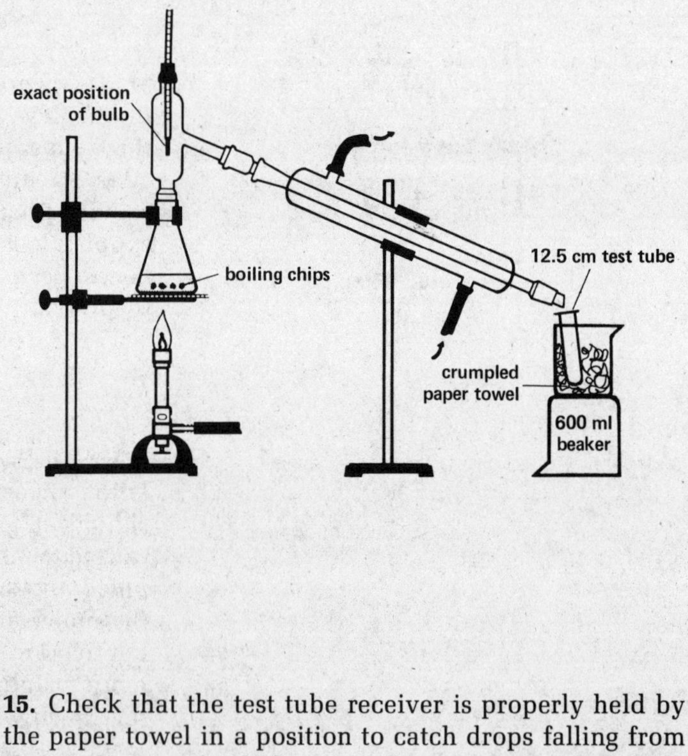

15. Check that the test tube receiver is properly held by the paper towel in a position to catch drops falling from the condenser. Check the water flow rate again, then light the burner and begin to heat the distilling flask carefully.

After boiling starts, you will observe a film of condensate moving gradually up the interior of the distilling head. Watch the thermometer reading carefully as the vapors approach the thermometer bulb. Record the temperature reading when the thermometer bulb is bathed with a thin

film of condensate and liquid just begins to collect inside the condenser. Adjust the heating rate so that drops fall from the condenser into the test tube receiver at the rate of about one drop every 2 to 3 sec. *Continue this heating rate* and check the thermometer regularly. When the temperature remains constant during collection of 10 to 12 drops, carefully remove the first test tube and replace it with a clean test tube. Continue the slow heating rate as long as the temperature reading remains within ±3°C of the "constant" reading. If the temperature changes by more than about ±3°C or if the receiver test tube is more than three-quarters full, change receivers. Keep a record of temperature ranges for all fractions collected and label the test tubes in order used. Continue distillation until nearly all of the ester has distilled. Under *no* circumstances should you heat the flask to dryness or allow the vapor temperature to exceed 160°C. (*n*-Amyl acetate boils around 149°C at 1.00 atm. The 1-pentanol boils around 137°C.) When distillation is completed, turn off the burner and water and let the apparatus cool to room temperature.

16. Weigh a 250-ml beaker on the triple beam balance. Place your first receiver test tube in the beaker and record the mass of the test tube and contents. Note and describe the odor of this fraction. Empty the test tube into the *specified* container and reweigh the empty tube. Repeat for each receiver test tube, recording the masses of each fraction collected. (Your instructor may specify separate containers for various fractions.)

17. After the distillation apparatus is cool, carefully disassemble the equipment. Wipe the grease film from all joints with a towel. Wash the distilling flask and the emptied test tubes thoroughly with soap and water. Rinse with tap water, then with distilled water (and, if directed, with acetone). *Do not* clean the distilling head or condenser unless you are given appropriate directions by your instructor.

CALCULATIONS AND REPORT

Use your notebook records and calculations to complete the Report Form. Attach separate pages containing details of your calculations and answers to the following questions:

1. What function was served by the zinc chloride used?
2. Why was the reaction performed under "reflux conditions"?
3. Why was excess acetic acid used?
4. On the basis of the relationship between molecular structure and water solubility, what was the purpose of the extraction procedure? Which components of the reaction mixture were probably removed in the aqueous phase?
5. Did you observe any evidence of 1-pentanol in the liquid placed in the distilling flask? Account for your answer in terms of the nature of the esterification reaction and the predistillation procedures.

GRADE / INSTRUCTOR'S INITIALS

Report Form

FOR INVESTIGATION 33:
Natural Flavoring Agent

NAME: ____________________ ID NO.: ____________________

LAB SECTION ____________________ INSTRUCTOR ____________________

Attach separate pages of calculations and answers to questions.

Calculated Quantities

0.10 mole of 1-pentanol corresponds to ________ g (________ ml)

0.17 mole of acetic acid corresponds to ________ g (________ ml)

The limiting reagent for the reaction used was ________

The theoretical* yield of *n*-amyl acetate was ________ g

The percentage† yield for this preparation was ________ %

Experimental Results

Volume of acetic acid actually used ________ ml

Odor of acetic acid ____________________

Volume of 1-pentanol actually used ________ ml

Odor of 1-pentanol ____________________

Odor of *n*-amyl acetate ____________________

distillation fraction number						
boiling range (°C)						
mass collected						
odor						

*Based on *actual amount used* of the limiting reagent.
†Based on fractions collected that appear to be the product ester.

(OVER)

On the basis of boiling range and odor, the fractions that appear to be principally the product ester were ______________________________

Total mass of fractions corresponding to the product ester ____________ g

_______________ DATE   _______________________ SIGNATURE

Investigation 34

THE KINETICS OF A PEROXIDE DECOMPOSITION

INTRODUCTION Hydrogen peroxide is an interesting reagent. Under appropriate conditions it may act as either an oxidizing agent or a reducing agent.

Oxidizing agent:

$$\overset{(-\mathrm{I})}{H_2O_2} + 2e^- \longrightarrow \overset{(-\mathrm{II})}{2OH^-}$$
$$H_2O_2 + 2e^- + 2H^+ \longrightarrow 2H_2O$$

Reducing agent:

$$\overset{(-\mathrm{I})}{H_2O_2} \longrightarrow \overset{(\mathrm{O})}{O_2} + 2H^+ + 2e^-$$
$$H_2O_2 + 2OH^- \longrightarrow O_2 + 2H_2O + 2e^-$$

Since oxygen in peroxides (as the −I oxidation state) is unstable with respect to both oxidation and reduction, hydrogen peroxide may *disproportionate* (undergo simultaneous oxidation and reduction):

$$2H_2O_2 \longrightarrow O_2 + 2H_2O$$

This disproportionation represents a hazard in terms of storing hydrogen peroxide, since the gas pressure resulting from formation of oxygen can cause a container to explode. The disproportionation reaction involves *free radicals* and these may be generated by exposure of H_2O_2 to light of the appropriate wavelength. To avoid the disproportionation problem, H_2O_2 is packaged in dark bottles, with directions to store the reagent in a dark, cool place. In some cases, inhibitors are added. These chemicals act to "scavenge" any free radicals formed before the disproportionation proceeds too far.

A number of methods are available for preparing hydrogen peroxide. A metal peroxide, such as Na_2O_2 or BaO_2, can be treated with a nonvolatile acid (typically H_2SO_4 or H_3PO_4) and the H_2O_2 generated can be obtained by careful distillation. For example,

$$Na_2O_2 + H_2SO_4 \longrightarrow Na_2SO_4 + H_2O_2$$
$$3BaO_2 + 2H_3PO_4 \longrightarrow Ba_3(PO_4)_2 + 3H_2O_2$$

A less expensive method, and one used commercially, involves the reaction of water with the peroxydisulfate ion formed by appropriate electrolysis of sulfuric acid:

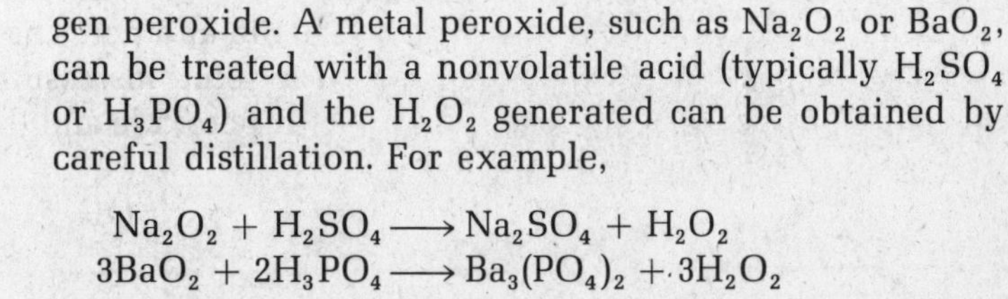

Hydrogen peroxide is sold as a 3-percent aqueous solution for medical use as an antiseptic, as a 6-percent solution for bleaching hair, and as 30-percent and 85-percent solutions for use in chemical laboratories.

The high chemical reactivity of H_2O_2 might suggest that this substance is unlikely to occur in living systems. On the contrary, hydrogen peroxide is a participant in a number of biochemical processes. Red blood cells, for example, may generate H_2O_2 under the stimulus of various absorbed oxidizing agents. Normally, the H_2O_2 is formed in only small concentrations and is rapidly reduced by a reaction utilizing a *peroxidase* and other enzyme systems. During the Korean War, a large number of black Americans developed acute hemolytic anemia as the result of H_2O_2 destruction of red blood cells. It was determined that a chemical used in the treatment of malaria was acting to stimulate excessive H_2O_2 production and that more than 10 percent of black males have a genetic deficiency in a particular enzyme, *glucose-6-phosphate dehydrogenase,* involved in the overall biochemical mechanism for removal of H_2O_2. As a result of the combination of high dosages of the antimalarial drug and the enzyme deficiency, accumulated H_2O_2 was destroying the red blood cell membrane.

In concentrated solution, H_2O_2 typically acts as a powerful oxidizing agent, capable of reacting with most organic materials. It would seem unlikely, then, that living organisms could "store" H_2O_2 in other than dilute solution. However, an insect, the *bombardier beetle,* does possess a suitable organic "storage chamber" for concentrated H_2O_2.

The bombardier beetle utilizes a rather unique chemical defense system. When alarmed or angered, the beetle forcefully ejects a spray propelled by oxygen gas from the decomposition of hydrogen peroxide:

$$2H_2O_2(aq) \longrightarrow 2H_2O(l) + O_2(g)$$

Although hydrogen peroxide will decompose spontaneously to oxygen and water, the reaction is very slow unless a catalyst is present.

The bombardier beetle stores hydrogen peroxide (about 7 *M*) in one small reservoir and hydroquinone (about 1 *M*) in another. The reaction involves the mixing of these two solutions, resulting in both the decomposition of hydrogen peroxide and the conversion of hydroquinone to quinone:

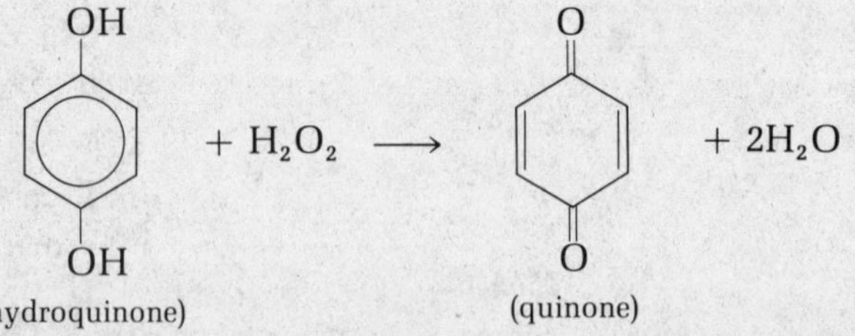

(hydroquinone) (quinone)

It has been demonstrated that quinone catalyzes both of the reactions, so that the "firing process" of the beetle is essentially autocatalytic. The quinone formed is not very soluble in water, so some of this product is expelled as a puff of smoke following the ejection of the spray. Some of the solid quinone apparently remains behind in the "mixing chamber" as a "primer" catalyst for the next time the beetle needs to mix its reagents.

In this investigation we shall study two reactions involving hydrogen peroxide, the I^--catalyzed disproportionation and the H^+-catalyzed oxidation of I^- by H_2O_2. Under conditions of pH 7, iodide ion acts to catalyze the disproportionation of H_2O_2, so I^- does not appear in the net equation for the process. It does, however, participate in the reaction and will appear in one or more of the steps (*elementary processes*) of the reaction mechanism. At low pH, acid-catalyzed oxidation of I^- by H_2O_2 occurs. For this case, a different mechanism of reaction must be suggested. By studying the effects of reactant concentrations on reaction rate, we may be able to obtain clues to plausible reaction mechanisms.

Before starting this investigation, you should be familiar with the concepts and calculations involved with kinetics data, rate laws, and reaction mechanisms, and with the principles of oxidation and reduction. In addition, you should review "gas law" calculations.

References

1. O'Connor, Rod. 1977; *Fundamentals of Chemistry* (2nd ed.). New York: Harper & Row (Units 10, 16 and 22).
2. Gardiner, W. C., Jr. 1969. *Rates and Mechanisms of Chemical Reactions.* Menlo Park, Calif: W. A. Benjamin.

SAFETY Wear approved eye protection at all times and be careful in working with reagents. Be especially cautious with the apparatus used for generating and collecting gases, checking that there is no possibility of generating a gas in a "closed system."

OBJECTIVES

1. Be able to accurately obtain appropriate kinetics data, including solution concentrations, gas volumes, and time intervals.
2. Be able to apply appropriate "correction factors" to gas volumes to account for temperature and pressure effects, and for the effect of collecting a gas over water.
3. Be able to interpret appropriate kinetics data in terms of *rate law* expressions.
4. Be able to suggest reaction mechanisms consistent with kinetics data and other experimental information.

You will be assigned a partner for this experiment. The two of you should go over the procedures involved and plan in advance how best to divide the work, particularly when time factors are critical. Before starting experimental work, prepare appropriate data tables in your notebook for procedures A and B.

PROCEDURES

A. The I^--catalyzed Disproportionation of H_2O_2

1. Prepare 100 ml of a 0.15 *M* KI solution.

2. Attach a buret clamp parallel to the direction of the base of a ringstand and about 30 cm above the base. Attach two other clamps near the top of the ringstand at angles of about 45 deg to the direction of the first clamp. Connect

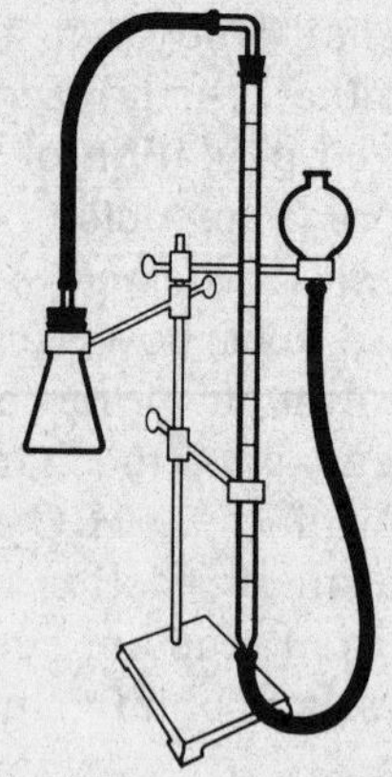

one end of a 60-cm length of rubber tubing to the lower end of a 50-ml buret and fasten the buret securely in the center clamp so that the lower end of the buret is about 15 cm above the ringstand base. Lubricate a size No. 00 one-hole rubber stopper and a previously prepared 90-deg-bent glass tube and, protecting your hands with a towel, insert the glass tube *carefully* into the stopper until the end of the tube just protrudes below the stopper. Repeat this procedure with a straight piece of glass tube and a one-hole stopper to fit the 125-ml Erlenmeyer flask.

If you are not thoroughly familiar with the proper technique for inserting glass tubing through a stopper (Appendix A), consult your instructor *before* attempting it.

Now fasten the 125-ml flask securely in the upper left clamp, connect the two stoppers by a 60-cm length of rubber tubing, and insert the stoppers into the top of the buret and the flask, respectively. Finally, connect the tubing from the bottom of the buret to a leveling bulb and fasten the bulb securely in the right-hand clamp, being sure that the rubber tubing is not constricted by the clamp.

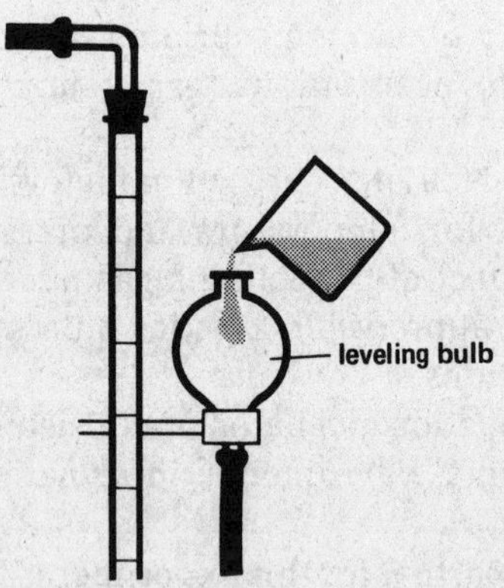

3. Remove the stopper from the flask. With the leveling bulb clamped securely, pour into it 125 ml of distilled water at room temperature. Loosen the clamp at the ringstand and alternately raise and lower the leveling bulb until all the air has been expelled from the lower tubing and no further bubbles rise in the buret. Clamp the bulb in its original position.

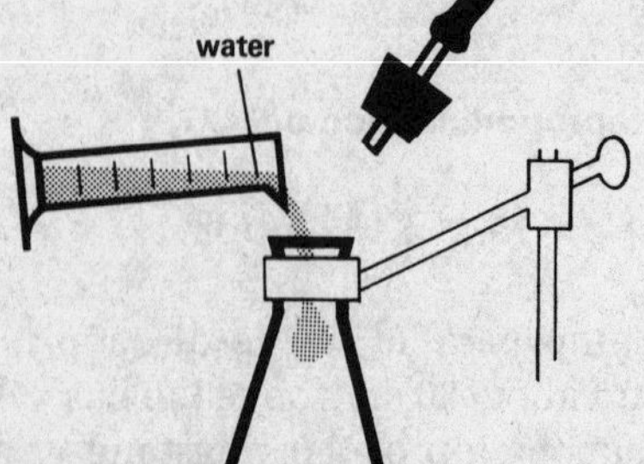

4. Carefully pour 20 ml of the KI solution into the Erlenmeyer flask, recording the volume added (to the nearest 0.5 ml). Then add 20 ml of distilled water at *room temperature*.

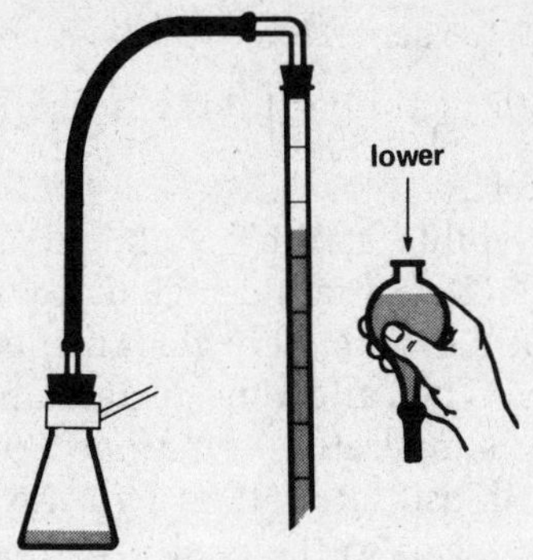

5. Stopper the Erlenmeyer flask and lower the leveling bulb to about 10 cm below its original position. If your system is gas tight, the liquid level in the buret should drop to somewhere *above* the level in the bulb. If this does not occur, check the system for leaks. (Check both stoppers carefully to be sure they fit snugly.) If you continue to have a gas leak, consult your instructor.

Plan the next steps carefully and perform them *quickly*.

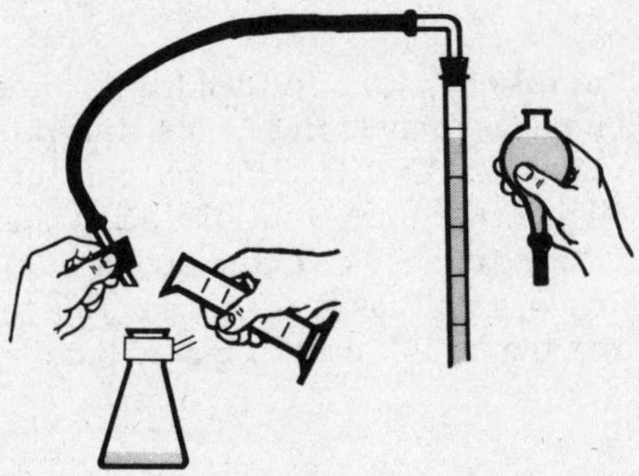

6. Unstopper the flask, add 10 ml of 3 percent H_2O_2 *all at once*, and *quickly* stopper the flask. While the stopper is being inserted, the leveling bulb must be held so that the top of the liquid column in the buret is just below the zero mark.

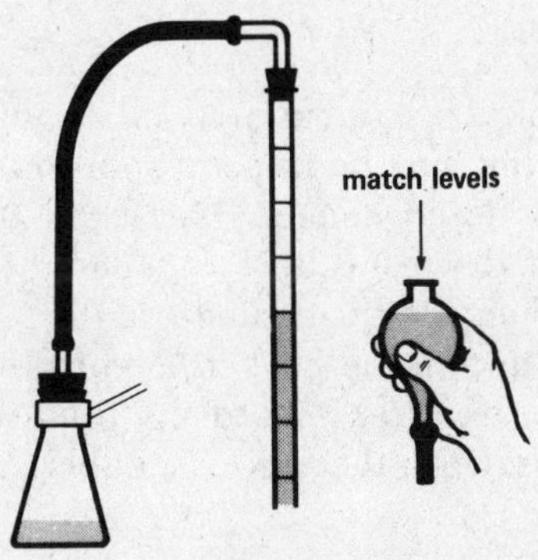

7. Remove the flask from its clamp and swirl the flask *gently*. As oxygen is evolved, the level of water in the buret will drop. To keep the pressure inside the system equal to that of the external atmosphere, adjust the height of the leveling bulb so that the level of water in the bulb is the same as that in the buret. When the pressure-equalized water level in the buret has dropped by 3 ml, record the *time* as "time zero." Continue recording time at 2-ml intervals until about 17 ml of O_2 have been collected. Label the record of time and volume measurements, in your notebook, as Trial 1.

8. Adjust the leveling bulb to permit you to disconnect the flask from the system without loss of water from the buret. Remove and rinse the flask thoroughly with distilled water. *Using the volumes specified for Trials 2 and 3*, repeat procedures A4–A7 two more times. Record all data as before.

	Procedure A.4		Procedure A.6
	Volume KI (ml)	Volume H_2O (ml)	Volume of 3% H_2O_2 (ml)
Trial 2	20	10	20
Trial 3	40	0	10

B. The H^+-Catalyzed Oxidation of I^- by H_2O_2

1. Prepare the following three solutions and label each with the letter indicated:

Solution A: 20 ml 3 percent H_2O_2 + 40 ml distilled water.

Solution B: 0.6 g KI + 59 ml distilled water + 15 drops of 3 *M* HCl.

Solution C: 0.1 g "soluble starch" + 10 ml distilled water. [Boil this mixture *gently*, with stirring, until a completely clear dispersion is obtained. Then allow the dispersion to cool to room temperature. Add one crystal of sodium thiosulfate (about half the size of a match head) to the cool dispersion and stir to dissolve the crystal.]

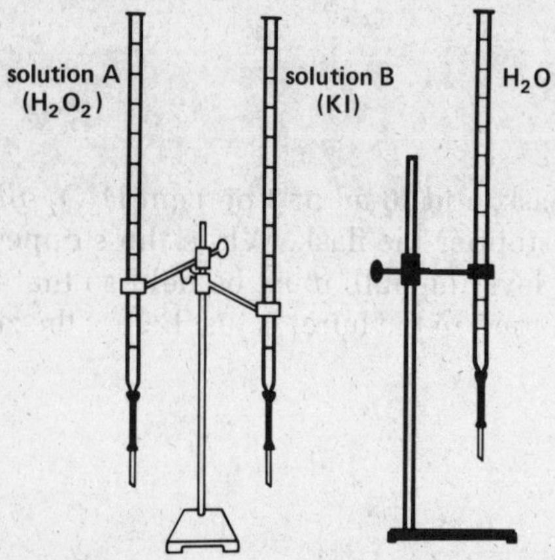

2. Obtain and, if necessary, clean (Appendix A) three burets. Clamp the burets securely to ringstands. Rinse and fill one buret with solution A (H_2O_2), the second buret with solution B (KI), and the third with distilled water. Label each buret as to its contents. Drain enough liquid from each buret into a "waste" flask to expel air from the buret tip and to drop the liquid level to below the zero mark.

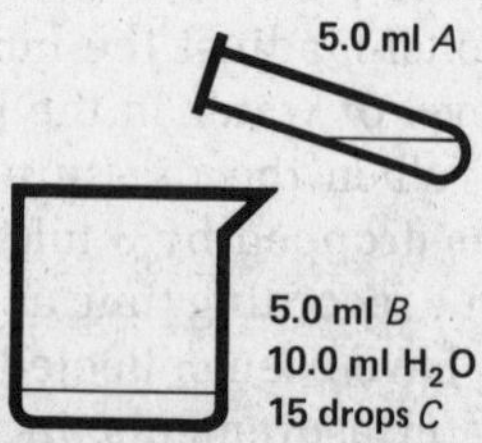

3. Into a *clean, dry* 100-ml beaker, add 5.0 ml of solution B, 10.0 ml of distilled water, and 15 drops of solution C. Stir the mixture until it is homogeneous. Measure 5.0 ml of solution A into a clean, dry test tube. Be prepared to measure a time interval. Then add the contents of the test tube *all at once* to the solution in the beaker. Record the time and begin stirring the mixture. Record the time when a blue color is *first* observed in the mixture. Label the data as Trial 1.

4. Rinse and wipe dry the beaker, test tube, and stirring rod used. *Using the volumes specified for Trials 2–5*, repeat procedure B3 four more times. Record all data as before.

	Volume Solution B (to beaker, first) (ml)	Volume dist. H_2O (to beaker, second) (ml)	Volume Solution C (to beaker, third)	Volume Solution A (all at once, from test tube) (ml)
Trial 2	10.0	5.0	15 drops	5.0
Trial 3	10.0	0	15 drops	10.0
Trial 4	5.0	5.0	15 drops	10.0
Trial 5	5.0	0	15 drops	15.0

CALCULATIONS AND REPORT The reaction studied in procedure A may be formulated as

$$2H_2O_2(aq) \xrightarrow{(I^- \text{ catalyst})} 2H_2O(l) + O_2(g)$$

Since we know this reaction involves I^- catalyst, it is probable that the reaction proceeds by a series of steps with the "slow" (rate-determining) step using I^- as a reactant and some subsequent, more rapid, step regenerating I^-. More sophisticated experiments suggest that molecular iodine (I_2) is a transient species in the overall process.

Using your data from procedure A, plot, for *each trial*, elapsed time (as the x-coordinate) versus volume of O_2 collected. Use the initial reading (time at 3 ml collected) as the zero time and zero-volume point. Construct the best straight line for the data in each trial, noting that kinetics data are generally more reliable during the initial part of the reaction than near the end of the reaction. Determine the slope of each line(y/x). Noting that these slopes are *rate expressions* (as milliliters O_2 per minute), determine the apparent reaction order in H_2O_2 and in I^-. Assuming that only these species appear in the rate-law expression, write the rate law for this reaction. Then suggest a plausible mechanism consistent with the overall reaction stoichiometry, the rate law, and the information that I_2 is a transient species in the process.

Analyze the data obtained in procedure B. Discuss the correlations, if any, of these data with the information obtained from procedure A. Given the following equations representing processes known to be involved in the "clock reaction" studied in procedure B, suggest a reason for the sudden ("alarm-clock") appearance of the blue color.

$$2H^+ + 2I^- + H_2O_2 \xrightarrow{(H^+ \text{ catalyst})} I_2 + 2H_2O$$

$$I_2 + 2S_2O_3^{2-} \xrightarrow{(\text{fast})} 2I^- + S_4O_6^{2-}$$

$$I_2 + \text{starch} \xrightleftharpoons{(\text{rapid, reversible})} \underset{(\text{blue complex})}{[I_2\ \text{starch}]}$$

Use your notebook records and calculations to complete the Report Form. Attach extra pages with the required calculation details and discussions.

GRADE	INSTRUCTOR'S INITIALS

Report Form

FOR INVESTIGATION 34:
The Kinetics of a Peroxide Decomposition

NAME: ______________________ ID NO.: ______________________

LAB SECTION ______________________ INSTRUCTOR ______________________

Prepare the required data graphs on the back of this page. Attach calculations and discussions on separate pages.

A. The I^--Catalyzed Disproportionation of H_2O_2

Trial No.	Initial $C_{H_2O_2}$ in reaction mixture*	Initial C_{I^-} in reaction mixture	Rate (from slope) [ml (O_2) per min]
1			
2			
3			

*3% H_2O_2 = 0.9 M H_2O_2

Suggested mechanism:

Suggested rate law
Rate of O_2 formation = $k \times$ ______________________

B. The H^+-Catalyzed Oxidation of I^- by H_2O_2

Trial No.	Initial $C_{H_2O_2}$ in reaction mixture	Initial C_{I^-} in reaction mixture	Time (min) for "completion"
1			
2			
3			
4			
5			

______________________ DATE ______________________ SIGNATURE

Investigation 35

INSTRUMENTAL DETERMINATION OF GLUCOSE

INTRODUCTION A major end product of photosynthesis is the compound glucose, a simple sugar. The glucose molecule ($C_6H_{12}O_6$) contains a definite pattern of carbon, hydrogen, and oxygen, as described in Investigation 5. All of these constituents come from a flow of raw materials, carbon dioxide, and water, absorbed by the plant from the atmosphere and the soil. Glucose is produced by plants and it is a most essential animal body chemical. The oxidation of glucose supplies the energy for muscle and gland activity and for body heat. The soluble glucose is maintained in the blood stream at a concentration of about 0.1 percent by the regulatory action of the liver. The liver converts excess glucose reversibly to "animal starch," glycogen $(C_6H_{10}O_5)_x$, which is a polymer of glucose and, as necessary, the liver reconverts the glycogen to glucose. Conversion of excess glucose to fatty acids and fats for more permanent storage in adipose tissue is another important biochemical process.

When there is some malfunction in the body's mechanisms for control of glucose concentration in the blood or of glucose metabolism, serious medical problems may result. In *hyperglycemia*, a condition resulting from improper regulation of the conversion of stored glycogen to glucose, the concentration of glucose in the blood is well above normal levels. A deficiency in the biochemical production of insulin leads to the improper regulation of glucose metabolism, resulting in the condition known as *diabetes*. In both diabetes and hyperglycemia, excess glucose accumulation shows up in elevated levels of "blood sugar" and when a sufficient glucose excess occurs, excreted glucose may be detected in the urine. Clinical tests for glucose concentrations in body fluids are important for clues in medical diagnosis and for monitoring the treatment of various conditions such as diabetes.

Although simple chemical tests are often satisfactory for detecting glucose and even for estimating approximate glucose concentrations in body fluids, instrumental methods are routinely employed when rapid and accurate analyses are needed. Spectrophotometric assays are particularly useful.

When light of appropriate wavelength is passed through a sample, the emerging beam of light will be of lower intensity than that which entered the sample. The loss in intensity (i.e., the *absorbance*) is directly related to the ability of the sample to absorb light of that wavelength, the concentration of the solution, and the distance the light beam must traverse through the solution. This is expressed by Beer's law as $A = \epsilon bC$, where A represents absorbance, ϵ represents the *molar absorptivity* (a measurement of the compound's ability to absorb light of the given wavelength), b represents path length through which the light beam must pass through the solution, and C is the molar concentration.

Beer's law is often used to determine the concentration of a sample by comparison of absorbance of a sample of unknown concentration to the absorbance of a sample of known concentration. This is accomplished by using a specific wavelength of light and a pair of matched sample holders (cuvets) that will have a fixed path length through which the light beam must pass. Under these conditions the two equations

$$A_{\mathrm{Kn}} = \epsilon b C_{\mathrm{Kn}} \qquad \text{and} \qquad A_{\mathrm{Unkn}} = \epsilon b C_{\mathrm{Unkn}}$$

can be combined, since ϵ and b are now constants, to yield

$$\frac{A_{\mathrm{Unkn}}}{C_{\mathrm{Unkn}}} = \frac{A_{\mathrm{Kn}}}{C_{\mathrm{Kn}}}$$

When the absorbance of the solution of known concentration and the absorbance of the solution of unknown concentration have been determined by the use of a spectrometer, the concentration of the unknown can be found.

Instruments used for such a determination vary in complexity and ease of operation. However, they all contain the following components.

Radiation source. The radiation source must generate a light beam with sufficient intensity for easy detection and measurement. This source must also produce radiation that contains all wavelengths over the region in which it is to be used. A tungsten lamp is the radiation source most often used for absorption studies within the visible region of the spectrum.

Dispersing device. Various dispersing devices are employed to select the particular wavelength or set of wavelengths desired. These devices may be filters, prisms, or gratings.

Sample holder. Sample holders may vary from test tubes, which have been matched and tested for uniformity of wall thickness and optical path length, to high quality cuvets constructed from quartz or other suitable material.

Radiation detector. Radiation detectors are devices that produce an electric current which is proportional to the amount of radiation reaching them. The electric current is then amplified and used to drive a pen on a chart recorder or to deflect the needle on a calibrated scale.

Controls, mirrors, and electronic devices. The instrument must also contain a number of devices that enable one to select radiation of a desired wavelength and to adjust the output of the signal obtained.

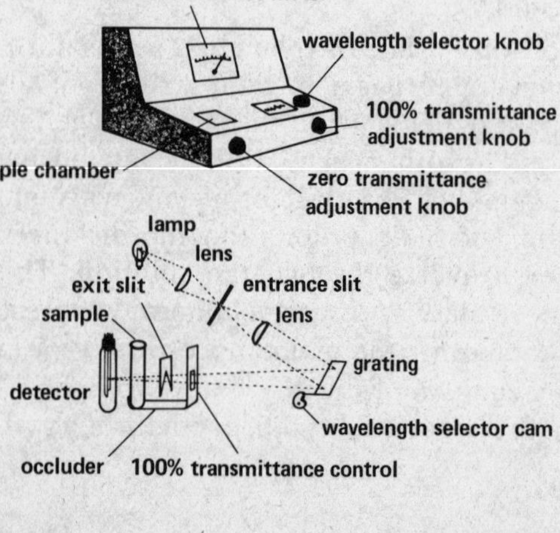

There are many different types of instruments in common use. This investigation has been designed for the Bausch and Lomb Spectronic 20. If you are to use a different instrument, your instructor will provide suitable modifications in the experimental procedures.

In this investigation, we will determine glucose concentration on the basis of a *decreased* absorption resulting from reaction of glucose with a light-absorbing species. Since glucose itself does not absorb light in the visible region of the spectrum, we must depend on an "indirect" determination. (Alternative procedures are available that use absorption *increase* resulting from the formation of a colored species by reaction of glucose with some appropriate colorless reagent.)

Before starting this investigation, you should be familiar with the fundamental principles of absorption spectroscopy, particularly with respect to the visible region of the spectrum, and with calculations based on Beer's law.

References

1. O'Connor, Rod. 1977. *Fundamentals of Chemistry* (2nd ed.). New York: Harper & Row (units 30 and 31).
2. Dyer, J. R. 1965. *Applications of Absorption Spectroscopy to Organic Compounds*. Englewood Cliffs, N.J.: Prentice-Hall.

SAFETY You will be working with some potentially dangerous reagents. Wear approved eye protection at all times. In case of accidental contact with chemicals, wash *immediately* with water. Exercise special precautions when heating solutions and when using the centrifuges.

OBJECTIVES

1. Be able to handle reagents and volumetric glassware properly.
2. Be able to make quantitative determinations using a spectrometer.
3. Be able to make calculations involving Beer's law.
4. Be able to prepare and interpret data graphs.

PROCEDURES

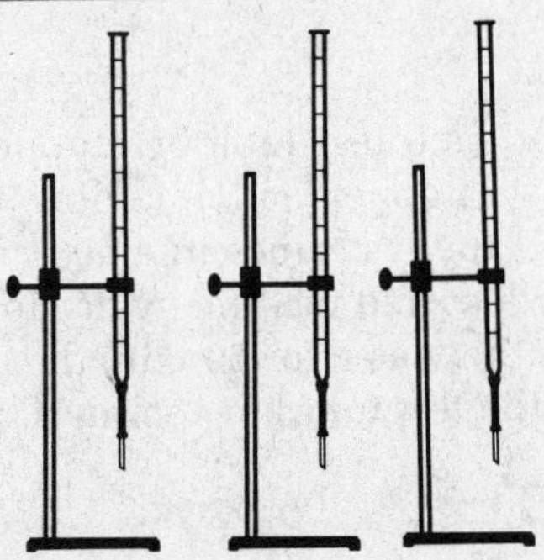

1. Working in groups designated by your instructor, set up three clean burets.

2. Weigh a 90 to 100-mg sample of glucose to the nearest 0.1 mg on the analytical balance. Record the mass of glucose used. Transfer the sample completely into a clean 100-ml volumetric flask already containing about 50 ml of distilled water. Use your squeeze bottle to rinse any glucose from inside the flask neck down into the solution. Swirl the flask carefully until all glucose has dissolved. Then fill the flask with distilled water to the 100-ml mark, stopper the flask snugly, and mix the solution thoroughly by inverting the flask several times.

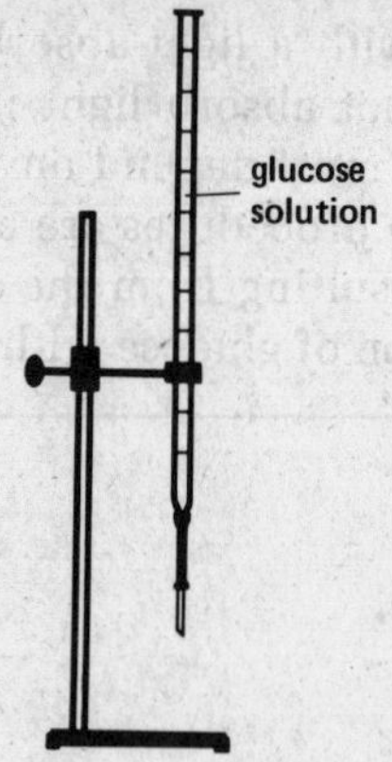

3. Rinse one of the burets twice with 5-ml portions of the glucose solution, then use the solution to fill this buret. Attach a label reading "Glucose." Rinse and fill a second buret with the Benedict's solution provided and label the buret appropriately. Rinse and fill the remaining buret with distilled water. Drain enough liquid from each buret into a "waste" flask to expel air from the buret tip and to drop the liquid level to below the zero mark.

4. Clean and flame dry seven test tubes of the appropriate size (volume >7 ml). After the test tubes have cooled, label them, near the top, with a grease pencil with numbers 1 through 7. Drain from the "glucose" buret the following volumes of solution into the respective test tubes, recording each volume in your notebook to the nearest 0.01 ml.

Tube no.	1	2	3	4	5	6	7
Volume glucose	0	0.5	1.0	2.0	2.5	3.0	4.0

Obtain two test tubes containing "unknown glucose" solutions from your instructor. Label these test tubes as directed and record the volume of each solution as indicated by your instructor. To each of the nine test tubes, add 1.0 ml of Benedict's reagent and a sufficient volume of distilled water to make a *total* volume of 7.0 ml in each test tube. Record all volumes added to the nearest 0.01 ml. Arrange the volume data in your notebook in a neat table format.

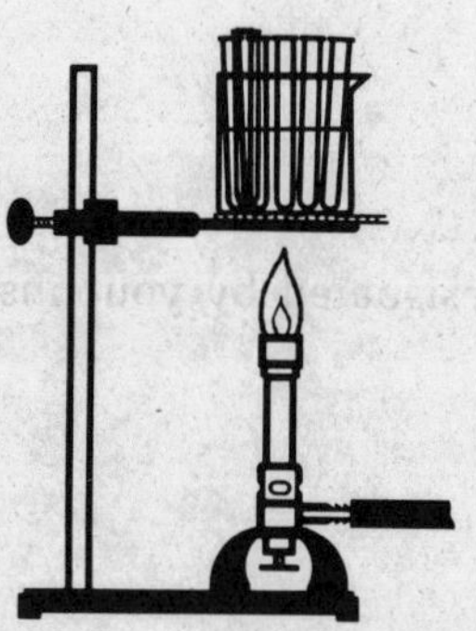

5. Mix the contents of each test tube by stoppering the tube with a *clean, dry* stopper and inverting the tube several times. Place all the test tubes in a 400-ml beaker containing 150 ml of water and heat the water to a gentle boil. Continue boiling the water for 30 min, then turn off the burner and allow the test tubes to cool until they can be handled comfortably.

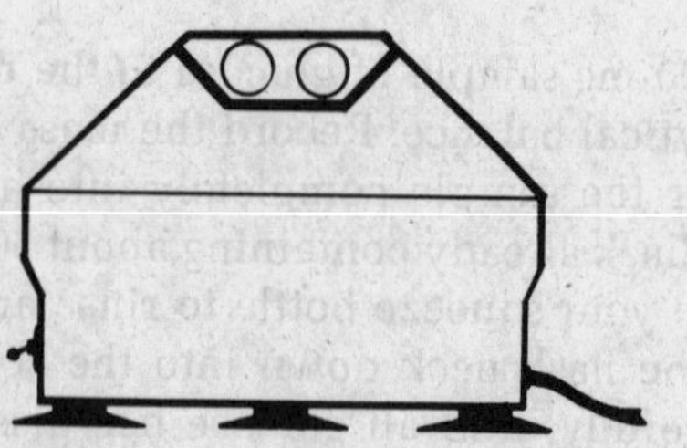

6. Centrifuge each test tube for 2 min. (Check Appendix A for proper centrifuge techniques.) Place the centrifuged tubes where they will not be disturbed.

Be sure you have received proper instructions for use of the spectrometer before proceeding further.

7. Prepare a data table in your notebook in the form shown below.

Tube No.	1	2	3	4	5	6	7	Unkn 1	Unkn 2
Absorbance									

8. Using the assigned spectrometer and a cuvet issued by your instructor, proceed as follows:

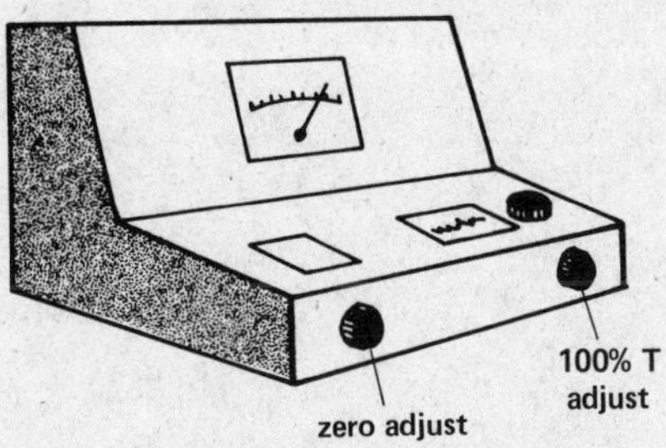

a. Adjust the wavelength selector knob for a wavelength setting of 735 mμ.

b. With the sample chamber empty and closed (unless directed otherwise by your instructor), adjust the "zero transmittance" knob for a meter reading of zero transmittance (100-percent absorbance).

c. Rinse and fill the cuvet with distilled water. Insert the cuvet into the sample chamber with the mark on the cuvet positioned toward the front of the instrument. Adjust the "100-percent transmittance" knob for a meter reading of 100-percent transmittance (zero absorbance).

d. Repeat the procedures for zero and 100-percent transmittance adjustment until no further adjustments are necessary when you switch from one to the other.

9. Now empty the cuvet and rinse it with a few drops of the supernatant solution from your test tube No. 1. Then fill the cuvet at least half full with this solution, being careful to decant the solution from any precipitate in the test tube without stirring up the precipitate. If bubbles adhere to the inside of the cuvet, shake the cuvet *gently* to dislodge the bubbles. Insert the cuvet properly into the sample chamber and record the *absorbance* reading shown on the meter dial. Repeat this procedure with each of your solutions, rinsing the cuvet each time with a few drops of the solution to be used. Record all absorbance data in the table prepared in your notebook. Show the final results to your instructor and ask for advice as to any data that should be discarded.

CALCULATIONS AND REPORT Use your notebook records to complete the Report Form. On the graph provided, plot absorbance versus molarity (of glucose) for your "known" solutions (omitting any points indicated by your instructor as "unreliable data"). Construct the straight line that best corresponds to those points representing absorbance greater than zero. Using this graph and the absorbance data for your two unknowns, determine the molarity of glucose in each unknown.

Attach extra pages showing all calculations.

Your instructor may describe the chemical reactions involved in this experiment. Alternatively, you may be assigned a "library project" to find these reactions.

GRADE / INSTRUCTOR'S INITIALS

Report Form

FOR INVESTIGATION 35:

Instrumental Determination of Glucose

NAME: ____________________ ID NO.: ____________________

LAB SECTION ____________________ INSTRUCTOR ____________________

Show all calculations on attached pages.

Glucose Standard

Glucose + weighing paper ________ g

Weighing paper ________ g

Glucose: ________ g

Concentration of standard glucose solution ________ *M*

Tube No.	1	2	3	4	5	6	7	Unknown 1	Unknown 2
Volume glucose (ml)	0.00	0.5_	1.0_	2.0_	2.5_	3.0_	4.0_	____	____
Volume Benedict's reagent (ml)	1.0_	1.0_	1.0_	1.0_	1.0_	1.0_	1.0_	1.0_	1.0_
Volume distilled water (ml)	6.0_	5.5_	5.0_	4.0_	3.5_	3.0_	2.0_	____	____
Absorbance									
Glucose concentration (*M*)								X	X

(OVER)

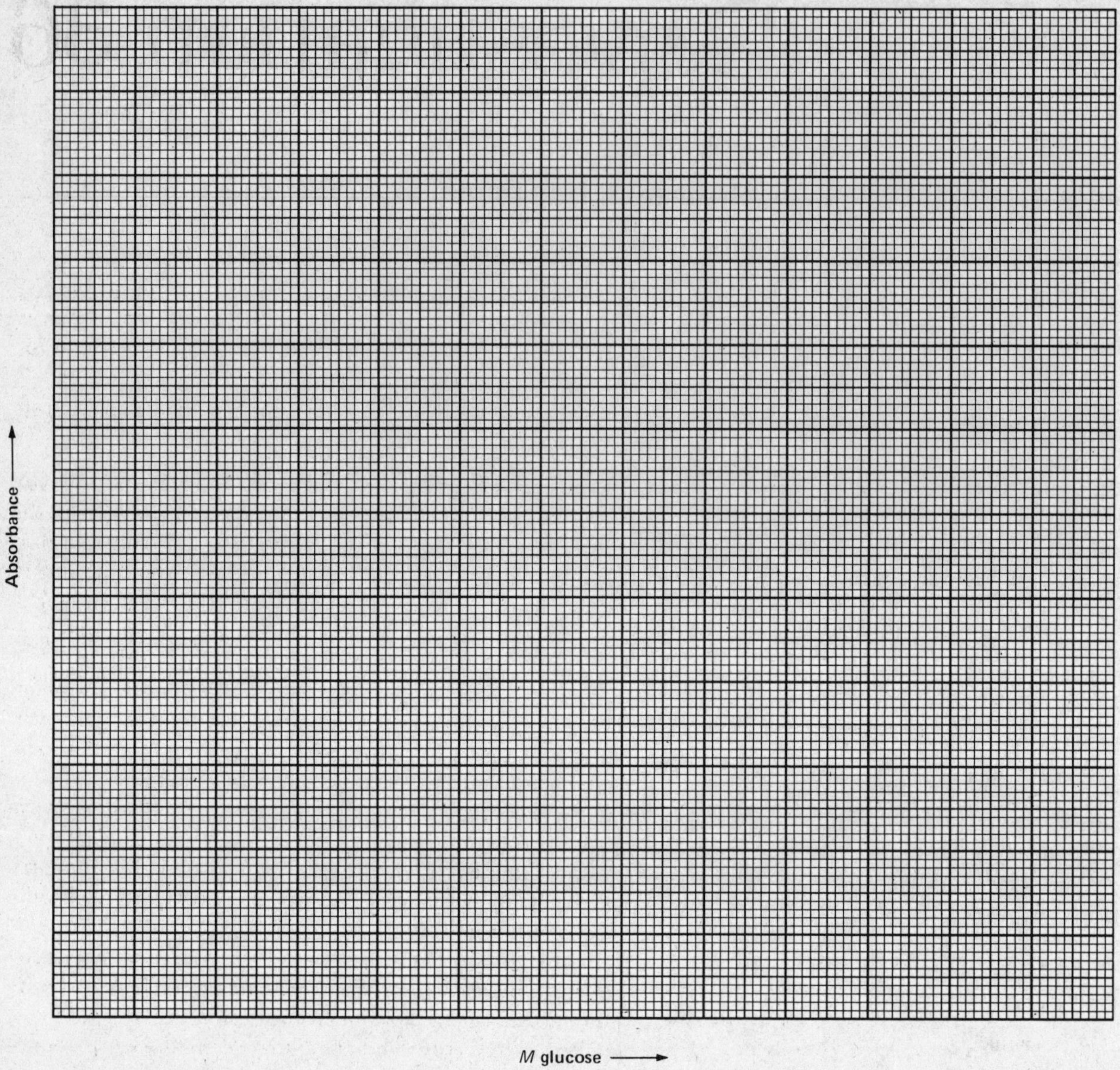

Unknown Solutions

From graph, for the total reaction mixture

Unknown #1 ____________ M

Unknown #2 ____________ M

By calculation, the original solution as obtained

Unknown #1 ____________ M

Unknown #2 ____________ M

DATE ____________ SIGNATURE ____________

Investigation 36

RAYON: A VALUABLE POLYMER

INTRODUCTION There are three types of fibrous polymers:

1. The natural polymers such as silk (a polyamide) or cellulose, as in cotton, (a polysaccharide).
2. The man-made polymers such as nylon (a polyamide) or Dacron (a polyester).
3. The man-modified polymers such as cellulose nitrate or rayon (both modified celluloses).

Until the mid-nineteenth century, fabrics were woven from natural fibers of cotton, wool, silk, or linen (from the celluloses of the flax plant). Although silk, from the harvesting of silkworm cocoons, was relatively abundant in the Orient, it was an expensive fiber for the Western world. Various attempts to make "synthetic silk" as competition for the high-priced imported fiber met with little success until 1855, when the Swiss chemist Audemars produced cellulose nitrate fibers from mulberry leaves. Unfortunately, cellulose nitrate proved to be highly flammable. Although various forms of the material, including a transparent sheet used for motion picture film stock, were produced commercially by the late nineteenth century, a series of disastrous fires eventually closed this industry and, by 1950, the cellulose nitrate was no longer profitable on any large scale.

In 1857 it had been discovered that cellulose could be dissolved by a solution of tetraaminecopper(II) hydroxide (Schweitzer's reagent). Further investigations of this system eventually revealed that unique fibers were reconstituted when the solution was squirted into acid. Fabrics woven from such fibers had the glossy sheen and soft texture of silk. A patent was granted for this process in 1890 and the "artificial silk" industry was born, with a German company as the first commercial manufacturer. In 1895, an English group obtained a patent for the production of similar fibers by a process involving the reaction of celluloses in an alkaline solution with carbon disulfide. This "viscose" process was soon highly competitive with the earlier "cuprammonium" process.

The rapid development of worldwide markets for the relatively inexpensive "artificial silk" soon resulted in manufacturing plants in every industrialized nation. With recognition of the chemical differences between the reconstituted cellulose fibers of "artificial silk" and the protein (polyamide) fibers of natural silk, various new names were suggested for the synthetic material. In 1924 *rayon* was adopted as the official U.S. name for the reconstituted cellulose fibers. Variations in process details and in further chemical modifications have led to many different types of *rayon* polymers, but all are referred to as *rayons* if they meet the U.S. Federal Trade Commission's definition as "A manufactured fiber composed of regenerated cellulose, as well as manufactured fibers composed of regenerated cellulose in which substituents have replaced not more than 15 percent of the hydrogens of the hydroxyl groups."

Today *rayons* account for nearly half of the world production of synthetic fibers. Both the "cuprammonium" and "viscose" processes are still in use, but the latter is most commonly employed because of its versatility and low cost. In addition to its extensive use in woven fabrics, rayon fiber is used as a reinforcing material in automobile tires and other rubber products, in production of specialty papers and "disposable" fabrics, and in a host of other materials. With improved forest management, it has become less expensive to produce rayons from the celluloses of wood pulp than from cotton. This should provide new areas of agricultural land for the raising of food crops.

In this investigation, we shall employ the early "cuprammonium" process to prepare a sample of rayon fiber. This process depends on the fragmentation of large cellulose polymers in an alkaline medium, the formation of a stable complex of Cu^{2+} with ammonia and cellulose units, and the regeneration of long cellulose polymers by extrusion of the complex solution into dilute sulfuric acid. Some of the $CuSO_4$ and $(NH_4)_2SO_4$ formed may initially adhere to the cellulose fibers, but these salts may be washed out by rinsing the fibers with distilled water. The fibers we will prepare will not have all the characteristics of commercial rayon, since we will omit the "aging" process and other treatments used industrially.

Before starting this investigation, you should be familiar with the general characteristics of natural and synthetic polymers and of complex ions such as tetraaminecopper(II).

References

1. O'Connor, Rod. 1977. *Fundamentals of Chemistry* (2nd ed.). New York: Harper & Row (Units 27, 31, and 33).
2. Margerison, D. and G. C. East. 1967. *Introduction to Polymer Chemistry.* New York: Pergammon Press.
3. *Encyclopedia of Polymer Science and Technology,* Volume 11. 1969. *Rayon.* New York: Interscience Publishers (pp. 810–845).

SAFETY Wear approved eye protection at all times. Be very careful in working with reagents, particularly concentrated ammonia and solutions of NaOH and of H_2SO_4. In case of accidental contact with chemicals, wash immediately and thoroughly with soap and water.

OBJECTIVES

1. Be able to work safely with hazardous reagents and with glass equipment.
2. Be able to prepare a sample of a synthetic polymer of good quality.

PROCEDURES

1. On a triple beam balance, weigh out about 7.5 g of $CuSO_4 \cdot 5H_2O$ crystals. Use a clean mortar and pestle to grind the crystals into a fine powder.

2. Transfer the powdered $CuSO_4 \cdot 5H_2O$ into a clean 250-ml beaker, add 50 ml of distilled water, and stir the mixture vigorously until a homogeneous solution is obtained.

Note: If a fume hood is available, perform all subsequent procedures involving ammonia in the hood. Ammonia vapors are penetrating and unpleasant, at best. They can also be harmful to your eyes. *Under no circumstances may you wear contact lenses while exposed to ammonia fumes.* (Cases of eye damage have resulted from this unsafe practice.) Avoid inhaling ammonia vapors.

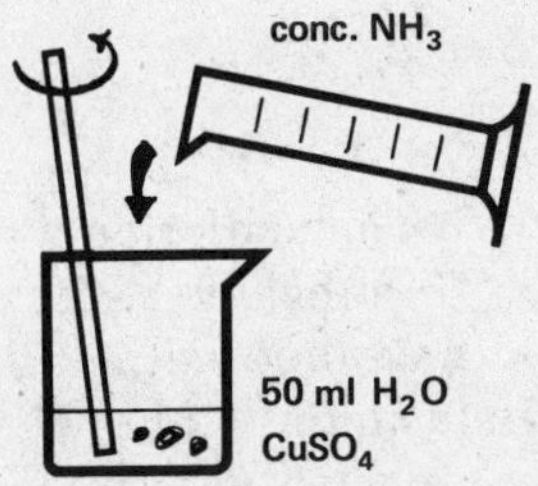

3. Measure 4 ml of concentrated (~15 *M*) aqueous ammonia with your graduated cylinder. Pour this solution *slowly, with gentle stirring,* into the copper(II) sulfate solution to precipitate the bluish-white $Cu(OH)_2$. Avoid adding excess ammonia.

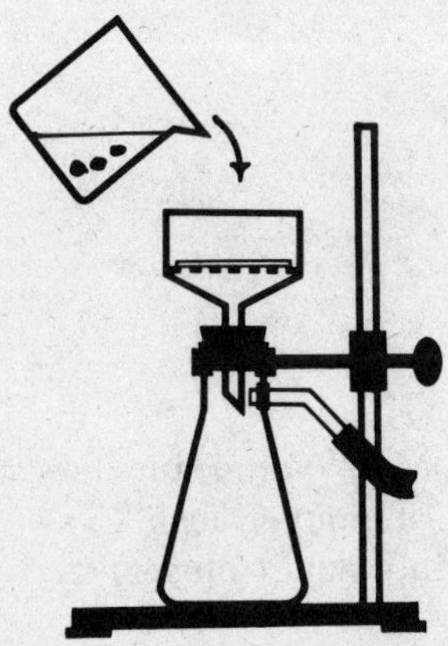

4. Allow the $Cu(OH)_2$ precipitate to settle for a few minutes. During this time, select a piece of filter paper that will fit your Buchner funnel and weigh the paper. Record the mass to the nearest 0.01 g. Assemble the vacuum filtration equipment. Filter the $Cu(OH)_2$ mixture by first decanting the liquid through the filter, under vacuum, and then transferring the solid to the filter.

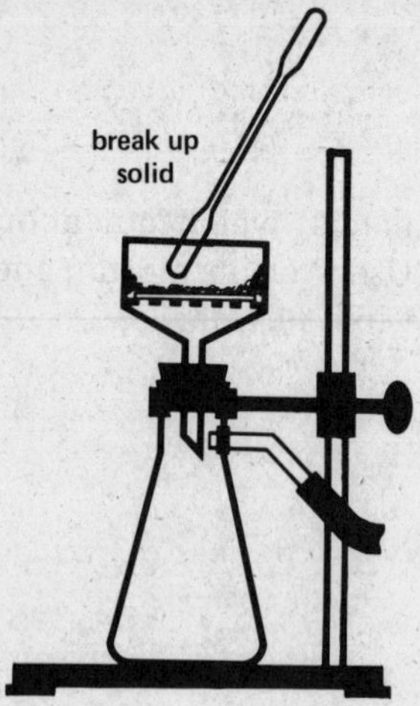

5. Rinse the beaker with small portions of distilled water to transfer any residual solid to the filter, then wash the solid on the filter with small portions of distilled water. (The total distilled water washes should be about 200 ml.) After each wash is poured into the filter, carefully break up the solid and stir it around. Then use strong suction to pull the wash water through the filter.

6. Disconnect the vacuum, carefully remove the Buchner funnel, and pour the wash liquid from the flask. Rinse the flask with a small portion of distilled water, then reassemble the filtration equipment. Wash the solid one more time with 2 to 3 ml of distilled water. Pour this rinsing from the flask into a test tube and add five or six drops of dilute HCl.

7. Add five or six drops of barium chloride solution to the acidified rinse solution. If a heavy white precipitate ($BaSO_4$) is observed, you have not washed the $Cu(OH)_2$ sufficiently free from sulfate anions and further washings must be performed until repeated tests indicate a negligible SO_4^{2-} contamination.

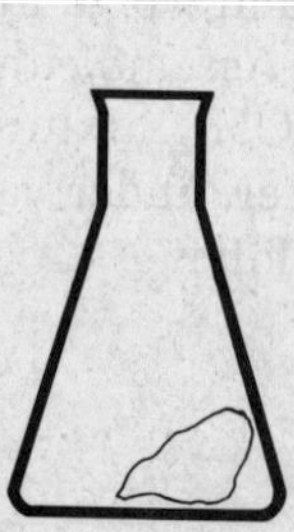

8. Transfer the filter paper containing the sulfate-free $Cu(OH)_2$ into a 125-ml Erlenmeyer flask. Use your spatula to tear the filter paper into small pieces.

9. Working at the hood, measure out 30 ml of concentrated aqueous ammonia and pour this into the flask containing the paper and $Cu(OH)_2$. Add to this mixture 0.5 g of additional paper (or 0.5 g of "untreated" cotton) that has been torn into tiny pieces. Stopper the flask securely and swirl the mixture *gently* for 5 min. Set the flask aside, returning to swirl it periodically while you work on procedures 10–12.

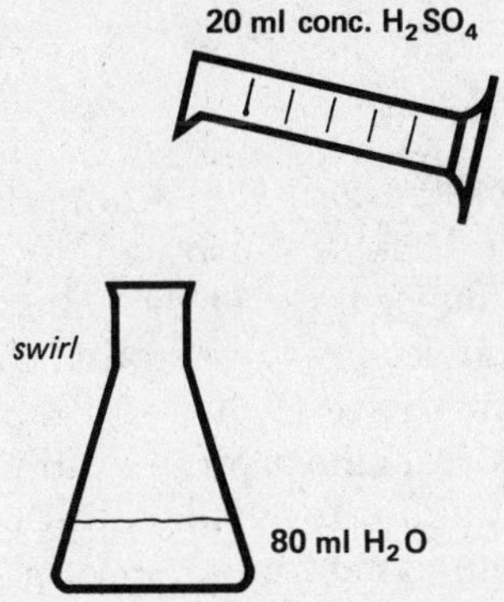

10. Prepare a dilute H_2SO_4 solution by *very carefully and slowly* pouring 20 ml of concentrated (18 *M*) H_2SO_4 into 80 ml of distilled water contained in a 250-ml Erlenmeyer flask, while swirling the flask gently to mix the contents. Cool the final solution to room temperature by running cold tap water over the outside of the flask while swirling the flask gently. (Be careful in washing the H_2SO_4 from the graduated cylinder.)

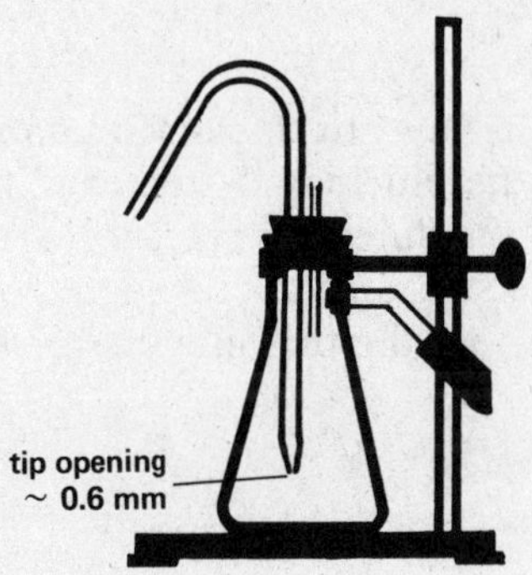

11. Using *safe* and correct techniques, cut, constrict, bend, and fire polish the glass tubing as needed for the apparatus shown. (If you have any questions on working with glass tubing, consult your instructor *before* starting work. Be careful to avoid burning yourself *or* the bench top with hot glass.) After the prepared tubing has cooled, lubricate the holes in the stopper and the part of the tubing to be inserted with glycerol. Protect your hands with a towel and insert the tubing through the stopper with a *gentle* twisting motion. *Do not exert undue force.* If you encounter problems, consult your instructor. Complete the assembly of the apparatus shown and have the setup checked by your instructor.

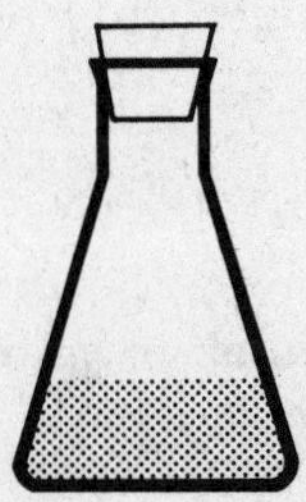

12. Check the mixture in your reaction flask. If it has not been recently disturbed, you should observe an intensely colored excess of Schweitzer's reagent (tetraaminecopper(II) hydroxide) above the more viscous mixture containing the complexed cellulose. The mixture should have "aged" for about 30 min and all paper (and cotton, if used) should have been dissolved before the mixture is ready for spinning the rayon fiber. Swirl the mixture gently. It should be fairly viscous ("syrupy"), but not so thick as to pose any problems by plugging the constricted glass tubing. If the mixture appears to be too "gummy," add a few more milliliters of concentrated ammonia.

Caution: Avoid inhalation or eye contact of vapors.

Swirl the mixture carefully during addition of the ammonia.

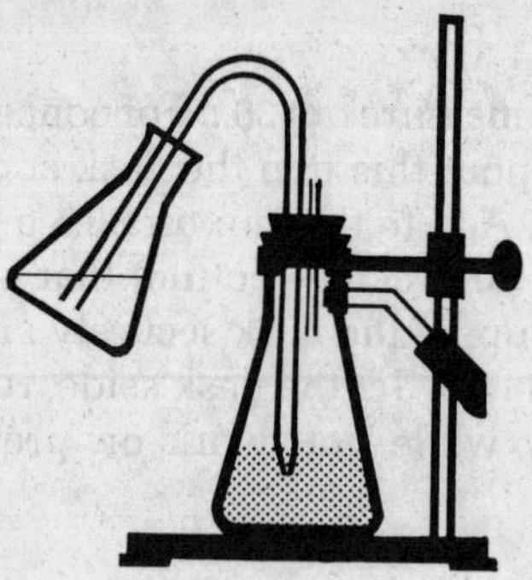

13. You should now be ready to spin your rayon fiber. Pour the prepared dilute H_2SO_4 solution into the suction flask to a depth such that the constricted glass tubing will extend below the surface of the acid.

Caution: Be *certain* you are using *diluted* H_2SO_4. *Concentrated* H_2SO_4 may react explosively with the rayon mixture.

Remove the stopper from the flask containing the rayon mixture and, holding the flask as illustrated, turn on the water aspirator. Place a finger tip over the short piece of glass tubing and adjust finger pressure so that the rayon mixture is drawn into the dilute H_2SO_4 at a rapid, even rate. Disconnect the flask from the aspirator tubing *immediately* on finishing the spinning to avoid pulling too much air through the mixture or having water "back up" into the flask.

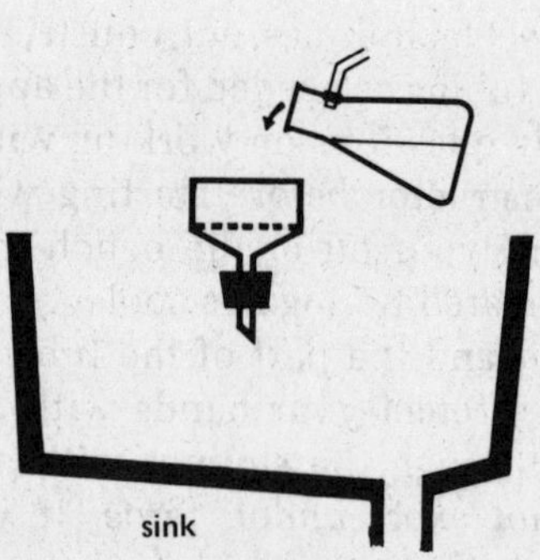

14. Allow the rayon "threads" to remain in contact with the dilute H_2SO_4 until the blue color is leached from the fibers. Then hold your Buchner funnel over a sink and pour the contents of the suction flask through the funnel. Rinse the collected fibers on the funnel with tap water.

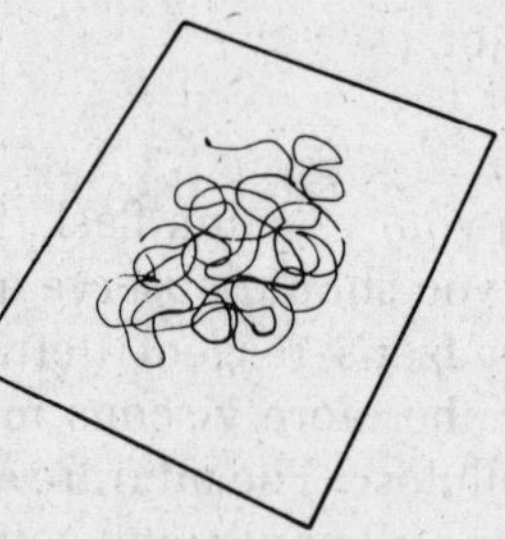

15. After the wash water has drained from the fibers, transfer the moist fibers onto a preweighed paper towel and weigh the towel and fibers on a triple beam balance. Record the mass of rayon obtained and have your instructor inspect your product.

CALCULATIONS AND REPORT Use your notebook records and observations to complete the Report Form. Show all calculations on the back of the Report Form.

GRADE

INSTRUCTOR'S INITIALS

Report Form

FOR INVESTIGATION 36:

Rayon: A Valuable Polymer

NAME: ______________________ ID NO.: ______________________

LAB SECTION ______________________ INSTRUCTOR ______________________

Show all calculations on the back of this form.

Experimental Data

Mass of $CuSO_4 \cdot 5H_2O$ used ____________ g

Mass of single filter paper ____________ g

Mass of additional paper ____________ g

Rayon + towel ____________ g

Towel ____________ g

Rayon (moist) ____________ g

Calculation

Assuming the conversion of paper (and cotton, if used) to rayon was 60 percent efficient, what percentage of the "moist rayon" collected was water?

____________ %

Observations

Instructor's evaluation of rayon ☐ good ☐ average ☐ poor ____________ Initials

Description of rayon obtained

____________ DATE ____________ SIGNATURE

Investigation 37

CAFFEINE IN NONPRESCRIPTION DRUGS

INTRODUCTION Caffeine is one of the most widely used nonprescription drugs, both in single-ingredient and multicomponent preparations. Among the more familiar products containing caffeine as the only "active" ingredient are the stimulants No-Doz and Vivarin, the latter also containing dextrose (glucose) as an energy source. Several well-known "analgesic" (pain-relieving) products contain caffeine in combination with aspirin (e.g., Modacin), phenacetin (e.g., Empirin), both aspirin and acetaminophen (e.g., Trigesic), both aspirin and cinnamedrine hydrochloride (e.g., Midol) or with a mixture of salicylamide, acetaminophen, pyrilamine maleate, and homatropine methylbromide (e.g., Femicin).

The properties and preparation of caffeine have already been described,* but it is worth reiterating that caffeine is a potent stimulant for which the effects of habitual use are not yet completely established.

When a drug is dispensed in tablet form, only a portion of the tablet is the drug itself. Inert additives, called "fillers" and "binders," are employed to maintain a proper consistency for the mixture to be pressed into stable tablets. In this experiment we will examine one of the nonprescription tablets containing caffeine by isolating the chloroform-soluble caffeine from chloroform-insoluble components such as aspirin or inert fillers. This procedure will enable us to determine the caffeine content of the tablet.

Before starting this investigation, you should be familiar with the principles and techniques of liquid–liquid extraction, distillation, and sublimation.

References

1. O'Connor, Rod. 1977. *Fundamentals of Chemistry,* (2nd ed.). New York: Harper & Row (Unit 18 and Excursion 4).
2. Bates, Robert B. and John P. Schaefer. 1971. *Research Techniques in Organic Chemistry.* Englewood Cliffs, N.J.: Prentice-Hall.

SAFETY Wear approved eye protection at all times. Do not inhale chloroform vapors. Exercise caution in use of the separatory funnel, the distillation equipment, and the sublimation apparatus. Be particularly careful with the hot oil bath. (*Vapors and smoke from the heated oil are easily ignited.*) Do not taste or swallow any of the drugs issued. They may be contaminated.

OBJECTIVES

1. Be able to perform properly and safely the techniques of liquid–liquid extraction, distillation, and sublimation.

*See "Caffeine in a Chloroform-Water System" (Investigation 16).

2. Be able to isolate and purify caffeine from a mixture in which the other components are insoluble in chloroform.

PROCEDURES

Obtain from your instructor three tablets of the drug to be assayed for caffeine. Weigh the tablets to the nearest 0.1 mg.

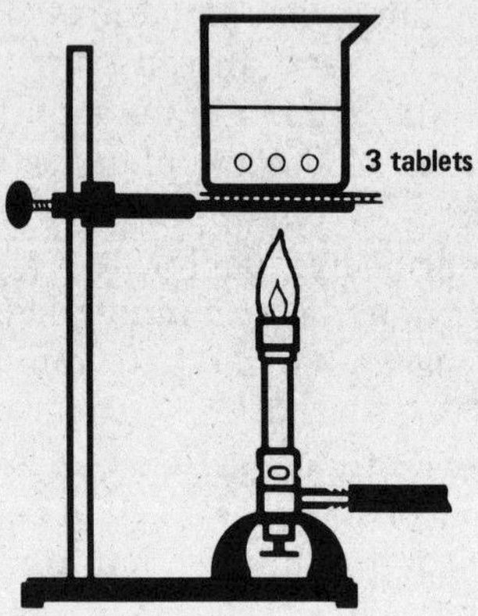

1. Place the weighed tablets in a clean 250-ml beaker and add 150 ml of distilled water and two or three boiling chips. Boil the mixture for 10 min to ensure solution of the caffeine. Then turn off the burner.

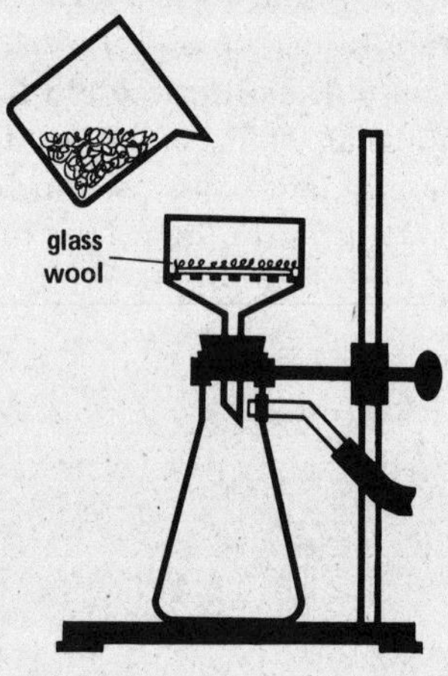

2. Using a folded paper towel for insulation, filter the hot mixture through a Buchner funnel containing a thin mat of glass wool to remove insoluble "filler."

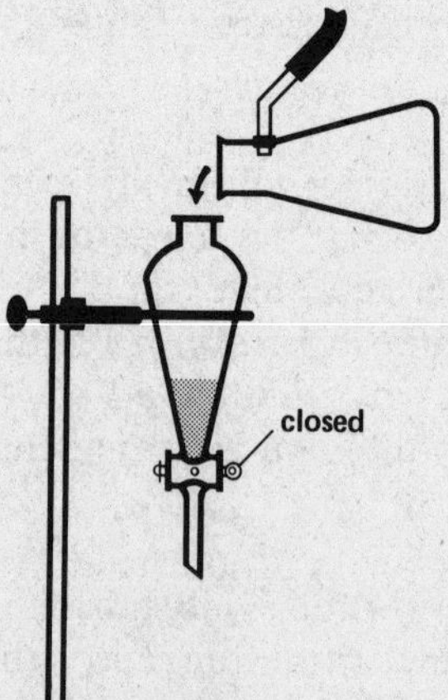

3. Allow the filtrate to cool to room temperature, then transfer it into a clean separatory funnel (stopcock closed). Rinse the vacuum flask twice with 5-ml portions of distilled water, transferring these rinses into the separatory funnel.

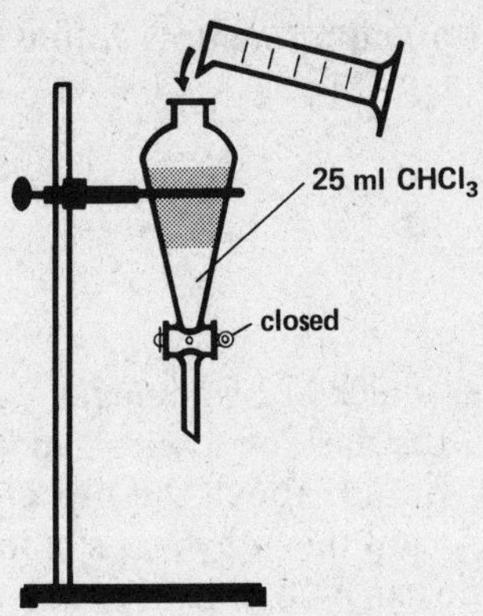

4. Add 25 ml of chloroform to the separatory funnel, then stopper the funnel. Remove the funnel from the ring and, holding the stopper with one hand and the stopcock with the other, invert the funnel and open the stopcock to vent pressure from solvent vapors.

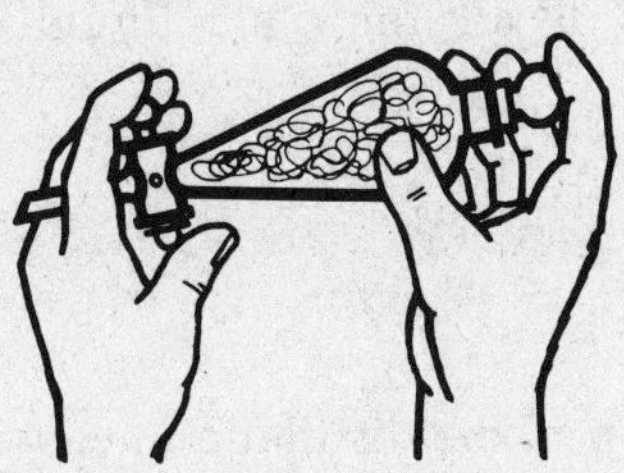

5. Rock the funnel up and down from a horizontal position for 5 min, venting periodically to relieve pressure. (Avoid vigorous shaking, which can produce an intractable emulsion.)

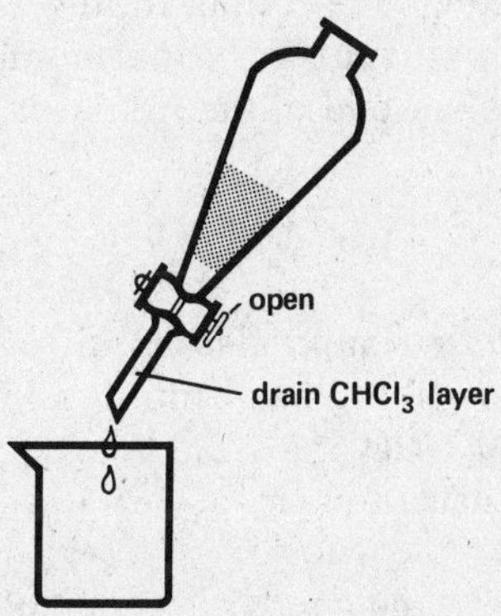

6. Return the funnel to the ring support and allow it to stand until the two liquid phases have separated. Then remove the stopper and set it on a clean watch glass. Drain the lower chloroform phase into a clean 250-ml beaker, being careful that none of the aqueous phase drains from the funnel.

Repeat procedures 4–6 twice more, collecting the total of 75 ml of chloroform extract in the same beaker.

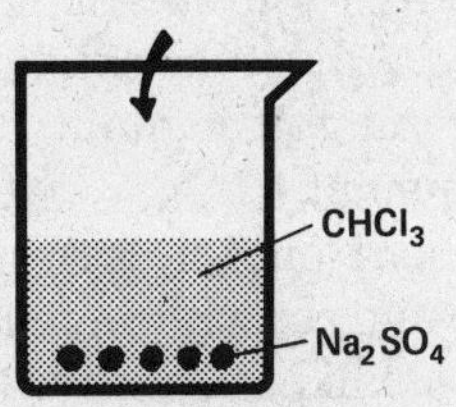

7. Add about 1 g of anhydrous sodium sulfate to the beaker containing the chloroform extracts. Swirl the mixture for 2 to 3 min. There should be no sign of water droplets in the beaker. If necessary, add a little more Na_2SO_4. Cover the beaker and allow it to stand while you begin assembling the distillation equipment.

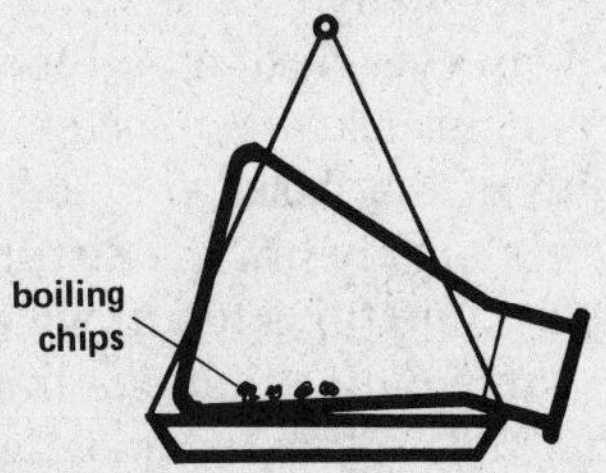

8. Weigh a clean, dry, ground-glass-necked 250-ml Erlenmeyer flask, containing three or four boiling chips, to the nearest 0.1 mg.

9. Assemble the distillation equipment as follows:

(Note: Neighboring distillation setups must be assembled "back to back" to avoid having one person's receiver flask near a neighbor's lighted burner.)

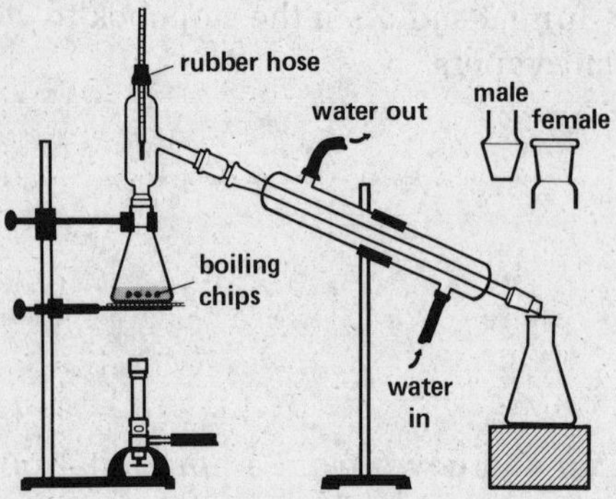

a. Clamp the ground-glass-necked Erlenmeyer flask to a ringstand and support the flask on a wire gauze set on a ring clamp about 2 to 3 in. above the top of an *unlighted* burner. Now decant the anhydrous chloroform extract away from the Na_2SO_4 into the Erlenmeyer flask through a funnel containing a small wad of cotton. Wash the sodium sulfate in the beaker with 5 ml of chloroform and also decant this wash liquid into the flask.

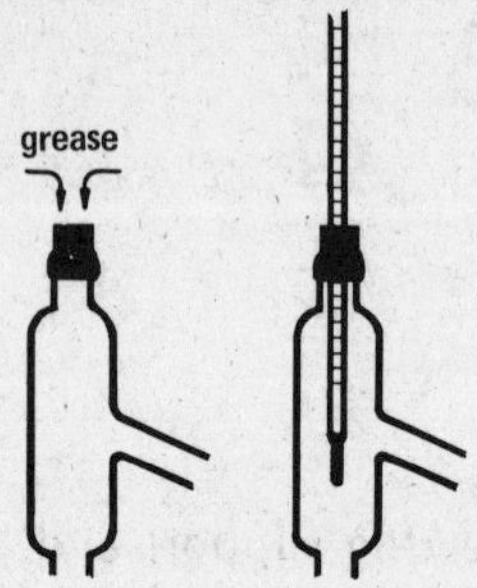

b. *Lightly* grease the top of the distillation head and slip on a short piece of rubber tubing. *Lightly* grease the lower end of a thermometer (−10° to 110°C or −10° to 260°C) and slip it through the rubber tubing, *carefully*, with a slow twisting motion until the thermometer bulb is just opposite the lower end of the side-arm opening.

c. Very lightly grease both ground-glass joints of the distilling head. Set the distilling head into the top of the Erlenmeyer flask and twist the head back and forth slowly until you observe a clear, uniform grease film inside the glass joint.

d. Attach rubber tubing to the condenser, then clamp the condenser to a second ringstand at an angle and height that aligns the female condenser joint with the male joint of the distilling head. The clamp holding the condenser to the ringstand should be tight, but the end clasping the condenser should be a bit loose.

e. Slide the condenser and ringstand toward the distillation head until the condenser joint fits snugly over the joint of the distilling head. Adjust the condenser clamp as necessary to ensure a snug fit *without placing any strain on the glass equipment.*

f. Connect the rubber tubing from the lower end of the condenser to a water tap and place the end of the other (water outlet) tubing in a sink. Turn on the water tap carefully and adjust the water flow rate to give a slow even discharge from the outlet hose. (You will need to check the flow rate periodically.)

g. Set a clean 500-ml Erlenmeyer flask on a block below the lower end of the condenser so that the condenser tip extends into the mouth of the flask.

h. Carefully lower the ring clamp below the distilling flask and place a 600-ml beaker one-third full of water on the ring clamp. Raise the clamp until about one-third of the distilling flask is immersed in the water bath.

Note: Have your distillation setup checked by your instructor before lighting the burner.

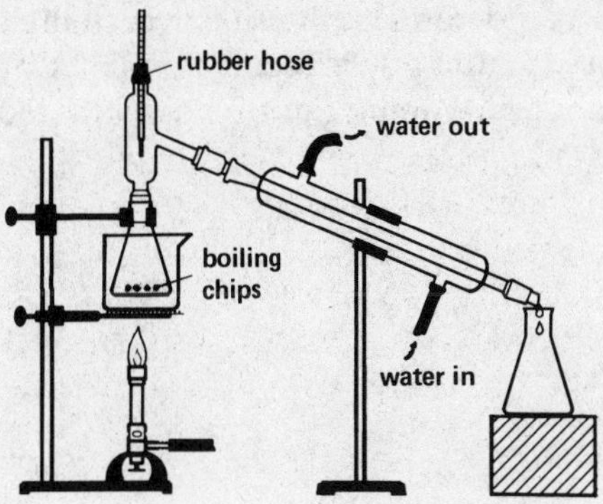

10. Now light the burner and adjust it to give a moderate flame. Begin heating the water bath carefully. When the chloroform begins to boil, adjust the heat to maintain a gentle boil and observe the thermometer readings. You should observe the condensation of chloroform vapor "creeping" up the neck of the distilling head and into the condenser. When liquid begins to drip from the condenser into the receiving flask, adjust the heating rate to give a rapid, but dropwise, distillation. Record the temperature at which the thermometer stabilizes as the boiling point of chloroform. Heat the water bath and continue heating until all the chloroform has been distilled, then turn off the burner. Stopper the receiver flask containing the distilled chloroform and set it out of the way after no more liquid drips from the condenser. When the distilling flask is cool, *carefully* disassemble the condenser and then the distilling head. Use a wad of towel to wipe off any grease remaining in the neck of the distilling flask.

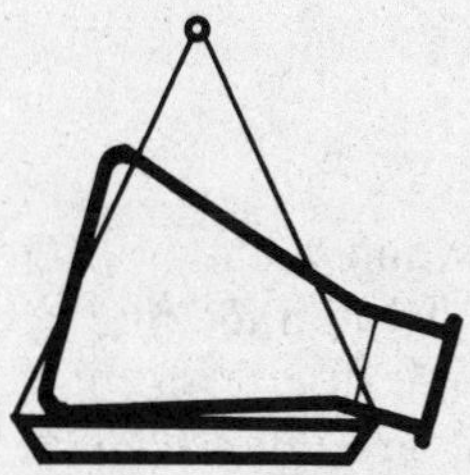

11. Weigh the distilling flask containing the solid residue of crude caffeine to the nearest 0.1 mg. Record the mass of crude caffeine in your notebook.

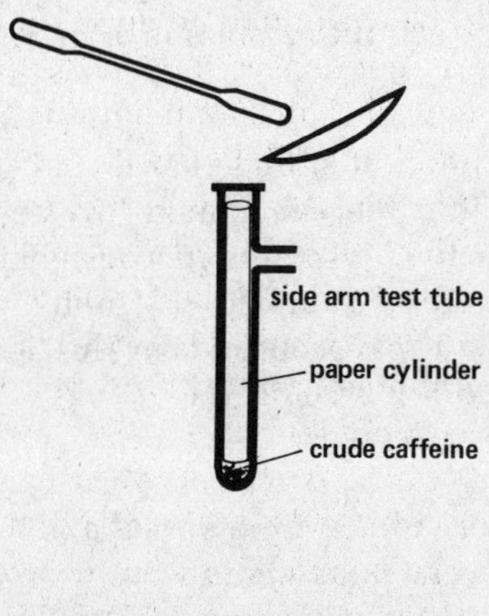

12. Scrape as much as possible of the crude caffeine from the flask onto a clean watch glass. Weigh the watch glass and caffeine to the nearest 0.1 mg. Roll a sheet of clean paper (about 12 cm × 12 cm) into a cylinder and insert this into a 150-mm side-arm test tube. Transfer the caffeine from the watch glass through the paper cylinder to the bottom of the test tube. Reweigh the emptied watch glass to the nearest 0.1 mg.

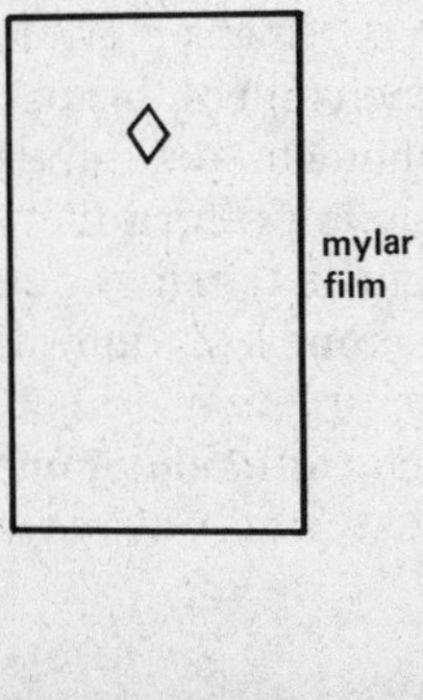

13. Remove the paper cylinder from the side-arm test tube, tapping the paper to dislodge any caffeine particles, and replace it with a cylinder prepared from a piece of Mylar film, 12 cm × 7 cm. The Mylar film will need to have a hole in it about 3.5 cm from its narrow side. Position the Mylar film so that the hole in it aligns with the side-arm opening.

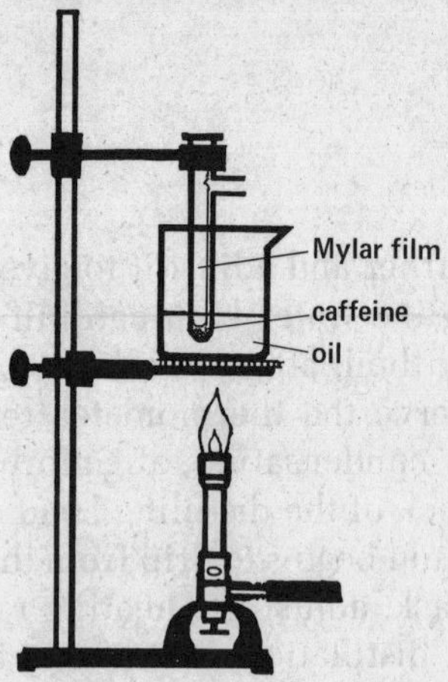

14. Set a 250-ml beaker containing 100 ml of heating oil on a wire gauze supported by a ring clamp above an unlighted burner. Clamp the side-arm test tube containing the crude caffeine and Mylar film so that the tube is immersed about 1 cm into the oil. Stopper the test tube with a No. 3 stopper and connect the side arm to the aspirator. (A trap in the vacuum line is recommended. Consult your instructor about how one might be constructed.)

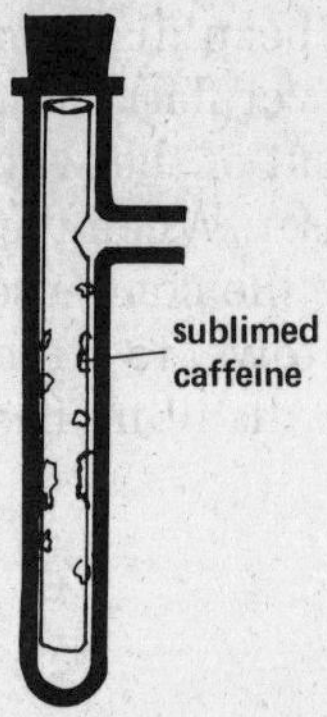

15. Turn on the aspirator and gently heat the oil. Crystals of sublimed caffeine will appear on the plastic film when the oil becomes sufficiently warm. *Do not heat the oil beyond the temperature required for sublimation of the caffeine.*

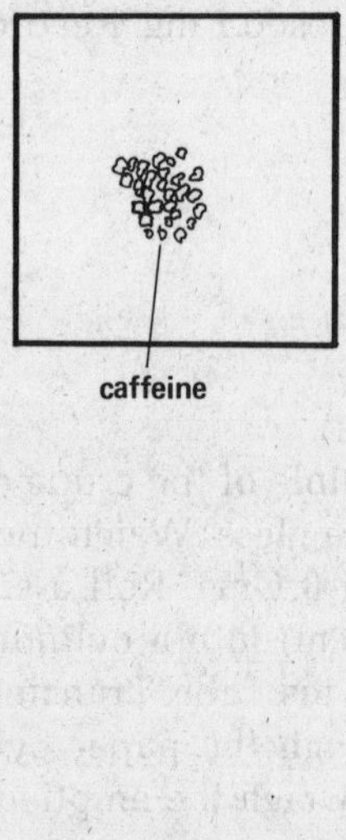

16. Discontinue heating when all of the crude caffeine appears to have sublimed. Disconnect the vacuum tubing before turning off the aspirator. Carefully remove the plastic sheet. Transfer the sublimed caffeine onto a preweighed sheet of paper (5 cm × 5 cm). The caffeine is easily removed from the plastic by flexing the Mylar sheet. Determine the mass of the sublimed caffeine recovered to the nearest 0.1 mg and record this mass in your notebook.

Pour the oil back into the oil storage bottle and pour the distilled chloroform into the designated "waste chloroform" container. Place the Mylar film in a designated container. Submit your caffeine sample, on a folded piece of paper, to your instructor and secure his evaluation of the caffeine quality on the Report Form.

CALCULATIONS AND REPORT On the basis of the mass of the tablets used and the mass of crude caffeine isolated, calculate the approximate caffeine content of the preparation assayed (in milligrams of caffeine per tablet). From your instructor obtain the caffeine content indicated on the label of this particular drug.

Using the distribution coefficient determined in Investigation 16, "Caffeine in a Chloroform-Water System," estimate the amount of caffeine left in the aqueous solution extracted, assuming 100 percent isolation of the caffeine extracted by the chloroform.

Use your notebook records and calculations to complete the Report Form.

GRADE

INSTRUCTOR'S INITIALS

Report Form

FOR INVESTIGATION 37:
Caffeine in Nonprescription Drugs

NAME: ______________________ ID NO.: ______________________

LAB SECTION ______________________ INSTRUCTOR ______________________

Experimental Results

Tablets used

Tablets + watch glass __________ g

Watch glass __________ g

Tablets __________ g

Crude caffeine isolated

Caffeine + distilling flask + boiling chips __________ g

Distilling flask + boiling chips __________ g

Crude caffeine __________ g

Boiling point of chloroform distilled __________ °C

Caffeine to be sublimed

Crude caffeine + watch glass __________ g

Watch glass __________ g

Crude caffeine __________ g

Sublimed caffeine recovered

Sublimed caffeine + paper __________ g

Paper __________ g

Sublimed caffeine __________ g

Instructor's Evaluation of Product

☐ reagent grade ☐ tech grade ☐ unacceptable

Instructor's initials __________

Information from Instructor

Name of preparation assayed ____________________

Label caffeine content __________ mg per tablet

Calculations (Show details on reverse side)

Assayed caffeine content __________ mg per tablet

Caffeine left in aqueous solution __________ mg per tablet

Percentage of caffeine recovered from sublimation __________%

DATE ______________ SIGNATURE ______________________

Investigation 38

AMINO ACID ANALYSIS

INTRODUCTION When protein foods are metabolized, the complex proteins are first broken down by hydrolysis of amide bonds to simpler species, the peptides, and these are eventually hydrolyzed to amino acids. The different amino acids undergo a variety of reactions in normal metabolic processes. One of the common amino acids, present in a relatively large amount in milk proteins, is phenylalanine:

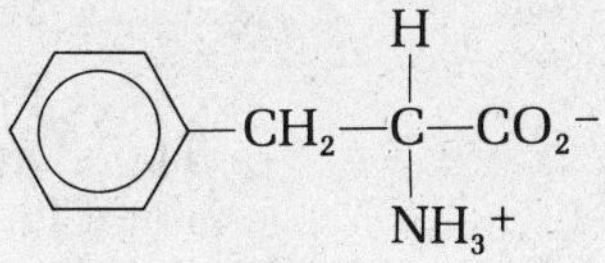

Some common amino acids

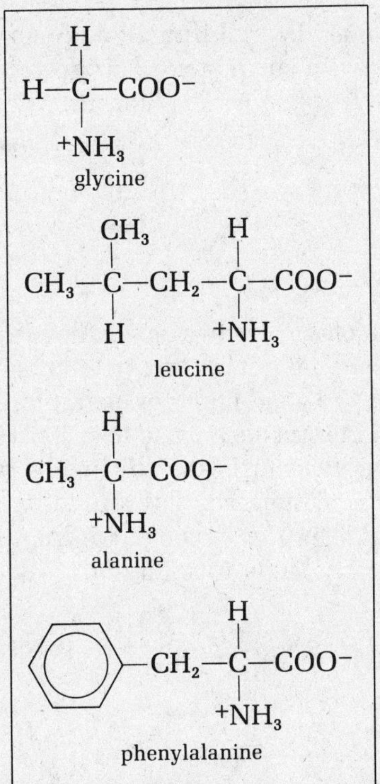

In the normal metabolic pathways, phenylalanine undergoes an oxidative degradation. A small fraction is converted to the keto acid, phenylpyruvic acid,

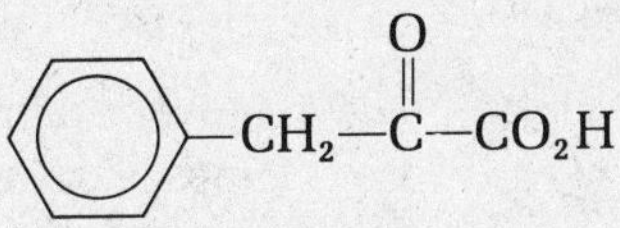

but most of the phenylalanine is changed to tyrosine,

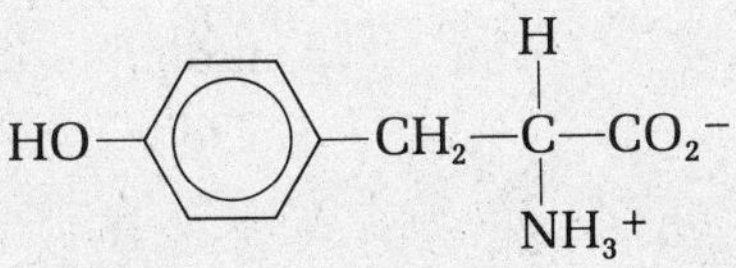

by a reaction catalyzed by the enzyme *phenylalanine hydroxylase*. The tyrosine formed undergoes further degradations, eventually being converted to simple aliphatic acids or, by an alternative pathway, into a component of the melanin pigments of skin or hair.

In the 1930s, Dr. Asbjorn Folling, a Swedish physician, discovered that the urine of two mentally retarded children contained abnormally high amounts of phenylpyruvic acid. Subsequent research showed that the mental retardation was the result of a metabolism error caused by deficiency in the phenylalanine hydroxylase enzyme system. The disease, called *phenylketonuria* (PKU), could be detected by a simple color test applied to the wet diaper of an infant. Like phenols, some keto acids react with iron(III) chloride to form a colored product. The reaction apparently detects the enol form of the keto acid:

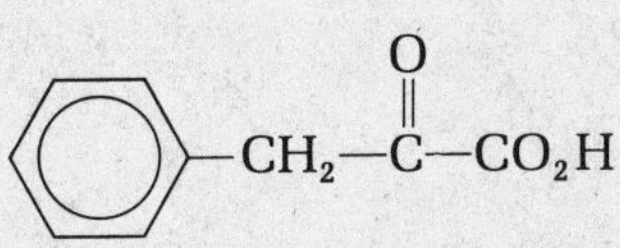

keto form

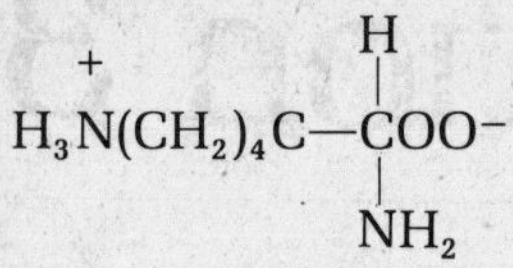

lysine

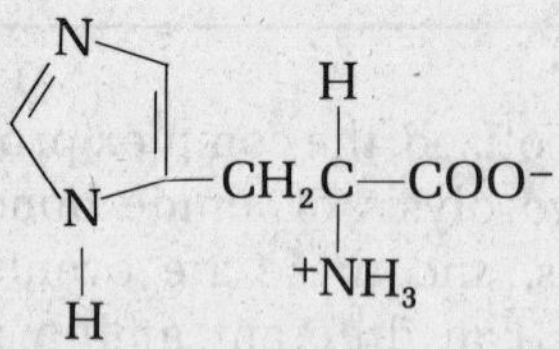

histidine

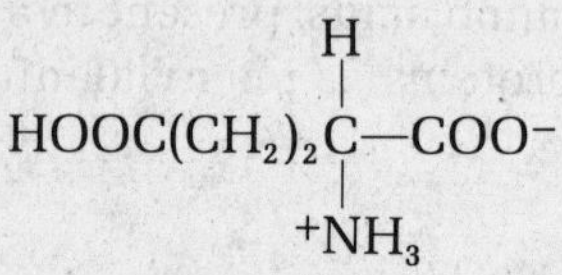

glutamic acid

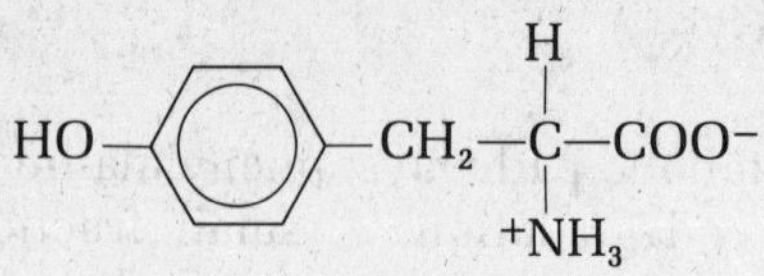

tyrosine

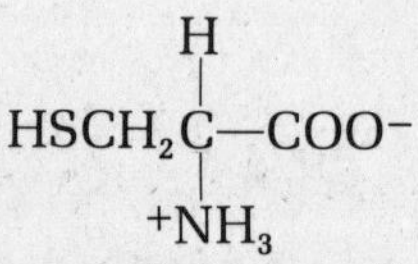

cysteine

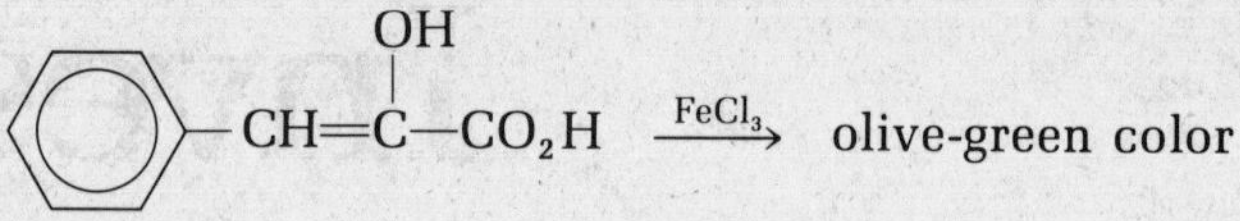

enol form

This simple test has now been largely replaced by a more sensitive and generally more reliable test* using a few drops of blood to detect abnormally high concentrations of phenylalanine in the blood serum.

If the disease is detected in early infancy, the treatment is quite simple. The diet of the baby is altered to reduce the intake of proteins rich in phenylalanine, and supplementary tyrosine is added to make up for that which would normally have been formed from phenylalanine.

Phenylketonuria (PKU) is only one of a number of molecular diseases, many of which involve errors in amino acid metabolism.

The simplest amino acids are bifunctional, containing a single basic group and a single acidic group. They are commonly represented by the general formula

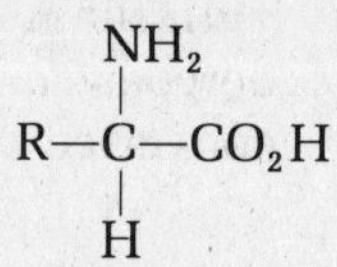

Now, we know enough chemistry to suspect that an amine and a carboxylic acid should not coexist without a proton exchange taking place. We would not, for example, expect a mixture of methylamine and acetic acid to remain stable. Rather, they should react to form the salt methylammonium acetate. Logic, then, suggests that amino acids are probably not very stable in the molecular form, and we might propose a more probable formulation:

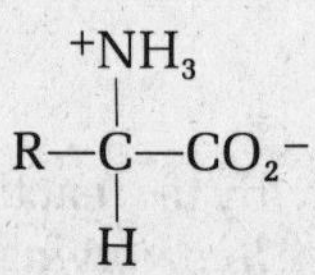

Such a structure represents a species containing both a positive and a negative charge. Ions of this type are called *Dipolarions* (or, in the older literature, *zwitterions*).

This is a good logical conclusion, but we have learned by now that we should not be satisfied with "logic" in science. Molecules are relatively immune to human thought processes and if we are to determine the correct formulation for amino acids, we must find experimental methods for testing our hypotheses. Such tests have been applied exhaustively to a variety of amino acids, and they confirm beyond any reasonable doubt that the dipolarionic formulation is the best description.

One of the necessary stages in the description of any macromolecule is the identification of the monomer units from which it is made. Starches, for example, can be hydrolyzed into smaller molecules, eventually glucose.

*The Guthrie test, developed by Dr. Robert Guthrie at the State University of New York at Buffalo.

Peptides and proteins, on the other hand, seldom yield a single hydrolysis product. Rather, a mixture of several amino acids is usually obtained.

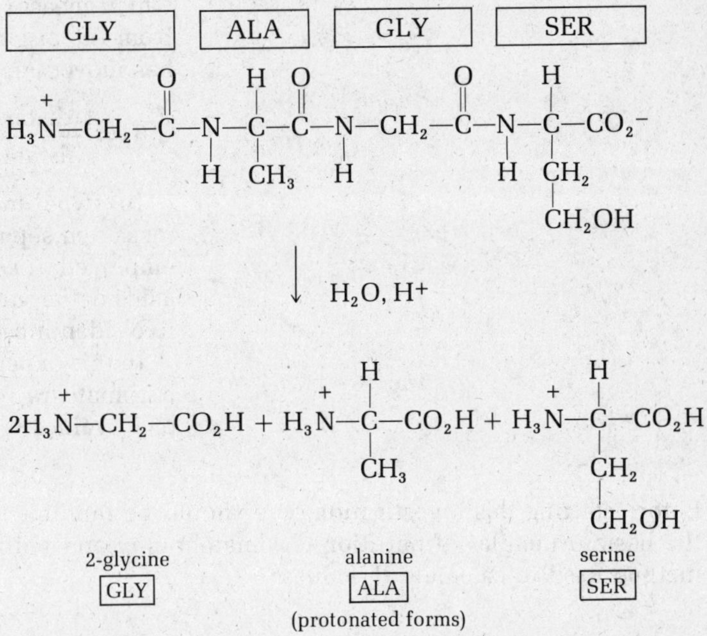

The most common method of hydrolysis of peptide bonds uses aqueous acid. Certain of the amino acids may, however, undergo structural changes under normal acid hydrolysis conditions so that other procedures (e.g., alkaline hydrolysis or hydrolysis catalyzed by appropriate enzymes) may be necessary to ensure that all component amino acids have been identified.

In order to identify the components of such a hydrolysis mixture, separation techniques must be employed that take advantage of the relatively minor differences in the structures of the amino acids. Two-dimensional chromatography is one such technique. By developing a chromatogram in one dimension with a given solvent system and then in another direction with a different solvent, a fairly effective separation of an amino acid mixture can be obtained. Comparison of a two-dimensional chromatogram, after detection with a color-producing reagent such as ninhydrin, with patterns obtained from known amino acids may be sufficient in some instances to identify the amino acids in the mixture.

Two-dimensional chromatography most commonly employs either a paper sheet or some powdered adsorbent adhering to a glass or plastic backing as the support medium. The latter technique is called *thin-layer chromatography*. Usually, a mixed solvent system is used so that separation by partitioning is the principal factor involved. For two-dimensional chromatography, the mixture is applied as a small spot near a lower corner of the support medium. The paper or thin-layer plate is then dipped into the solvent mixture to a line just below the applied spot, and the solvent moves up through the support, which acts as a wick. After the solvent front reaches a desirable height, the chromatogram is removed, dried, turned 90 deg, and dipped into a second solvent system. The second development is followed by drying and spraying with some detection reagent to produce color at the locations of the separated amino acids.

Spots may be characterized by so-called R_f values, which measure the distance the component* has moved from the origin relative to the distance the solvent front has moved (as measured from the origin).

$$R_f = \frac{\text{distance of spot center from origin}}{\text{distance of solvent front from origin}}$$

In two-dimensional chromatography, R_f values are measured separately for each of the two solvent systems employed. A comparison with R_f values for known amino acids in the same chromatography system permits a tentative† identification of each amino acid.

In this experiment we will use simple one-dimensional chromatography to identify amino acids by comparison of R_f values with those of known samples.

Before starting this investigation, you should be familiar with the basic principles of partition chromatography and with the method used to calculate R_f values.

References

1. O'Connor, Rod, 1977. *Fundamentals of Chemistry* (2nd ed.). New York: Harper & Row (Units 37 and 40).
2. Kopple, K. D. 1966. *Peptides and Amino Acids*. Menlo Park, Calif.: Benjamin.

SAFETY Wear approved eye protection. Avoid inhaling vapors of the solvents used or of the ninhydrin spray. In case of contact with any reagents, wash thoroughly with *soap* and water. *All of the solvents, as well as the ninhydrin aerosol mist, are highly flammable.*

OBJECTIVES
1. Be able to prepare and use "spotting capillary tubes."
2. Be able to develop simple chromatograms accurately.
3. Be able to calculate and use R_f values.

PROCEDURES

1. You and your neighbor will each prepare a chromatography chamber containing a different solvent system.

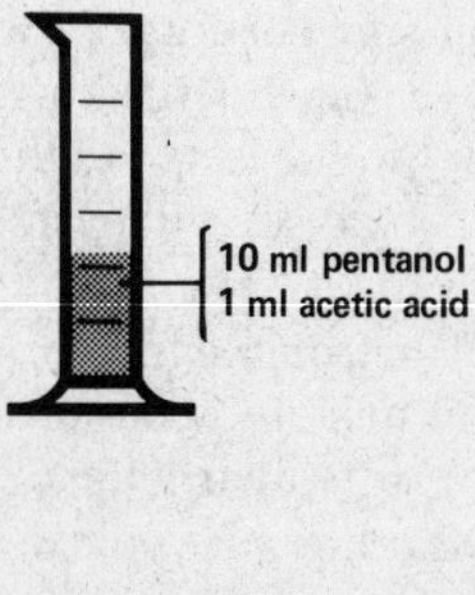

STUDENT 1 Measure 10 ml of 1-pentanol and 1.0 ml of glacial acetic acid into a clean, *dry* 600-ml beaker. Cover the beaker with a 15-cm × 15-cm piece of Saran Wrap, secured with a rubber band. Swirl the beaker several times to mix the solvents.

*Measurement is made from the center of the spot to the origin.

†R_f values may vary with a number of experimental conditions such as thickness of the support or temperature. As a result, chromatography patterns usually require confirmation by other analytical methods before unambiguous results can be expected, except in relatively simple cases.

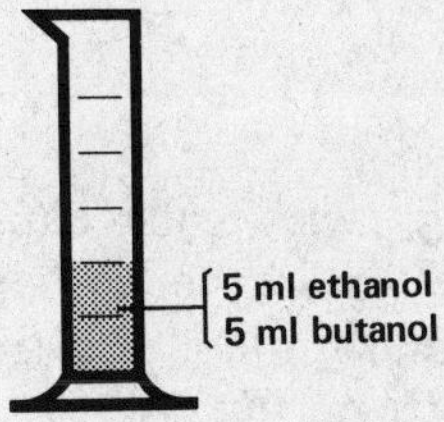

STUDENT 2 Measure 5 ml of ethanol and 5 ml of 1-butanol into your beaker. Seal and swirl the beaker.

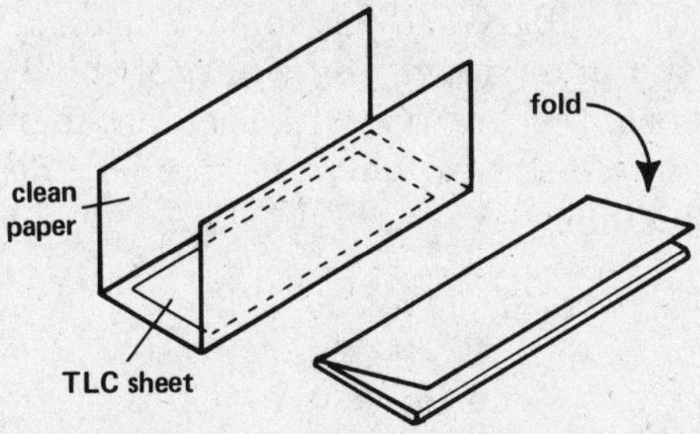

2. Fold a clean piece of paper in thirds and obtain in this, from your instructor, 2 thin-layer chromatography (TLC) sheets, impregnated with silica gel. Never touch the surface of a TLC sheet with your fingers. Handle a TLC sheet carefully (it is quite fragile).

3. Carefully clean and dry six small test tubes and seal them with clean stoppers labeled I (for an ink sample*), G (for glycine), A (for alanine), L (for leucine), M (for a mixture of these three amino acids), and U (for your unknown). Obtain from your instructor three to five drops of each of these solutions.

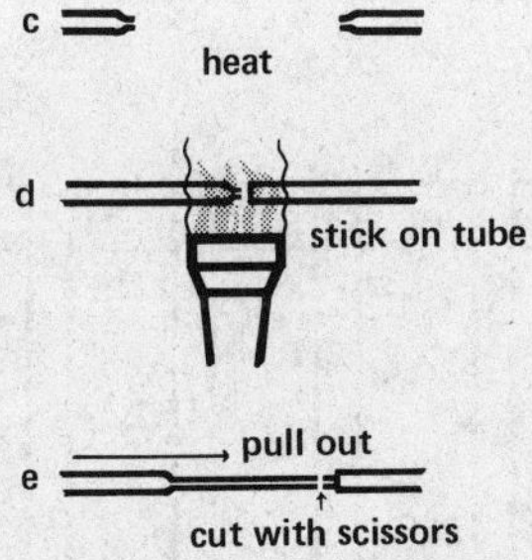

4. Prepare six chromatographic spotting capillaries, as follows (repeat the sequence three times):

a. Fire polish both ends of a 20-cm length of 5 mm (O.D.) soft glass tubing. *When the ends have cooled,* rotate the center segment of the tubing over a hot flame until the tubing softens.
b. Remove the tubing from the flame and quickly pull on both ends.
c. Break off the center capillary section.
d. Seal one end of the capillary in the edge of a flame.
e. Soften one of the larger pieces from step (c) and touch this to the sealed end of the capillary in the flame until the two pieces fuse together. Then pull a very fine capillary and snip it off with scissors. Do not touch the ends of the capillaries with your fingers.

*The ink sample is used so that you can see one of the chromatograms developing. Amino acids are colorless, so their spots are not visible until after reaction with ninhydrin (Procedure 13).

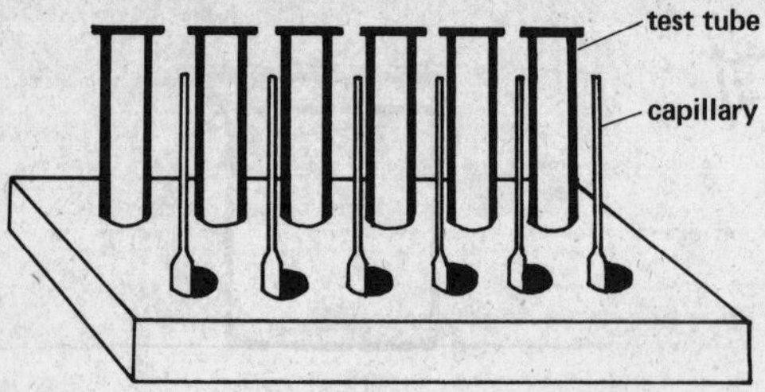

5. Store the capillaries, large ends down, in the test tube block opposite the solutions to be used.

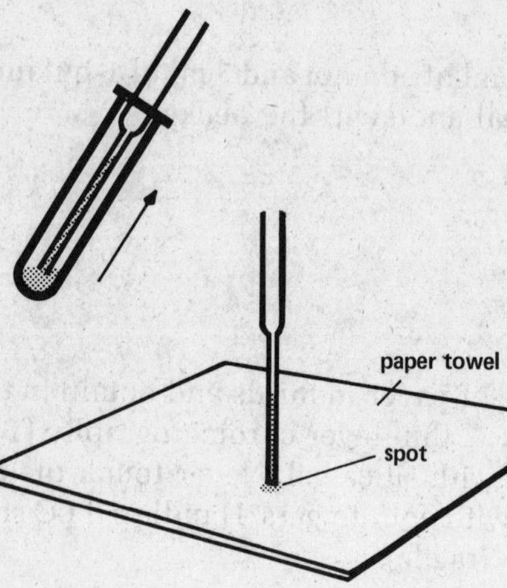

6. Now practice "spotting" with each capillary onto a double-folded paper towel. Hold the "I" capillary so the fine tip extends into the ink sample and a tiny column of ink rises in the capillary. (If no ink rises, *snip off a short segment* of the capillary and try again with the remaining piece.) Remove the capillary and *just touch it* against the paper towel. (If you press down, you will probably snap the capillary.) Repeat this test run with each of the other solutions, being careful to use only the capillary designated for each solution.

Caution: *Never* touch the TLC sheets with your fingers. Handle them only with tweezers to avoid contamination from amino acids of the skin.

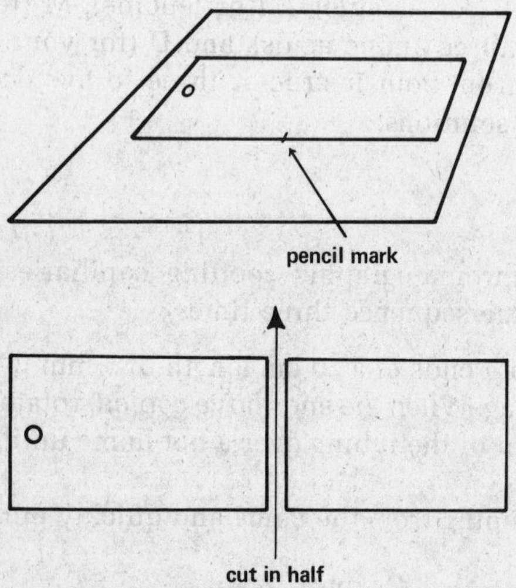

7. Carefully unfold the paper containing the TLC sheets. Transfer one sheet, with tweezers, onto a flat piece of clean paper. With a ruler, determine the midpoint of the TLC sheet and mark it lightly with a pencil dot. Cut the sheet into two 10-cm × 5-cm halves. Mark the bottom of each piece with a pencil dot each 16 mm. Draw a light pencil line parallel to the marked edge and 1 cm from the bottom of the sheet. Draw a second line 7 cm above the first. With scissors cut three 16-mm wide strips from each half of the TLC sheet. Repeat this procedure with the other TLC sheet.

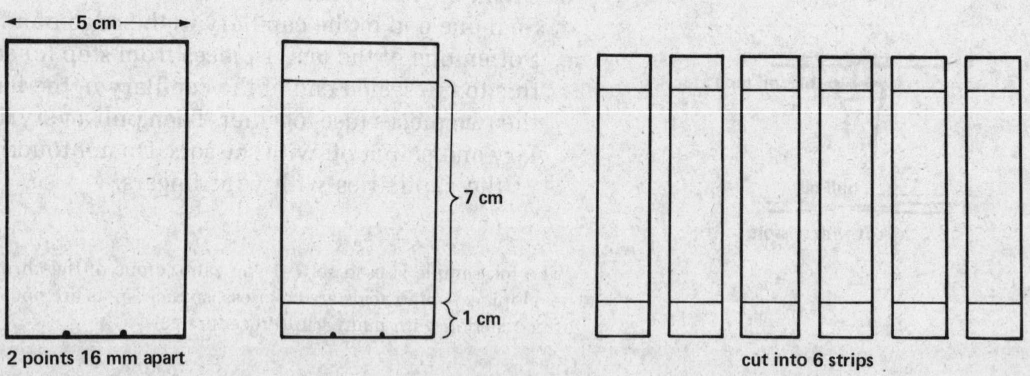

Now you are ready for the critical step of spotting the chromatograms. It is important that you apply only a *tiny* spot of solution.

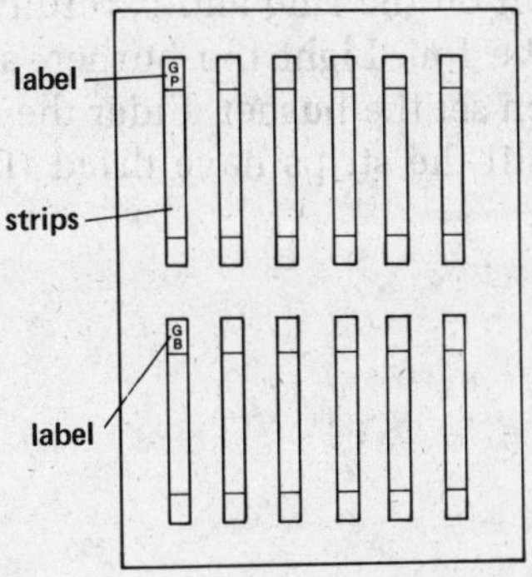

8. Line up the TLC strips in two groups of six each. With a pencil, mark above the top line of one strip in the upper row IP (for ink, to be run in the 1-pentanol-acetic acid system). Similarly mark the top of the strip just below this IB (for ink, to be run in the 1-butanol–ethanol system). Mark the remaining strips in the top row GP, AP, LP, MP, and UP, respectively; and those in the bottom row GB, AB, LB, MB, and UB. Hold the top of the TLC strip flat with a spatula blade while marking.

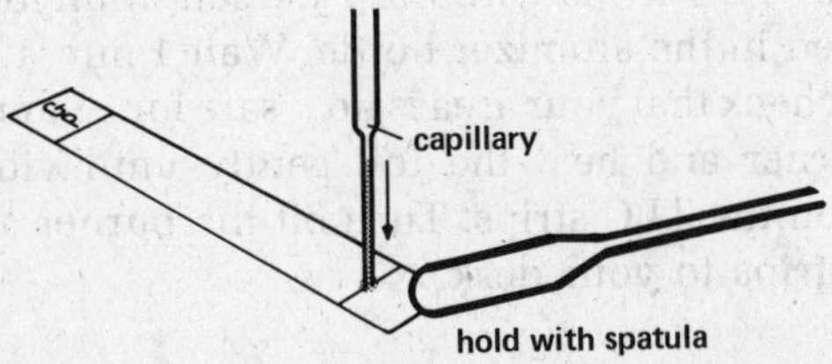

9. Now, while holding the bottom of a strip flat with a spatula blade, use a spotting capillary to apply a tiny spot (1 mm diameter or less) at the center of the lower pencil line, using the solution corresponding to the code marked on the TLC strip.

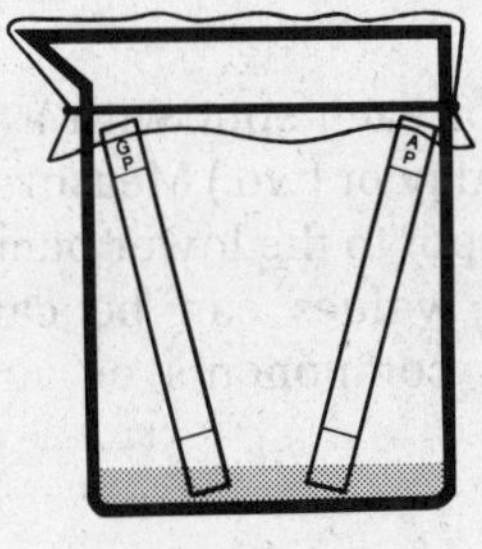

10. STUDENT 1 With tweezers, carefully stand your six strips marked with a P (e.g., IP) in your beaker (containing the 1-propanol–acetic acid), leaning the strips against the side of the beaker so that no two strips touch each other. The TLC strip must *not* dip into the solvent far enough to immerse the applied spot. Reseal the Saran Wrap over the beaker.

STUDENT 2 Follow the procedure outlined for Student 1, but placing the six strips marked with a B (e.g., IB) in your beaker (containing the 1-butanol–ethanol).

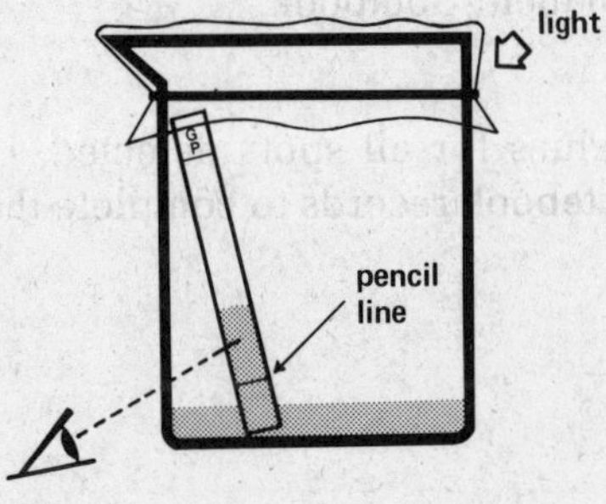

11. As the solvent migrates up the TLC strip, you can see the solvent front as a border between a light and dark region on the strip. This is best observed by *carefully* holding the beaker toward the light and looking at the back side of the strips. When the solvent front reaches the upper pencil line, remove the strips from the beaker, allowing any excess solvent to drip into the beaker. Place the strips carefully on a clean piece of paper. Reseal the beaker.

After you and your partner have both removed your first set of strips, exchange beakers and develop your second set of TLC strips.

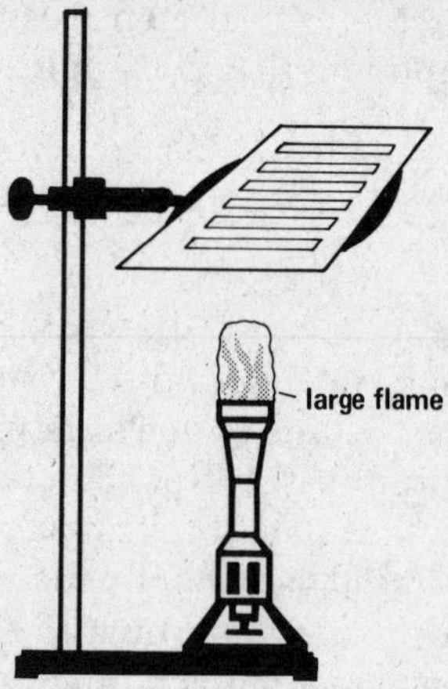

12. When you have developed both sets of strips, assemble in the fume hood a ringstand with the ring clamp set 34 cm above the base of the stand.* Place a 10-cm × 15-cm piece of aluminum foil on the ring and carefully arrange your TLC strips on the foil. Light the burner, adjust the flame to 6 cm, and then set the burner under the foil. Heat for a few minutes until the strips have dried. Then *turn off the burner*.

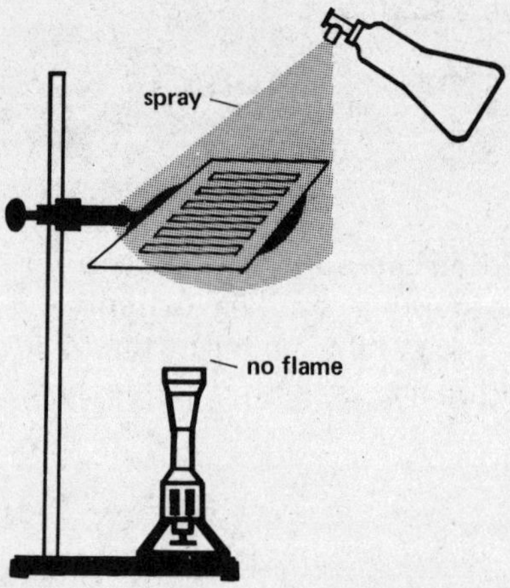

13. *Be sure there are no flames in your working area.* Then spray your TLC strips with the 1-percent ninhydrin reagent provided in the atomizer bottle. Wait 1 min after spraying, then check that your area is now safe for a flame.* Relight the burner and heat the foil gently until violet spots develop on the TLC strips. Turn off the burner and take the TLC strips to your desk.

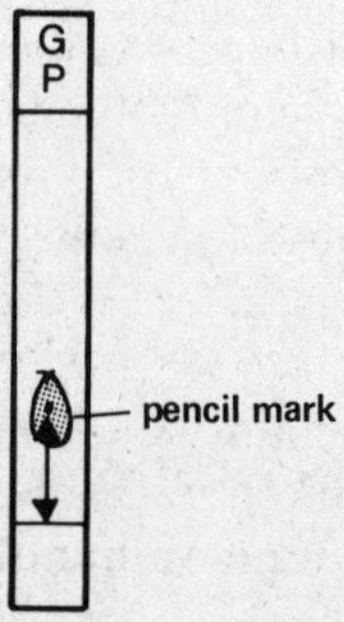

14. Outline the border of each spot with a blunt pencil. (The spots will fade in a day or two.) Measure the distance from the center of each spot to the lower pencil line. From these measurements, R_f values can be calculated and used in identifying the components of your unknown solution.

Enclose your TLC strips in an envelope and paper clip this into your permanent notebook.

CALCULATIONS AND REPORT Calculate the R_f values for all spots detected. Use your calculations and notebook records to complete the Report Form.

*Instead of the burner flame, a hot plate or chromatogram oven may be used. Ask your instructor for details of these alternatives.

GRADE	INSTRUCTOR'S INITIALS

Report Form

FOR INVESTIGATION 38:
Amino Acid Analysis

NAME: ______________________ ID NO.: ______________________

LAB SECTION ______________________ INSTRUCTOR ______________________

R_f Values

Show a sample calculation, with a sketch of one of the chromatograms, on the reverse side.

Sample	1-Pentanol–acetic acid solvent system	1-Butanol–ethanol solvent system
ink		
glycine		
alanine		
leucine		
mixture		
unknown		

Composition of unknown ______________________

DATE ______________ SIGNATURE ______________________

Investigation 39

CHEMICAL OXYGEN DEMAND

INTRODUCTION Many organic, and some inorganic, wastes are degraded in natural waters by oxidation processes. In some cases these involve direct reaction of chemical wastes with molecular oxygen dissolved in the water. More commonly, aerobic bacteria in the water degrade the chemicals while utilizing dissolved oxygen. In either case, oxygen is consumed and if waste degradation demands abnormally high amounts of O_2, fish and other aquatic life cannot survive in the oxygen-depleted waters.

The *biological oxygen demand* $(BOD)_5$ is defined as the amount of oxygen (in milligrams) consumed per liter of solution over a five day period, and is determined by mixing a measured sample of polluted water with a known volume of oxygen-saturated pure water. The total O_2 concentration in the mixture is measured just after mixing and again after five days incubation at 20°C. It is the *change* in O_2 concentration over this period that is used to calculate the BOD.

Although the BOD value provides an important indication of the capability of a natural water system to degrade domestic and industrial wastes, the BOD alone does not provide enough information on *total* pollutant levels since *biological* oxygen demand applies *only* to those pollutants degraded by aerobic aquatic organisms. A better measure of the total *oxidizable* pollutants, including those inert to biological degradation, is given by the *chemical oxygen demand* (COD).

With rare exceptions, organic compounds, both those resistant and those nonresistant to biological degradation, are readily attacked by suitable chemical oxidizing agents. Historically, several different reagents have been employed for determinations of "oxidizable pollutants," but dichromate ion in acidic solution is now the accepted standard. Except for a few species such as pyridine or certain aromatic hydrocarbons, the organic compounds typically found in domestic or industrial wastes are rapidly oxidized by acidic $Cr_2O_7^{2-}$. A few species, such as the smaller fatty acids, are attacked slowly, but it has been found that silver ion catalyzes the dichromate oxidation of fatty acids, so this problem has been solved. Carbon and hydrogen in the organic compounds are oxidized completely to CO_2 and H_2O, respectively. Nitrogen in the −III oxidation state (e.g., in amino groups) is liberated as ammonia, but nitrogen in higher oxidation states is converted to nitrate ion.

The rapidity of dichromate oxidations, catalyzed as necessary, permits the determination of chemical oxygen demand in just a few hours. As a result, COD values are more useful for systems, such as waste treatment plants, in which pollutant levels must be routinely determined over short time periods because of the rapid turnover of the system. In order for $Cr_2O_7^{2-}$ oxidation to be complete, the solution must have a high concentration of H^+ and must be maintained at an elevated temperature. To prevent the loss of volatile pollutants or partial-oxidation products, the reaction is best performed under reflux conditions. Some inorganic species typically present in waste waters, such as chloride ion, are oxidized by $Cr_2O_7^{2-}$. Various methods are available for removing such species when it is desired to exclude them from COD determinations. The addition of Hg^{2+}, for example, effectively ties up Cl^- as a complex or the weakly ionized $HgCl_2$ so that oxidation of Cl^- is avoided.

The COD test is the best available simple method for determining the "total" oxidizable pollutants in a water sample. It is, of course, limited to "oxidizable" pollutants and it cannot distinguish between biologically degradable and biologically inert pollutants, nor does it provide any clues as to the rate of pollutant degradation by the natural aquatic ecosystem. The best overall picture of a water pollution problem is one that combines both COD and BOD data, since this combination provides some level of prediction of long-range pollution effects.

The COD determination does not involve a *direct* measurement of dichromate consumption, since an excess of $Cr_2O_7^{2-}$ is necessary to ensure complete oxidation. It is more feasible to use an *indirect* determination by titrating the *excess* $Cr_2O_7^{2-}$ with a reducing agent (typically Fe^{2+}). The amount of $Cr_2O_7^{2-}$ *consumed* (and, hence, the "oxidation demand") by the sample is then equal to the *difference* between the quantity of dichromate added to the sample and the quantity remaining as excess after consumption of all oxidizable species. Since it is not practical to exclude *all* oxidizable material from the reagents and equipment used in the COD determinations, a *blank* determination is made using all conditions except for the "polluted" sample. The COD values are reported in the same units as those used for BOD determinations, ppm O_2 (i.e., as milligram O_2 consumed per liter of sample). Since the reaction employed uses $Cr_2O_7^{2-}$ as the oxidizing agent, rather than molecular oxygen, we must have some way of relating the "chemical equivalence" of $Cr_2O_7^{2-}$ to O_2. The clues to this "equivalence" include the knowledge that during the course of the type of oxidation involved, O_2 is reduced to the $-$II oxidation state (e.g., as in H_2O or CO_2) and dichromate is reduced to chromium(III) ion. We may write some hypothetical half-reaction equations, from which the necessary relationships can be calculated:

$$O_2 + 4H^+ + 4e^- \longrightarrow 2H_2O$$
$$Cr_2O_7^{2-} + 14H^+ + 6e^- \longrightarrow 2Cr^{3+} + 7H_2O$$

Since any oxidation-reduction reaction may be considered in terms of electron transfer, we may write

$$\frac{1 \text{ mole } O_2}{4 \text{ moles } e^-} \times \frac{32 \text{ g } (O_2)}{1 \text{ mole } O_2} \times \frac{10^3 \text{ mg}}{1 \text{ g}} \times \frac{6 \text{ moles } e^-}{1 \text{ mole } Cr_2O_7^{2-}} \times \frac{1 \text{ mole}}{10^3 \text{ mmol}} = 48 \frac{\text{mg } (O_2)}{\text{mmol } Cr_2O_7^{2-}}$$

If we can now find a method for determining the number of millimoles of dichromate consumed by a measured volume of a water sample, we then have all the information needed for calculation of the COD. To find the quantity of $Cr_2O_7^{2-}$ consumed, we can measure the volume of dichromate solution (of known molarity) added to the measured volume of sample, to find

$$\text{mmol}_{Cr_2O_7^{2-}} \text{ added} = M_{Cr_2O_7^{2-}} \times V_{(ml)} \text{ added}$$

After oxidation of the sample is complete, we can determine the quantity of dichromate remaining as excess (as $\text{mmol}_{Cr_2O_7^{2-}}$ remaining) from a knowledge of the molarity and volume of reducing agent used in titration of the excess $Cr_2O_7^{2-}$ and a knowledge of the stoichiometry of the oxidation-reduction reaction. The difference (allowing for the *blank* determination) then gives us the quantity of dichromate consumed:

$$\text{mmol}_{Cr_2O_7^{2-}} \text{ consumed} = (\text{mmol}_{Cr_2O_7^{2-}} \text{ added}) - (\text{mmol}_{Cr_2O_7^{2-}} \text{ remaining})$$

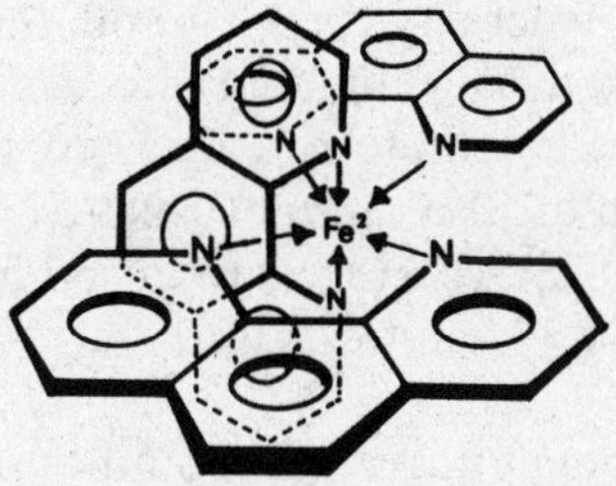

The endpoint of the oxidation-reduction reaction between $Cr_2O_7^{2-}$ and Fe^{2+} is readily visualized by the indicator *ferroin,* iron(II) 1,10-phenanthroline sulfate, which changes to a distinct red-brown color as soon as all $Cr_2O_7^{2-}$ has been consumed. This color change is not masked by the purple-green color of Cr^{3+}. The color change of the ferroin indicator results from the difference in colors of the Fe^{2+} and Fe^{3+} complexes with 1,10-phenanthroline. The original ferroin indicator is an iron(II) complex.

When added to dichromate solution, the iron(II) is quickly oxidized to iron(III), for which the complex is a pale blue, almost colorless at the concentration used. The 1,10-phenanthroline has a much greater affinity for Fe^{2+} than for Fe^{3+}, so the accumulation of a slight excess of Fe^{2+} titrant (after consumption of the $Cr_2O_7^{2-}$ is complete) quickly forms the iron(II) complex. The intense red-brown color of this complex is readily observed as an indication of the endpoint of the $Fe^{2+}/Cr_2O_7^{2-}$ titration. The 1,10-phenanthroline is stable to oxidation, so no problem is encountered by having the indicator present in the dichromate solution, but both complexes are decomposed at temperatures much above 50°C. It is therefore essential that the indicator not be added until the solution is cool.

In this investigation, we shall use a slightly abbreviated version of the standard COD determination for an "unknown" sample.

Before starting this investigation, you should be familiar with the concepts of oxidation and reduction as they apply to both organic and inorganic species and with calculations involving titrations (solution stoichiometry and solution preparation). In addition, prior to coming to the laboratory, you must set up (and complete, when appropriate) the calculations indicated in the Procedures section of this investigation.

References

1. O'Connor, Rod. 1977. *Fundamentals of Chemistry* (2nd ed.). New York: Harper & Row (Units 13, 22, and 22).

2. Sawyer, Clair and Perry McCarty. 1967. *Chemistry for Sanitary Engineers* (2nd Ed.). New York: McGraw-Hill.

SAFETY Wear approved eye protection at all times and exericse *extra* precautions in working with reagents. A hot, acidic solution of dichromate is a *very powerful* oxidizing agent. In case of contact with *any* of the reagents, wash *immediately* and thoroughly with water. Be careful in assembling and using glass equipment, checking particularly that clamps are arranged for proper support without placing any strain on glassware. Be sure to use glass bead "boiling aids" properly to avoid superheating and "bumping" during reflux procedures.

OBJECTIVES

1. Be able to make appropriate calculations involving solution concentrations and solution stoichiometry.
2. Be able to weigh accurately a *specified amount* of a solid reagent.
3. Be able to perform safely and accurately the techniques of solution preparation, reflux, and titration.
4. Be able to determine the *chemical oxygen demand* for an "unknown solution."

PROCEDURES

You will be assigned a partner for this investigation. Each of you must independently perform all calculations. Compare your results and correct any errors found, then have your calculations checked by your instructor.

Before starting work, calculate the mass of $K_2Cr_2O_7$ required for preparation of 100.0 ml of 0.04250 *M* solution and the mass of $Fe(NH_4)_2(SO_4)_2 \cdot 6H_2O$ required for preparation of 100 ml of 0.10 *M* solution.

1. Clean two burets (Appendix A) and mount them properly on a ringstand. Attach a label reading "$Cr_2O_7^{2-}$" to one buret and label the other "Fe^{2+}." These burets will be used for the $K_2Cr_2O_7$ and $Fe(NH_4)_2(SO_4)_2$ solutions, respectively.

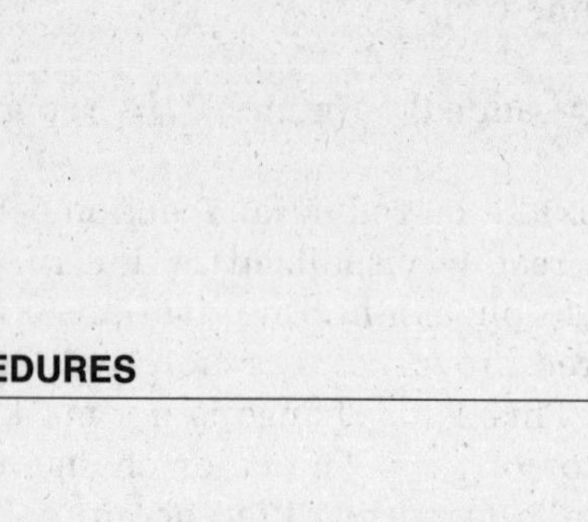

2. Review the techniques for weighing an "exact amount" of solid (as described in Appendix A) and then weigh out the previously calculated amount of $K_2Cr_2O_7$ on a small piece of creased, smooth paper. You must weigh out within ±2 mg of the required amount to an accuracy of ±0.1 mg. Record the mass used.

3. Transfer the $K_2Cr_2O_7$ completely into a clean 100-ml volumetric flask. Add about 70 ml of distilled water to the flask, rinsing any crystals within the flask neck down into the solution. Swirl the flask until a homogeneous solution is obtained, then add distilled water to the 100-ml mark, stopper the flask securely, and mix the contents thoroughly by inverting the flask several times.

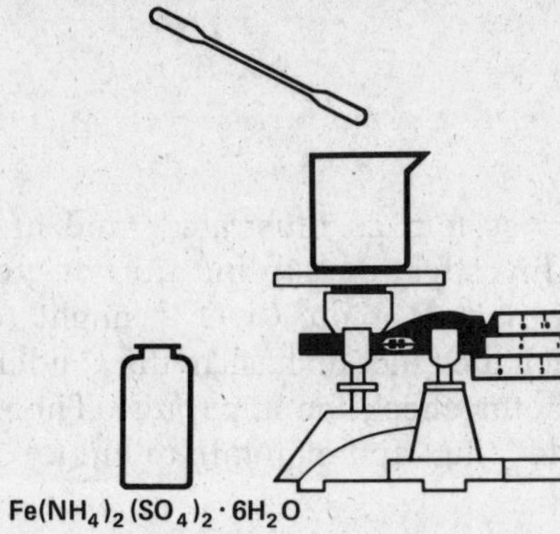

4. On the triple beam balance, weigh into a clean, preweighed 250-ml beaker the calculated amount of $Fe(NH_4)_2(SO_4)_2 \cdot 6H_2O$. Remove the beaker from the balance and pour into it 85 ml of distilled water.

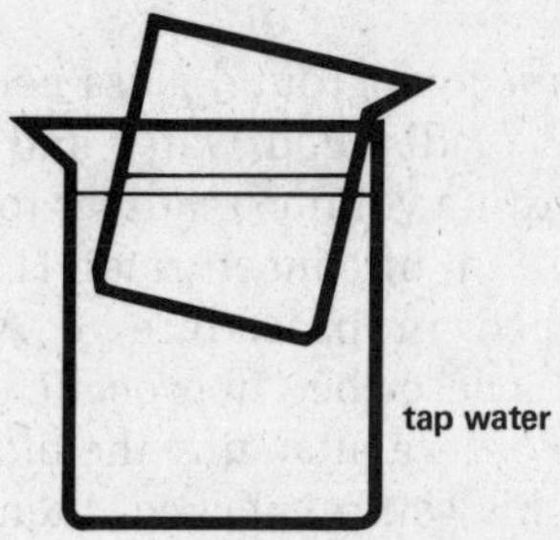

5. To the beaker containing the iron(II) salt and water, slowly and carefully add 20 ml of concentrated (18 *M*) H_2SO_4, stirring the mixture constantly during addition of the acid. (*Be careful.* The solution will become quite hot.) Then set the beaker cautiously into a larger beaker containing cold tap water. Leave the beaker there to cool until you need it, but be certain the beakers are placed where they cannot be accidentally knocked over.

Take a *clean,* ground-glass-necked 250-ml Erlenmeyer flask to your instructor and obtain 10.0 ml of the assigned "unknown" solution.

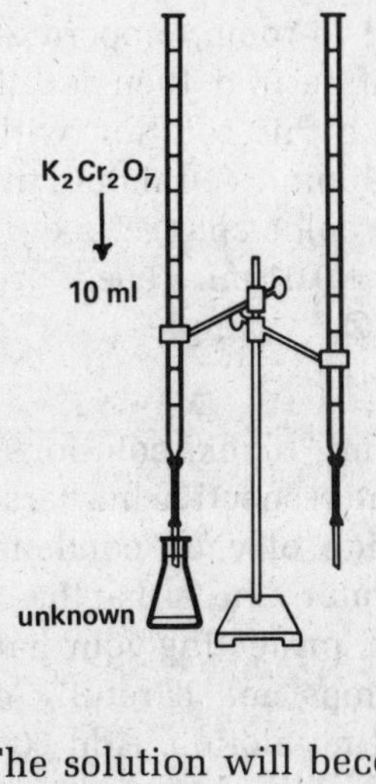

6. Rinse and fill the $Cr_2O_7^{2-}$-labeled buret with the dichromate solution. Drain enough liquid into a "waste" flask to expel air from the buret tip and to drop the liquid level below the zero mark. Record the buret reading to the nearest 0.01 ml, then drain 10.0 ml of the $Cr_2O_7^{2-}$ solution into the ground-glass-necked flask containing your "unknown." Record the final buret reading to the nearest 0.01 ml. Add four glass beads (as "boiling aids") to the flask. *Carefully and slowly,* while swirling the flask, pour in 30 ml of concentrated sulfuric acid.

Caution: The solution will become quite hot.

Continue swirling until the solution is homogeneous.

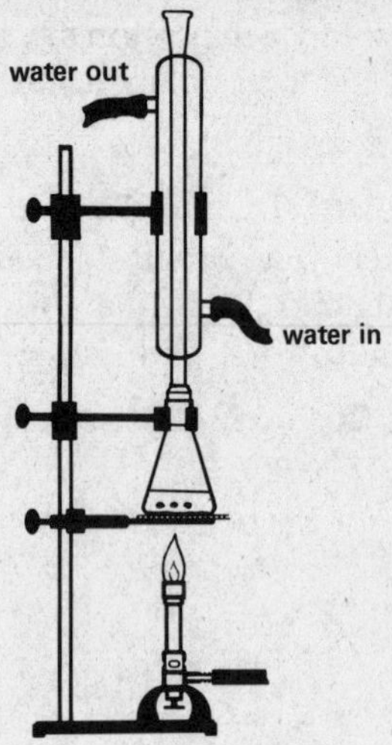

7. Assemble a reflux system as illustrated (and as described in detail in Investigation 33), but *do not grease* the ground-glass joint (since acidic $Cr_2O_7^{2-}$ might react with the grease). Attach the reaction flask to the condenser and have your instructor check the apparatus. Then adjust the heat for gentle reflux and maintain reflux for 1 hr.

8. Prepare a *blank* in a second ground-glass-necked 250 ml flask by mixing 10 ml of distilled water and 10.0 ml of $Cr_2O_7^{2-}$ solution (recording volume readings to ± 0.01 ml), and *carefully* adding 30 ml of concentrated H_2SO_4, using the technique described in procedure 6. Assemble a second reflux system, as described in procedure 7. Be sure to add the glass beads before attaching the *blank* flask to the condenser. Have this setup checked, then reflux the *blank* solution for 1 hr.

9. When the $Fe(NH_4)_2(SO_4)_2/H_2SO_4$ solution has cooled, rinse and fill the Fe^{2+}-labeled buret with this solution. Drain enough liquid into a "waste" flask to expel air from the buret tip and to drop the liquid level to below the zero mark.

10. Drain 10.0 ml of the $Cr_2O_7^{2-}$ solution into a clean 250-ml Erlenmeyer flask. Add 90 ml of distilled water then, carefully with swirling, add 30 ml of concentrated H_2SO_4. Allow this solution to cool to room temperature, using a cooling bath of tap water if desired, then add three drops of ferroin indicator. Titrate this solution with the Fe^{2+} solution to a red-brown endpoint. You should titrate rather slowly since the endpoint color change is quite sudden. Record all volume data to ± 0.01 ml. Repeat procedure 10 with a second 10.0-ml $Cr_2O_7^{2-}$ sample.

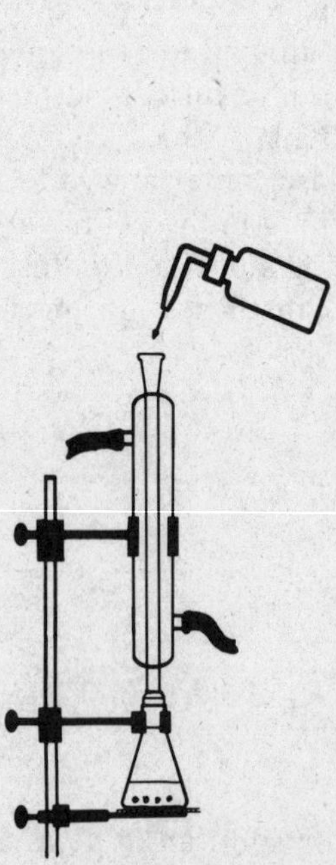

11. After the "unknown" and "blank" solutions have each refluxed for 1 hr, turn off the respective burners and water taps. Rinse down the inside of each condenser with a small portion of distilled water. Check that the condensers are securely clamped, then, protecting your hands with a towel, lower the ring clamps and carefully detach the flasks, using a gentle twisting motion. Add 90 ml of distilled water to each flask, cool the flasks to room temperature, add three drops of ferroin indicator to each flask, and titrate the respective solutions with the Fe^{2+} solution to a red-brown (sudden) endpoint. Record all volume data to ± 0.01 ml. When you have finished your titrations, clean all equipment and dispose of solutions and excess reagents as directed by your instructor.

CALCULATIONS AND REPORT Use your notebook records and calculations to complete the Report Form. You will need to balance the equation for the titration reaction,

$$Fe^{2+} + Cr_2O_7^{2-} \longrightarrow Fe^{3+} + Cr^{3+}$$

(in acidic solution)

and you will need to assume that the potassium dichromate used was a *primary standard,* that is, that the concentration of the $Cr_2O_7^{2-}$ was accurately determined by the solute mass and solution volume used.

Attach separate pages showing all calculations.

GRADE	INSTRUCTOR'S INITIALS

Report Form

FOR INVESTIGATION 39:

Chemical Oxygen Demand

NAME: ____________________ ID NO.: ____________________

LAB SECTION ____________________ INSTRUCTOR ____________________

Attach separate pages with all calculations.

Description of Reagents and Solutions

$K_2Cr_2O_7$

Calculated mass required ________ g

$K_2Cr_2O_7$ + weighing paper ________ g

Weighing paper ________ g

Mass of $K_2Cr_2O_7$ used ________ g

$Fe(NH_4)_2(SO_4)_2 \cdot 6H_2O$

Calculated mass required ________ g

Salt + beaker ________ g

Beaker ________ g

Fe(II) salt used ________ g

Concentration of $Cr_2O_7^{2-}$ solution ____________ M

Identification number (or code) of "unknown" ____________________

Volume of "unknown" used ____________ ml

	Fe^{2+} standardization		Blank	Unknown
	Trial 1	Trial 2		
Final volume $Cr_2O_7^{2-}$				
Initial volume $Cr_2O_7^{2-}$				
Volume $Cr_2O_7^{2-}$ used				
Final volume Fe^{2+}				
Initial volume Fe^{2+}				
Volume Fe^{2+} used				

Calculated Values

Balanced equation for the titration reaction

______ Fe^{2+} + ______ $Cr_2O_7^{2-}$ + ______ $H^+ \longrightarrow$ ______ Cr^{3+} + ______ Fe^{3+} + ______ H_2O

Concentration of Fe^{2+} solution (based on the balanced equation, the $M_{Cr_2O_7^{2-}}$, and the "standardization" data)

Trial 1 ____________ *M* Trial 2 ____________ *M* Average ____________ *M*

(OVER)

	Blank	"Unknown"
mmol $Cr_2O_7^{2-}$ added ($M_{Cr_2O_7^{2-}}$ × No. ml added)		
mmol $Cr_2O_7^{2-}$ remaining (from titration data, $M_{Fe^{2+}}$, and the balanced equation)		
mmol $Cr_2O_7^{2-}$ consumed [(mmol added) − (mmol remaining)]		

$Cr_2O_7^{2-}$ *actually* consumed by "unknown" ____________ mmol
[("Unknown's" apparent consumption) − (mmol consumed by *blank*)]

COD of "unknown" ____________ ppm (O_2)

PROBLEM: Assuming that glucose ($C_6H_{12}O_6$) is completely oxidized by $Cr_2O_7^{2-}$ to CO_2 and H_2O, with reduction of $Cr_2O_7^{2-}$ to Cr^{3+}, what is the COD of a solution containing 0.684 g of glucose per 100 ml with no other solutes?

____________________ ppm (O_2)

DATE ____________ SIGNATURE ____________

Appendix A

SOME LABORATORY TECHNIQUES

Contents

USE OF A LABORATORY CENTRIFUGE

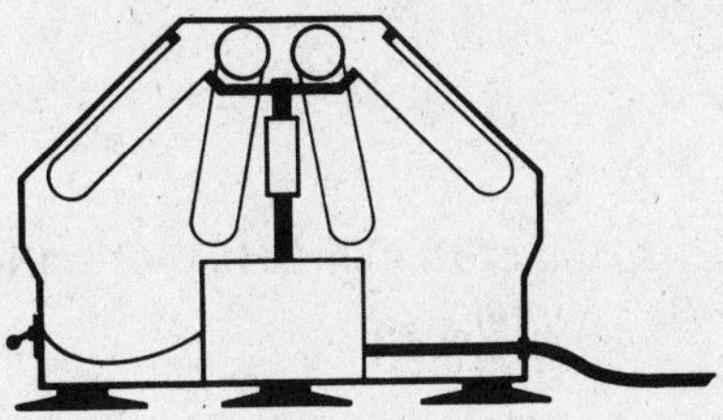

A centrifuge is useful for the rapid separation of an insoluble solid from a liquid medium. The centrifuge consists of a housing inside of which are located the electromotor and gear system used to spin the rotor head. Typical instructional laboratory centrifuges contain two, four, or six openings for 10-cm test tubes in the rotor head. There is a power cord and an on-off switch. To use the centrifuge, you should follow a regular sequence:

1. Check that the centrifuge is plugged in to a 110-V line.

2. Check the rotor and remove any test tubes left in the rotor head.

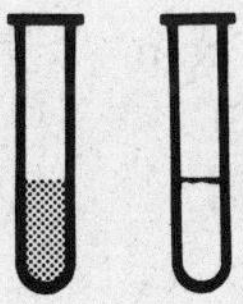

3. Fill an empty test tube with distilled water to the level of the mixture in your sample tube. This will allow you to balance the centrifuge.

4. Insert the two test tubes into the metal thimbles of *opposite* holes in the rotor head.

5. Turn on the centrifuge.

6. Wait until the rotor has attained full speed.

7. When full speed has been attained, switch off the centrifuge.

8. Let the rotor come to a *complete* stop on its own.

9. Remove both test tubes.

SAFETY PRECAUTIONS

Do not touch the rotor while it is in motion.
Do not attempt to stop the rotor motion, even when the motor is switched off.
Do not come close to the moving rotor with long hair or loose clothing.

PRACTICE USING A CENTRIFUGE

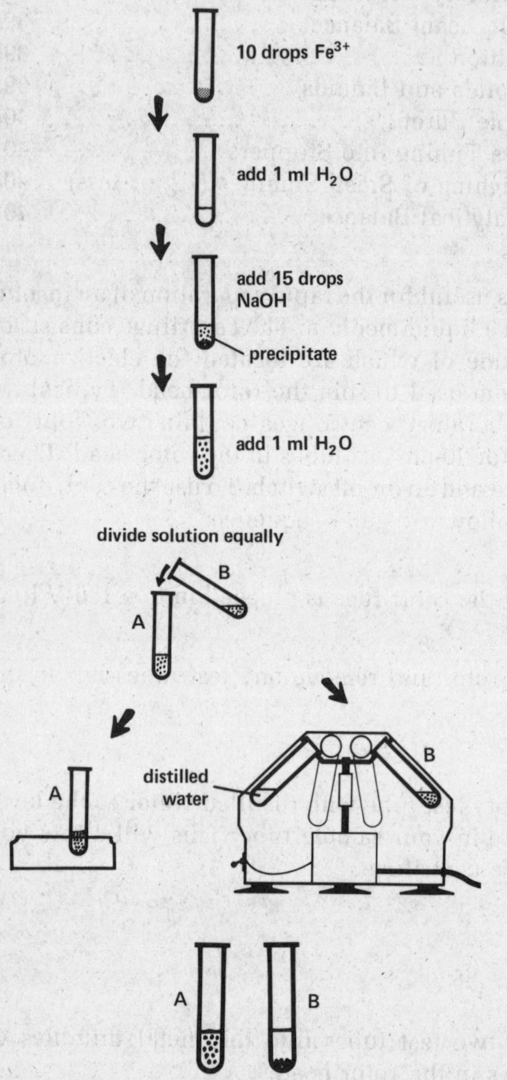

1. Place 10 drops of an Fe^{3+} solution in a clean centrifuge tube.

2. Add 20 drops of distilled water.

3. Add dilute NaOH solution, dropwise, until a flocculent brown precipitate appears.

4. Add 10 additional drops of the NaOH solution and 20 drops (1 ml) of distilled water.

5. Mix the contents of the tube thoroughly and pour half into a clean test tube. Set this tube in the test tube block.

6. Centrifuge the other tube, using the procedures described.

7. Compare the appearance of the centrifuged and non-centrifuged tubes. Attempt with both tubes to decant (pour off) the clear solution without transferring solid $Fe(OH)_3$.

USE OF A TRIPLE BEAM BALANCE

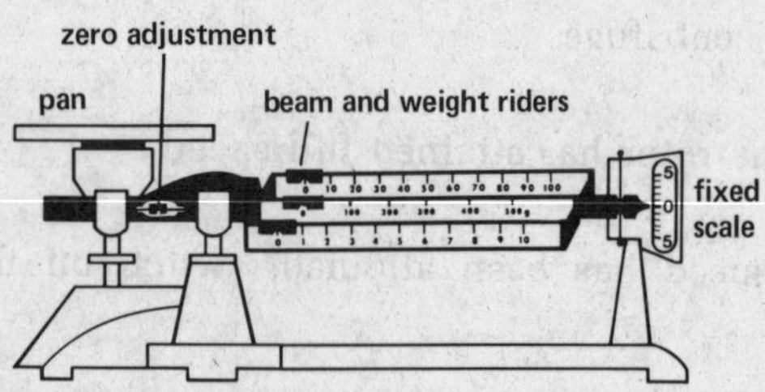

A triple beam balance (used when weighings within ±0.05 g are appropriate) consists of a pan, a supporting base system, a fulcrum, a balance beam, a balance-adjustment screw, three beams with appropriate scales and mass riders, a pointer, and a fixed scale. Examine the balance in your laboratory to identify the parts.

The following procedure is recommended.

Remember: Never place chemicals directly on the balance pan! Always weigh chemicals in a flask or beaker or on a watch glass, aluminum foil, or weighing paper.

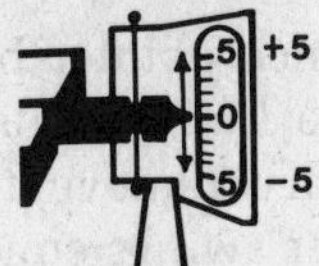

First check the balance "zero" position. Slide the 100-g and 10-g mass riders into the zero notches and the 1-g rider to its exact zero position. Then *lightly* push the balance beam to cause it to swing* between the (+5) and (−5) pointer positions on the fixed scale. If the balance is properly zeroed, successive pointer readings should include (+4)–(−4), (+3)–(−3). If you observe readings such as (+5)–(−3), (+4)–(−2), carefully turn the zero adjustment screw until the balance is properly zeroed.

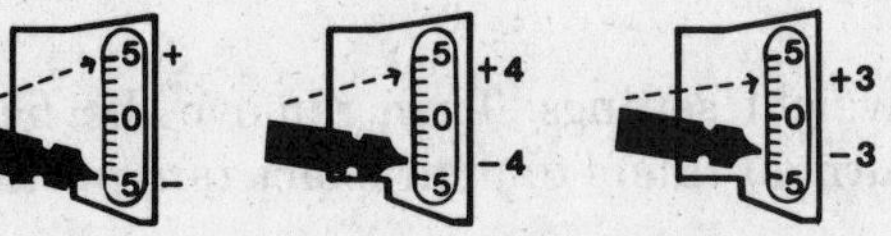

Now, as an example, let us trace through the steps for weighing out 7.8 g of a solid to the nearest 0.1 g, after having zeroed the balance.

1. Place a clean dry beaker onto the balance pan and set the 100-g rider in the 100-g notch. If the balance pointer goes down, the beaker is lighter than 100 g and you will have to set that rider back in its zero notch. If the balance pointer goes up against the stop, the beaker is heavier than 100 g. In this case, advance the 100-g rider one notch at a time until the pointer goes down, then set it back one notch.

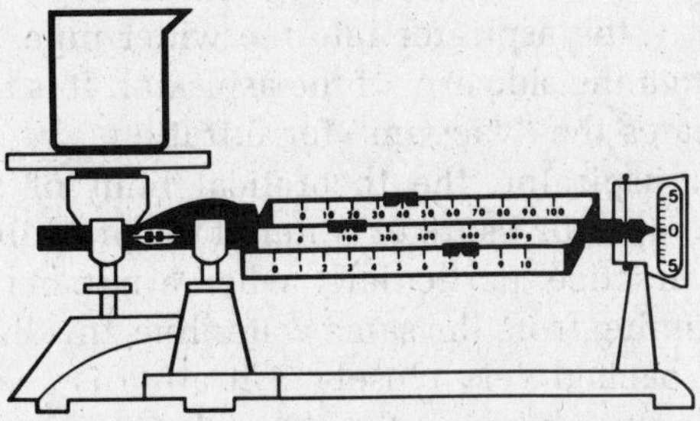

2. Repeat the above procedure with the 10-g rider until the pointer goes down, then set this rider back one notch.

3. Now advance the 1-g rider to around the 5-g position. Slide it back or forward from this position, in decreasing increments, until the balance pointer swings as it did for the zero position.

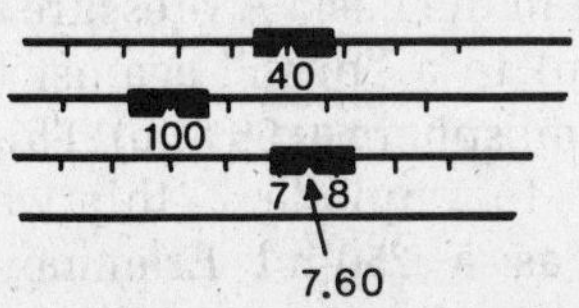

4. Now read and record the total weight settings. The illustration shows a combination of rider positions amounting to 147.6 g. You can learn to estimate the position of the 1-g rider to about ±0.01 g, but this balance is not reliable (usually) to better than ±0.05 g.

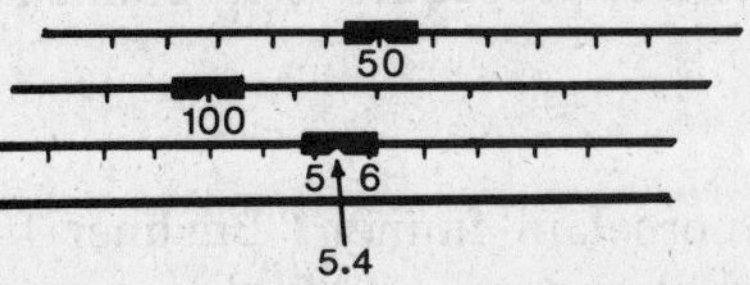

5. Now, to obtain the desired 7.8-g sample, change the rider positions to give a total reading of 7.8 g more than that of the empty beaker (i.e., since 147.6 + 7.8 = 155.4, reset the 10-g rider in the 50 notch and slide the 1-g rider to the 5.4 position).

6. Add small increments of your sample, being careful not to spill any, into the beaker. From time to time, push gently on the balance pan. If it seems reluctant to swing, you are not yet close to the desired mass of sample. If it starts to swing easily, reduce the size of the sample increments

*If the balance you are to use is magnetically damped, consult your instructor.

being added. When the balance pointer begins to swing freely,* carefully observe the limits of its swing on the fixed scale. Add (or, if necessary, remove) tiny increments of the sample until the pointer swings equally above and below its zero position. (This is referred to as "returning the balance to equilibrium.")

7. Record the weight settings. Then remove the beaker (gently) and return all riders to their exact zero positions.

If you spilled any sample on the balance pan, it will reduce the accuracy of your sample mass. In any event, check before leaving the balance that both the balance and the surrounding area are left completely clean.

VACUUM FILTRATION

Although centrifugation is an effective separation method for many solid-liquid mixtures, it is limited (with typical instructional laboratory centrifuges) to small samples. Filtration is normally used for larger mixtures and, except for gelatinous or very fine precipitates, the application of a vacuum is a useful aid to rapid filtration.

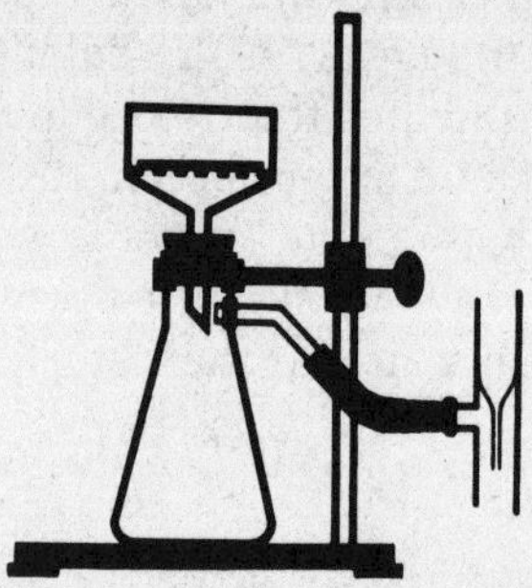

A simple vacuum filtration apparatus consists of a suction flask with a side arm and thick glass construction, a funnel (containing a rigid support for the filter) inserted through a rubber stopper, a thick-walled vacuum hose, and a water aspirator. When water rushes through the constricted tube of the aspirator into the wider pipe, air is pulled in through the side arm of the aspirator. It is this air flow that creates the "vacuum" for filtration.

With a water aspirator, the theoretical limit of the "vacuum" is the vapor pressure of water at the prevailing temperature. In practice, particularly when a number of aspirators are running from the same water line, this limit may not be approached very closely. Filtration is accelerated by the difference between the atmospheric pressure against the surface of the liquid in the funnel and the "vacuum" pressure within the flask. A pressure difference of 730 torr is equivalent to a "push" against the liquid surface of almost one atmosphere (0.96 atm). This pressure difference is sufficient to implode a thin-walled flat-bottomed flask (such as a 250-ml Erlenmeyer). Flat-bottomed flasks (unless they have special thick glass) that are larger than 125 ml must never be evacuated.

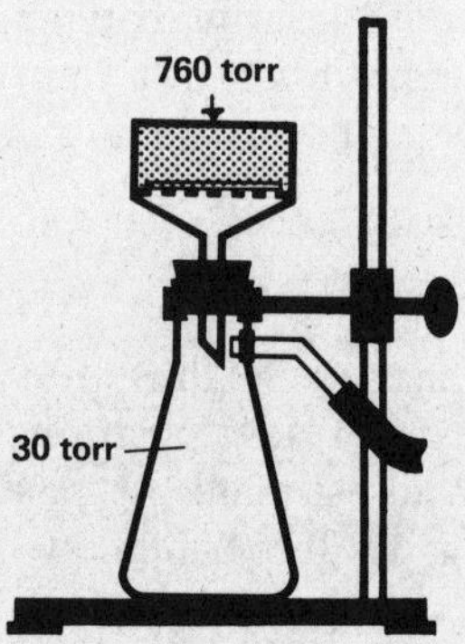

The following general procedure is recommended for vacuum filtrations.

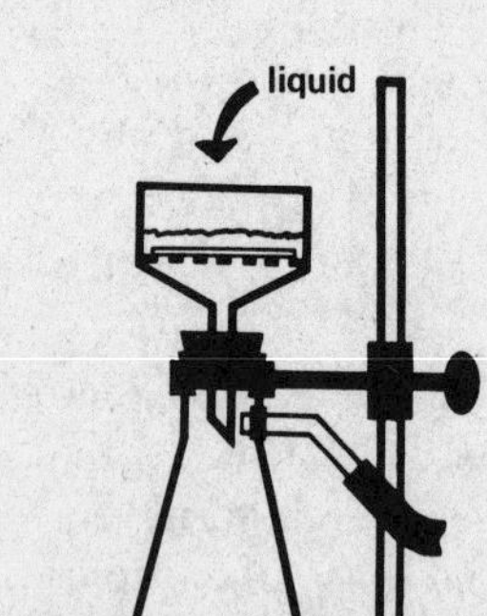

1. Secure a clean porcelain funnel ("Buchner funnel") inserted through a rubber stopper to fit the vacuum flask. Set the funnel and stopper securely in the neck of the flask. Connect the flask to a water aspirator by a length of vacuum hose. (To keep the flask stable, clamp it to a ringstand.) Place a filter paper (of a size to fit easily in the funnel while covering all the holes in the porcelain base of the funnel) into the funnel and moisten the paper. Turn on the aspirator to pull the moistened paper snugly against the filter support.

*Unless magnetically damped.

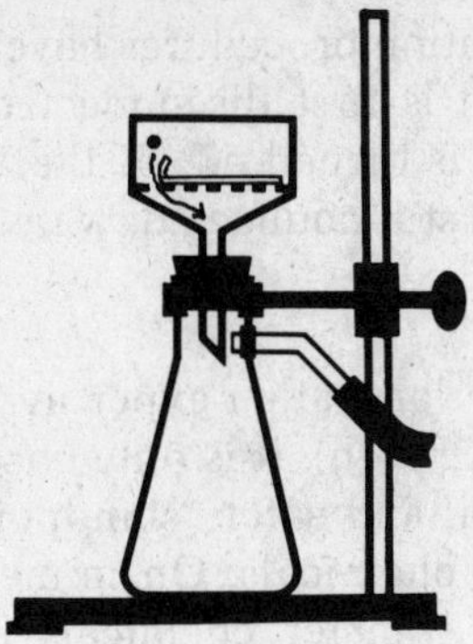

If the paper is not moistened, it might curl up when a mixture is poured into the funnel, permitting part of the mixture to escape filtration.

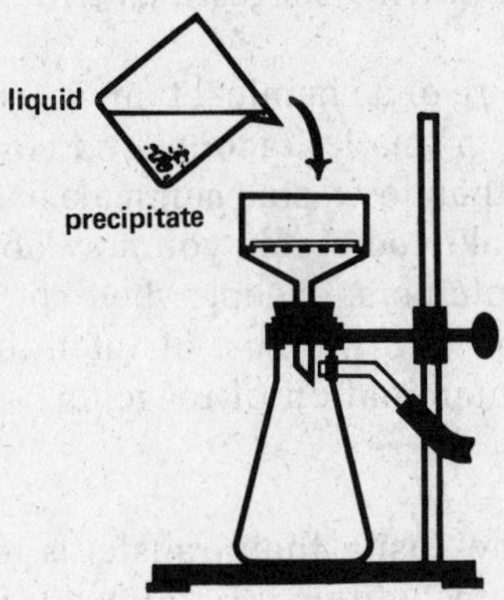

2. Now decant as much liquid as possible away from settled solids into the funnel. Leave as much of the solid in the beaker as possible during the initial part of the filtration. The liquid will filter more rapidly before accumulated solid begins to clog the pores of the filter paper. (Note: Do not fill the funnel more than two-thirds full at any time.)

3. After most of the liquid has passed through the filter, transfer the remaining liquid and solid into the funnel. Rinse the original container, with small volumes of the liquid used, to transfer residual solid into the funnel. Continue with the vacuum at least until no further liquid drops are coming from the funnel stem. The further treatment of the solid in the funnel will vary, depending on particular experimental requirements. For example:

a. It is generally desirable to wash collected crystals with a small amount of liquid (usually the original solvent). To do this, first disconnect the vacuum hose carefully from the side arm of the flask. Then pour the desired amount of wash liquid into the funnel and stir the mixture with a clean spatula, being very careful not to disturb or tear the filter paper. Reconnect the vacuum hose to pull the wash liquid through the filter. Washing may be repeated as necessary.

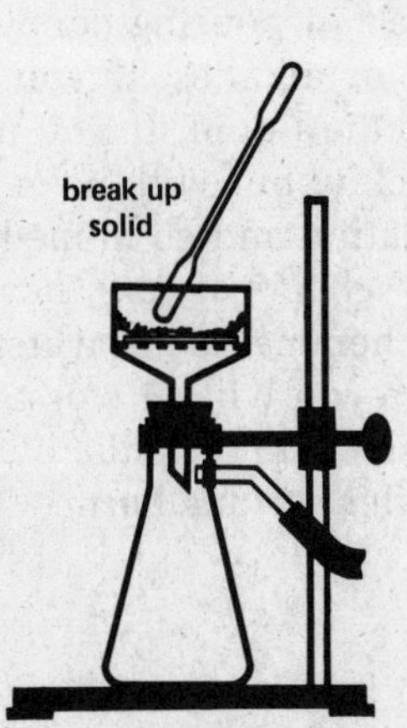

b. If it is desired to air dry the collected solid, after all liquid has been removed, the solid cake may be carefully broken up into finer pieces with a spatula or glass rod. The aspirator is allowed to pull air through the solid until it is dry.

4. When the desired filtration procedures have been completed, the vacuum hose is first disconnected from the flask. *Then* the aspirator is turned off. (If the aspirator is shut off while the hose is still connected, water may back into the suction flask.)

TRANSFER OF SOLIDS AND LIQUIDS

Most laboratory chemicals are rather expensive. Many are corrosive or poisonous (or both). It is only good common sense, from both financial and safety standpoints, not to spill or otherwise waste chemicals. On numerous occasions, you will have to transfer chemicals from stock bottles into beakers or flasks, or from one piece of equipment to another. The following suggestions will facilitate such operations.

(Note: Never take more of a chemical than you actually need. It is better to take a smaller amount and come back for more, if necessary, than to waste chemicals needed by other students. If you take too much, you may not return excess to the stock containers, except when specifically directed to do so. Even with the best of intentions, this often results in the contamination of reagents.)

Liquid Transfer

Since liquids usually flow easily, their transfer is generally quite simple if a few precautions are followed. Usually, you will be pouring the liquid from a large container into a smaller one. This becomes a problem only if there is a *major* difference in size of the containers (e.g., pouring from a 5-liter jug into a 10-cm test tube). Should this appear necessary, pour the liquid first into a small beaker. Then pour from this into the test tube.

To pour from a large screw-cap bottle (or one having a flat-topped glass cap), first remove the cap and place it *upside down* on a clean paper towel. Then slowly and carefully tilt the bottle, holding your collection vessel at the mouth of the bottle, until liquid pours slowly and evenly into the container. Avoid rapid, jerky movements that might splatter the liquid. Replace the cap when you have finished. If any liquid has run down the outside of the container, wipe it off carefully with a damp paper towel.

When pouring from a glass-stoppered bottle having a "vertical disc" top, hold the stopper between two fingers (as shown). Never place this type of stopper on the bench top.

Special care is required in pouring concentrated reagents, particularly acids or alkalies. If you are careful, you will not spill any of the liquid. If you do, however, you must react very quickly, but without panic. Set the bottle down in a safe location and go immediately to the nearest sink. Wash off the chemical very *thoroughly* with water (using soap if the chemical is not water soluble). If you follow this procedure you will, at worst, get a minor skin burn. Should you panic and drop the bottle, you and your neighbors could suffer serious harm.

Caution: If you use a pipet for liquid transfer, *never use mouth suction.* Always use a mechanical suction device, such as a pipet bulb.

Solid Transfer

Before attampting to transfer a solid, be sure the contents of the bottle are loose and in small pieces. If necessary, tap the *closed* bottle against the palm of your hand to loosen the contents. If this does not work, consult your instructor. Never use a spatula, scoopula, or glass rod to break up the solid in a reagent container. You might inadvertently contaminate the reagent. (In special cases you may be directed to use a scoopula or spatula for solid transfer. If so, be sure the instrument is *completely* clean and dry.)

The trickiest type of solid transfer is one in which a specific weight of solid is required. During such a transfer, any spillage will not only contaminate the working area or balance, it will also, if the spill is onto the balance pan, invalidate the weighing.

Solids are normally stored in wide-mouth screw-cap reagent bottles. After you have checked that the solid is of appropriate particle size, remove the screw-cap and set it *upside down* on a clean surface. If others are working with different reagents in the same area, be careful to return the correct cap to the bottle.

Let us consider some typical solid transfer situations.

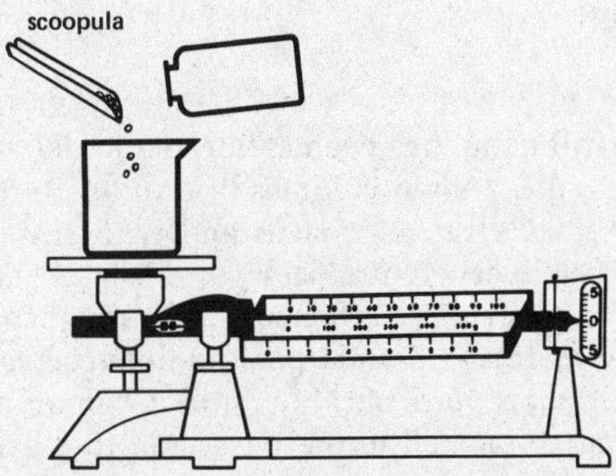

1. *Obtaining an approximate mass of solid, using a balance and a scoopula assigned to the particular reagent:* Weigh the container to be used (e.g., a 100-ml beaker). Then obtain the reagent bottle and the *assigned* scoopula. Check that the solid is loose and that the scoopula is clean. Advance the balance masses by the amount of reagent required. Tilt the opened bottle and insert the scoopula to obtain a small amount of reagent. While the scoopula is inside the bottle, tap it gently to shake off any solid that might otherwise fall off when the scoopula is removed. Carefully move the scoopula to inside the beaker and allow the solid to slide off. If any adheres to the scoopula, tap the hand holding the scoopula with your other hand. When you are approaching the desired mass, the solid must be added in very small increments. To do this, hold the scoopula at a slight angle in the beaker and tap the hand holding the scoopula with your other hand. Watch the balance pointer carefully and stop adding solid exactly at the desired mass. In this special situation, return any unused solid to the reagent bottle. Recap the bottle, wipe the scoopula with a clean paper towel, and return the bottle and scoopula to their assigned locations.

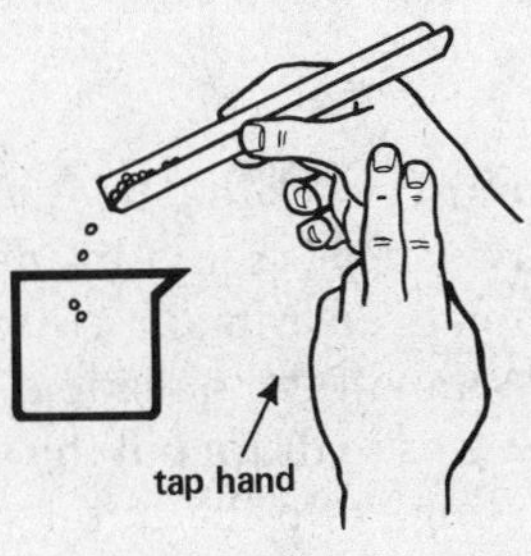

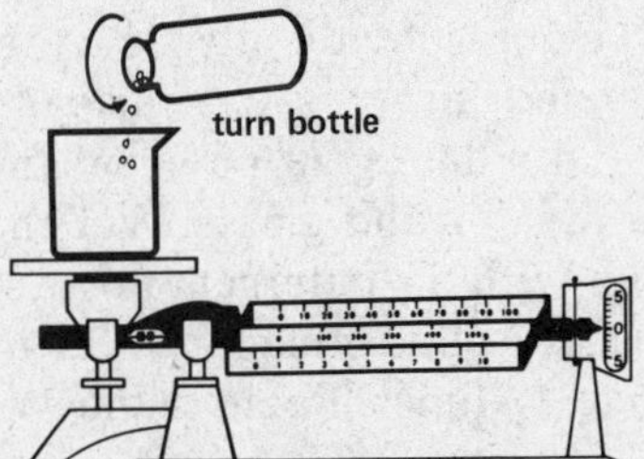

2. *Obtaining an approximate mass of solid without the use of a scoopula:* Solid transfer is usually accomplished by use of a creased piece of paper or aluminum foil of appropriate size. To pour the solid from the bottle, hold the bottle at a slight angle and rotate it slowly back and forth to deliver solid slowly onto the paper or foil. This method can also be used to pour solid directly into a preweighed beaker or wide-mouth flask set on a laboratory balance. Care must be exercised to deliver small increments as the required mass is approached. This method takes a bit of practice.

3. *Obtaining a small sample of solid, which must be accurately weighed:* It should be noted that a spatula is only suited for transferring small amounts of solid. The possibility of spilling solid from a spatula is considerably greater than from a scoopula or creased paper (or foil). Usually, for small samples, the required mass is specified by a limiting range (e.g., 0.2 to 0.3 g), but once the approximate mass has been transferred, its mass may have to be measured accurately (e.g., to ±0.1 mg). Spillage inside the analytical balance *must* be avoided. If it should occur, the spilled sample must be brushed out carefully and immediately. It is best if a corrosive reagent is being used to perform addition or removal of sample with a preweighed watch glass or other container *outside* the balance. With "safer" reagents, the sample can be added to the container on the balance pan, but spillage must still be avoided. Addition of small increments is facilitated by a careful "hand tapping" similar to that described earlier, but with the arm holding the sample resting on the balance table. Before attempting any use of the analytical balance, study the sections of Appendix A describing precision weighing and use of an analytical balance.

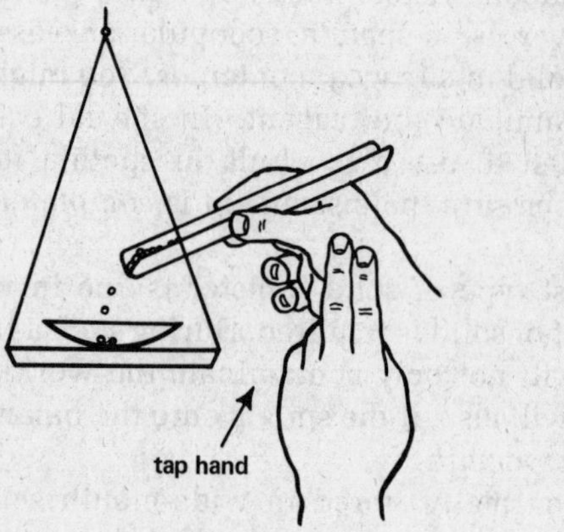

Note: In all chemical transfer operations, be very careful to avoid spillage. If spillage does accidentally occur, clean it up as directed by your instructor.

USE OF A SIMPLE BURET

A buret is a glass tube that has been calibrated to deliver, in controlled amounts, known volumes of liquids. Burets are available in a wide range of sizes and with several different types of delivery control devices, such as glass or teflon stopcocks or a simple pinchcock. The pinchcock is the most easily maintained and operated delivery control. It consists of a short piece of rubber tubing containing a glass sphere and connected to the buret and to a glass tip. To deliver liquid from the buret, the rubber tubing is squeezed between the thumb and middle finger to form an ellipsoidal shape which allows liquid to flow past the glass sphere.

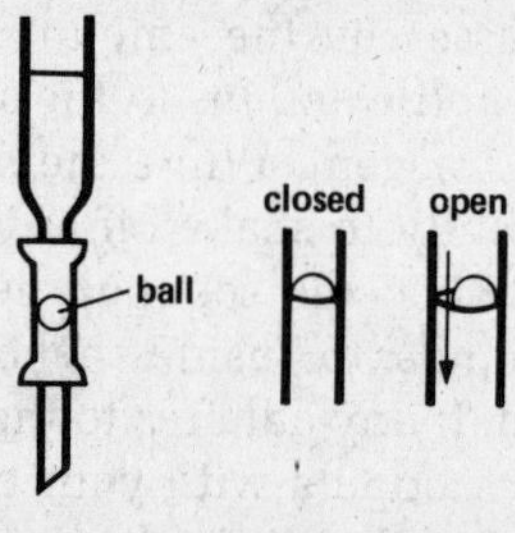

Before using a buret, you must check that it is clean, paying particular attention to any dirt or grease inside the buret which causes water droplets to remain when water is drained from the buret. Wipe off the outside of the buret, fill it with tap water, allow this to drain out the tip, and see if droplets are observed clinging to the buret walls. If so, you must clean the buret thoroughly.

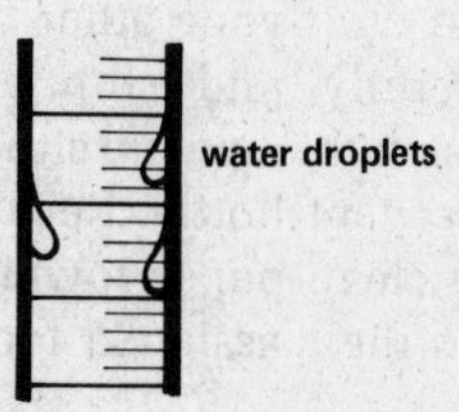

Various special cleaning mixtures are sometimes employed, but most of these are dangerously corrosive and some ("chromic acid") can react violently with the rubber tubing. It is recommended that you use instead a long-handled buret brush and a detergent powder* (*not* soap, which often leaves a film on the glass). Wet the brush and sprinkle the bristles with the detergent powder. Insert the brush into the buret and slide it back and forth the full length of the buret 10 to 12 times. Remove the brush and

*Abrasive cleansers must not be used, since repeated use will scratch the interior of the buret.

rinse the buret with tap water several times. If it still does not drain clean, repeat the cleaning. Once the buret drains clean, rinse it several times with distilled water. There is no need to use a lot of distilled water. Use the plastic squeeze bottle to squirt water around and around the inside of the buret top so that a thin film rinses down the entire inner surface. Drain the rinses through the pinchcock so that this and the tip are also rinsed.

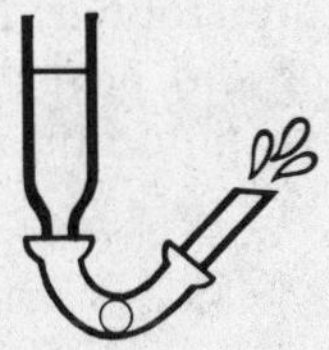

After the buret has been cleaned, rinsed, and clamped to a ringstand, use two successive 5-ml portions of the titrant solution to rinse the buret again. Pour these in with a circular motion so that a thin film of the titrant rinses down the complete inner surface and is expelled through the tip. Then fill the buret with the titrant to above the zero mark. Drain out enough titrant into a waste flask to expel air from the buret tip and to drop the liquid level to just below the zero mark. Read and record the liquid level.

When you have finished with the buret, drain out any remaining titrant and rinse the buret two or three times with distilled water as before. Either return the buret to the storage area or, if it is to be left clamped to the ringstand, fill it with distilled water and cover it with an inverted test tube or a small piece of aluminum foil.

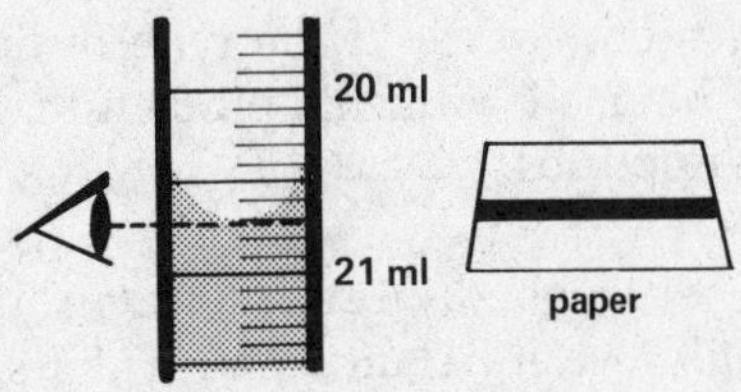

To read the buret, have your eye at the same level as the meniscus (curved surface) of the liquid. Record the position of the bottom of the meniscus with respect to the graduation marks on the buret. A black-and-white marked card, held so that the dark line is just below (and reflected by) the bottom of the meniscus, will help improve the accuracy of your readings. With practice you can learn to read a buret to ±0.01 ml.

If you are to use burets equipped with plastic or teflon stopcocks, ask your instructor for a description of their use and care.

INSERTING GLASS TUBING INTO STOPPERS

One of the most common laboratory accidents involves the breakage of glass tubing or other glass equipment while attempting its insertion into a stopper. Two rules are *essential*.

Always protect your hands with a towel and never apply leverage or excessive force. Before attempting to insert glass tubing, be sure both ends of the tubing are smooth and rounded. If not, heat the end of the tubing at the edge of the hot inner cone of a burner flame until the glass softens to a smooth, rounded edge. *Let the glass cool* before working with it again.

The following procedure is recommended:

1. Insert glass tubing (or other equipment) only into a stopper having the proper size hole. The hole must be just *slightly* smaller than the tubing to give a snug fit, but *not* too small.

2. Lubricate the stopper hole and the part of the tubing to be inserted with a few drops of glycerin or water.

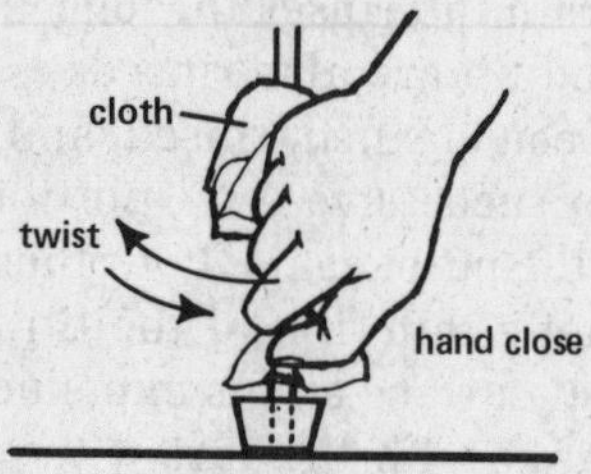

3. Protect both hands with a towel. Grasp the tubing *close to the end to be inserted* and apply a twisting motion while pushing the tube gently into the stopper. Never push at the end away from the stopper and never use excessive force. If you have problems, consult your instructor.

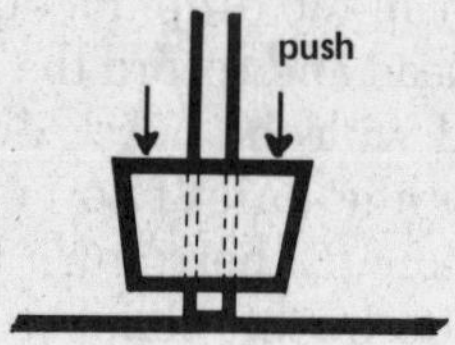

If you should have to remove tubing (or other glass equipment) from a stopper, protect your hands with a towel and push down *on the stopper* while the tubing is set *at a 90 deg angle* against a paper towel on the bench top. If the tubing does not loosen readily from the stopper, consult your instructor. *Do not use excessive force.*

PRECISION WEIGHING OF SMALL SAMPLES (1 g or less)

Check with your instructor to see if the type of balance available to you can be used with this procedure. If not, request alternative directions.

If samples are to be weighed more accurately than is possible with the triple beam balance *and* if the sample does not weigh more than 1 g, an analytical balance may be used in a simplified manner not requiring detailed balance instructions. The balance must, however, have an optical scale with a range of 1.2 g. If the balance is in good operating condition the zero point of the balance need not be adjusted.

The following steps should be used for such weighing.

1. Place a small piece of paper (5 × 5 cm) on the balance pan. Close the balance doors. The fractional mass dial must be at 00. Return it to this position if necessary.

2. Turn the arrest knob to the left stop (the released position).

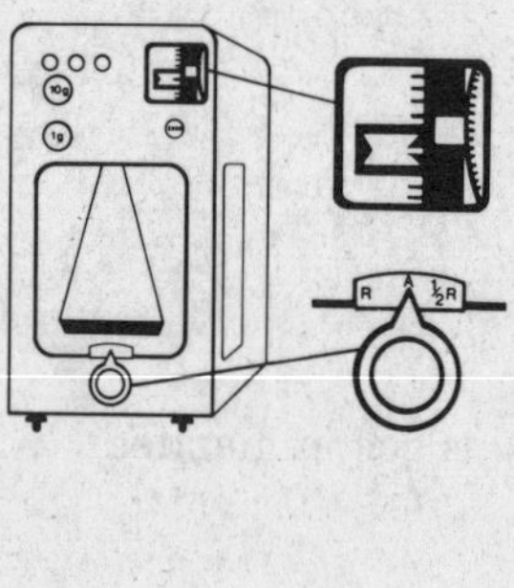

3. Wait until the optical scale in the light window stops moving.

4. Turn the fractional mass dial until the line just below the indicator gap is centered in the gap.

5. Read the masses. For example, when the line representing 22 is centered in the gap, and if the tape window shows 56; the mass is 0.2256 g or 225.6 mg.

6. Arrest the balance by turning the arrest knob to the 12 noon position and return the fractional mass dial to 00.

7. Place your object or sample on the paper.

8. Repeat steps 2 through 6 and record the new mass. Arrest the balance and remove paper and object from the pan.

9. Calculate the difference between the two readings to find the mass of your object or sample.

10. Make sure balance doors are closed, pan is arrested, and all dials read zero.

Note: Should you have spilled anything inside the balance, for example, a few crystals, remove them all with a brush.

USE OF AN ANALYTICAL BALANCE

Although all single-pan analytical balances operate on the same basic principles, the procedures for their use will vary. The description included here applies to the Ainsworth Model 21N. If you are to use a different model balance, ask your instructor for the necessary modifications of the procedures described.

Presented here are only the operations that have to be performed in order to weigh a sample to the nearest 0.1 mg. It is not absolutely necessary to know what goes on inside the balance during a weighing operation to obtain correct results. It might, however, be worthwhile to satisfy your curiosity and learn what occurs "inside the white box." If you are interested, consult your instructor.

The parts of the balance that you have to know are as follows.

1. The Leveling Bubble. The air bubble must be centered in the black circle. If it is some place else, call it to your instructor's attention.

2. The Two Balance Doors. When you approach the balance the doors should be closed. If you find them open, close them.

3. The Balance Pan. The balance pan *must not* be littered with chemicals. If it is, clean it carefully with a brush and remove all traces of chemicals from the balance compartment.

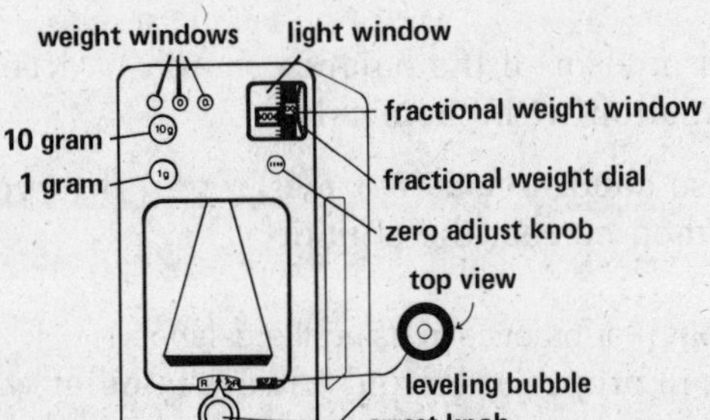

4. The Arrest Knob. This knob has three positions. When the balance is not in operation, the pointer on the knob must *always* point to the arrested position. Turning the knob to the right, to the stop, will partially release the balance beam. The beam will be *entirely* free when the knob is turned to the left.

5. The 1G (1-g) Dial. The 1-g dial moves masses from the balance beam in units of 1 g up to 9 g. The total number of grams removed will automatically be shown in the right circular weight indicator window. *Always* turn the mass dial very slowly and carefully.

6. The 10G (10-g) Dial. The 10-g dial removes masses from the balance beam in units of 10 g up to 150 g. The total number of 10-g units removed will be automatically recorded in the middle and left circular weight indicator window. Always turn the mass dial very slowly and carefully.

7. The Light Window. The recessed light window will show part of a mass scale, which consists of 120 divisions (00 to 120), each division representing 0.01 g. The division scale is lighted when the arrest knob is turned to the partial released or released position. With the pan empty, you will see in the light window the first 10 positions. Painted onto the matted glass in the light window is an "indicator gap." The glass with the gap can be moved up and down by turning the "zero adjust knob."

8. The Fractional-Mass Dial. To the right of the light window is a dial coupled with a tape with numbers 00 to 100, which appear in the small window to the left of the dial. The numbers being read in this small window represent the thousandth and 10-thousandth gram units (milligrams and 0.1 mg). This dial should be set to have 00 in the small window. Dial to 00 if you see another number in this window.

9. The Zero Adjust Knob. The zero adjust knob moves up or down the matted glass with the indicator gap in the light window. The knob is used to center the 00 mark in the indicator gap when the balance is in the released position when unloaded.

When you step up to the balance you should find it in the following condition:

a. Both balance doors should be closed.
b. The arrest knob should be in the arrested position.
c. The right and the middle circular weight indicator windows should show 0, the left one should be black.
d. No light should be showing in the light window.
e. The fractional mass dial should have been returned to the position that made the 00 mark appear in the small window.
f. The pan and interior of the balance should be free of any contamination with chemicals.

If one of these conditions is not met, report it to your instructor and then correct the situation.

Weighing of a Coin (For practice in using the balance.)
If the balance is in proper order and you are familiar with the functions of the parts of the balance necessary for its operation, practice the weighing procedure using a coin. Proceed in the following way.

1. With all dials at zero, nothing on the pan, and both doors closed, turn the arrest knob to the released position.

2. Center the 00 mark projected onto the light window in the indicator gap by turning the zero adjust knob. As soon as you have done this, do not touch the zero knob again.

3. Turn the arrest knob to the arrested position.

4. Open one balance door, place a coin on a spatula and transfer the coin onto the center of the balance pan, *gently*. Close the door.

5. Turn the arrest knob to the partial released position. Should the balance pan begin to swing like a pendulum, move the arrest knob gently several times between the arrested and partially released positions until the pan stops. Bring the arrest knob to the partially released position.

6. Check the light window. A mark somewhere between 25 and 35 will be inside the indicating gap. Watching this scale, turn the 1G dial gently counterclockwise. When the sign "Remove Weight" appears on the scale, turn the dial clockwise to the preceding setting. Should the "Remove Weight" sign not have appeared with 9 g in the weight window, turn the 1G dial one more notch and begin with the 10G dial. (This will not happen with coins, unless you have a silver dollar.) When the "Remove Weight" sign appears, back up one 10-g unit. Now repeat this procedure with the 1G dial, and again back up one unit as soon as the "Remove Weight" sign appears. Should you suspect that your object is heavier than 10 g, begin with the 10G dial first.

7. Turn the arrest knob to the arrested position and then to the released position. When the scale in the light window comes to rest, turn the fractional mass dial until the line marking just below the gap is exactly centered in the gap.

8. Read the mass from left to right. For example,

Circular mass windows:	05.
From scale in light window:	03
From fractional weight dial window:	69
Total mass:	5.0369 g

9. Turn the arrest knob to the arrested position. Open the door and remove the coin with your spatula. Close the door. Turn all dials to zero.

The balance is now ready for another weighing.

Determination of the Mass of an Unknown

(Optional, at discretion of instructor)

When you are ready to show that you can weigh an object, ask your instructor for an unknown. Determine the mass of this unknown, recording in your notebook the steps you took to accomplish this. The time limit is 1 min. Report the result to your instructor.

IMPORTANT NOTES

a. Never place any hot object inside the balance or onto the balance pan.

b. Never place any chemical directly on the balance pan. It is strongly recommended that you use a watch glass, a beaker, a piece of paper or foil, or a flask to contain a sample being weighed.

c. Weigh your container first to the nearest 0.1 mg. Then add your substance and reweigh. The difference between these two masses is the mass of your chemical.

d. It is convenient and time saving to use the same watch glass or other container for weighings. If you record their masses in your notebook, you can dial the approximately correct masses without delay.

Weighing Out an "Exactly-Specified" Amount of Solid

For most experimental work, it is more important to determine the mass of a substance with suitable precision and to the number of significant figures *appropriate to* the experimental requirements than it is to weigh out some "exact" amount. Of course, no physical measurement is "exact" anyway. There is always some degree of uncertainty. Nor is it feasible to weigh to the number of significant figures suggested by a calculation on one of our modern electronic calculators.

In most cases a weighing is made by adding increments of the substance to a preweighed container or weighing paper until the amount is within a few percent of that calculated or indicated by experimental directions. *Then* the mass of the substance is determined (and recorded) to the appropriate number of significant figures. Generally, this approach is recommended for three reasons.

a. It is, at best, tedious (and, in fact, often impossible) to *add* increments in such a way as to obtain a desired amount to within less than ±0.1 g. In some cases, as with large crystals or drops of a liquid, there is a physical limit on increment size available.

b. If a slight excess above the desired amount is added, it is usually much easier to take this into account in subsequent work and calculations than it is to *remove* some of the substance.

c. There is always the possibility of a "subconscious" tendency to "make the answer right" by recording the desired mass when that actually measured is slightly different.

In rare cases, however, you will need to weigh an "exactly specified" amount of solid. This cannot, of course, mean a "zero uncertainty" in the weighing. Rather, this phrase refers to extending the limitations on weighing to beyond those normally employed ("a few percent"). For our purposes the technique required for an "exactly specified" amount is one that permits us to weigh out a substance to *within ±2 mg* of a calculated mass. When this is necessary, the following steps must be used.

1. Use standard procedures to weigh the container, watch glass, or weighing paper to be used, recording the mass to the nearest 0.1 mg.

2. With the balance pan *arrest knob* in the "arrested" position, add an amount of the solid estimated to be somewhat less than that required. Determine the mass added. Continue this procedure until less than 1 g of additional solid is required. (Note: Never turn the 1G dial while the

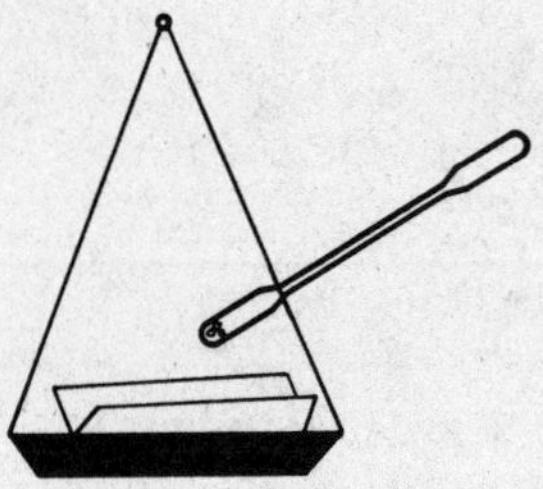

arrest knob is in "released" position.) If, at any stage, too much solid was added, *arrest* the balance, then use a clean spatula to remove excess. (*Do not return excess* to a reagent bottle. Place it instead in a specified waste container.)

3. When less than 1 g of additional solid is required, adjust the mass dials to the settings corresponding to the total mass of the container + the desired amount of solid. *Then* open the balance door and turn the *arrest knob* to the "released" position. *Slowly,* by tapping the spatula handle with your finger, drop on increments until the "exact" reading (within ±2 mg) is registered in the mass window. If too much is added, *arrest* the balance and remove excess as described in step 2. Then continue with step 3 until the desired mass is obtained to within ±2 mg.

4. Determine and record the actual mass used* to the nearest 0.1 mg.

5. When your weighing is completed, *arrest* the balance, return *all* dial settings to their zero positions, and be sure the balance and surrounding area are *clean.*

*If any solid is spilled *on the balance pan,* your mass determination will not be reliable unless the spilled material is removed. *With the balance pan arrested,* use a balance brush to remove spilled solid from the pan and from inside the weighing chamber.

Appendix B

LOGARITHM TABLE

Natural numbers	0	1	2	3	4	5	6	7	8	9	Proportional parts								
											1	2	3	4	5	6	7	8	9
10	0000	0043	0086	0128	0170	0212	0253	0294	0334	0374	4	8	12	17	21	25	29	33	37
11	0414	0453	0492	0531	0569	0607	0645	0682	0719	0755	4	8	11	15	19	23	26	30	34
12	0792	0828	0864	0899	0934	0969	1004	1038	1072	1106	3	7	10	14	17	21	24	28	31
13	1139	1173	1206	1239	1271	1303	1335	1367	1399	1430	3	6	10	13	16	19	23	26	29
14	1461	1492	1523	1553	1584	1614	1644	1673	1703	1732	3	6	9	12	15	18	21	24	27
15	1761	1790	1818	1847	1875	1903	1931	1959	1987	2014	3	6	8	11	14	17	20	22	25
16	2041	2068	2095	2122	2148	2175	2201	2227	2253	2279	3	5	8	11	13	16	18	21	24
17	2304	2330	2355	2380	2405	2430	2455	2480	2504	2529	2	5	7	10	12	15	17	20	22
18	2553	2577	2601	2625	2648	2672	2695	2718	2742	2765	2	5	7	9	12	14	16	19	21
19	2788	2810	2833	2856	2878	2900	2923	2945	2967	2989	2	4	7	9	11	13	16	18	20
20	3010	3032	3054	3075	3096	3118	3139	3160	3181	3201	2	4	6	8	11	13	15	17	19
21	3222	3243	3263	3284	3304	3324	3345	3365	3385	3404	2	4	6	8	10	12	14	16	18
22	3424	3444	3464	3483	3502	3522	3541	3560	3579	3598	2	4	6	8	10	12	14	15	17
23	3617	3636	3655	3674	3692	3711	3729	3747	3766	3784	2	4	6	7	9	11	13	15	17
24	3802	3820	3838	3856	3874	3892	3909	3927	3945	3962	2	4	5	7	9	11	12	14	16
25	3979	3997	4014	4031	4048	4065	4082	4099	4116	4133	2	3	5	7	9	10	12	14	15
26	4150	4166	4183	4200	4216	4232	4249	4265	4281	4298	2	3	5	7	8	10	11	13	15
27	4314	4330	4346	4362	4378	4393	4409	4425	4440	4456	2	3	5	6	8	9	11	13	14
28	4472	4487	4502	4518	4533	4548	4564	4579	4594	4609	2	3	5	6	8	9	11	12	14
29	4624	4639	4654	4669	4683	4698	4713	4728	4742	4757	1	3	4	6	7	9	10	12	13
30	4771	4786	4800	4814	4829	4843	4857	4871	4886	4900	1	3	4	6	7	9	10	11	13
31	4914	4928	4942	4955	4969	4983	4997	5011	5024	5038	1	3	4	6	7	8	10	11	12
32	5051	5065	5079	5092	5105	5119	5132	5145	5159	5172	1	3	4	5	7	8	9	11	12
33	5185	5198	5211	5224	5237	5250	5263	5276	5289	5302	1	3	4	5	6	8	9	10	12
34	5315	5328	5340	5353	5366	5378	5391	5403	5416	5428	1	3	4	5	6	8	9	10	11
35	5441	5453	5465	5478	5490	5502	5514	5527	5539	5551	1	2	4	5	6	7	9	10	11
36	5563	5575	5587	5599	5611	5623	5635	5647	5658	5670	1	2	4	5	6	7	8	10	11
37	5682	5694	5705	5717	5729	5740	5752	5763	5775	5786	1	2	3	5	6	7	8	9	10
38	5798	5809	5821	5832	5843	5855	5866	5877	5888	5899	1	2	3	5	6	7	8	9	10
39	5911	5922	5933	5944	5955	5966	5977	5988	5999	6010	1	2	3	4	5	7	8	9	10
40	6021	6031	6042	6053	6064	6075	6085	6096	6107	6117	1	2	3	4	5	6	8	9	10
41	6128	6138	6149	6160	6170	6180	6191	6201	6212	6222	1	2	3	4	5	6	7	8	9
42	6232	6243	6253	6263	6274	6284	6294	6304	6314	6325	1	2	3	4	5	6	7	8	9
43	6335	6345	6355	6365	6375	6385	6395	6405	6415	6425	1	2	3	4	5	6	7	8	9
44	6435	6444	6454	6464	6474	6484	6493	6503	6513	6522	1	2	3	4	5	6	7	8	9
45	6532	6542	6551	6561	6571	6580	6590	6599	6609	6618	1	2	3	4	5	6	7	8	9
46	6628	6637	6646	6656	6665	6675	6684	6693	6702	6712	1	2	3	4	5	6	7	7	8
47	6721	6730	6739	6749	6758	6767	6776	6785	6794	6803	1	2	3	4	5	5	6	7	8
48	6812	6821	6830	6839	6848	6857	6866	6875	6884	6893	1	2	3	4	4	5	6	7	8
49	6902	6911	6920	6928	6937	6946	6955	6964	6972	6981	1	2	3	4	4	5	6	7	8
50	6990	6998	7007	7016	7024	7033	7042	7050	7059	7067	1	2	3	3	4	5	6	7	8
51	7076	7084	7093	7101	7110	7118	7126	7135	7143	7152	1	2	3	3	4	5	6	7	8
52	7160	7168	7177	7185	7193	7202	7210	7218	7226	7235	1	2	2	3	4	5	6	7	7
53	7243	7251	7259	7267	7275	7284	7292	7300	7308	7316	1	2	2	3	4	5	6	6	7
54	7324	7332	7340	7348	7356	7364	7372	7380	7388	7396	1	2	2	3	4	5	6	6	7

Natural numbers	0	1	2	3	4	5	6	7	8	9	Proportional parts								
											1	2	3	4	5	6	7	8	9
55	7404	7412	7419	7427	7435	7443	7451	7459	7466	7474	1	2	2	3	4	5	5	6	7
56	7482	7490	7497	7505	7513	7520	7528	7536	7543	7551	1	2	2	3	4	5	5	6	7
57	7559	7566	7574	7582	7589	7597	7604	7612	7619	7627	1	2	2	3	4	5	5	6	7
58	7634	7642	7649	7657	7664	7672	7679	7686	7694	7701	1	1	2	3	4	4	5	6	7
59	7709	7716	7723	7731	7738	7745	7752	7760	7767	7774	1	1	2	3	4	4	5	6	7
60	7782	7789	7796	7803	7810	7818	7825	7832	7839	7846	1	1	2	3	4	4	5	6	6
61	7853	7860	7868	7875	7882	7889	7896	7903	7910	7917	1	1	2	3	4	4	5	6	6
62	7924	7931	7938	7945	7952	7959	7966	7973	7930	7987	1	1	2	3	3	4	5	6	6
63	7993	8000	8007	8014	8021	8028	8035	8041	8048	8055	1	1	2	3	3	4	5	5	6
64	8062	8069	8075	8082	8089	8096	8102	8109	8116	8122	1	1	2	3	3	4	5	5	6
65	8129	8136	8142	8149	8156	8162	8169	8176	8182	8189	1	1	2	3	3	4	5	5	6
66	8195	8202	8209	8215	8222	8228	8235	8241	8248	8254	1	1	2	3	3	4	5	5	6
67	8261	8267	8274	8280	8287	8293	8299	8306	8312	8319	1	1	2	3	3	4	4	5	6
68	8325	8331	8338	8344	8351	8357	8363	8370	8376	8382	1	1	2	3	3	4	4	5	6
69	8388	8395	8401	8407	8414	8420	8426	8432	8439	8445	1	1	2	2	3	4	4	5	6
70	8451	8457	8463	8470	8476	8482	8488	8494	8500	8506	1	1	2	2	3	4	4	5	6
71	8513	8519	8525	8531	8537	8543	8549	8555	8561	8567	1	1	2	2	3	4	4	5	5
72	8573	8579	8585	8591	8597	8603	8609	8615	8621	8627	1	1	2	2	3	4	4	5	5
73	8633	8639	8645	8651	8657	8663	8669	8675	8681	8686	1	1	2	2	3	4	4	5	5
74	8692	8698	8704	8710	8716	8722	8727	8733	8739	8745	1	1	2	2	3	4	4	5	5
75	8751	8756	8762	8768	8774	8779	8785	8791	8797	8802	1	1	2	2	3	3	4	5	5
76	8808	8814	8820	8825	8831	8837	8842	8848	8854	8859	1	1	2	2	3	3	4	5	5
77	8865	8871	8876	8882	8887	8893	8899	8904	8910	8915	1	1	2	2	3	3	4	4	5
78	8921	8927	8932	8938	8943	8949	8954	8960	8965	8971	1	1	2	2	3	3	4	4	5
79	8976	8982	8987	8993	8998	9004	9009	9015	9020	9025	1	1	2	2	3	3	4	4	5
80	9031	9036	9042	9047	9053	9058	9063	9069	9074	9079	1	1	2	2	3	3	4	4	5
81	9085	9090	9096	9191	9106	9112	9117	9122	9128	9133	1	1	2	2	3	3	4	4	5
82	9138	9143	9149	9154	9159	9165	9170	9175	9180	9186	1	1	2	2	3	3	4	4	5
83	9191	9196	9201	9206	9212	9217	9222	9227	9232	9238	1	1	2	2	3	3	4	4	5
84	9243	9248	9253	9258	9263	9269	9274	9279	9284	9289	1	1	2	2	3	3	4	4	5
85	9294	9299	9304	9309	9315	9320	9325	9330	9335	9340	1	1	2	2	3	3	4	4	5
86	9345	9350	9355	9360	9365	9370	9375	9380	9385	9390	1	1	2	2	3	3	4	4	5
87	9395	9400	9405	9410	9415	9420	9425	9430	9435	9440	0	1	1	2	2	3	3	4	4
88	9445	9450	9455	9460	9465	9469	9474	9479	9484	9489	0	1	1	2	2	3	3	4	4
89	9494	9499	9504	9509	9513	9518	9523	9528	9533	9538	0	1	1	2	2	3	3	4	4
90	9542	9547	9552	9557	9562	9566	9571	9576	9581	9586	0	1	1	2	2	3	3	4	4
91	9590	9595	9600	9605	9609	9614	9619	9624	9628	9633	0	1	1	2	2	3	3	4	4
92	9638	9643	9647	9652	9657	9661	9666	9671	9675	9680	0	1	1	2	2	3	3	4	4
93	9685	9689	9694	9699	9703	9708	9713	9717	9722	9727	0	1	1	2	2	3	3	4	4
94	9731	9736	9741	9745	9750	9754	9759	9763	9768	9773	0	1	1	2	2	3	3	4	4
95	9777	9782	9786	9791	9795	9800	9805	9809	9814	9818	0	1	1	2	2	3	3	4	4
96	9823	9827	9832	9836	9841	9845	9850	9854	9859	9863	0	1	1	2	2	3	3	4	4
97	9868	9872	9877	9881	9886	9890	9894	9899	9903	9908	0	1	1	2	2	3	3	4	4
98	9912	9917	9921	9926	9930	9934	9939	9943	9948	9952	0	1	1	2	2	3	3	4	4
99	9956	9961	9965	9969	9974	9978	9983	9987	9991	9996	0	1	1	2	2	3	3	3	4

Appendix C

ACCURACY AND PRECISION

During your study of chemistry you will employ a number of measuring devices such as a buret, a graduated cylinder, a thermometer, and a balance. Whenever you set out to make a measurement, you must use a device capable of doing the job properly. The particular measuring instrument selected will depend both on the accuracy required and on the "sensitivity" of the instruments available.

If, for example, you wished to measure a 10-mg sample of a drug, you might think first of a triple beam balance, since such an instrument could easily give a two-place weighing accuracy. However, the triple beam balance is not sufficiently "sensitive." At best, it could give a mass to within ±0.05 g (50 mg). You would therefore require the more sensitive analytical balance, capable of weighing to ±0.1 mg. In general, the simplest measuring device having both the accuracy and sensitivity required is selected. If you had needed to weigh "about 125 g" of a salt, for example, the triple beam balance would have been a logical choice. For such a measurement, the analytical balance would have been unnecessary (and more time consuming).

The results of measurements are influenced by the limitations inherent in the instruments used. Each measurement will have a certain *error* (e.g., the ±0.05 g tolerance of the triple beam balance) arising from this limitation. If you misread the mass settings on a balance, this is not referred to as an "error." We might call it a "blunder." It is assumed that your results will be limited only by instrument error, that is, "blunders" are not expected of you.

Another factor that must be considered is precision in measurement. This is reflected in the differences among the individual results of a series of measurements made in the same way. Accuracy, on the other hand, is expressed by the difference between the average result from a series of measurements and the "true," or accepted, value. Rarely will you know the "true" value of a measurement, so you will learn to place the greatest trust in your measurements when they represent good precision. One simple way of expressing precision is by a *range,* the difference between the highest and lowest result of a series of measurements. A "narrow range" suggests a more reliable measurement than a "broad range," although, as we shall see, good precision does not *guarantee* good accuracy.

There are two types of errors: *systematic* and *random.* Systematic errors may result from incorrect calibration of instruments, from incorrect instrument design or manufacture, or from use of a method not well suited to the problem. Such errors affect all measurements the same way so that results may be precise, but inaccurate. Random errors result from the difficulty of exactly repeating a measuring procedure. Even though skill and practice can reduce random error, it is not possible to eliminate it completely. In some cases random error occurs for reasons beyond the control of the experimenter, as in a line voltage fluctuation during the use of a sensitive electric measuring device.

Some comparisons of precision and accuracy are represented schematically in the following diagram.

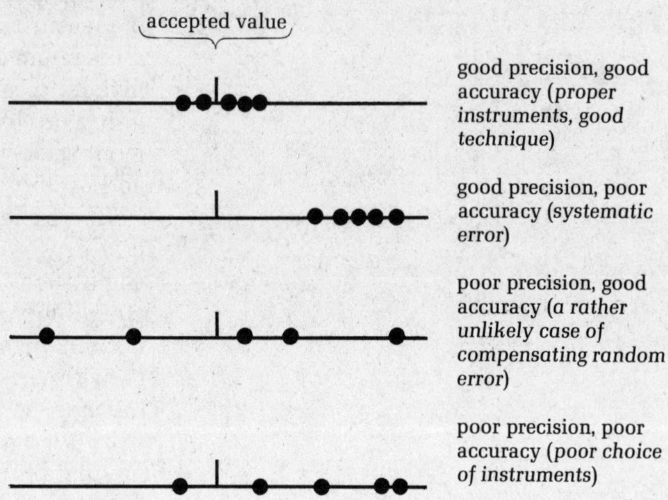

Every result of a series of measurements should be reported in such a way as to indicate some idea of the error involved. There are various ways of doing this, some rather exotic. We will limit our work to reporting either the *range* or the average result with its *average deviation*. The average result is obtained by summing the individual results and dividing this sum by the number of individual values, for example, for a series of five weighings with an analytical balance,

1st mass	1.6752
2nd mass	1.6748
3rd mass	1.6754
4th mass	1.6747
5th mass	1.6753
sum:	8.3754

$$\text{average mass} = \frac{8.3754}{5} = 1.67508$$

(rounded* to: 1.6751)

The average deviation is determined by first finding the differences between each individual value and the average value, then dividing the sum of these differences (irrespective of algebraic sign) by the number of individual values. For example, for the preceding weighings,

*If a calculation yields more figures than are significant (as discussed later), the result is rounded to the proper number of significant figures. When the first nonsignificant digit is 0 through 4, omit it and all following digits. When the first nonsignificant digit is 5 through 9, add one unit to the preceding digit. For example,

$$\frac{1.6751}{5} = 0.33502 \qquad \text{(round to 0.3350)}$$

$$\frac{1.6751}{2} = 0.83755 \qquad \text{(round to 0.8376)}$$

$$1.6752 - 1.6751 = +0.0001$$
$$1.6748 - 1.6751 = -0.0003$$
$$1.6754 - 1.6751 = +0.0003$$
$$1.6747 - 1.6751 = -0.0004$$
$$1.6753 - 1.6751 = +0.0002$$

sum of differences: 0.0013

Average deviation

$$= \frac{0.0013}{5} = 0.00026$$

(rounded to 0.0003)

The result is then reported as

1.6751 ± 0.0003 g

All measured or calculated experimental results should indicate* the error involved, as by the use of an average deviation. Results should be reported in such a way that the last figure given in the average value corresponds to the magnitude of "doubt" expressed by the average deviation. A result reported in this way is said to contain only *significant figures* (that is, those having experimental validity). For example:

A mass obtained with an average deviation of 0.0003 g cannot be meaningfully reported to 0.01 mg (0.00001 g). There is already a range of six units in the 0.1-mg place (from 1.6748 to 1.6754). An added figure to the 0.01-mg position does not add any *meaningful* information.

Incorrect: 1.67508 ± 0.0003 g
Correct: 1.6751 ± 0.0003 g

There is generally no problem in reporting the proper number of significant figures for the direct result of a series of experimental measurements. Additional figures sometimes appear, however, when calculations are performed, especially with one of the now popular "pocket electronic calculators." There is a method for determining the proper number of significant figures for any particular situation. It is, however, more sophisticated and complex than we really need at the level of our work. If we remember that no *mathematical operation* can increase accuracy in an experimental result, we can get by with two simple rules.

1. If the operation involves multiplication or division, or both, round the result to the number of significant figures in the least accurately known quantity used, for example:

$$\frac{1.6751\ \text{g}}{0.332\ \text{ml}} = 5.0454819\ \text{g ml}^{-1}$$ (calculator result)

(rounded to 5.05 g ml^{-1})

2. If the operation involves only addition or subtraction, or addition and subtraction, round the result to the *decimal place* of the least accurately known quantity used, for example,

16.5
21.2
68.34
40.1
10.327
156.467 (rounded to 156.5)

*When no average deviation is given, the last figure reported is assumed to be in doubt by ±1 unit.

It is important in laboratory work to be aware of the appropriate limitations on experimental results. It is equally important to learn to use the measuring devices best suited for particular needs, and to use them skillfully.

Appendix D

EQUIPMENT AND SUPPLIES

Equipment needed per laboratory station or per student, if every student has a drawer.

2 aluminum foil sheets, 5 × 3 cm; 10 × 10 cm
1 asbestos pad, 6″ × 6″
1 beaker, 100 ml, with graduation
1 beaker, 250 ml, with graduation
1 beaker, 400 ml, with graduation
1 beaker, 600 ml, with graduation
1 Buchner funnel, 7 cm OD
2 burets, 50 ml
1 buret funnel
3 clamps, utility
1 clay triangle, 2 cm
1 crucible, 35 mm high, 22 mm bottom ϕ, 42 mm top ϕ
1 crucible cover, to fit crucible
1 crucible tongs
1 Erlenmeyer flask, 125 ml, with graduation
1 Erlenmeyer flask, 250 ml, with graduation
1 Erlenmeyer flask, 500 ml, with graduation
1 file, triangular

Glass Tubing
- 1 right-angle bend, 5 mm OD, 5 cm long
- 6 glass tubing, soft glass, 1 foot, 5 mm OD

1 graduated cylinder, 25 ml
1 leveling bulb, 125 ml
1 mortar, 8 cm OD, and pestle
4 medicine droppers

Rubber Tubing
- 2 rubber tubing, 60 cm, 10 mm OD, 5 mm ID
- 2 rubber tubing, 3 mm long, 10 mm OD, 5 mm ID
- 1 vacuum tubing, 60 cm, 15 mm OD, 6 mm ID
- 1 vacuum tubing, 1 cm long, 15 mm OD, 6 mm ID

1 powder funnel
1 ruler, plastic, 6″, with metric and inch scale
1 scissors
1 scoopula
1 screw cap vial (OD 2.5 cm, 7 cm high)
1 separatory funnel, 250 ml, with plastic stopper and Teflon stopcock
1 spatula
1 squeeze bottle, plastic, 250 ml
1 stirring rod, 6″
1 stirring rod, 10″

Stoppers
1 stopper, #6, 1 hole, 14 mm ϕ
- 2 stoppers, #6, 1 hole, 11 mm ϕ
- 1 stopper #6, solid
- 1 stopper #7, solid
- 1 stopper #7, 1 hole, 4 mm ϕ
- 8 stoppers, #5, solid
- 1 stopper #3, 1 hole, 4 mm ϕ
- 1 stopper #00, 1 hole, 4 mm ϕ
- 5 stoppers, #00, solid

3 Styrofoam cups (can be used repeatedly), 8 cm high
1 suction flask, 250 ml
10 test tubes, 10 cm
5 test tubes, 12.5 cm
8 test tubes, 20 cm
1 test tube block
1 test tube (side arm) 12.5 cm
1 test tube brush
1 thermometer −200° to +30°C
1 thermometer −10° to +260°C
1 thermometer −5° to +80°C (0.2°C divisions)
1 volumetric flask, 100 ml, with plastic stopper
1 watch glass, 9 cm ϕ

MISCELLANEOUS

aluminum foil (1 box per 25 students)
atomizer (for ninhydrin solution; 4 per 25 students)
cotton (1 roll per 25 students)
glasswool (1 package per 25 students)
gummed labels (5 per student)
melting-point capillaries (5 per student; not required if students make their own)
pennies, (80 per student) or same size disks
plastic spheres 1½ in. (13 per student)
roll of tape (5 rolls per 25 students)
rubber bands (1 box per 25 students)
salt, sugar, Kool-Aid (unsweetened), cocoa mix, yeast, packaged (5 per student)
Saran Wrap (1 box per 25 students)
Sheaffer's blue-black #22 ink (1 bottle for 25 students)
string (thin, 40 cm per student)
Styrofoam spheres 1½ in. (30 per student)
toothpicks (1 box per 3 students)
wooden spheres 1.575 cm, 0.856 cm, 1.9 cm, 1.1 cm, and 0.5 cm (6 of each per 30 students)

GENERAL LABORATORY EQUIPMENT

It is recommended that the following equipment be made available for each group of 25 students:

6 analytical balances, single pan, 160-g capacity
12 buret brushes
8 centrifuges
8 graduated cylinders, 250 ml
6 magnetic stirrers
6 pH meters
12 sets of concentrated reagents: HCl conc., HNO_3 conc., H_2SO_4 conc., glac. acetic acid, NaOH 6 *M*, NH_3 6 *M*
6 spectrometers
12 triple beam balances
waste containers, appropriately labeled, for collection of organic liquids

EQUIPMENT PER DESK (Stored on Bench Top)

1 Fisher burner with hose
1 ring 6.5 cm ID
1 ring, 12.5 cm OD
2 ringstands
1 wire gauze

CHEMICALS IN DESK

Each of the following solutions are contained in dropping bottles of 30- or 50-ml capacity.

$CaCl_2$, 5%
CH_3COOH 6 *M*
$Cu(NO_3)_2$ 10 mg ml^{-1}
$Fe(NO_3)_3$ 10 mg ml^{-1}
HCl 6 *M*
HNO_3 6 *M*
H_2SO_4 6 *M*
$K_4Fe(CN)_6$ 0.05 *M*
$MgCl_2$, 5%
NaOH 6 *M*
NH_3 6 *M*
NH_4CNS 1 *M*
$Pb(NO_3)_2$ 10 mg ml^{-1}
phenolphthalein, 1% in ethanol
K_2CrO_4 0.3 *M*
1 vial blue litmus paper
1 vial red litmus paper

GROUND-GLASS EQUIPMENT

1 condensor, reflux, ₮24/40 (male, female), 300 mm length
1 connecting tube, with thermometer opening, 3 way, 75 deg with 2 ₮24/40 joints (both male)
1 Erlenmeyer flask, 250 ml ₮24/40 (female)

MINERALS FOR DISPLAY PER SECTION (and Others, as Available)

1 cerussite
1 chalcopyrite
1 fluorite
1 pyrite

CHEMICALS PER 100 STUDENTS

acetamide, 0.3 kg
acetophenone, 0.5 kg
p-anisaldehyde, 0.3 kg
alanine, 1 g
ammonia, conc., 3 liters
ammonium thiocyanate, 0.5 kg
acetic acid, glacial, 3 liters
acetone, 1.5 liters
barium nitrate, 50 g
benzaldehyde, 0.5 liter
benzoic acid, 0.2 kg
bromophenol blue, 10 g
buffer solution, 0.5 liter, pH 7
butanol, 0.5 liter
2-butanone, 0.3 kg
butyric acid, 0.5 liter
cadmium nitrate, 50 g
caffeine, 0.5 kg
calcium carbonate, reagent, 0.5 kg
calcium chloride, 0.5 kg
calcium nitrate, 25 g
calcium oxide, 0.1 kg
camphor, 0.5 kg
carbon black, 0.5 kg
carbon tetrachloride, 0.1 liter
cerussite, $PbCO_3$, 10 g
chalcopyrite, $CuFeS_2$, 10 g
chlorobenzene, 0.1 liter
chloroform, 15 liters
cinnamaldehyde, 0.3 kg
cobalt chloride hexahydrate, 70 g
cobalt nitrate, 0.2 kg
copper(II) nitrate, 0.5 kg
copper(II) sulfate, $CuSO_4 \cdot 5H_2O$, 0.7 kg
cyclohexanol, 0.5 liter
cyclohexanone, 0.3 kg
detergent, synthetic, 0.5 kg
dimethylglyoxime, 5.0 lb
2,4-dinitrophenylhydrazine, 0.1 kg
disodium ethylenediamine tetraacetate dihydrate (EDTA), 0.2 kg
dry ice, 15 kg
Eriochrome Black T, 5 g
ethanol, 95%, 45 liters
ethylene glycol, 5 liters
filter paper, Whatman #1, 5.5 cm ϕ, 600 circles
formamide, 0.5 kg
galena, PbS, 1 kg

glass wool, ½ lb
glucose, 0.2 kg
glycine, 1 g
hydrochloric acid, conc., 9 liters
hydrogen peroxide, 3%, 0.1 liter
iron(II) ammonium sulfate,
  $Fe(NH_4)_2SO_4 \cdot 6H_2O$, 0.2 kg
iron(II) nitrate, 0.2 kg
iron(III) nitrate, 0.5 kg
iron(II) sulfate, $FeSO_4 \cdot 7H_2O$, 0.3 kg
"Ivory" soap flakes, 2.0 lb
lead, metal, 0.2 kg
lead nitrate, 0.5 kg
leucine, 1 g
magnesium chloride, 0.5 kg
manganese(II) nitrate, 0.1 kg
manganese(II) sulfate, $MnSO_4 \cdot 4H_2O$, 0.5 kg
methanol, 1 liter
methoxybenzoic acid, 0.2 kg
mercuric chloride, 0.2 kg
mercuric nitrate, 50 g
mercurous nitrate, 50 g
mineral oil, white, 7 liters
Mylar film (10-in. × 10-in. sheets), 15 sheets
1-naphthylamine hydrochloride, 0.2 kg
nickel(II) nitrate hexahydrate, 0.1 kg
ninhydrin, 5 g
nitric acid, conc., 4 liters
3-nitrophthalic anhydride, 0.1 kg
oyster shells, 0.1 lb
palmitic acid, 1.0 kg
pentanol, 0.5 liter
1,10-phenathroline monohydrate, 10 g
phenolphthalein, 0.1 kg
phenylacetamide, 0.3 kg
phenylacetic acid, 0.3 kg
phosphoric acid, conc., 0.5 liter
potassium aluminum sulfate,
  $KAl(SO_4)_2 \cdot 12H_2O$, 0.5 kg
potassium chromate, 0.5 kg
potassium chromium sulfate,
  $KCr(SO_4)_2 \cdot 12H_2O$, 0.5 kg
potassium dichromate, 0.5 kg
potassium hexacyanoferrate(II), 0.5 kg
potassium hydrogen phthalate, reagent, 1 kg
potassium hydroxide, 1 kg
potassium nitrate, 0.2 kg
potassium iodide, 0.1 kg
potassium permanganate, 0.1 kg
2-propanol, 0.1 liter
pyridine, 0.5 liter
pyrite, FeS_2, 150 g
salicylaldehyde, 0.3 kg
salicylamide, 0.3 kg
shortening, 2.5 lb
silver nitrate, 50 g
sodium acetate, 1.0 kg
sodium bicarbonate, 0.2 kg
sodium carbonate, 0.5 kg
sodium chloride, 4 kg
sodium citrate, 1.0 kg
sodium dodecylbenzenesulfonate, 50 g
sodium hydroxide, 3.5 kg
sodium oxalate, reagent, 0.5 kg
sodium sulfate, anhydrous, 0.3 kg
sodium thiosulfate, $Na_2S_2O_3 \cdot 5H_2O$, 2.5 kg
stearic acid, 0.5 kg
strontium nitrate, 0.1 kg
sugar beets, 5 lb
sulfanilic acid, 0.2 kg
sulfuric acid, conc., 6 liters
thin-layer chromatography paper, type
  silica gel 5 × 20 cm (S/P 4943-22), 3 boxes
tin(II) chloride, $SnCl_2 \cdot 2H_2O$, 0.5 kg
vanillin, 0.3 kg
zinc, powder, 0.2 kg
zinc chloride, anhydrous, 0.3 kg
zinc nitrate, 0.1 kg
thioacetamide, 50 g

DRUGS

Alka-Seltzer tablets, 100
Camalox tablets, 50
magnesia tablets, 50
No-Doz tablets,* 300
Rolaid tablets, 50
Tums tablets, 50

*May be supplemented or replaced by Midol tablets, Modacin tablets, Vivarin tablets, (or other caffeine preparations containing no additional chloroform-soluble components).

Printer and Binder: The Murray Printing Company

79 80 9 8 7 6 5 4 3